Design of Concrete Structures

Eighth Edition

George Winter
Professor of Engineering
(The Class of 1912 Chair)
Cornell University

Arthur H. Nilson
Professor of Structural Engineering
Cornell University

A textbook originated by

L. C. Urquhart and C. E. O'Rourke
Late Professors of Structural Engineering
Cornell University

McGraw-Hill Book Company

New York St. Louis San Francisco Düsseldorf Johannesburg
Kuala Lumpur London Mexico Montreal New Delhi
Panama Rio de Janeiro Singapore Sydney Toronto

Design of Concrete Structures

0 FGRFGR 7 9 8

This book was set in Modern by The Maple Press Company.
The editors were B. J. Clark and Michael Gardner;
the designer was Barbara Ellwood;
and the production supervisor was Ted Agrillo. The drawings were done
by John Cordes, J & R Technical Services, Inc.

Library of Congress Cataloging in Publication Data

Winter, George, 1907–
    Design of concrete structures.

    (McGraw-Hill civil engineering series)
    "A textbook conceived by L. C. Urquhart and
C. E. O'Rourke."
    Includes bibliographical references.
    1. Reinforced concrete construction. 2. Prestressed
concrete construction. I. Nilson, Author H., joint
author. II. Urquhart, Leonard Church, 1886–   Design
of concrete structures. III. Title.
TA683.2.W48 1973   624'.1834   72–6312
ISBN 0-07-071115-1

Contents

Preface

This edition maintains the dual orientation first introduced in its immediate predecessor: to lay the foundation of a thorough understanding of materials' behavior and the basic mechanics and performance of reinforced- and prestressed-concrete structures and, at the same time, to present ample material on up-to-date design practices and procedures as preparation for, and aid in, professional practice.

It is now firmly recognized that mere training in specialized skills and codified design procedures is inadequate as a basis for successful professional practice. These skills and procedures undergo frequent and sweeping changes. To keep abreast of these rapid developments, the engineer needs a thorough grounding in the basic performance of concrete and steel as structural materials and of the behavior, elastic and inelastic, of reinforced-concrete members and structures. On the other hand, the final goal of the structural engineer is to design structures safely, economically, and efficiently. Hence, with this basic understanding as a firm foundation, familiarity with current design procedures and skill in employing them is of the essence. This edition, like its predecessor, serves both these purposes.

The general arrangement of the seventh edition has been retained, with two exceptions. For one, a separate new chapter on precast construction has been included. It is generally recognized that on-site construction alone can no longer satisfy the vast demand for new buildings for two reasons: disproportionately increased cost and an increasingly inadequate supply of site labor. It is for this reason that so-called systems building is assuming increasing importance. The essence of such construction is prefabrication, that is, quantity production of prefabricated components in factories or yards. This reduces cost, minimizes on-site construction, and shortens construction time. It is for these reasons that pre-cast concrete construction during the last several years has developed at an ever-increasing pace. The new chapter on this method, though of necessity brief, will serve as a first introduction to this field and, through extensive references, guide the reader to a more thorough understanding.

The other change in arrangement is the omission of the chapter on retaining walls. The problems specific to retaining walls lie more in the field of soil mechanics than of reinforced-concrete design. It was thought more important to enlarge the treatment of footings and other spread foundations.

Apart from these major changes, new and updated material permeates all portions of the book. To mention only the more important ones: Lightweight concrete has received extensive new treatment, in consonance with its increasing practical importance; entirely new material on design for torsion, and of two-way slab systems is included; because of the greater importance of performance under service loads when modern high-strength materials are used, the coverage of control of deflections and of cracking has been expanded and updated; likewise expanded and updated is the treatment of slender columns and of prestressed concrete. Most of these changes, as well as many minor ones, are occasioned by the publication of the 1971 edition of the "ACI Building Code Requirements for Reinforced Concrete" (henceforth referred to as the Code).[1] All design procedures, examples, and problems throughout the book have been adapted to this new Code edition.

Other new or improved material not directly connected with the Code includes a more fundamental discussion of structural safety, and expanded treatment of cylindrical shells and folded plates, and of approximate analysis of indeterminate beams and frames, at the expense of eliminating standard methods of indeterminate analysis which can be found in any number of other books. Finally, a variety of new design aids should make the volume more useful, not only to the student but also to the practicing engineer.

The first two chapters present the structurally relevant *properties and modes of behavior of steel and concrete* and the *basic mechanics and performance of reinforced- and prestressed-concrete members*. These two chapters are written without reference to codes and specifications. The mechanics and behavior in compression, tension, bending, eccentric compression, shear, torsion, bond, and anchorage

---

[1] This ACI standard 318-71 can be obtained from the American Concrete Institute, P.O. Box 4754 Redford Station, Detroit, Michigan 48219. For more detail on the Institute and its Code see the footnote in Art. 3.2.

are developed systematically through all stages of loading, from low loads to failure.

The next four chapters (beams, slabs, columns, and footings) present the design methods of these basic reinforced-concrete members in close conformity with the 1971 edition of the Code. The design methods are fully illustrated with detailed examples. In many cases, these examples and an ample number of problems for the student are designed, not only to illustrate specific design methods, but also to develop more general points of interest (e.g., the comparative use of high- vs. low-strength materials). Apart from practical design methods, these four chapters also contain a sizable amount of fundamental information not included in the first two. For instance, new design methods for continuous strip, grid, and mat foundations have been introduced in the chapter on Footings.

The last six chapters, as in previous editions, cover the major design aspects of various types of structures. The chapter on continuous beams and frames features an expanded treatment of inelastic moment redistribution and of approximate analysis as employed in early design stages, at the expense of omitting standard methods of indeterminate analysis. The chapter on reinforced-concrete buildings, though somewhat reduced to eliminate duplication of material, still contains fairly complete design examples of various types of framing. As in previous editions, a chapter on arches, cylindrical shells, folded plates, hyperbolic paraboloids, and spherical domes represents a first introduction to these types of structures, updated and slightly expanded. As already mentioned, the new chapter on prefabricated construction reflects the important new tendencies toward systems building. The chapter on prestressed concrete has been updated in regard to the Code, but also expanded and enriched with partially original design methods. Finally, the last chapter on bridges has been updated in regard to the latest edition of the AAHSO specification.

These features, plus an updated Appendix, which contains a variety of new design aids, tables, and graphs, should continue to make this book a valuable desk aid *for the practicing engineer*.

For the teacher it may be important that the text is designed to accommodate itself to a variety of course organizations. If time limitations permit only an introduction to reinforced concrete, Chapters 1 and 2 plus selected parts of Chapters 3 and 5 will serve this purpose. If the curriculum contains a three- or four-hour course in concrete, Chapters 1 to 5 and parts of Chapter 6 will provide thorough coverage. In either of these two arrangements, time will probably not permit any classroom coverage of Chapters 8 and 9 on buildings and on precast construction. It is suggested, however, that portions of Chapter 8 and all of the brief Chapter 9 could well be assigned as required or recommended reading. This will provide the student with some design-oriented information of a more descriptive type on how structures are actually put together, complementing the more analytical topics in the other chapters. Also, in the writers' experience, such complementary independent reading tends to enhance student motivation. Finally, for a two-course sequence, or for one undergraduate and one graduate course, the book in its entirety should provide a solid foundation for professional design practice.

The authors wish to thank their colleague, Professor R. G. Sexsmith, for contributing the material on structural safety in Chapter 2. They also wish to thank Cornell University for essential secretarial and other assistance, and Messrs. John T. DeWolf and Riadh Al-Mahaidi, graduate students, for calculating many examples.

George Winter

Arthur H. Nilson

1
Concrete and Reinforcing Steel

1.1 INTRODUCTION: CONCRETE, REINFORCED CONCRETE, PRESTRESSED CONCRETE

Concrete is a stonelike material obtained by permitting a carefully proportioned mixture of cement, sand and gravel or other aggregate, and water to harden in forms of the shape and dimensions of the desired structure. The bulk of the material consists of fine and coarse aggregate. Cement and water interact chemically to bind the aggregate particles into a solid mass. Additional water, over and above that needed for this chemical reaction, is necessary to give the mixture the workability that enables it to fill the forms and surround the embedded reinforcing steel prior to hardening. Concretes in a wide range of strength properties can be obtained by appropriate adjustment of the proportions of the constituent materials. Special cements (such as high-early-strength cements), special aggregates (such as various lightweight or heavyweight aggregates), and special curing methods (such as steam-curing) permit an even wider variety of properties to be obtained.

These properties depend to a very substantial degree on the proportions of the mix, on the thoroughness with which the various constituents are intermixed, and on the conditions of humidity and temperature in which the mix is maintained from the moment it is placed in the forms until it is fully hardened. The process of controlling these conditions is known as *curing*. To protect against the unintentional production of substandard concrete, a high degree of skillful control and supervision is necessary throughout the process, from the proportioning by weight of the individual components, through mixing and placing, until the completion of curing.

The factors which make concrete a universal building material are so pronounced that it has been used, in more primitive kinds and ways than at present, for thousands of years, probably beginning in Egyptian antiquity. The facility with which, while plastic, it can be deposited and made to fill forms or molds of almost any practical shape is one of these factors. Its high fire and weather resistance are evident advantages. Most of the constituent materials, with the possible exception of cement, are usually available at low cost locally or at small distances from the construction site. Its compressive strength, as that of natural stones, is high, which makes it suitable for members primarily subject to compression, such as columns and arches. On the other hand, again as in natural stones, it is a relatively brittle material whose tensile strength is small compared with its compressive strength. This prevents its economical use in structural members which are subject to tension either entirely (such as tie rods) or over part of their cross sections (such as beams or other flexural members).

To offset this limitation, it has been found possible, in the second half of the nineteenth century, to use steel with its high tensile strength to reinforce concrete, chiefly in those places where its small tensile strength would limit the carrying capacity of the member. The reinforcement, usually round steel rods with appropriate surface deformations to provide interlocking, is placed in the forms in advance of the concrete. When completely surrounded by the hardened concrete mass, it forms an integral part of the member. The resulting combination of two materials, known as *reinforced concrete*, combines many of the advantages of each: the relatively low cost, good weather and fire resistance, good compressive strength, and excellent formability of concrete and the high tensile strength and much greater ductility and toughness of steel. It is this combination which allows the almost unlimited range of uses and possibilities of reinforced concrete in the construction of buildings, bridges, dams, tanks, reservoirs, and a host of other structures.

In more recent times it has been found possible to produce steels, at relatively low cost, whose yield strength is of the order of 4 times and more that of ordinary reinforcing steel. Likewise, it is possible to produce concrete 2 and 3 times as strong in compression as the more ordinary concretes. There are limits to the strengths of the constituent materials beyond which they can no longer be combined effectively in one member. To be sure, the strength of such a member would increase roughly in proportion to those of the constituent materials. However, the high strains which result from such high stresses, particularly in the steel, would make for large deformations and deflections of such members under load. Equally, if not more important, the large strains in such high-strength reinforcing steel would induce large cracks in the surrounding low-tensile-strength concrete, cracks which would not only be unsightly, but which would expose the steel reinforcement to corrosion by moisture and other chemical action. This limits the useful yield strength of reinforcing steel to about 90 ksi, about twice that of conventional reinforcing steels.

A special way has been found, however, to use steels and concretes of very high strength in combination. This type of construction is known as *prestressed concrete*. The steel, mostly in the shape of wires or strands, but sometimes as bars, is embedded in the concrete under high tension which is held in equilibrium by compression stresses in the surrounding concrete after hardening. Because of this precompression, the concrete in a flexural member will crack on the tension side at a much larger load than when not so precompressed. This reduces radically both the deflections and the tensile cracks at service loads in such structures and thereby enables these high-strength materials to be used effectively. Prestressed concrete is particularly suited to prefabrication on a mass-production basis, although it is being used as well without such prefabrication. Its introduction has extended, to a very significant degree, the range of structural uses of the combination of these two materials.

Figure 1.1 shows simplified sketches of some of the *principal structural forms* of reinforced concrete; pertinent design methods for many of them are discussed later in this volume. A slab, beam, and girder floor, as it is used in many multistory buildings, is shown in (a), and a ribbed floor which permits savings of concrete and reuse of standardized metal forms in (b). The flat-slab floor of (c), distinct by the absence of beams and girders, is frequently used for more heavily loaded buildings, such as warehouses. The rigid-frame structure of (d) can be made to span large openings for auditoriums, gymnasiums, or industrial buildings. Figure 1.1e and f shows a cylindrical shell roof and a folded-plate roof, respectively, forms which, like the other shell roofs in (h), (l), and (m), have been developed in recent years and permit the use of extremely thin surfaces, often thinner, relatively, than an eggshell. Two of the many bridge forms are shown in (g) and (j), the former a multiarch bridge for large spans, the latter a rigid-frame bridge typical of highway overpasses. The tank or reservoir of (i) combines a cylindrical and a spherical shell, the former frequently prestressed. Representative of the extensive use of reinforced concrete in foundation work is the counterfort retaining wall of (k).

The forms of Fig. 1.1 do not constitute a complete range, but are merely typical of shapes appropriate to the properties and possibilities of concrete structures. They illustrate the adaptability of the material to a great variety of one-dimensional (beams, girders, columns), two-dimensional (slabs, arches, rigid frames), and three-dimensional (shells) structures and structural components. This variability allows the shape of the structure to be adapted to

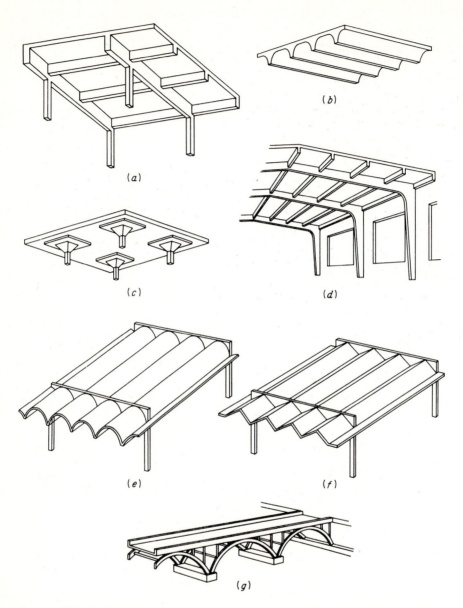

Fig. 1.1 Principal structural forms in reinforced concrete. (*a*) Beam-and-girder floor; (*b*) ribbed floor; (*c*) flat-slab floor; (*d*) rigid-frame building; (*e*) cylindrical-shell roof; (*f*) folded-plate roof; (*g*) multiple-arch bridge.

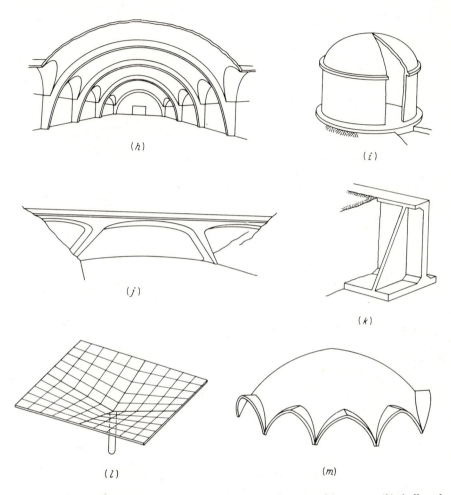

Fig. 1.1 (Continued)   Principal structural forms in reinforced concrete: (*h*) shell roof; (*i*) storage tank; (*j*) rigid-frame bridge; (*k*) counterfort retaining wall; (*l*) hyperbolic-paraboloid umbrella shell; (*m*) point-supported dome roof.

its function in an economical manner and furnishes the architect with a wide range of possibilities for esthetically satisfying structural solutions.

1.2 CEMENT

A cementitious material is one which has the adhesive and cohesive properties necessary to bond inert aggregates into a solid mass of adequate strength and durability. This technologically important category of materials includes not only cements proper, but also limes, asphalts, and tars as they are used in road building, and others. For making structural concrete, so-called *hydraulic* cements are used exclusively. Water is needed for the chemical process (hydration) in which the cement powder sets and hardens into one solid mass. Of the various hydraulic cements which have been developed, *portland cement*, which was first patented in England in 1824, is by far the most common.

Portland cement is a finely powdered, grayish material which consists chiefly of calcium and aluminum silicates. The common raw materials from which it is made are limestones, which provide CaO, and clays or shales, which furnish SiO_2 and Al_2O_3. These are ground, blended, fused to clinkers in a kiln, cooled, and ground to the required fineness. The material is shipped in bulk or in bags containing 94 lb of cement. Concretes made with portland cement generally need about two weeks to reach sufficient strength so that forms of beams and slabs can be removed and reasonable loads applied; they reach their design strength after 28 days and continue to gain strength thereafter at a decreasing rate. To speed construction when needed, *high-early-strength cements* have been developed; they are more costly than ordinary portland cement, but reach, within 1 to 3 days, the strength a portland cement would have after 28 days. They have the same basic composition as portland cements, but are more carefully blended and more finely ground, both before and after clinkering.

When cement is mixed with water to form a soft paste, it gradually stiffens until it becomes a solid. This process is known as *setting* and *hardening;* The cement is said to have set when it has gained sufficient rigidity to support an arbitrarily defined pressure, after which it continues for a long time to harden, i.e., to gain further strength. The water in the paste dissolves material at the surfaces of the cement grains and forms a gel which gradually increases in volume and stiffness. This leads to a rapid stiffening of the paste 2 to 4 hr after water has been added to the cement. *Hydration* continues to proceed deeper into the cement grains, at decreasing speed, with continued stiffening and hardening of the mass. In ordinary concrete the cement is probably never completely hydrated. The gel structure of the hardened paste seems to be the chief reason for the volume changes which are caused in concrete by variations in moisture, such as the *shrinkage* of concrete as it dries.

For complete hydration of a given amount of cement, according to H. Rüsch, an amount of water equal to about 25 per cent of that of cement, by weight, is needed chemically. An additional 10 to 15 per cent must be present, however, to provide mobility for the water in the cement paste during the hydration process so that it can reach the cement particles. This makes for a total minimum *water-cement ratio* of 0.35 to 0.40 by weight. This corresponds to 4 to 4.5 gal of water per sack of cement, the more customary way of expressing the water-cement ratio. Water-cement ratios in concretes are generally considerably larger than this minimum, to provide the necessary workability of the concrete mix. Any amount of water above the 25 per cent consumed in the chemical reaction produces pores in the cement paste. The strength of the hardened paste decreases in inverse proportion to the fraction of the total volume occupied by pores. Put differently, since only the solids, and not the voids, resist stress, strength increases directly as the fraction of the total volume occupied by the solids. This is why the strength of the cement paste depends primarily on, and decreases directly with, increasing water-cement ratio.

The chemical process involved in the setting and hardening liberates heat, known as *heat of hydration*. In large concrete masses, such as dams, this heat is dissipated very slowly and results in a temperature rise and volume expansion of the concrete during hydration, with subsequent cooling and contraction. To avoid the serious cracking and weakening which may result from this process, special measures must be taken for its control.

1.3 AGGREGATES

In ordinary structural concretes the aggregates occupy about 70 to 75 per cent of the volume of the hardened mass. The remainder consists of hardened cement paste, uncombined water (i.e., water not involved in the hydration of the cement), and air voids. The latter two evidently do not contribute to the strength of the concrete. In general, the more densely the aggregate can be packed, the better are the strength, weather resistance, and economy of the concrete. For this reason the gradation of the particle sizes in the aggregate, to produce close packing, is of considerable importance. It is also important that the aggregate have good strength, durability, and weather resistance; that its surface be free from impurities such as loam, silt, and organic matter which may weaken the bond with the cement paste; and that no unfavorable chemical reaction take place between it and the cement.

Natural aggregates are generally classified as fine and coarse. *Fine aggregate* or *sand* is any material which will pass a No. 4 sieve, i.e., a sieve with four openings per linear inch. Material coarser than this is classified as *coarse aggregate* or *gravel*. When favorable gradation is desired, aggregates are separated by sieving into two or three size groups of sand and several size groups of coarse aggregate. These can then be combined according to grading charts to result in a densely packed aggregate. The *maximum size of coarse aggregate* in reinforced concrete is governed by the requirement that it shall easily fit into the forms and between the reinforcing bars. For this purpose it should not be larger than one-fifth of the narrowest dimension of the forms or one-third of the depth of slabs, nor two-thirds (for columns) to three-quarters (for other members) of the minimum distance between reinforcing bars. Authoritative information on aggregate properties and their influence on concrete properties and on selection, preparation, and handling of aggregate is found in the report of ACI Committee 621 (Ref. 1.1).[1]

[1] References are listed at the end of the chapter.

The *unit weight* of so-called *stone concrete*, i.e., concrete with natural stone aggregate, varies from about 140 to 152 pcf and can generally be assumed as 145 pcf. For special purposes, lightweight concretes, on the one hand, and heavy concretes, on the other, are being used with increasing frequency.

A variety of *lightweight* aggregates is available. Some unprocessed aggregates, such as pumice or cinders, are suitable for insulating concretes, but for structural lightweight concrete, processed aggregates are used because of their better control. These consist of expanded shales, clays, slates, slags, or pelletized fly ash. They are light in weight because of the porous, cellular structure of the individual aggregate particle, which is achieved by gas or steam formation in processing the aggregates in rotary kilns at high temperatures (generally in excess of 2000°F). Requirements for satisfactory lightweight aggregates are found in ASTM C330, Specifications for Lightweight Aggregates for Structural Concrete.

Three classes of lightweight concrete are distinguished (Ref. 1.2): low-density concretes, which are chiefly employed for insulation and whose unit weight rarely exceeds 50 pcf; moderate-strength concretes, with unit weights from about 60 to 85 pcf and compressive strengths of 1000 to 2500 psi, which are chiefly used as "fill," e.g. over light-gage steel floor panels; and structural concretes, with unit weights from 90 to 120 pcf and compressive strengths equal to those of stone concretes. Similarities and differences in structural characteristics of lightweight as compared with stone concretes are discussed in Art. 1.11.

Heavyweight concrete is frequently required for shielding against gamma and X radiation in nuclear-reactor and similar installations, for protective structures, and for special purposes, such as counterweights of lift bridges. Heavy aggregates are used for such concretes. These consist of heavy iron ores or barite (barium sulfate) rock crushed to suitable sizes. Steel in the form of scrap, punchings, or shot (as fines) is also used. Unit weights of heavyweight concretes with natural heavy rock aggregates range from about 200 to 230 pcf; if iron punchings are added to high-density ores, weights as high as 270 pcf are achieved. The weight may be as high as 330 pcf if ores are used for the fines only, and steel for the coarse aggregate.

1.4 PROPORTIONING AND MIXING OF CONCRETE

The various components of a mix are proportioned so that the resulting concrete has adequate strength, proper workability for placing, and low cost. The latter calls for use of the minimum amount of cement (the most costly of the components) which will achieve adequate properties. The better the gradation of aggregates, i.e., the smaller the volume of voids, the less cement paste is needed to fill these voids. In addition to the water required for hydration (see Art. 1.2), water is needed for wetting the surface of the aggregate. As water is added, the plasticity and fluidity of the mix increases (i.e., its workability

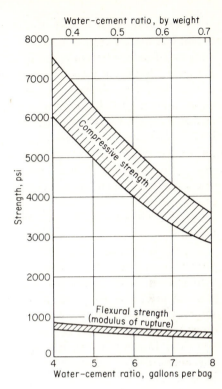

Fig. 1.2 Effect of water-cement ratio on 28-day compressive and flexural strength. (*Adapted from Ref.* 1.4.)

improves), but the strength decreases because of the larger volume of voids created by the free water. To reduce the free water while retaining the workability, cement must be added. Therefore, as for the cement paste, the *water-cement ratio* is the chief factor which controls the strength of the concrete. For a given water-cement ratio, one selects that minimum amount of cement, in sacks per cubic yard, which will secure the desired workability.

Figure 1.2 shows the decisive influence of the water-cement ratio on the compressive strength of concrete. Its influence on the tensile strength, as measured by the nominal flexural strength or modulus of rupture, is seen to be pronounced, but much smaller than its effect on compression strength. This seems to be so because, in addition to the void ratio, the tensile strength depends strongly on the strength of bond between coarse aggregate and cement mortar (i.e., cement paste plus fine aggregate). According to tests at Cornell University, this bond strength is relatively slightly affected by the water-cement ratio (Ref. 1.6).

It has been customary to define the *proportions* of a concrete mix by the ratio, by volume or weight, of cement to sand to gravel, for example, 1:2:4. This method refers to the solid components only and, unless the water-cement ratio is specified separately, is insufficient to define the properties of the resulting concrete either in its fresh state or when set and hardened. For a complete definition of proportions it is now customary to specify, per 94-lb bag of cement, the weight of water, sand, and coarse aggregate. Thus, a mix may be defined as containing (per 94-lb bag of cement) 45 lb of water, 320 lb of sand, and 500 lb of coarse aggregate. Alternatively, batch quantities are often defined in terms of the total weight of each component needed to make up one cubic yard of wet concrete (e.g., 517 lb of cement, 300 lb of water, 1270 lb of dry sand, and 1940 lb of dry coarse aggregate).

Various methods of proportioning are used to obtain mixes of the desired properties from the cements and aggregates at hand. One is the so-called *trial-batch method*. Selecting a water-cement ratio from information such as that in Fig. 1.2, one produces several small trial batches with varying amounts of aggregate to obtain the required strength, consistency, and other properties with a minimum amount of paste. Concrete *consistency* is most frequently measured by the *slump test*. A metal mold in the shape of a truncated cone 12 in. high is filled with fresh concrete in a carefully specified manner. Immediately upon being filled, the mold is lifted off, and the slump of the concrete is measured as the difference in height between the mold and the pile of concrete. The slump is a good measure of the total water content in the mix and should be kept as low as is compatible with workability. Slumps for concretes in building construction generally range from 2 to 6 in.

The so-called ACI method of proportioning makes use of the slump test in connection with a set of tables which, for a variety of conditions (types of structures, dimensions of members, degree of exposure to weathering, etc.), permit one to estimate proportions which will result in the desired properties (Ref. 1.4). These preliminarily selected proportions are checked and adjusted by means of trial batches to result in concrete of the desired quality. Inevitably, strength properties of a concrete of given proportions scatter from batch to batch. It is therefore necessary to select proportions which will furnish an average strength sufficiently greater than the specified design strength so that even the accidentally weaker batches are of adequate quality (for details, see Art. 1.6). Discussion in detail of practices for proportioning concrete is beyond the scope of this volume; this topic is treated fully in Refs. 1.3 to 1.5, both for stone concrete and for lightweight aggregate concrete.

In addition to the main components of concretes, *admixtures* are often used for special purposes. There are admixtures to improve workability, to accelerate or retard setting and hardening, to aid in curing, to improve durability, to add color, and to impart other properties. While the beneficial effects of some admixtures are well established, the claims of others should be viewed with caution. *Air-entraining agents* at present are the most important and most widely used admixtures. They cause the entrainment of air in the form of small well-dispersed bubbles in the concrete. These improve workability and durability, chiefly resistance to freezing and thawing, and reduce segregation during placing. They decrease density because of the increased void ratio and thereby decrease strength; however, this decrease can be partially offset by reduction of mixing water without loss of workability. The chief use of air-entrained concretes is in pavements, but they are also used for structural work, particularly for outdoor structures.

On all but the smallest jobs, *batching* is carried out in special batching plants. Separate hoppers contain cement and the various fractions of aggregate. Proportions are controlled, by weight, by means of manually operated or automatic dial scales connected to the hoppers. The mixing water is batched either by measuring tanks or by water meters.

The principal purpose of *mixing* is to produce an intimate mixture of cement, water, fine and coarse aggregate, and possible admixtures of uniform consistency throughout each batch. This is achieved in machine mixers of the revolving-drum type. Minimum mixing time is 1 min for mixers of not more than 1 yd^3 capacity, with an additional 15 sec for each additional $\frac{1}{2}$ yd^3. Mixing can be continued for a considerable time without adverse effect. This fact is particularly important in connection with ready-mixed concrete.

On large projects, particularly in the open country where ample space is available, movable mixing plants are installed and operated at the site. On the other hand, in construction under congested city conditions, on smaller jobs, and frequently in highway construction, *ready-mixed concrete* is used. Such concrete is batched in a stationary plant and then hauled to the site in trucks in one of three ways: (1) mixed completely at the stationary plant and hauled in a truck agitator, (2) transit-mixed, i.e., batched at the plant but mixed in a truck mixer, or (3) partially mixed at the plant with mixing completed in a truck mixer. Concrete should be discharged from the mixer or agitator within at most 1$\frac{1}{2}$ hr after the water is added to the batch.

1.5 CONVEYING, PLACING, COMPACTING, CURING

Conveying of most building concrete from the mixer or truck to the form is done in bottom dump buckets or in wheelbarrows or buggies or by pumping through steel pipelines. The chief danger during conveying is that of *segregation*. The individual components of concrete tend to segregate because of their dissimilarity. In overly wet concrete standing in containers or forms, the heavier gravel components tend to settle, and the lighter materials, particularly water, to rise. Lateral movement, such as flow within the forms, tends to separate the coarse gravel from the finer components of the mix. The danger of segregation has caused the discarding of some previously common means of conveying, such as chutes and conveyor belts, in favor of methods which minimize this tendency.

Placing is the process of transferring the fresh concrete from the conveying device to its final place in the forms. Prior to placing, loose rust must be removed from reinforcement, forms must be cleaned, and hardened surfaces of previous concrete lifts must be cleaned and treated appropriately. Placing and compacting are critical in their effect on the final quality of the concrete.

Proper placement must avoid segregation, displacement of forms or of reinforcement in the forms, and poor bond between successive layers of concrete. Immediately upon placing, the concrete should be *compacted* by means of hand tools or vibrators. Such compacting prevents honeycombing, assures close contact with forms and reinforcement, and serves as partial remedy to possible prior segregation. Compacting is achieved by hand tamping with a variety of special tools, but now more commonly and successfully with high-frequency, power-driven *vibrators*. These are of the *internal* type, immersed in the concrete, or of the *external* type, attached to the forms. The former are preferable, but must be supplemented by the latter where narrow forms or other obstacles make immersion impossible (Ref. 1.7).

Fresh concrete gains strength most rapidly during the first few days and weeks. Structural design is generally based on the 28-*day strength*, about 70 per cent of which is reached at the end of the first week after placing. The final concrete strength depends greatly on the conditions of moisture and temperature during this initial period. The maintenance of proper conditions during this time is known as *curing*. Thirty per cent of the strength or more can be lost by premature drying out of the concrete; similar amounts may be lost by permitting the concrete temperature to drop to 40°F or lower during the first few days unless the concrete is maintained continuously moist for a long time thereafter. Freezing of fresh concrete may reduce its strength by as much as 50 per cent.

To prevent such damage, concrete should be protected from loss of moisture for at least 7 days and, in more sensitive work, up to 14 days. When high-early-strength cements are used, curing periods can be cut in half. Curing can be achieved by keeping exposed surfaces continually wet through sprinkling, ponding, covering with wet burlap, or the like. Recent methods include the use of sealing compounds, which, when properly used, form evaporation-retarding membranes, and waterproof papers. In addition to improved strength, proper moist-curing provides better shrinkage control. To protect the concrete against low temperature during cold weather, the mixing water and, occasionally, the aggregates are heated, temperature insulation is used where possible, and special admixtures, particularly calcium chloride, are employed. When air temperatures are very low, external heat may have to be supplied in addition to insulation.

1.6 TESTS, QUALITY CONTROL, INSPECTION

The quality of mill-produced materials, such as structural or reinforcing steel, is guaranteed by the producer, who must exercise systematic quality controls, usually specified by pertinent ASTM (American Society for Testing and Materials) standards. Concrete, in contrast, is produced at or close to the site, and its final qualities are affected by a number of factors, which have been briefly discussed. Thus, systematic quality control must be instituted at the construction site.

The main measure of the structural quality of concrete is its *compression strength*. Tests for this property are made on cylindrical specimens of height equal to twice the diameter, usually 6 × 12 in. Impervious molds of this shape are filled with concrete during the operation of placement as specified by ASTM C172, Method of Sampling Fresh Concrete, and ASTM C31, Method of Making and Curing Concrete Specimens in the Field. The cylinders are moist-cured at 70 ± 5°F, generally for 28 days, and then tested in the laboratory at a specified rate of loading. The compression strength obtained from such tests is known as the *cylinder strength* f'_c and is the main property specified for design purposes.

To provide structural safety, continuous control is necessary to ensure that the strength of the concrete as furnished is in satisfactory agreement with the value called for by the designer. The Building Code Requirements for Reinforced Concrete of the American Concrete Institute, ACI 318-71 (henceforth referred to as the ACI Code), specify that a pair of cylinders shall be tested for each 150 yd³ of concrete or for each 5000 ft² of surface area placed, but not less than once a day. As mentioned in Art. 1.4, the results of strength tests of different batches mixed to identical proportions show inevitable scatter. The scatter can be reduced by closer control, but occasional tests below the cylinder strength specified in the design cannot be avoided. To assure adequate concrete strength in spite of such scatter, the ACI Code stipulates that concrete quality is satisfactory if (1) no individual strength-test result (the average of a pair of cylinder tests) falls below the required f'_c by more than 500 psi, and (2) the average of any three consecutive strength tests is equal to or exceeds the required f'_c.

It is evident that if concrete were proportioned so that its mean strength were just equal to the required f'_c, it would not pass these quality requirements, because about half of its strength-test results would fall below the required f'_c. It is therefore necessary to proportion concrete so that its mean strength exceeds the required design strength f'_c by an amount sufficient to ensure that the two quoted requirements are met. This minimum amount by which the mean strength must exceed f'_c can be determined only by statistical methods because of the random nature of test scatter. To do so, requirements have been derived to limit the probability of strength deficiencies to the following amounts:

1. A probability of 1 in 100 that an individual strength test will fall below f'_c by more than 500 psi.
2. A probability of 1 in 100 that an average of three consecutive strength tests will fall below f'_c.
3. A probability of 1 in 10 that a single random strength test will fall below f'_c.

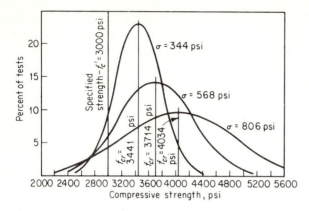

Fig. 1.3 Frequency curves and average strengths for various degrees of control of concretes with $f_c' = 3000$ psi. (*Adapted from Ref. 1.15.*)

The situation is illustrated in Fig. 1.3, which shows three normal frequency curves of distribution of strength-test results, all of which will satisfy requirement (3), above. The required design strength is $f_c' = 3000$ psi. The curves correspond to three different degrees of quality control, the narrowest curve representing the best control, i.e., the least scatter, and the widest curve the worst control. The degree of control, i.e., the amount of scatter, is measured by the standard deviation σ, which is shown for each of the curves. It is seen that in order to satisfy requirement (3) for a closely controlled concrete with a standard deviation of only 344 psi, the mean strength f_{cr} must be at least 441 psi greater than the design strength f_c'. On the other hand, for the most poorly controlled of the three concretes, with a standard deviation of 806 psi, the mean strength must exceed the specified design strength by at least 1034 psi. Conditions 1 and 2 can be similarly evaluated by statistical methods.

On the basis of such studies the ACI Code requires that concrete production facilities maintain records from which the standard deviation achieved in the particular facility can be determined. It then stipulates the minimum amount by which the average strength aimed at when selecting concrete proportions shall exceed the specified design strength f_c', depending on the standard deviation σ, as follows:

400 psi if $\sigma < 300$ psi
550 psi if $\sigma = 300$ to 400 psi
700 psi if $\sigma = 400$ to 500 psi
900 psi if $\sigma = 500$ to 600 psi

If σ exceeds 600 psi or if no adequate record of concrete plant performance is available, the average strength must exceed f_c' by at least 1200 psi.

It is seen that this method of control recognizes the fact that occasional deficient batches are inevitable. Requirements 1 to 3 ensure, in order, a small probability that such strength deficiencies as are bound to occur will be so large that they represent serious danger; an equally small probability that a sizable portion of the structure, as represented by three consecutive strength tests, will be made of below-par concrete; and a total fraction of concrete with strength below f_c' no larger than 10 per cent of all concrete placed.

In spite of scientific advances, building in general, and concrete making in particular, retain some elements of an art; they depend on many skills and imponderables. It is the task of systematic *inspection* to ensure close correspondence between plans and specifications and the finished structure. Inspection during construction should be carried out by a competent engineer, preferably the one who produced the design or one who is responsible to the design engineer. The inspector's main functions in regard to materials quality control are sampling, examination, and field testing of materials, control of concrete proportioning, inspection of batching, mixing, conveying, placing, compacting, and curing, and supervision of preparation of specimens for laboratory tests. In addition, he must inspect foundations, formwork, placing of reinforcing steel, and other pertinent features of the general progress of work; keep records of all the inspected items; and prepare periodic reports. The importance of thorough inspection to the correctness and adequate quality of the finished structure cannot be emphasized too strongly.

This brief account of concrete technology represents the merest outline of an important subject. Anyone in practice who is actually responsible for any of the phases of producing and placing concrete must familiarize himself with the details in much greater depth. In addition to the literature quoted in the preceeding articles, Refs. 1.8 to 1.11 are relevant in this connection.

1.7 STRENGTH AND DEFORMATION OF CONCRETE IN COMPRESSION

a. Short-time loading   Performance of a structure under load depends to a large degree on the stress-strain relationship of the material from which it is made, under the type of stress to which the material is subjected in the structure. Since concrete is used mostly in compression (see Art. 1.1), its compressive stress-strain curve is of primary interest. Such a curve is obtained by appropriate strain measurements in cylinder tests (Art. 1.6) or on the compression side in beams. Figure 1.4 shows a typical set of such curves, obtained at normal, moderate testing speeds on concretes 28 days old, for various cylinder strengths f_c'.

All the curves have somewhat similar character. They consist of an initial relatively straight elastic portion in which stress and strain are closely proportional, then begin to curve to the horizontal, reaching the maximum stress, i.e., the compressive strength, at a strain of approximately 0.002 in. per in., and finally show a descending branch. It is also seen that concretes of lower strength are less brittle, i.e., fracture at a larger maximum strain, than high-strength concretes.

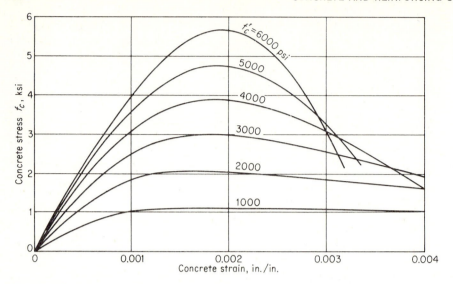

Fig. 1.4 Typical concrete stress-strain curves.

The *modulus of elasticity* E_c (in psi units), that is, the slope of the initial straight portion of the stress-strain curve, is seen to be larger the higher the strength of the concrete. It can be computed with reasonable accuracy from the empirical equation

$$E_c = 33w^{3/2} \sqrt{f_c'} \tag{1.1}$$

where w is the unit weight of the hardened concrete in pcf, and f_c' its cylinder strength in psi. Equation (1.1) has been obtained by testing structural concretes with values of w from 90 to 155 pcf. For normal sand-and-stone concretes, with $w = 145$ pcf, one obtains

$$E_c = 57,000 \sqrt{f_c'} \tag{1.2}$$

Information on concrete strength properties such as those discussed is usually obtained through tests made 28 days after pouring. However, cement continues to hydrate, and consequently concrete continues to harden, long after this age, at a decreasing rate. Figure 1.5 shows a typical curve of the gain of concrete strength with age.

In present practice, 28-day cylinder strengths in the range of

$$f_c' = 2500 \text{ to } 6000 \text{ psi}$$

are usually specified for reinforced-concrete structures, values between 3000 and 4000 being the most common. For prestressed concrete, higher strengths

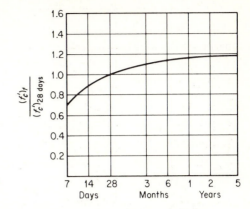

Fig. 1.5 Effect of age on compression strength f'_c. (*From Ref. 1.4.*)

are designated, from about 4000 to 8000 psi, with $f'_c = 5000$ to 6000 psi most customary.

It should be noted that the shape of the stress-strain curve for various concretes of the same cylinder strength, and even for the same concrete under various conditions of loading, varies considerably. An example of this is shown in Fig. 1.6, where different specimens of the same concrete are loaded at different rates of strain, from one corresponding to relatively fast loading (0.001 in. per in. per min) to one corresponding to an extremely slow applica-

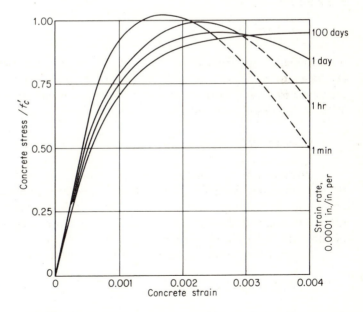

Fig. 1.6 Stress-strain curves at various strain rates, concentric compression. (*Adapted from Ref. 1.13.*)

tion of load (0.001 in. per in. per 100 days). It is seen that the descending branch of the curve, probably indicative of internal disintegration of the material, is much more pronounced at fast than at slow rates of loading. It is also seen that the peaks of the curves, i.e., the maximum strength reached, is somewhat smaller at slower rates of strain.

When compressed in one direction, concrete, like other materials, expands in the direction transverse to that of the applied stress. The ratio of the transverse to the longitudinal strain is known as *Poisson's ratio* and depends on composition and other factors. At stresses lower than about $0.7f'_c$, Poisson's ratio for concrete falls within the limits of 0.15 and 0.20, with 0.17 the most representative value.

b. Longtime loading   In some engineering materials, such as steel, strength and the stress-strain relationship are independent of rate and duration of loading, at least within the usual ranges of rate of stress, temperature, and other variables. In contrast, Fig. 1.6 illustrates the fact that the influence of time, in this case of rate of loading, on the behavior of concrete under load is pronounced. The main reason for this is that concrete creeps under load, while steel does not exhibit creep under conditions prevailing in buildings, bridges, and similar structures.

Creep is the property of many materials by which they continue deforming over considerable lengths of time at constant stress or load. The nature of the creep process is shown schematically in Fig. 1.7. This particular concrete was loaded at age 28 days with resulting instantaneous strain ϵ_{inst}.

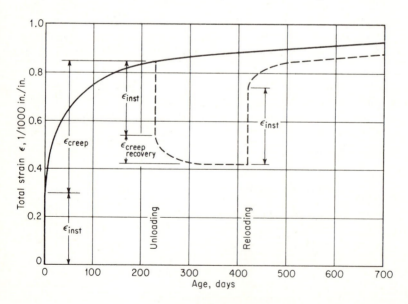

Fig. 1.7   Typical creep curve (concrete loaded to 600 psi at age 28 days).

The load was then maintained for 230 days, during which time creep is seen to have increased the total deformation to almost 3 times its instantaneous value. If the load were maintained, the deformation would follow the solid curve. If the load is removed, as shown by the dashed curve, the elastic instantaneous strain ϵ_{inst} is of course recovered, and some creep recovery is seen to occur. If the concrete is reloaded at some later date, instantaneous and creep deformations develop again, as shown.

Creep deformations for a given concrete are practically proportional to the magnitude of the applied stress; at any given stress, high-strength concretes show less creep than lower-strength concretes. As seen in Fig. 1.7, with elapsing time creep proceeds at a decreasing rate and ceases after 2 to 5 years at a final value which, depending on concrete strength and other factors, attains 1.5 to 3 times the magnitude of the instantaneous strain. If, instead of being applied quickly and thereafter kept constant, the load is increased slowly and gradually, as is the case in many structures during and after construction, instantaneous and creep deformations proceed simultaneously. The effect is that shown in Fig. 1.6; that is, the previously discussed difference in the shape of the stress-strain curve for various rates of loading is chiefly the result of the creep deformations of concrete.

For stresses not exceeding about one-half of the cylinder strength, creep strains are directly proportional to stress. Creep also depends on the average ambient relative humidity, being about twice as large for 50 per cent as for 100 per cent humidity. This is so because part of the reduction in volume under sustained load is caused by an outward migration of free pore water which evaporates into the surrounding atmosphere. Other things being equal, creep is larger for low-strength than for high-strength concretes. Table 1.1 refers to average humidity conditions. If ϵ'_{creep} is the final asymptotic value of the creep strain ϵ_{creep}, and ϵ_{inst} the initial, instantaneous strain (Fig. 1.7), the specific creep is ϵ'_{creep} per psi, and the creep coefficient $C_c = \epsilon'_{creep}/\epsilon_{inst}$.

To illustrate, if the concrete in a column, with $f'_c = 4000$ psi, is subject to a longtime load which causes sustained stress of 1200 psi, then after several years under load the final value of the creep strain will be of the order of $1200 \times 0.80 \times 10^{-6} = 0.00096$ in. per in. Thus, if the column were 20 ft long, creep would shorten it by about $\frac{1}{4}$ in.

Table 1.1   Creep parameters

Compressive strength, psi	Specific creep, 10^{-6} per psi	Creep coefficient C_c
3000	1.0	3.1
4000	0.80	2.9
6000	0.55	2.4
8000	0.40	2.0

Sustained loads affect not only the deformation, but also the strength of concrete. The cylinder strength f'_c is determined at normal rates of test loading (about 35 psi per sec). Tests by Rüsch and others have shown that for concentrically loaded, unreinforced-concrete prisms or cylinders, the *strength under sustained load* is significantly smaller than f'_c, of the order of 75 per cent of f'_c for loads maintained for a year or more.

When concrete is subject to fluctuating rather than sustained loading, its *fatigue strength*, as for all other materials, is considerably smaller than its static strength. When plain concrete in compression is stressed cyclically from zero to maximum stress, its fatigue limit is from 50 to 60 per cent of the static compression strength, for 2,000,000 cycles. A reasonable estimate can be made for other stress ranges using the modified Goodman diagram. For other types of applied stress, such as flexural compression stress in reinforced-concrete beams or flexural tension stress in unreinforced beams or on the tension side of reinforced beams, the fatigue limit likewise appears to be about 55 per cent of the corresponding static strength (Ref. 1.14).

1.8 TENSION STRENGTH

While concrete is best employed in a manner which utilizes its favorable compression strength, its tension strength is also of consequence in a variety of connections. Thus, the shear and torsion resistance of reinforced-concrete beams appears to depend primarily on the tension strength of the concrete. Also, the conditions under which cracks form and propagate on the tension side of reinforced-concrete flexural members depend strongly on the tension strength.

There are considerable experimental difficulties in determining true tensile strengths for concrete. In direct tension tests, minor misalignments and stress concentrations in the gripping devices are apt to mar the results. For many years, tension properties have been measured in terms of the *modulus of rupture*, i.e., that computed flexural tension stress Mc/I at which a test beam of plain concrete would fracture. Because this nominal stress is computed on the assumption that concrete is an elastic material, and because this bending stress is localized in the outermost fibers, it is apt to be larger than the strength of concrete in uniform axial tension. It is thus a measure of, but not identical with, the real axial tension strength.

In recent years the result of the so-called *split-cylinder test* has established itself as a measure of the tensile strength of concrete. A 6- $\times$ 12-in. concrete cylinder, the same type as is used for compression tests, is inserted in a compression-testing machine in the horizontal position, so that compression is applied uniformly along two opposite generatrices. Plywood pads are inserted between the compression platens of the machine and the cylinder in order to equalize and distribute the pressure. It can be shown that in an elastic cylinder so loaded, a nearly uniform tensile stress of magnitude $2P/\pi dL$ exists at right angles to the plane of load application. Correspondingly, such cylinders, when tested, split into two halves along that plane, at a stress which can be computed from the above expression. P is the applied compression load at failure, and d and L are the diameter and length of the cylinder. Because of local stress conditions at the load lines and the presence of stresses at right angles to the aforementioned tension stresses, the results of the split-cylinder tests likewise are not identical with (but are believed to be a good measure of) the true axial tensile strength. The results of all types of tensile tests show considerably more scatter than those of compression tests.

Tensile strength, whichever way determined, does not correlate well with the compression strength f'_c. It appears that for sand-and-gravel concrete the tensile strength depends primarily on the strength of bond between hardened cement paste and aggregate, whereas for lightweight concretes it depends largely on the tensile strength of the porous aggregate. The compression strength, on the other hand, is much less determined by these particular characteristics.

It appears that a reasonable estimate for the split-cylinder strength f'_{sp} is given by 6 to 7 times $\sqrt{f'_c}$ for sand-and-gravel concretes, and by 4 to 5 times $\sqrt{f'_c}$ for lightweight concretes.[1] The true tensile strength f'_t for the former appears to be of the order of 0.5 to 0.7 times f'_{sp}, and the flexural tension strength f_r (modulus of rupture) from 1.25 to 1.75 times f'_{sp}. The smaller of the foregoing factors apply to higher-strength concretes, and the larger to lower-strength concretes. These approximate expressions show that tension and compression strengths are by no means proportional and that any increase in compression strength, such as that achieved by lowering the water-cement ratio, is accompanied by a much smaller percentage increase in tensile strength (see also Fig. 1.2).

1.9 STRENGTH UNDER COMBINED STRESS

In many structural situations concrete is subjected simultaneously to various stresses acting in various directions. For instance, in beams much of the concrete is subject simultaneously to compression and shear stresses, in slabs and footings to compression in two perpendicular directions plus shear. By methods well known in the study of strengths of materials, any such state of combined stress, no matter how complex, can be reduced to three principal stresses acting at right angles to each other on an appropriately oriented elementary cube in the material. Any or all of the principal stresses can be either tension or compression. If one of them is zero, a state of *biaxial* stress is said to exist; if two of them are zero, the state of stress is *uniaxial*, either simple compression or simple tension. In most cases only the uniaxial strength prop-

---

[1] In these expressions, f'_c is expressed in psi, and the resulting f'_{sp} is obtained in psi. For example, for $f'_c = 3600$ psi, $f'_{sp} = 6\sqrt{f'_c}$ to $7\sqrt{f'_c} = 6 \times 60$ to $7 \times 60 = 360$ to 420 psi.

erties of a material are known from simple tests, such as the cylinder strength f_c' and the tensile strength f_t'. For predicting the strengths of structures in which concrete is subject to biaxial or triaxial stress, it would be desirable to be able to calculate the strength of concrete in such states of stress, knowing from tests only either f_c' or f_c' and f_t'.

In spite of extensive and continuing research, no general theory of the strength of concrete under combined stress has yet emerged. Modifications of various strength theories, such as the maximum-tension-stress, the Mohr, and the octahedral-stress theories, all of which are discussed in strength-of-materials texts, have been adapted with varying partial success to concrete (Refs. 1.16 to 1.19). At this time none of them is free from internal contradictions or limitations in applicability. One of the reasons for this lies in the highly nonhomogeneous nature of concrete and in the degree to which its behavior at high stresses and at fracture is influenced by microcracking and other discontinuity phenomena (Refs. 1.20 and 1.20a).

One current theory, successively developed by Cowan, Zia, and Mattock (Refs. 1.19, 1.21, and 1.22) and applied by them to a variety of practical

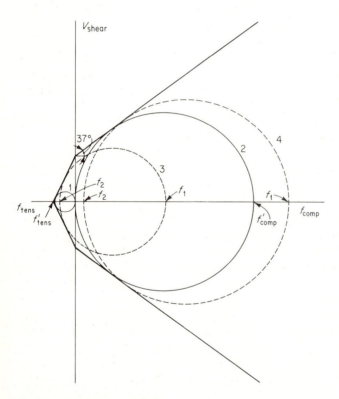

Fig. 1.8 Strength of concrete under combined stress. (*According to Refs.* 1.19 *and* 1.21.)

situations, is shown in Fig. 1.8. It represents a special form of the Mohr theory which defines a failure envelope such that any Mohr stress circle which is tangent to the envelope represents a combination of stresses that will cause failure of the material. In Fig. 1.8 circle 1 represents failure in simple tension at a stress f'_{tens}, circle 2 failure in compression at a stress f'_{comp}. If the cylinder strength f'_c is known, these stresses can be taken as $f'_{tens} = 6 \sqrt{f'_c}$ psi and $f'_{comp} = f'_c$, or more conservatively, $f'_{comp} = 0.85 f'_c$. The envelope, then, consists of the two straight lines shown. On the compression side the line has an inclination of 37° and is tangent to circle 2. This implies that the resistance of concrete to compression consists of a cohesive portion given by the intercept with the ordinate axis and by a friction contribution, 37° being a reasonable average value of the angle of internal friction of concrete. On the tension side the line is laid from the intercept tangent to circle 1.

The dashed circles show two states of combined stress. In circle 3 a principal tension stress f_2 acts at right angles to a principal compression stress f_1. It is seen that $f_1 < f'_{comp}$; that is, when tension acts at right angles to compression, the compression strength of concrete is reduced compared with its strength f'_{comp} in simple compression. In circle 4 a smaller principal compression stress f_2 acts at right angles to the larger principal compression stress f_1. It is seen that here $f_1 > f'_{comp}$; that is, when compression acts at right angles to compression, the compressive strength of concrete is increased compared with that in uniaxial compression. Similar relations can be read off the figure for concrete strength in shear as it is affected by simultaneous compression or tension.

This theory suffers from the same defect as the general Mohr theory in implying that the intermediate of the three principal stresses has no effect on strength. In particular, for the frequent case of biaxial stress, it predicts that if both principal stresses are of the same sign, there is no effect of biaxiality on strength. Recent tests have shown this to be a rather crude approximation, as seen from Fig. 1.9, which presents test evidence on concrete strength under biaxial stress (Ref. 1.23). It is seen that when equal compression stresses act in both perpendicular directions, the strength is increased by about 15 per cent over the uniaxial cylinder strength and that this increase rises to about 23 per cent when the smaller applied compression stress is about 70 per cent of the larger. Much larger strength increases obtain if compression stresses act in all three principal directions, as can be read from Fig. 1.8.

Information of the type presented in Figs. 1.8 and 1.9 permits one to visualize correctly the effects of bi- and triaxial stresses qualitatively and to obtain reasonable quantitative approximations in simple situations. However, the fact that the strength of concrete under combined stress cannot yet be calculated rationally and, equally important, that in many situations in concrete structures it is nearly impossible to calculate all the acting stresses and their directions is one of the main reasons for continued reliance on tests. Because of this, the design of reinforced-concrete structures continues to be

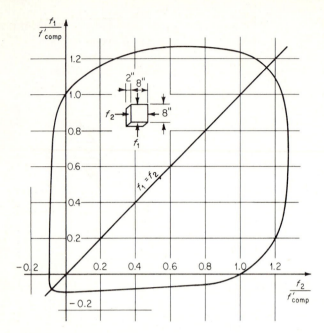

Fig. 1.9 Strength of concrete in biaxial stress. (*Adapted from Ref.* 1.23.)

based more on extensive experimental information than on consistent analytical theory, particularly in the many situations where combined stresses occur.

1.10 VOLUME CHANGES: SHRINKAGE, TEMPERATURE

The deformations discussed in Art. 1.7 were induced by stresses caused by external loads. Influences of a different nature cause concrete, even when free of any external loading, to undergo deformations and volume changes. The most important of these are shrinkage and the effects of temperature variations.

a. Shrinkage   As was discussed in Arts. 1.2 and 1.4, any workable concrete mix contains more water than is needed for hydration. If the concrete is exposed to air, the larger part of this free water evaporates in time, the rate and completeness of drying depending on ambient temperature and humidity conditions. As the concrete dries, it shrinks in volume, probably due to the capillary tension which develops in the water remaining in the concrete. Conversely, if dry concrete is immersed in water, it expands, regaining much of the volume loss from prior shrinkage. Shrinkage, which continues at a decreasing rate for several months, depending on the configuration of the member, is a detrimental property of concrete in several respects. When not

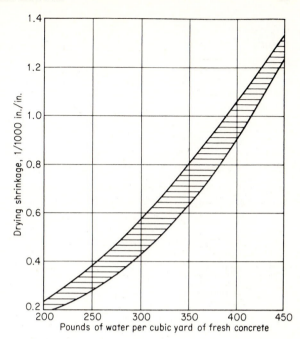

Fig. 1.10 Effect of water content on drying shrinkage. (*From Ref.* 1.4.)

adequately controlled, it will cause unsightly and often deleterious cracks, as in slabs, walls, etc. In structures which are statically indeterminate (and most concrete structures are), it can cause large and harmful stresses. In prestressed concrete it leads to partial loss of initial prestress. For these reasons it is essential that shrinkage be minimized and controlled.

As is clear from the nature of the process, the chief factor which determines the amount of final shrinkage is the unit water content of the fresh concrete. This is clearly illustrated in Fig. 1.10, which shows the amount of shrinkage in units of 0.001 in. per in. for varying amounts of mixing water. The same aggregates were used for all tests, but in addition to and independently of water content, the amount of cement was also varied, from 4 to 11 sacks per cubic yard of concrete. This very large variation of cement content had only very minor effects on the amount of shrinkage, as compared with the effect of water content; this is evident from the narrowness of the band which comprises the test results for the widely varying cement contents. It is evident from this that the chief means of reducing shrinkage is to reduce the water content of the fresh concrete to the minimum compatible with the required workability. In addition, prolonged and careful curing is beneficial for shrinkage control.

Values of final shrinkage for ordinary concretes are generally of the order of 0.0002 to 0.0007 in. per in., depending on initial water content, ambient temperature and humidity conditions, and the nature of the aggregate. Highly absorptive aggregates, such as some sandstones and slates, result in shrinkage values two and more times those obtained with less absorptive materials, such as granites and some limestones. Some lightweight aggregates, in view of their great porosity, easily result in much larger shrinkage values than ordinary concretes.

b. Temperature changes Like most other materials, concrete expands with increasing temperature and contracts with decreasing temperature. The effects of such volume changes are similar to those caused by shrinkage. That is, temperature contraction can lead to undue cracking, particularly when superposed on shrinkage; in indeterminate structures, deformations due to temperature changes can cause large and occasionally harmful stresses.

The coefficient of expansion varies somewhat, depending on the type of aggregate and richness of the mix. It is generally within the range of 0.000004 to 0.000006 in. per in. per °F. A value of 0.0000055 is generally accepted as satisfactory for calculating stresses and deformations caused by temperature changes.

1.11 LIGHTWEIGHT CONCRETE

As was discussed in Art. 1.3, lightweight aggregates are widely used to produce lightweight concretes for a variety of purposes: purely for insulation, for moderate-strength fills, and for structural use. The latter are known as *structural lightweight aggregate concretes* and are defined as "concretes having 28-day compressive strength in excess of 2500 psi and 28-day air-dry unit weight not exceeding 115 pcf." Most lightweight concretes in structures weigh between 100 and 110 pcf. Specified compressive strengths usually range from 3000 to 4000 psi, but strengths up to and even exceeding 6000 psi can be obtained by appropriate mix proportioning. These strengths are adequate for the entire range of structural usage, for cast-in-place, precast, or prestressed concrete construction. Because the closer and more critical control required in producing structural lightweight concretes is more easily achieved in precasting plants (Ref. 1.3), their use has become particularly widespread in precast and prestressed construction.

The modulus of elasticity of structural lightweight concrete is from one-half to three-quarters of that of stone concretes of the same strength and can be calculated satisfactorily from Eq. (1.1). Creep and shrinkage deformations are generally larger in lightweight than in stone concrete by amounts ranging from about 10 to 100 per cent. The split-cylinder tensile strength of air-dried lightweight concrete ranges from 70 to about 90 per cent of that of stone concrete of the same compressive strength, while it is about the same in both kinds of concrete when they are continually moist-cured. The modulus of rupture of lightweight concrete is about the same as that of stone concrete when moist-cured; for air-dried conditions it is probably smaller, but adequate data seem to be lacking. More detailed information on these and other performance characteristics of lightweight concretes are found in Ref. 1.2.

It follows from these comparisons of structural performance characteristics that structural lightweight aggregate concrete is a perfectly satisfactory material in concrete construction. However, it does require special considerations in design where performance depends on creep and shrinkage, as in determining deflections under longtime loads, and where tensile strength plays a significant role, as in shear and bond strength and for crack control.

1.12 REINFORCING STEELS

a. General   As compared with concrete, steel is a high-strength material. The useful strength of ordinary reinforcing steels in tension as well as in compression, i.e., the yield strength, is of the order of 10 times the compression strength of common structural concrete, or of the order of 100 times its tensile strength. On the other hand, steel is a high-cost material as compared with concrete. It follows that the two materials are best used in combination if the concrete is made to resist the compression stresses, and the steel the tension stresses. Thus, in reinforced-concrete beams the concrete resists the compression force, longitudinal steel rods are located close to the tension face to resist the tension force, and frequently additional steel bars are so disposed that they resist the inclined tension stresses which are caused by the shear force in the webs of beams. However, reinforcement is also used for resisting compression forces primarily where it is desired to reduce the cross-sectional dimensions of compression members, as in the lower-floor columns of multistory buildings. Even if no such necessity exists, a minimum amount of reinforcement is placed in all compression members to safeguard them against the effects of small accidental bending moments which might crack and even fail an unreinforced member.

For most effective reinforcing action, it is essential that steel and concrete deform together, i.e., that there be a sufficiently strong *bond* between the two materials so that no relative movements of the steel bars and the surrounding concrete occur. This bond is provided by the relatively large *chemical adhesion* which develops at the steel-concrete interface, by the *natural roughness* of the mill scale of hot-rolled reinforcing bars, and by the closely spaced rib-shaped *surface deformations* with which reinforcing bars are furnished in order to provide a high degree of interlocking of the two materials.

Additional features which make for the satisfactory joint performance of steel and concrete are the following: (1) The *thermal expansion coefficients* of the two materials, about 0.0000065 for steel vs. an average of 0.0000055 for concrete, are sufficiently close to forestall cracking and other undesirable effects of differential thermal deformations. (2) While the *corrosion resistance* of bare steel is poor, the concrete which surrounds the steel reinforcement provides excellent corrosion protection, minimizing corrosion problems and corresponding maintenance costs. (3) The *fire resistance* of unprotected steel is impaired by its high thermal conductivity and by the fact that its strength decreases sizably at high temperatures. Conversely, the thermal conductivity of concrete is relatively low. Thus, damage caused by even prolonged fire exposure, if any, is generally limited to the outer layer of concrete, and a moderate amount of concrete cover provides sufficient thermal insulation for the embedded reinforcement.

b. Types of reinforcement   Steel is used in two different ways in concrete structures: as reinforcing steel and as prestressing steel. Reinforcing steel is placed in the forms prior to casting of concrete. Stresses in the steel, as in the hardened concrete, are caused only by the loads on the structure, except for possible parasitic stresses from shrinkage or similar causes. In contrast, in prestressed-concrete structures large tension forces are applied to the reinforcement prior to letting it act jointly with the concrete in resisting external loads. The steels for these two usages are very different and will be discussed separately.

The most common type of *reinforcing steel* (as distinct from prestressing steel) is in the form of round bars. They are available in a large range of diameters, from about $\frac{1}{4}$ to about $1\frac{3}{8}$ in. for ordinary applications, and in two heavy bars of about $1\frac{3}{4}$ and $2\frac{1}{4}$ in. These bars, with the exception of the $\frac{1}{4}$-in. size, are furnished with surface deformations for the purpose of increasing the bond strength between steel and concrete. Minimum requirements for these deformations (spacing, projection, etc.) have been developed in lengthy experimental research and are specified in ASTM Specification A305. Different bar producers use different patterns, all of which satisfy these requirements. Figure 1.11 shows a variety of current types of deformations.

Bar sizes are designated by numbers, Nos. 2 to 11 being commonly used,

Fig. 1.11   Types of deformed bars.

and Nos. 14 and 18 representing the two special large-size bars previously mentioned. Designation by number, instead of by diameter, has been introduced because the surface deformations make it impossible to define a single easily measured value of the diameter. The numbers are so arranged that the unit in the number designation corresponds closely to $\frac{1}{8}$ in. of diameter size. That is, a No. 5 bar has a nominal diameter of $\frac{5}{8}$ in., and similarly for the other sizes. Table 1 of the Appendix gives areas, perimeters, and weights of standard bars. Tables 2 to 4 give similar information for groups of bars.

Apart from single reinforcing bars, *welded wire fabric* is often used for reinforcing slabs and other surfaces, such as shells. This consists of a series of longitudinal and transverse cold-drawn steel wires at right angles to each other and welded together at all points of intersection. Size and spacing of wires may be the same in both directions or may be different, depending on the requirements of the design. Detailed data can be found in Ref. 1.24 and in commercial catalogues.

Prestressing steel is used in three forms: wire strands, single wires, and high-strength bars. Wire strands are of the seven-wire type in which a center wire is enclosed tightly by six helically placed outer wires with a pitch of 12 to 16 times the nominal diameter of the strand. Strand diameters range from $\frac{1}{4}$ to $\frac{1}{2}$ in. Prestressing wire ranges in diameter from 0.192 to 0.276 in. It is made by cold-drawing from high-carbon steel, just as the individual wires for wire strand. High-strength alloy-steel bars for prestressing are available in diameters from $\frac{3}{4}$ to $1\frac{3}{8}$ in.

Table 1.2   Standardized reinforcing steels

Type	ASTM Standard No.	Grade	Specified minimum Yield strength, ksi	Specified minimum Tensile strength, ksi
Deformed billet steel bars and	A615	40	40	70
deformed axle steel bars	A617	60	60	90
		75*	75	100
Deformed rail steel bars	A616	50	50	80
		60	60	90
Cold-drawn wire	A82		70	80
Welded wire fabric	A185 and	Size W1.5	65	75
	A82	and larger		
		Smaller than W1.5	56	70
Deformed steel wire	A496		70	80
Welded deformed wire fabric	A497		70	80

* Grade 75 furnished only in sizes 11, 14, 18; not produced from axle steel.

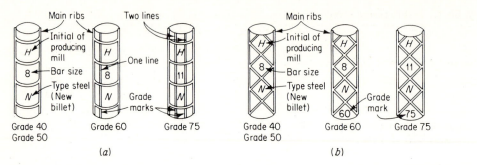

Fig. 1.12   Marking system for reinforcing bars. (*Adapted from Ref.* 1.24.)

c. Grades and strengths   In reinforced concrete a longtime trend is evident toward the use of higher-strength materials, both steel and concrete. Reinforcing bars with 33-ksi yield points, which were widespread not too long ago, are no longer made, and steels with 60-ksi yield point are in the process of replacing the 40-ksi steel which was prevalent until quite recently. Large-diameter bars with 75-ksi yield points are available, particularly for use in columns. Table 1.2 lists all presently available reinforcing steels, their designations, the ASTM specifications which define their properties in detail, and their two main minimum specified strength values.

In order for bars of various grades and sizes to be easily distinguished, which is necessary to avoid accidental use of lower-strength or smaller-size bars than called for in the design, all deformed bars are furnished with rolled-in markings. These identify the producing mill (usually an initial), the bar size number (3 to 18), the type of steel (N for billet, A for axle, and a rail sign for rail steel), and an additional marking for identifying the higher-strength steels. These markings consist either of one or two longitudinal lines for 60- and 75-ksi yield steels, respectively, or the numbers 60 and 75. The identification marks are shown in Fig. 1.12.

Prestressing is practical only when steels of very high strength are used. Table 1.3 summarizes the essential characteristics of these steels.

While the modulus of elasticity E_s of reinforcing steels can be taken as 29×10^6 psi, the effective modulus of prestressing steels varies, depending on type of steel (e.g., strand vs. bars) and type of usage, and should be ascertained by test or supplied by the manufacturer. For unbonded strand (i.e., strand not embedded in concrete) the modulus may be as low as 26×10^6 psi.

d. Stress-strain curves   The two chief numerical characteristics which determine the character of reinforcement are its yield point (generally identical in tension and in compression) and its *modulus of elasticity E*. The latter is practically the same for all reinforcing steels (but not for prestressing steels) and is taken as $E = 29,000,000$ psi.

Table 1.3 Prestressing steels

| Type | Grade or size | Specified minimum | | Elongation |
		Tensile strength, ksi	Yield strength, ksi	
Seven-wire stress-relieved strand ASTM A416	Grade 260 Grade 270	250 270	212 230	3.5% in 24-in. gage length
Uncoated stress-relieved wire ASTM A421	0.192-in. diam 0.276-in. diam	250 235	200 188	4% in 10-in. gage length
High-strength alloy-steel bars* AISI 5160 and 9260	Regular Special	145 160	130 130	4% in 10-in. gage length

* The ACI Code specifies that such bars shall be prestressed to 87 per cent of the guaranteed tensile strength f'_s, after which they shall show a minimum yield strength of $0.87f'_s$, a minimum elongation of 4 per cent in a gage length of 20 diam, and a minimum reduction in area at rupture of 20 per cent.

In addition, however, the shape of the *stress-strain curve*, and particularly of its initial portion, has significant influence on the performance of reinforced-concrete members. Typical stress-strain curves of standardized American reinforcing steels are shown in Fig. 1.13. The curve for a low-alloy reinforcing steel with 90-ksi specified yield point, which at this writing is in the process of development, is also shown in that figure. The complete stress-strain curves are shown in the left part of the figure; the right part gives the initial portions of the curves magnified 10 times.

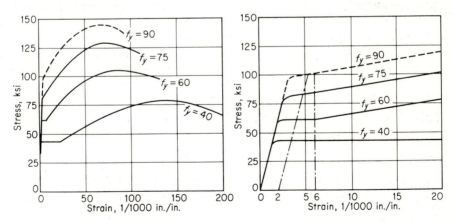

Fig. 1.13 Typical stress-strain curves for current reinforcing steels with minimum specified yield points f_y from 40 to 90 ksi.

Low-carbon steels, typified by the 40-ksi curve, show an elastic portion followed by a "yield plateau," i.e., by a horizontal portion of the curve where strain continues to increase at constant stress. For such steels the yield point is that stress at which the yield plateau establishes itself. With further strains the stress begins to increase again, though at a slower rate, a process which is known as *strain-hardening*. The curve flattens out when the tensile strength is reached; it then turns down until fracture occurs. Higher-strength carbon steels, such as those with 60- or 75-ksi yield points, either have a yield plateau of much shorter length or enter strain-hardening immediately without any continued yielding at constant stress. In the latter case the yield point is defined as the stress at which the total strain under load is 0.5 to 0.6 per cent, as shown. Low-alloy, high-strength steels rarely show any yield plateau and usually enter strain-hardening immediately upon beginning to yield.

Figure 1.14 shows stress-strain curves for prestressing strand and for high-strength prestressing bars. For comparison, the initial portion of a stress-strain curve for a 60-ksi yield-strength reinforcing bar is also shown. It is seen that, in contrast to reinforcing bars, these prestressing steels do not have a definite yield plateau; i.e., they do not yield at constant or nearly constant stress. Yielding develops gradually, and in the inelastic range the curve continues to rise smoothly until the tensile strength f_{pu} is reached, with a corresponding total elongation at rupture ϵ'_s. Because discontinuous yielding is not observed in these steels, the yield strength is somewhat arbitrarily defined as the stress at a total elongation of 1 per cent for strand and wire, and at 0.2 per cent permanent strain for high-strength bars. Figure 1.14 shows that the yield strengths so defined represent a good limit below which stress and strain are fairly proportional, and above which strain increases much more rapidly with increasing stress. It is also seen that the spread between tensile strength

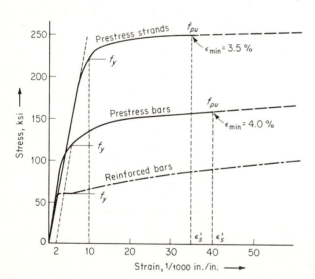

Fig. 1.14 Typical stress-strain curves for prestressing steels.

f_{pu} and yield strength f_y is much smaller in prestressing than in reinforcing steels. Lastly, by comparison with Fig. 1.13, it is seen that, in spite of the much greater strength, the elongation at fracture, ϵ'_s, is considerably smaller in prestressing than in reinforcing steels; that is, the former possess less ductility than the latter. (The ϵ_{min} values shown in Fig. 1.14 are the specified minimum elongations given in Table 1.3. Actual measured elongations usually exceed these minimum values, as indicated in the figure.)

In contrast to reinforcing steels, prestressing steels under service conditions (i.e., as stressed by the initial prestressing and by the loads acting on the member) exhibit creep somewhat analogous to that of concrete. While creep is defined as increase of strain under constant stress, for prestressing steels it is more convenient to describe the same phenomenon in terms of *stress relaxation*, i.e., the decrease of stress at constant strain. To be specific, if a length of high-strength prestressing steel is stressed to a sizable fraction of its yield strength (say 80 to 90 per cent) and held at a constant strain (e.g., between immovable jaws), the steel stress f_s will gradually decrease from its initial value f_{si}. In prestressed concrete members this stress relaxation is important in regard to distribution of internal stresses and magnitude of deflections at some time after initial prestress was applied.

The amount of relaxation varies, depending on type and grade of steel. However, a satisfactory estimate can be obtained from an equation which was derived from more than 400 relaxation tests of up to nine years duration (Ref. 1.25):

$$\frac{f_s}{f_{si}} = 1 - \frac{\log t}{10}\left(\frac{f_{si}}{f_y} - 0.55\right) \tag{1.3}$$

where f_s is the steel stress t hr after the application of the initial stress f_{si}, and $\log t$ is to the base 10. It is seen that the relaxation is greater the larger the initial stress relative to the yield stress f_y. Thus, it is seen that there is no stress relaxation if f_{si} is less than 55 per cent of f_y.

REFERENCES

1.1. Selection and Use of Aggregates for Concrete, reported by ACI Committee 621, *J. ACI*, vol. 58, p. 515, 1961.
1.2. Guide for Structural Lightweight Aggregate Concrete, reported by ACI Committee 213, *J. ACI*, vol. 64, p. 422, 1967.
1.3. Selecting Proportions for Structural Lightweight Concrete, ACI Standard 211.2-69.
1.4. G. E. Troxell, H. E. Davis, and J. W. Kelly: "Composition and Properties of Concrete," 2d ed., McGraw-Hill Book Company, New York, 1968.
1.5. Selecting Proportions for Concrete, ACI Standard 613-54.
1.6. T. C. Hsu and F. O. Slate: Tensile Bond Strength between Aggregate and Cement Paste or Mortar, *J. ACI*, vol. 60, p. 465, 1963.
1.7. Consolidation of Concrete, reported by ACI Committee 609, *J. ACI*, vol. 57, p. 985, 1961.
1.8. Measuring, Mixing and Placing of Concrete, ACI Standard 614-59.

1.9. Cold Weather Concreting, ACI Standard 306-66.

1.10. Hot Weather Concreting, ACI Standard 605-59.

1.11. A. M. Neville: "Properties of Concrete," John Wiley & Sons, Inc., New York, 1963.

1.12. E. Hognestad, N. W. Hansson, and D. McHenry: Concrete Stress Distribution in Ultimate Strength Design, *J. ACI*, vol. 55, p. 455, 1955.

1.13. H. Rüsch: Research toward a General Flexural Theory for Structural Concrete, *J. ACI*, vol. 57, p. 1, 1960.

1.14. G. M. Nordby: Fatigue of Concrete: A Review of Research, *J. ACI*, vol. 55, p. 191, 1958.

1.15. Evaluation of Compression Test Results of Field Concrete, ACI Standard 214-65.

1.16. S. Timoshenko: "Strength of Materials," part II, 3d ed., D. Van Nostrand Company, Inc., Princeton, N.J., 1956.

1.17. B. Bresler and K. S. Pister: Strength of Concrete under Combined Stress, *J. ACI*, vol. 55, p. 321, 1958.

1.18. D. McHenry and J. Karni: Strength of Concrete under Combined Tensile and Compressive Stress, *J. ACI*, vol. 54, p. 829, 1958.

1.19. H. J. Cowan: The Strength of Plain, Reinforced and Prestressed Concrete under the Action of Combined Stresses, *Mag. Concrete Res.*, vol. 5, p. 75, 1953.

1.20. T. C. Hsu, F. O. Slate, G. M. Sturman, and G. Winter: Microcracking of Plain Concrete and the Shape of the Stress-Strain Curve, *J. ACI*, vol. 60, p. 209, 1963.

1.20a. S. P. Shah and G. Winter: Inelastic Behavior and Fracture of Concrete, in "Causes, Mechanism and Control of Cracking in Concrete," ACI Publ. SP-20, p. 5, 1968.

1.21. P. Zia: Torsional Strength of Prestressed Concrete Members, *J. ACI*, vol. 57, p. 1337, 1961.

1.22. J. F. Hofbek, I. O. Ibrahim, and A. H. Mattock: Shear Transfer in Reinforced Concrete, *J. ACI*, vol. 66, p. 119, 1969.

1.23. H. Kupfer: Performance of Concrete in Biaxial Stress (in German), Lehrstuhl für Massivbau, *Tech. Univ. Munich Rept.* 78, 1969.

1.24. Concrete Reinforcing Institute: Manual of Standard Practice, 19th ed., Chicago, 1968.

1.25. W. G. Corley, M. A. Sozen, and C. P. Siess: Time-dependent Deflections of Prestress Concrete Beams, *Highway Res. Board Bull.* 307, pp. 1–25, 1961.

2

Mechanics and Behavior of Reinforced Concrete

2.1 FUNDAMENTALS

The chief task of the structural engineer is the design of structures. By *design* is meant the determination of the general shape and all specific dimensions of a particular structure so that it will perform the function for which it is created and will safely withstand the influences which will act on it throughout its useful life. These influences are primarily the loads and other forces to which it will be subjected, as well as other detrimental agents, such as temperature fluctuations, foundation settlements, and corrosive influences. *Structural mechanics* is one of the main tools in this process of design. As here understood, it is the body of scientific knowledge which permits one to predict with a good degree of certainty how a structure of given shape and dimensions will behave when acted upon by known forces or other mechanical influences. The chief items of behavior which are of practical interest are (1) the strength of the structure, i.e., that magnitude of loads of a given distribution which will cause the structure to fail, and (2) the deformations, such as deflections and extent of cracking, which the structure will undergo when loaded under service conditions.

35

In this chapter, the structural mechanics of reinforced-concrete members will be developed in general terms. Specific methods of design and analysis, based on the information provided by structural mechanics but frequently simplified and conventionalized to various degrees, will be taken up in subsequent chapters.

The fundamental propositions on which the mechanics of reinforced concrete is based are as follows:

1. The internal forces, such as bending moments, shear forces, and normal and shear stresses, at any section of a member are in equilibrium with the effects of the external loads at that section. This proposition is not an assumption but a fact, because any body or any portion thereof can be at rest only if all forces acting on it are in equilibrium.

2. The strain in an embedded reinforcing bar (unit extension or compression) is the same as that of the surrounding concrete. Expressed differently, it is assumed that perfect bonding exists between concrete and steel at the interface, so that no slip can occur between the two materials. Hence, as the one deforms, so must the other. With modern deformed bars (see Art. 1.12) a high degree of mechanical interlocking is provided in addition to the natural surface adhesion, so that this assumption is very close to correct.

3. Cross sections which were plane prior to loading continue to be plane in the member under load. Accurate measurements have shown that when a reinforced-concrete member is loaded close to failure, this assumption is not accurate. However, the deviations are usually minor, and the results of theory based on this assumption check well with extensive test information.

4. In view of the fact that the tensile strength of concrete is only a small fraction of its compressive strength (see Art. 1.8), the concrete in that part of a member which is in tension is usually cracked. While these cracks, in well-designed members, are generally so narrow as to be hardly visible (they are known as *hairline* cracks), they evidently render the cracked concrete incapable of resisting tension stress. Correspondingly, it is assumed that concrete is not capable of resisting any tension stress whatever. This assumption is evidently a simplification of the actual situation because, in fact, concrete prior to cracking, as well as the concrete located between cracks, does resist tension stresses of small magnitude. Also, in discussions of the resistance of reinforced-concrete beams to shear, it will become apparent that under certain conditions this particular assumption is dispensed with and advantage is taken of the modest tension strength which concrete can develop.

5. The theory is based on the actual stress-strain relationships and strength properties of the two constituent materials (see Arts. 1.7 and 1.10) or some reasonable simplifications thereof, rather than on some assumptions of ideal material behavior. This last feature is a relatively new develop-

ment. It supplants a method of analysis which was based on the assumption that both concrete and steel behave in an elastic manner at all stress levels relevant in calculations. This latter assumption, while leading to relatively simple computational methods, is very far from true (see Figs. 1.4, 1.6, 1.13, and 1.14). Correspondingly, the newer methods of analysis which recognize the departure from elasticity at the higher stress levels are in much closer agreement with the actual behavior of structures and with very extensive experimental information. At the same time the facts that nonelastic behavior is reflected in modern theory, that concrete is assumed to be ineffective in tension, and that the joint action of the two materials is taken into consideration result in analytical methods which are considerably more complex, but also more challenging, than those which are adequate for members made of a single, substantially elastic material.

These five assumptions permit one to predict by calculation the performance of reinforced-concrete members only for some simple situations. Actually, the joint action of two materials as dissimilar and complicated as concrete and steel is so complex that it has not yet lent itself to purely analytical treatment. For this reason, methods of design and analysis, while utilizing these assumptions, are very largely based on the results of extensive and continuing experimental research. They are modified and improved as additional test evidence becomes available.

2.2 AXIAL COMPRESSION

In members which sustain chiefly or exclusively axial compression loads, such as building columns, it is economical to make the concrete carry most of the load. Still, some steel reinforcement is always provided for various reasons. For one, very few members are truly axially loaded; steel is essential for resisting any bending that may exist. For another, if part of the total load is carried by steel with its much greater strength, the cross-sectional dimensions of the member can be reduced, the more so the larger the amount of reinforcement.

The two chief forms of reinforced-concrete columns are shown in Fig. 2.1. In the square column, the four longitudinal bars serve as main reinforcement. They are held in place by transverse small-diameter steel ties which prevent displacement of the main bars during construction operations and counteract any tendency of the compression-loaded bars to buckle out of the concrete by bursting the thin outer cover. On the left is shown a round column with eight main reinforcing bars. These are surrounded by a closely spaced spiral which serves the same purpose as the more widely spaced ties and acts to confine the concrete within it, thereby increasing its resistance to axial compression. The discussion which follows applies to tied columns.

When axial load is applied, the compression strain is the same over the entire cross section, and in view of the bonding between concrete and steel,

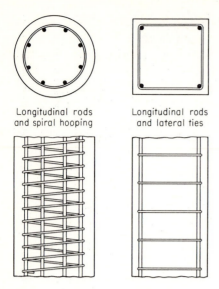

Longitudinal rods   Longitudinal rods
and spiral hooping   and lateral ties

Fig. 2.1   Reinforced-concrete column.

is the same in the two materials (see propositions 2 and 3 on page 36). To
illustrate the action of such a member as load is applied, Fig. 2.2 shows two
typical stress-strain curves, one for a concrete with $f_c' = 3000$ psi (3 ksi) and
the other for a steel with $f_y = 40,000$ psi (40 ksi). The curves for the two
materials are drawn on the same graph, using different vertical stress scales.
Curve b has the shape which would be obtained in a concrete cylinder test.
The rate of loading in most structures is considerably slower than that in a
cylinder test, and this affects the shape of the curve (see Art. 1.7). Curve c,
therefore, is drawn as being characteristic of the performance of concrete
under slow or sustained loading. Under these conditions the maximum available compression strength off reinforced concrete is about $0.85f_c'$, as shown (see
also Fig. 1.6).

a. Elastic behavior (working stresses)   At low stresses, up to about $f_c'/2$, the
concrete is seen to behave nearly elastically, i.e., stresses and strains are quite
closely proportional; the straight line d represents this range of behavior with
little error for both rates of loading. For the given concrete the range extends
to a strain of about 0.00045. The steel, on the other hand, is seen to be elastic
nearly to its yield point of 40 ksi, or to the much greater strain of about 0.0013.
    Because the compression strain in the concrete, at any given load, is
equal to the compression strain in the steel,

$$\epsilon_c = \frac{f_c}{E_c} = \epsilon_s = \frac{f_s}{E_s}$$

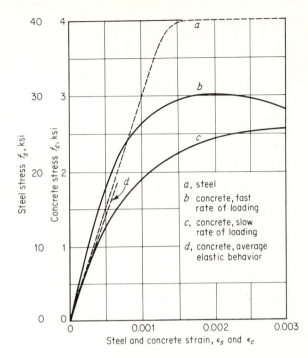

Fig. 2.2 Concrete and steel stress-strain curves.

from which the relation between the steel stress f_s and the concrete stress f_c is obtained as

$$f_s = \frac{E_s}{E_c} f_c = n f_c \qquad (2.1)$$

where $n = E_s/E_c$ is known as the *modular ratio*.

Let A_c = net area of concrete, i.e., gross area minus area occupied by reinforcing bars

$\quad A_g$ = gross area

$\quad A_s$ = area of reinforcing bars

$\quad P$ = axial load

Then

$$P = f_c A_c + f_s A_s = f_c A_c + n f_c A_s$$

or

$$P = f_c (A_c + n A_s) \qquad (2.2)$$

The term $(A_c + n A_s)$ can be interpreted as the area of a fictitious concrete cross section, the so-called *transformed area* which, when subjected to the particular concrete stress f_c, results in the same axial load P as the actual section composed of both steel and concrete. This transformed concrete area is seen to consist of the actual concrete area plus n times the area of the rein-

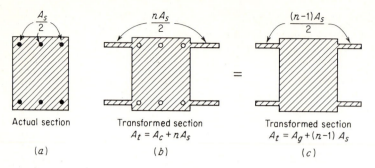

Fig. 2.3 Transformed section in axial compression.

forcement. It can be visualized as shown in Fig. 2.3. That is, in Fig. 2.3*b* the three bars along each of the two faces are thought of as removed and replaced, at the same distance from the axis of the section, with added areas of fictitious concrete of total amount nA_s. Alternatively, as shown in Fig. 2.3*c*, one can think of the area of the steel bars as replaced with concrete, in which case one has to add to the gross concrete area A_g so obtained only $(n - 1)A_s$ in order to obtain the same total transformed area. Therefore, alternatively,

$$P = f_c[A_g + (n - 1)A_s] \tag{2.2a}$$

If load and cross-sectional dimensions are known, the concrete stress can be found by solving Eq. (2.2) or (2.2a) for f_c, and the steel stress can be calculated from Eq. (2.1). These relations hold in the range in which the concrete behaves nearly elastically, i.e., up to about 50 to 60 per cent of f'_c. For reasons of safety and serviceability, concrete stresses in structures under design loads are kept within this range. Therefore, these relations permit one to calculate *working stresses*, i.e., stresses which act in a member when it carries the working or design loading.

Example   A column of the materials defined in Fig. 2.2 has a cross section of 16 × 20 in. and is reinforced by six No. 9 bars disposed as shown in Fig. 2.3. Determine the axial load which will stress the concrete to 1000 psi.
     One finds $A_g = 16 \times 20 = 320$ in.², and from Table 2 of the Appendix, $A_s = 6.00$ in. From Eq. (1.2), $E_c = 57,000 \sqrt{3000} = 3,120,000$ psi, and $n = 29,000,000/3,120,000 = 9$. (In view of the scatter inherent in E_c, it is customary and satisfactory to round off the value of n to the nearest integer.) The load on the column, from Eq. (2.2a), is $P = 1000[320 + (9 - 1)6.00] = 368,000$ lb. Of this total load the concrete is seen to carry $P_c = f_c A_c = f_c(A_g - A_s) = 1000(320 - 6) = 314,000$ lb, and the steel $P_s = f_s A_s = (nf_c)A_s = 9000 \times 6 = 54,000$ lb, which is 14.7 per cent of the total axial load.

b. Inelastic range   Inspection of Fig. 2.2 shows that the elastic relationships which have been utilized so far cannot be used beyond a strain of about 0.0005, for the given concrete. To obtain information on the behavior of the member at larger strains and, correspondingly, at larger loads, it is therefore necessary to make direct use of the information of Fig. 2.2.

By way of example one may want to calculate the magnitude of the axial load which will produce a strain or unit shortening $\epsilon_c = \epsilon_s = 0.0010$ in the column of the preceding example. At this strain the steel is seen to be still elastic, so that the steel stress $f_s = \epsilon_s E_s = 0.001 \times 29,000,000 = 29,000$ psi. The concrete is in the inelastic range, so that its stress cannot be directly calculated, but it can be read from the stress-strain curve for the given value of strain. (1) If the member has been loaded at a fast rate, curve b holds at the instant when the entire load is applied. The stress for $\epsilon = 0.001$ can be read as $f_c = 2660$ psi. Consequently, the total load can be obtained from

$$P = f_c A_c + f_s A_s \tag{2.3}$$

which evidently applies in the inelastic as well as in the elastic range. Hence, $P = 2660(320 - 6) + 29,000 \times 6 = 835,000 + 174,000 = 1,009,000$ lb. Of this total load, the steel is seen to carry 174,000 lb, or 17.3 per cent. (2) For slowly applied or sustained loading, curve c represents the behavior of the concrete. Its stress at a strain of 0.001 can be read as $f_c = 1900$ psi. Then $P = 1900 \times 314 + 29,000 \times 6 = 596,000 + 174,000 = 770,000$ lb. Of this total load, the steel is seen to carry 22.6 per cent.

Comparison of the results for fast and slow loading shows the following: Owing to creep of concrete, a given shortening of the column is produced by a smaller load when slowly applied or sustained over some length of time than when quickly applied. More important, the farther the stress is beyond the proportional limit of the concrete, and the more slowly the load is applied or the longer it is sustained, the smaller is the share of the total load carried by the concrete, and the larger that carried by the steel. In the sample column the steel was seen to carry 14.7 per cent of the load in the elastic range, 17.3 per cent for a strain of 0.001 under fast loading, and 22.6 per cent at the same strain under slow or sustained loading.

c. Ultimate strength   The one quantity of chief interest to the structural designer is the *ultimate strength*, i.e., the maximum load which his structure or member will carry. Information on stresses, strains, and similar quantities serves chiefly as a tool for determining carrying capacity. The performance of the column discussed so far indicates two things: First, in the range of large stresses and strains which precede attainment of ultimate strength and subsequent failure, elastic relationships cannot be utilized. Second, the member behaves differently under fast than under slow or sustained loading and shows less resistance to the latter than to the former. In usual construction, many types of loads, such as the weight of the structure and any permanent equipment housed therein, are sustained, and others are applied at slow rates. For this reason, to calculate a reliable magnitude of ultimate strength, curve c of Fig. 2.2 must be used as far as the concrete is concerned.

As for the steel, Fig. 1.13 shows that it reaches its ultimate strength (peak of the curve) at strains of the order of 0.08. Concrete, on the other hand, fails by crushing at the much smaller strain of about 0.003 and, as seen from Figs. 1.4 and 2.2 (curve c), reaches its ultimate strength in the strain range of 0.002 to 0.003. Because the strains in steel and concrete are equal in axial compression, the load at which the steel begins to yield can be calculated from the information of Fig. 2.2.

If the small knee prior to yielding of the steel is disregarded, i.e., if the steel is assumed to be sharp-yielding, the strain at which it yields is

$$\epsilon_y = \frac{f_y}{E_s} \tag{2.4}$$

or

$$\epsilon_y = 40,000/29,000,000 = 0.00138$$

At this strain, curve c of Fig. 2.2 indicates a stress of 2200 psi in the concrete; therefore, by Eq. (2.3), the load in the member when the steel starts yielding is $P_y = 2200 \times 314 + 40,000 \times 6 = 931,000$ lb. At this load the concrete has not yet reached its ultimate strength, which, as mentioned before, can be assumed as $0.85f'_c = 2550$ psi for slow or sustained loading, and therefore the load on the member can be further increased. During this stage of loading, the steel keeps yielding at constant stress. Finally, the ultimate load[1] of the member is reached when the concrete crushes while the steel yields, i.e.,

$$P'_u = 0.85f'_c A_c + f_y A_s \tag{2.5}$$

Numerous careful tests have shown the reliability of Eq. (2.5) in predicting the ultimate strength of a concentrically loaded reinforced-concrete column, provided its slenderness ratio is small so that buckling will not reduce its strength (Ref. 2.1).

For the particular numerical example, $P'_u = 2550 \times 314 + 40,000 \times 6 = 800,000 + 240,000 = 1,040,000$ lb. At this stage the steel carries as much as 23 per cent of the load.

Summary   In the elastic range of low stresses the steel carries a relatively small portion of the total load of an axially compressed member. As the ultimate strength is approached, there occurs a redistribution of the relative shares of the load resisted, respectively, by concrete and steel, the latter taking an increasing amount. The ultimate load at which the member is on the point of

---

[1] Throughout this book, quantities which refer to the ultimate strength of a member are primed and furnished with the subscript u, for example P'_u. The prime is used to distinguish these quantities from those which the ACI Code prescribes for use in design. These ultimate design strengths are designated in the Code by the subscript u without prime, for example P_u. To ensure structural safety, the codified design values such as P_u are smaller than the actual ultimate-strength values, such as P'_u, by percentages which are specified in the Code and explained in Chaps. 3 and 5.

failure consists of the contribution of the steel when it is stressed to the yield point plus that of the concrete when its stress has attained the ultimate strength of $0.85f_c'$, as reflected in Eq. (2.5).

PROBLEM

The same 16- $\times$ 20-in. column as in the preceding examples is made of the same concrete and reinforced with the same six No. 9 bars, except that a steel with yield strength $f_y = 60$ ksi is used. The stress-strain diagram of this reinforcing steel is that shown in Fig. 1.13 for $f_y = 60$ ksi. For this column determine: (a) the axial load which will stress the concrete to 1000 psi; (b) the load at which the steel starts yielding; (c) the ultimate strength; (d) the share of the total load carried by the reinforcement at these three stages of loading. Compare results with those calculated in the preceding examples for $f_y = 40$ ksi, keeping in mind, in regard to relative economy, that the price per pound for reinforcing steels with 40-ksi and with 60-ksi yield points is about the same.

2.3 AXIAL TENSION

It was pointed out in Art. 1.8 that the tension strength of concrete is only a small fraction of its compressive strength. It follows that reinforced concrete is not well suited for use in tension members because the concrete will contribute little, if anything, to their strength. Still, there are situations in which reinforced concrete is stressed in tension, chiefly in tie rods in structures such as arches. Such members consist of one or more bars embedded in concrete in a symmetrical arrangement similar to compression members (cf. Figs. 2.1 and 2.3).

When the tension force in the member is sufficiently small so that the stress in the concrete is considerably below its tensile strength, both steel and concrete behave elastically. In this situation all the expressions derived for compression in Art. 2.2a are identically valid for tension. In particular, Eq. (2.2) becomes

$$P = f_{ct}(A_c + nA_s) \tag{2.6}$$

where f_{ct} is the tensile stress in the concrete.

However, when the load is further increased, the concrete reaches its tensile strength at a stress and strain of the order of one-tenth of what it could sustain in compression. At this stage the concrete cracks across the entire cross section. When this happens, it ceases to resist any part of the applied tension force, since, evidently, no force can be transmitted across the air gap in the crack. At any load larger than that which caused the concrete to crack, the steel is called upon to resist the entire tension force. Correspondingly, at this stage,

$$P = f_s A_s \tag{2.7}$$

With further increased load, the tensile stress f_s in the steel reaches the yield point f_y. When this occurs, the tension members cease to exhibit small, elastic deformations, but instead, stretch a sizable and permanent amount at substantially constant load. This does not impair the strength of the member. Its elongation, however, becomes so large (of the order of 1 per cent or more of its length) as to render it useless. Therefore, the maximum useful strength P'_{ut} of a tension member is that force which will just cause the steel stress to reach the yield point. That is,

$$P'_{ut} = f_y A_s \tag{2.8}$$

To provide adequate safety, the force permitted in a tension member under normal service loads should be of the order of $\frac{1}{2} P'_{ut}$. Because the concrete has cracked at loads considerably smaller than this, it does not contribute to the carrying capacity of the member in service. It does serve, however, as a fire and corrosion proofing and often improves the appearance of the structure.

There are situations, though, in which reinforced concrete is used in axial tension under conditions in which the occurrence of tension cracks must be prevented. A case in point is a circular tank (see Fig. 1.1i). To provide water-tightness, the hoop tension caused by the fluid pressure must be prevented from causing the concrete to crack. In this case, Eq. (2.6) can be used to determine a safe value for the axial tension force P by using, for the concrete tension stress f_{ct}, an appropriate fraction of the tensile strength of the concrete, i.e., of that stress which would cause the concrete to crack.

PROBLEM

A reinforced-concrete water pipe has an inside diameter of 5 ft 0 in. and a wall thickness of $3\frac{1}{2}$ in. The circumferential hoop reinforcement consists of No. 4 bars at 3 in. distance, center to center; the bars are located midway between inside and outside surfaces. The concrete has a compression strength $f'_c = 5000$ psi and a split cylinder strength $f'_{sp} = 450$ psi; for the reinforcement, $f_y = 40,000$ psi. (a) Determine the allowable water pressure such that there is a safety factor of 1.5 against leakage caused by tension cracks; (b) for this allowable water pressure, determine the safety factor against failure of the pipe. Question: Would it be advantageous to use higher-strength reinforcement? Hint: To expedite calcula-tions, analyze a 1-ft length of pipe and make use of Table 2 or 4 of the Appendix. Regarding tension strength of concrete, see Art. 1.8.

2.4 BENDING

a. Bending of homogeneous beams Reinforced-concrete beams are nonhomo-geneous in that they are made of two entirely different materials. The methods used in the analysis of reinforced-concrete beams are therefore different from those used in the design or investigation of beams composed entirely of steel, wood, or any other structural material. The fundamental principles involved are, however, essentially the same. Briefly, these principles are as follows:

At any cross section there exist internal forces which may be resolved into components normal and tangential to the section. Those components which are normal to the section are the *bending* stresses (tension on one side of the neutral axis and compression on the other). Their function is to resist the bending moment at the section. The tangential components are known as the *shear* stresses, and they resist the transverse or shear forces.

FUNDAMENTAL ASSUMPTIONS

1. A cross section which was plane before loading remains plane under load. This means that the unit strains in a beam, above and below the neutral axis, are proportional to the distance from that axis.
2. The bending stress f at any point depends on the strain at that point in a manner given by the stress-strain diagram of the material. If the beam is made of a homogeneous material whose stress-strain diagram in tension and compression is that of Fig. 2.4a, the following holds: If the maximum strain at the outer fibers is smaller than the strain ϵ_p up to which stress and strain are proportional for the given material, then the compression and tension stresses on either side of the axis are proportional to the distance from the axis, as shown in Fig. 2.4b. However, if the maximum strain at the outer fibers is larger than ϵ_p, this is no longer true. The situation which then obtains is shown in Fig. 2.4c; that is, in the outer portions of the beam, where $\epsilon > \epsilon_p$, stresses and strains are no longer proportional. In these regions the magnitude of stress at any level, such as f_2 in Fig. 2.4c, depends on the strain ϵ_2 at that level in the manner

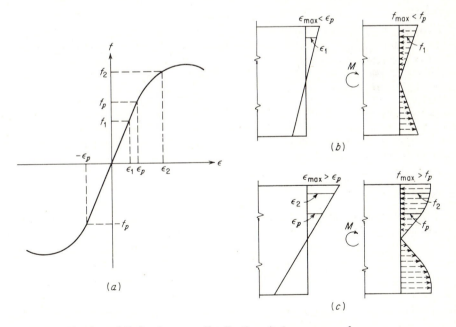

Fig. 2.4 Elastic and inelastic stress distributions in homogenous beam.

given by the stress-strain diagram of the material. In other words, for a given strain in the beam, the stress at a point is the same as that given by the stress-strain diagram.

3. The distribution of the shear stresses v over the depth of the section depends on the shape of the cross section and of the stress-strain diagram. These shear stresses are largest at the neutral axis and equal to zero at the outer fibers. The shear stresses on horizontal and vertical planes through any point are equal.

4. Owing to the combined action of shear stresses (horizontal and vertical) and flexure stresses, at any point in a beam there are inclined stresses of tension and compression, the largest of which form an angle of 90° with each other. The intensity of the inclined maximum or principal stress at any point is given by the equation

$$t = \frac{f}{2} \pm \sqrt{\frac{f^2}{4} + v^2} \qquad\qquad (2.9)$$

where f = intensity of normal fiber stress
      v = intensity of tangential shearing stress

The inclined stress makes an angle α with the horizontal such that $\tan 2\alpha = 2v/f$.

5. Since the horizontal and vertical shearing stresses are equal and the flexural stresses are zero at the neutral plane, the inclined tensile and compressive stresses at any point in that plane form an angle of 45° with the horizontal, the intensity of each being equal to the unit shear at the point.

6. When the stresses in the outer fibers are smaller than the proportional limit f_p, the beam behaves *elastically* as shown in Fig. 2.4b. In this case the following obtains:

    a. The neutral axis passes through the center of gravity of the cross section.

    b. The intensity of bending stress normal to the section increases directly with the distance from the neutral axis and is a maximum at the extreme fibers. The stress at any given point in the cross section is represented by the equation

$$f = \frac{My}{I} \qquad\qquad (2.10)$$

where f = bending stress at a distance y from neutral axis
     M = external bending moment at section
      I = moment of inertia of cross section about neutral axis

The maximum bending stress occurs at the outer fibers and is equal to

$$f_{max} = \frac{Mc}{I} = \frac{M}{S} \qquad\qquad (2.11)$$

where c = distance from neutral axis to outer fiber
$S = I/c$ = section modulus of cross section

c. The shear stress (longitudinal equals transverse) v at any point in the cross section is given by

$$v = \frac{VQ}{Ib} \tag{2.12}$$

where V = total shear at section
Q = statical moment about neutral axis of that portion of cross section lying between a line through point in question parallel to neutral axis and nearest face (upper or lower) of beam
I = moment of inertia of cross section about neutral axis
b = width of beam at given point

d. The intensity of shear along a vertical cross section in a rectangular beam varies as the ordinates of a parabola, the intensity being zero at the outer fibers of the beam and a maximum at the neutral axis. The maximum is $\frac{3}{2}V/ba$, since at the neutral axis $Q = ba^2/8$ and $I = ba^3/12$ in Eq. (2.12).

The remainder of this article deals only with bending stresses and their effects on reinforced-concrete beams. Shear stresses and their effects are discussed separately in Art. 2.5.

b. Reinforced-concrete beams: behavior under load   Plain concrete beams are inefficient as flexural members because the tension strength in bending (modulus of rupture, see Art. 1.8 and Fig. 1.2) is a small fraction of the compression strength. In consequence, such beams fail on the tension side at low loads, long before the strength of the concrete on the compression side has been fully utilized. For this reason, steel reinforcing bars are placed on the tension side as close to the extreme tension fiber as is compatible with proper fire and corrosion protection of the steel. In such a reinforced-concrete beam the tension which is caused by the bending moments is chiefly resisted by the steel reinforcement, while the concrete alone is usually capable of resisting the corresponding compression. Such joint action of the two materials is assured if relative slip is prevented. This is achieved by using deformed bars with their high bond strength at the steel-concrete interface (see Art. 1.12) and, if necessary, by special anchorage of the ends of the bars. A simple example of such a beam, with the customary designations for the cross-sectional dimensions, is shown in Fig. 2.5. For the sake of simplicity the discussion which follows will deal with beams of rectangular cross section, even though members of other shapes are very common in most concrete structures.

When the load on such a beam is gradually increased from zero to that magnitude which will cause the beam to fail, several different stages of behavior

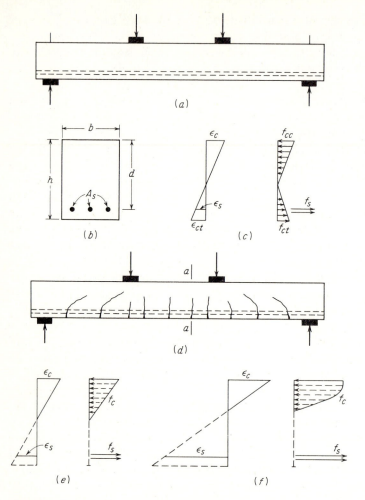

Fig. 2.5 Behavior of reinforced-concrete beam under increasing load.

can be clearly distinguished. At low loads, as long as the maximum tension stress in the concrete is smaller than the modulus of rupture, the entire concrete is effective in resisting stress, in compression on one side and in tension on the other side of the neutral axis. In addition, the reinforcement, deforming the same amount as the adjacent concrete, is also subject to tension stresses. At this stage all stresses in the concrete are of small magnitude and are proportional to strains. The distribution of strains and stresses in concrete and steel over the depth of the section is as shown in Fig. 2.5c.

When the load is further increased, the tension strength of the concrete is soon reached, and at this stage tension cracks develop. These propagate quickly upward to or close to the level of the neutral plane, which in turn shifts upward with progressive cracking. The general shape and distribution of these tension cracks is shown in Fig. 2.5d. In well-designed beams the width of these cracks is sufficiently small (hairline cracks) so that they are not objectionable from the viewpoint of either corrosion protection or appearance. Their presence, however, affects profoundly the behavior of the beam under load. Evidently, in a cracked section, i.e., in a cross section located at a crack such as a-a in Fig. 2.5d, the concrete does not transmit any tension stresses. Hence, just as in tension members (Art. 2.3), it is the steel which is called upon to resist the entire tension. At moderate loads, if the concrete stresses do not exceed approximately $f'_c/2$, stresses and strains continue to be closely proportional (see Fig. 2.2). The distribution of strains and stresses at or near a cracked section is then that shown in Fig. 2.5e. When the load is still further increased, stresses and strains rise correspondingly and are no longer proportional. The ensuing nonlinear relation between stresses and strains is that given by the concrete stress-strain curve. Therefore, just as in homogeneous beams (see Fig. 2.4), the distribution of concrete stresses on the compression side of the beam is of the same shape as the stress-strain curve. Figure 2.5f shows the distribution of strains and stresses close to the ultimate load.

Eventually, the carrying capacity of the beam is reached. Failure can be caused in one of two ways. When relatively moderate amounts of reinforcement are employed, at some value of the load the steel will reach its yield point. At that stress the reinforcement yields suddenly and stretches a large amount (see Fig. 1.13), and the tension cracks in the concrete widen visibly and propagate upward, with simultaneous significant deflection of the beam. When this happens, the strains in the remaining compression zone of the concrete increase to such a degree that crushing of the concrete, the so-called *secondary compression failure*, ensues at a load only slightly larger than that which caused the steel to yield. Effectively, therefore, attainment of the yield point in the steel determines the carrying capacity of moderately reinforced beams. Such yield failure is gradual and is preceded by visible signs of distress, such as the widening and lengthening of cracks and the marked increase in deflection.

On the other hand, if large amounts of reinforcement or normal amounts of steel of very high strength are employed, the compression strength of the concrete may be exhausted before the steel starts yielding. Concrete fails by crushing when strains become so large that they disrupt the integrity of the concrete. Exact criteria for this occurrence are not yet known, but it has been observed that rectangular beams fail in compression when the concrete strains reach values of about 0.003 to 0.004 for concretes with f'_c varying from 5000 to 2000 psi, respectively. Compression failure through crushing of the concrete is sudden, of an almost explosive nature, and occurs without warning. For this reason it is good practice to dimension beams in such a manner that, should they be overloaded, failure would be initiated by yielding of the steel rather than by crushing of the concrete.

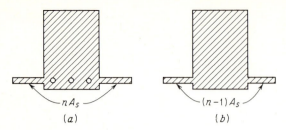

(a)   (b)

Fig. 2.6 Uncracked transformed section.

The analysis of stresses and strength in the different stages just described will be discussed in the next several sections.

c. Stresses elastic, section uncracked As long as the tensile stress in the concrete is smaller than the modulus of rupture so that no tension cracks develop, the strain and stress distribution as shown in Fig. 2.5c is essentially the same as in an elastic, homogeneous beam (Fig. 2.4b). The only difference is the presence of another material, the steel reinforcement. As was shown in Art. 2.2a, in the elastic range, for any given value of strain, the stress in the steel is n times that of the concrete [Eq. (2.1)]. In the same article it was shown that one can take account of this fact in calculations by replacing the actual steel-and-concrete cross section with a fictitious section thought of as consisting of concrete only. In this "transformed section" the actual area of the reinforcement is replaced with an equivalent concrete area equal to nA_s located at the level of the steel. The transformed, uncracked section pertaining to the beam of Fig. 2.5b is shown in Fig. 2.6.

Once the transformed section has been obtained, the usual methods of analysis of elastic homogeneous beams apply. That is, the section properties (location of neutral axis, moment of inertia, section modulus, etc.) are calculated in the usual manner, and in particular, stresses are computed with Eqs. (2.10) to (2.12).

Example A rectangular beam has the dimensions (see Fig. 2.5b) $b = 10$ in., $h = 25$ in., and $d = 23$ in. and is reinforced with three No. 8 bars so that $A_s = 2.35$ in.^{2} The concrete cylinder strength f_c' is 4000 psi, and the tensile strength in bending (modulus of rupture) is 475 psi. The yield point of the steel f_y is 60,000 psi, the stress-strain curves of the materials being those of Fig. 2.2. Determine the stresses caused by a bending moment $M = 45$ ft-kips.

With a value $n = E_s/E_c = 29,000,000/3,640,000 = 8$, one has to add to the rectangular outline an area $(n - 1)A_s = 7 \times 2.35 = 16.45$ in.^{2}, disposed as shown on Fig. 2.7, in order to obtain the uncracked, transformed section. Conventional calculations show that the location of the neutral axis of this section is given by $\bar{y} = 13.2$ in., and its moment of inertia about this axis is 14,710 in.^{4} For $M = 45$ ft-kips $= 540,000$ in.-lb, the concrete compression stress at the top fiber is, from Eq. (2.11),

$$f_{cc} = 540,000 \times 13.2/14,710 = 485 \text{ psi}$$

and similarly, the concrete tension stress at the bottom fiber is

$$f_{ct} = 540,000 \times 11.8/14,710 = 433 \text{ psi}$$

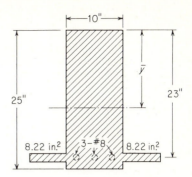

Fig. 2.7

Since this value is below the given tensile bending strength of the concrete, 475 psi, no tension cracks will form, and calculation by the uncracked, transformed section is justified. The stress in the steel, from Eqs. (2.1) and (2.10), is

$$f_s = n\frac{My}{I} = 8(540,000 \times 9.8/14,710) = 2880 \text{ psi.}$$

By comparing f_{cc} and f_s with the cylinder strength and the yield point, respectively, it is seen that at this stage the actual stresses are quite small as compared with the available strengths of the two materials.

d. Stresses elastic, section cracked   When the tension stress f_{ct} exceeds the modulus of rupture, cracks form, as shown in Fig. 2.5d. If the concrete compression stress is less than approximately $\frac{1}{2}f_c'$ and the steel stress has not reached the yield point, both materials continue to behave elastically, or very nearly so. This situation generally obtains in structures under normal service conditions and loads, since at these loads the stresses are generally of the order of magnitude just discussed. At this stage, for simplicity and with little if any error, it is assumed that tension cracks have progressed all the way to the neutral axis, and that sections plane before bending are plane in the bent member. The situation in regard to strain and stress distribution is then that shown in Fig. 2.5e.

   To compute stresses, and strains if desired, the device of the transformed section can still be used. One need only take account of the fact that all the concrete which is stressed in tension is assumed cracked, and therefore effectively absent. As shown in Fig. 2.8a, the transformed section then consists of the concrete in compression on one side of the axis, and n times the steel area on the other. The distance to the neutral axis, in this stage, is conventionally expressed as a fraction kd of the effective depth d. (Once the concrete is cracked, any material located below the steel is ineffective, which is why d is the effective depth of the beam.) To determine the location of the neutral axis, the moment of the tension area about the axis is set equal to the moment of the compression area, which gives

$$b\frac{(kd)^2}{2} - nA_s(d - kd) = 0 \qquad\qquad (2.13)$$

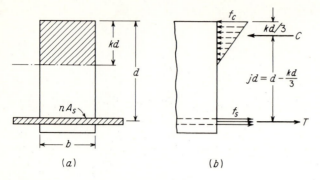

Fig. 2.8   Cracked transformed section.

Having obtained kd by solving this quadratic equation, one can determine the moment of inertia and other properties of the transformed section as in the preceding case. Alternatively, one can proceed from basic principles by accounting directly for the forces which act in the cross section. These are shown in Fig. 2.8b. The concrete stress, with maximum value f_c at the outer edge, is distributed linearly as shown. The entire steel area A_s is subject to the stress f_s. Correspondingly, the total compression force C and the total tension force T are

$$C = \frac{bkd}{2} f_c \quad \text{and} \quad T = A_s f_s \tag{2.14}$$

The requirement that these two forces must be equal numerically has been taken care of by the manner in which the location of the neutral axis has been determined.

Equilibrium requires that the couple constituted by the two forces C and T be equal numerically to the external bending moment M. Hence, taking moments about C,

$$M = Tjd = A_s f_s jd \tag{2.15}$$

from which the steel stress is

$$f_s = \frac{M}{A_s jd} \tag{2.16}$$

Conversely, taking moments about T,

$$M = Cjd = \frac{bkd}{2} f_c jd = \frac{bd^2}{2} kjf_c \tag{2.17}$$

from which the concrete stress is

$$f_c = \frac{M}{\frac{1}{2} bd^2 kj} \tag{2.18}$$

From Fig. 2.8*b* it is seen that $jd = d - kd/3$, or

$$j = 1 - \frac{k}{3} \tag{2.19}$$

Example The beam of the preceding example is subject to a bending moment $M = 90$ ft-kips (rather than 45 ft-kips as previously). Calculate the relevant properties and stresses.

    If the section were to remain uncracked, the tension stress in the concrete would now be twice its previous value, that is, 866 psi. Since this exceeds by far the modulus of rupture of the given concrete (475 psi), cracks will have formed, and the analysis must be adapted appropriately. Equation (2.13), with the known quantities b, n, and A_s inserted, gives the distance to the neutral axis $kd = 7.6$ in., or $k = 7.6/23 = 0.33$. From Eq. (2.19), $j = 1 - 0.33/3 = 0.89$. With these values the steel stress is obtained from Eq. (2.16) as $f_s = 22{,}400$ psi, and the maximum concrete stress from Eq. (2.18) as $f_c = 1390$ psi.

    Comparing the results with the pertinent values for the same beam when subject to one-half the moment, as previously calculated, one notices: (1) The neutral plane has migrated upward so that its distance from the top fiber has changed from 13.2 to 7.6 in.; (2) even though the bending moment has only been doubled, the steel stress has increased from 2880 psi to 22,400 psi, or about 7.8 times, and the concrete compression stress has increased from 485 to 1390 psi, or 2.9 times; (3) the moment of inertia of the cracked transformed section is easily computed to be 5910 in.[4], as compared with 14,710 in.[4] for the uncracked section. This affects the magnitude of the deflection, as discussed in Art. 2.7. Thus it is seen how radical is the influence of the formation of tension cracks on the behavior of reinforced-concrete beams.

e. Ultimate strength, general analysis It is of interest in structural practice to calculate those stresses and deformations which occur in a structure in service under design load. For reinforced-concrete beams this can be done by the methods just presented, which assume elastic behavior of both materials. It is equally, if not more, important that the structural engineer be able to predict with satisfactory accuracy the ultimate strength of his structure or structural member. By making this strength larger by an appropriate amount than the largest loads which can be expected during the lifetime of the structure, an adequate margin of safety is assured. Until recent times, methods based on elastic analysis, such as those just presented or variations thereof, have been used for this purpose. It is clear, however, that at or near the ultimate load, stresses are no longer proportional to strains. In regard to axial compression this has been discussed in detail in Art. 2.2, and in regard to bending, it has been pointed out that at high loads, close to the ultimate, the distribution of stresses and strains is that of Fig. 2.5*f* rather than the elastic distribution of Fig. 2.5*e*. More realistic methods of analysis, based on actual inelastic rather than assumed elastic behavior of the materials and on results of extremely extensive experimental research, have been developed in recent years to predict the ultimate strength. They are now used almost exclusively in structural-design practice.

    If the distribution of concrete compression stresses at or near ultimate load (Fig. 2.5*f*) had a well-defined and invariable shape—parabolic, trapezoidal, or otherwise—it would be possible to derive a completely rational theory of ultimate bending strength, just as the theory of elastic bending with its known triangular shape of stress distribution (Figs. 2.4b and 2.5c and e) is straightforward and rational. Actually, inspection of Figs. 1.4, 1.6, and 2.2, and of many more concrete stress-strain curves which have been published, shows that the geometrical shape of the stress distribution is quite varied and depends on a number of factors, such as the cylinder strength and the rate and duration of loading. For this and other reasons, a wholly rational flexural theory for reinforced concrete has not yet been developed (Refs. 2.2 and 2.3). Present methods of analysis, therefore, are based in part on known laws of mechanics, supplemented, where needed, by extensive test information.

Let Fig. 2.9 represent the distribution of internal stresses and strains when the beam is about to fail. One desires a method to calculate that moment M'_u (ultimate moment) at which the beam will fail either by tension yielding of the steel or by crushing of the concrete in the outer compression fiber. For the first mode of failure the criterion is that the steel stress equal the yield point, $f_s = f_y$. It has been mentioned before that an exact criterion for concrete compression failure is not yet known, but that for rectangular beams strains of 0.003 to 0.004 in. per in. have been measured immediately preceding failure. If one assumes, slightly conservatively, that the concrete is about to crush when the maximum strain reaches 0.003, then comparison with a great many tests of beams and columns of a considerable variety of shapes and conditions of loading shows that a satisfactorily accurate and safe prediction of ultimate strength can be made (Ref. 2.4). In addition to these two criteria (yielding of the steel at a stress of f_y and crushing of the concrete at a strain of 0.003), it is not really necessary to know the exact shape of the concrete stress distribution in Fig. 2.9. What is necessary is to know, for a given distance c of the neutral axis, (1) the total resultant compression force C in the concrete and (2) its vertical location, i.e., its distance from the outer compression fiber.

In a rectangular beam the area which is in compression is bc, and the

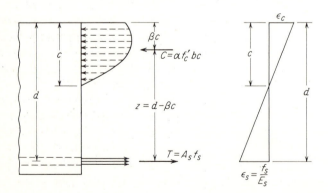

Fig. 2.9   Stress distribution at ultimate load.

total compression force on this area can be expressed as $C = f_{avg}bc$, where f_{avg} is the average compression stress on the area bc. Evidently, the average compression stress which can be developed before failure occurs becomes larger the higher the cylinder strength f'_c of the particular concrete. Let

$$\alpha = \frac{f_{avg}}{f'_c} \qquad\qquad (2.20)$$

Then

$$C = \alpha f'_c bc \qquad\qquad (2.21)$$

For a given distance c to the neutral axis, the location of C can be defined as some fraction β of this distance. Thus, as indicated in Fig. 2.9, for a concrete of given strength it is only necessary to know α and β in order to define completely the effect of the concrete compression stresses.

Extensive direct measurements, as well as indirect evaluation of numerous beam tests, have shown that the following values for α and β are satisfactorily accurate (see Ref. 2.4, where α is designated as k_1k_3, and β as k_2) :

α equals 0.72 for $f'_c \leq 4000$ psi and decreases by 0.04 for every 1000 psi above 4000.
β equals 0.425 for $f'_c \leq 4000$ psi and decreases by 0.025 for every 1000 psi above 4000.

The decrease in α and β for high-strength concretes is related to the fact that such concretes are more brittle, i.e., show a more sharply curved stress-strain plot with a smaller near-horizontal portion (see Fig. 1.3). Figure 2.10 shows these simple relations.

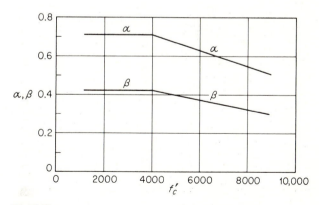

Fig. 2.10

If this experimental information is accepted, the ultimate strength can be calculated from the laws of equilibrium and from the assumption that plane cross sections remain plane. Equilibrium requires that

$$C = T \quad \text{or} \quad \alpha f'_c b c = A_s f_s \tag{2.22}$$

Also, the bending moment, being the couple of the forces C and T, can be written as either

$$M = Tz = A_s f_s (d - \beta c) \tag{2.23}$$

or

$$M = Cz = \alpha f'_c b c (d - \beta c) \tag{2.24}$$

For *tension failure* by yielding of the steel, $f_s = f_y$. Substituting this value in Eq. (2.22), one obtains the distance to the neutral axis:

$$c = \frac{A_s f_y}{\alpha f'_c b} \tag{2.25}$$

It is often convenient to express the steel area nondimensionally as a fraction of the effective area of the section, the so-called *reinforcement ratio:*

$$\rho = \frac{A_s}{bd} \tag{2.26}$$

Using this quantity, one can also write

$$c = \frac{\rho f_y}{\alpha f'_c} d \tag{2.25a}$$

as the distance to the neutral axis when tension failure occurs. The ultimate moment M'_u is then obtained from Eq. (2.23), with the value for c just determined, and $f_s = f_y$; that is,

$$M'_u = A_s f_y d \left(1 - \frac{\beta f_y}{\alpha f'_c} \rho \right) \tag{2.27}$$

With the specific, experimentally obtained values for α and β given previously, this becomes, for $f'_c \leq 4000$ psi,

$$M'_u = A_s f_y d \left(1 - 0.59 \frac{f_y}{f'_c} \rho \right) \tag{2.27a}$$

On the other hand, for *compression failure* the criterion is that the compression strain in the concrete become $\epsilon_c = 0.003$, as previously discussed. The steel stress f_s, not having reached the yield point, is proportional to the steel strain ϵ_s; that is, according to Hooke's law,

$$\epsilon_s = \frac{f_s}{E_s}$$

From the strain distribution of Fig. 2.9, the distance to the neutral axis is
obtained by evaluating the similar triangles and is

$$c = \frac{\epsilon_c}{\epsilon_s + \epsilon_c} d = \frac{0.003}{f_s/E_s + 0.003} d \qquad (2.28)$$

Calculating f_s from Eq. (2.28) and substituting it in Eq. (2.23), one equates the
right sides of Eqs. (2.23) and (2.24) to obtain a quadratic equation for c in
terms of the cross-sectional dimensions f_c', E_s, and α. Substitution of this value
of c in either Eq. (2.23) or (2.24) determines the ultimate moment of a beam so
heavily reinforced that failure occurs by crushing of the concrete.

It was noted that compression failure occurs explosively and without
warning. For this reason it is good practice to keep the amount of reinforcement
sufficiently small so that, should the member be overstressed, it will give
adequate warning before failing in a gradual manner by yielding of the steel,
rather than by crushing of the concrete. This can be done by keeping the
reinforcement ratio $\rho = A_s/bd$ below a certain limiting value. This value, the
so-called *balanced steel ratio* ρ_b, represents that amount of reinforcement necessary to make the beam fail by crushing of the concrete at the same load which
causes the steel to yield. But this means that the neutral axis must be so located
that at the load at which the steel starts yielding, the concrete reaches its
assumed breaking strain of 0.003. Correspondingly, equating the right sides of
Eqs. (2.25a) and (2.28) and solving for ρ, one obtains for the balanced steel
ratio

$$\rho_b = \alpha \frac{0.003}{f_y/E_s + 0.003} \frac{f_c'}{f_y} \qquad (2.29)$$

In a well-designed member the actual steel ratio $\rho = A_s/bd$ is kept well below
the balanced ratio ρ_b.

Example Determine the ultimate moment M_u' at which the beam of the preceding example
will fail.

For this beam the steel ratio $\rho = A_s/bd = 2.35/10 \times 23 = 0.0102$. The balanced steel ratio [Eq. (2.29)] is found to be 0.284. Since the amount of steel in the
beam is less than that which would cause failure by crushing of the concrete, the beam
will fail in tension by yielding of the steel. Its ultimate moment, from Eq. (2.27a), is
found to be

$$M_u' = 2.35 \times 60,000 \times 23[1 - 0.59(60,000/4000)0.0102]$$
$$= 2,950,000 \text{ in.-lb} = 246 \text{ ft-kips}$$

It is interesting to note that when the beam reaches its ultimate strength, the distance
to its neutral axis, from Eq. (2.25a), is

$$c = \frac{0.0102 \times 60,000}{0.72 \times 4000} 23 = 4.90 \text{ in.}$$

In previous calculations it was found that at low loads, when the concrete had not yet cracked in tension, the neutral axis was located at a distance of 13.2 in. from the compression edge; at higher loads, when the tension concrete was cracked but stresses were still sufficiently small to be elastic, this distance was 7.6 in. Immediately before the beam fails, as has just been shown, this distance has further decreased to 4.9 in. This migration of the neutral axis toward the compression edge as load is increased is a graphic illustration of the differences among the various stages of behavior through which a reinforced-concrete beam passes as its load is increased from zero to that value which causes it to fail.

f. Ultimate strength, equivalent rectangular stress distribution   The preceding method of calculating the ultimate strength, derived from basic concepts of structural mechanics and pertinent experimental research information, also applies to situations other than the case of rectangular beams reinforced on the tension side. It can be used and gives valid answers for beams of other cross-sectional shapes, reinforced in other manners, and for members subject not only to simple bending, but also to the simultaneous action of bending and axial force (compression or tension). However, the pertinent equations for these more complex cases become increasingly cumbersome and lengthy. What is more important, it becomes increasingly difficult for the designer to visualize the physical basis for his design methods and formulas; this could lead to a blind reliance on formulas, with a resulting lack of actual understanding. This is not only undesirable on general grounds but, practically, is more likely to lead to numerical errors in design work than when the designer at all times has a clear picture of the physical situation in the member which he is dimensioning or analyzing. Fortunately, it is possible, essentially by a conceptual trick, to formulate the ultimate-strength analysis of reinforced-concrete members in a different manner, which gives the same answers as does the general analysis just developed, but which is much more easily visualized and much more easily applied to cases of greater complexity than that of the simple rectangular beam. Its consistency is shown below, and its application to more complex cases has been checked against the results of a vast number of tests on a great variety of types of members and conditions of loading (Ref. 2.4).

It was noted in the preceding section that the actual geometrical shape of the concrete compression-stress distribution varies considerably and that, in fact, one need not know this shape exactly, provided one does know two things: (1) the magnitude C of the resultant of the concrete compression stresses and (2) the location of this resultant. Information on these two quantities was obtained from the results of experimental research and expressed in the two parameters α and β.

Evidently, then, one can think of the actual complex stress distribution as replaced by a fictitious one of some simple geometrical shape, provided that this fictitious distribution results in the same total compression force C applied

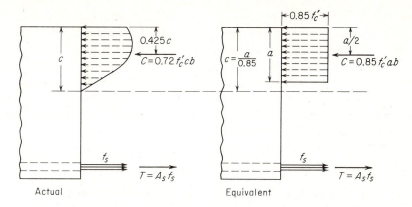

Fig. 2.11 Actual and equivalent rectangular stress distributions at ultimate load (for f'_c less than 4000 psi).

at the same location as in the actual member when it is on the point of failure. Historically, a number of simplified, fictitious equivalent stress distributions have been proposed by various investigators in various countries. The one generally accepted in this country, and increasingly abroad, was first proposed by C. S. Whitney and subsequently elaborated and checked experimentally by others (see, for example, Ref. 2.4). For concrete with cylinder strength f'_c not exceeding 4000 psi, an actual stress distribution and the fictitious equivalent distribution are shown in Fig. 2.11.

It is seen that the actual distribution is replaced by a fictitious equivalent one of simple rectangular outline. The intensity of this uniform and constant equivalent compression stress is $0.85f'_c$, and it is thought of as acting over a part of the section of depth $a = 0.85c$, where c is the distance to the actual neutral axis. It is in no way maintained that the compression stresses are actually distributed in this most unlikely manner. It is maintained, however, that this equivalent distribution gives the answers just derived for the actual situation. Of this one can easily convince oneself. In a rectangular beam of width b the equivalent uniform compression stress $0.85f'_c$ acts on an area ab, so that the total compression force is

$$C = 0.85f'_c ab \tag{2.30}$$

If one expresses the depth a of the rectangular stress block in terms of the distance c to the neutral axis as assumed, that is, $a = 0.85c$, one obtains for the compression force $C = 0.85f'_c \times 0.85c \times b = 0.72f'_c cb$, which is the same as Eq. (2.21) for $\alpha = 0.72$. Further, since the centroid of a rectangle is at mid-depth, the distance from the compression edge to C is evidently $a/2$. Again expressing a in terms of c as above, one obtains the distance to the compression resultant as $0.5 \times 0.85c = 0.425c$, which is the correct distance for $\beta = 0.425$.

Since the chosen equivalent rectangular stress distribution satisfies the stated requirements 1 and 2 above, it will give exactly the same answers as were derived previously. This is so for f_c' not exceeding 4000 psi because, in this range of concrete strengths, the parameters α and β are constants equal, respectively, to 0.72 and 0.425. For concretes of higher strength it has been stated that α and β decrease because of the greater brittleness of such concretes. In the equivalent rectangular stress distribution, this is reflected in the following stipulation:

$$a/c = \beta_1 = 0.85 \text{ for } f_c' \leq 4000 \text{ psi and decreases by 0.05 for}$$
$$\text{every 1000 psi above 4000} \tag{2.31}$$

This equivalent rectangular stress distribution can be used for deriving the equations which have been developed in the preceding section. The failure criteria, of course, are the same as before: yielding of the steel at $f_s = f_y$ or crushing of the concrete at $\epsilon_c = 0.003$. More important, however, because the rectangular stress block is easily visualized and its geometrical properties are extremely simple, many calculations are conveniently carried out directly without reference to formally derived equations.

Example Using the equivalent rectangular stress distribution, calculate directly the ultimate strength of the previously analyzed beam.

The distribution of stresses, internal forces, and strains is as shown in Fig. 2.12. The depth a of the equivalent stress block is found from the equilibrium condition $C = T$. Hence $ab \times 0.85f_c' = A_s f_y$, or $a \times 10 \times 3400 = 2.35 \times 60,000$, from which $a = 4.15$ in. The distance to the neutral axis, by the definition of the rectangular stress block, is $c = a/0.85 = 4.15/0.85 = 4.88$ in. One way of determining whether the strength of the beam is governed by yielding of the steel or by crushing of the concrete is to calculate whether the concrete strain in the top fiber is smaller or larger than 0.003 at the instant the steel starts yielding. At that point the strain in the steel is $\epsilon_s = f_y/E_s = 60,000/29,000,000 = 0.00207$. From the similar triangles of the strain distribution, $\epsilon_c = \epsilon_s c/(d - c) = 0.00207(4.88/18.12) = 0.000557$. This value is smaller

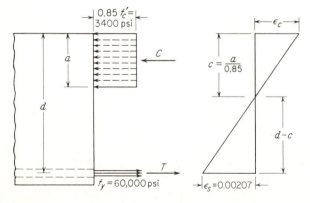

Fig. 2.12

than the assumed crushing strain of the concrete, and hence failure will be initiated by yielding of the steel. The ultimate moment, then, is $M'_u = T(d - a/2) = 2.35 \times 60,000(23.0 - 2.07) = 2,950,000$ in.-lb $= 246$ ft-kips.

If one wanted to determine the balanced steel ratio p_b for the given beam, one could proceed as follows: At simultaneous yielding of the steel and crushing of the concrete, it is known that the two pertinent strains are $\epsilon_s = 0.00207$ and $\epsilon_c = 0.003$. Hence, from the similar triangles of the strain distribution, $c/(d - c) = 0.003/0.00207$, from which $c = (3/5.07)d$, and $a = 0.85c = 0.85(3/5.07)23 = 11.6$ in. Hence the tensile force, equal to the compression force, is $T = C = 0.85f'_c ab = 3400 \times 11.6 \times 10 = 394,000$ lb. From this the balanced steel ratio is $p_b = A_s/bd = (T/f_y)/bd = 394,000/(60,000 \times 10 \times 23) = 0.0284$. The results of this simple and direct numerical analysis, based on the equivalent rectangular stress distribution, are identical with those previously determined from the general ultimate-strength analysis of beams.

PROBLEM

A simple beam of 20-ft span has the cross section of the preceding examples, 10×25 in., and carries a uniform load of 2450 lb per ft in addition to its own weight. (a) Check whether this beam, if reinforced with three No. 8 bars as in the preceding examples, is adequate to carry this load with a minimum factor of safety against failure of 1.85. A different way of stating the same requirement is to stipulate that the ultimate strength of the beam shall be at least equal to 1.85 times the design load. If this requirement is not met, select a three-bar reinforcement of diameter adequate to provide this safety. (b) Determine the maximum stress in the steel and in the concrete under working load, i.e., when the beam carries its own weight and the specified uniform design load. (c) Will the beam show hairline cracks on the tension side under working load? As before, $f'_c = 4000$ psi, $f_y = 60,000$ psi. Use Table 2 of the Appendix for steel areas; assume a unit weight of 150 lb per ft³ for reinforced concrete.

(*Note:* Throughout the remainder of this chapter, additional problems frequently refer to this same beam and loading. To avoid repetitive calculations, the student is advised to retain his computation sheets until reaching the end of the chapter.)

2.5 SHEAR AND DIAGONAL TENSION

a. Homogeneous elastic beams   The stresses acting in homogeneous beams were briefly reviewed in Art. 2.4a. It was pointed out that when the material is elastic (stresses proportional to strains), shear stresses

$$v = \frac{VQ}{Ib} \tag{2.12}$$

act in any section in addition to the bending stresses

$$f = \frac{My}{I} \tag{2.10}$$

except for those locations at which the shear force V happens to be zero.

The role of shear stresses is easily visualized by the performance under load of the laminated beam of Fig. 2.13; it consists of two rectangular pieces bonded together along their contact surface. If the adhesive is strong enough, the member will deform as one single beam, as shown in Fig. 2.13a. On the

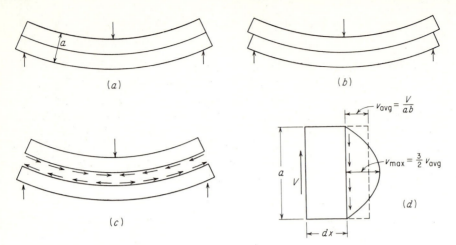

Fig. 2.13 Shear in homogeneous rectangular beam.

other hand, if the adhesive is weak, the two pieces will separate and slide relative to each other, as shown in Fig. 2.13b. Evidently, then, when the adhesive is effective, there are forces or stresses acting in it which prevent this sliding or shearing. These horizontal shear stresses are shown in Fig. 2.13c as they act, separately, on the top and bottom pieces. The same stresses occur in horizontal planes in single-piece beams; they are different in intensity at different distances from the neutral axis.

Figure 2.13d shows a differential length of a single-piece rectangular beam acted upon by a shear force of magnitude V. Upward translation is prevented, i.e., vertical equilibrium is provided, by the vertical shear stresses v. Their average value is equal to the shear force divided by the cross-sectional area, $v_{avg} = V/ab$, but their intensity varies over the depth of the section. As is easily computed from Eq. (2.12), the shear stress is zero at the outer fibers and has a maximum of $1.5v_{avg}$ at the neutral axis, the variation being parabolic as shown. Other values and distributions are found for other shapes of the cross section, the shear stress always being zero at the outer fibers and of maximum value at the neutral axis.

If a small square element located at the neutral axis of such a beam is isolated as in Fig. 2.14b, the vertical shear stresses on it, equal and opposite on the two faces for reasons of equilibrium, act as shown. However, if these were the only stresses present, the element would not be in equilibrium; it would spin. Therefore, on the two horizontal faces there exist equilibrating horizontal shear stresses of the same magnitude. That is, at any point in the beam, the horizontal shear stresses of Fig. 2.14b are equal in magnitude to the vertical shear stresses of Fig. 2.13d.

It is proved in any strength-of-materials text that, on an element cut at 45°, these shear stresses combine in such a manner that their effect is as

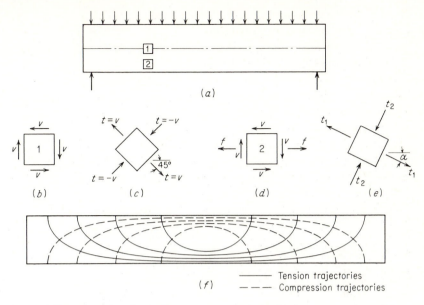

Fig. 2.14   Stress trajectories in homogeneous rectangular beam.

shown in Fig. 2.14c. That is, the action of the two pairs of shear stresses on the vertical and horizontal faces is the same as that of two pairs of normal stresses, one tension and one compression, acting on the 45° faces and of numerical value equal to that of the shear stresses. If an element of the beam is considered which is located neither at the neutral axis nor at the outer edges, then its vertical faces are subject not only to the shear stresses, but also to the familiar bending stresses whose magnitude is given by Eq. (2.10) (Fig. 2.14d). The six stresses which now act on the element can again be combined into a pair of inclined compression stresses and a pair of inclined tension stresses which act at right angles to each other. They are known as *principal* stresses (Fig. 2.14e). Their value, as was mentioned in Art. 2.4a, is given by

$$ t = \frac{f}{2} \pm \sqrt{\frac{f^2}{4} + v^2} \tag{2.9} $$

and their inclination α by $\tan 2\alpha = 2v/f$.

Since the magnitudes of the shear stresses v and the bending stresses f change both along the beam and vertically with distance from the neutral axis, the inclinations as well as the magnitudes of the resulting principal stresses t also vary from one place to another. Figure 2.14f shows, for a rectangular beam uniformly loaded, the inclinations of these principal stresses. That is, these stress trajectories are lines which, at any point, are drawn in that direction in which the particular principal stress, tension or compression, acts at that point. It is seen that at the neutral axis the principal stresses in a beam are always inclined at 45° to the axis. In the vicinity of the outer fibers they are horizontal near midspan.

An important point which follows from this discussion is the following: Tension stresses, which are of particular concern in view of the low tensile strength of the concrete, are not confined to the horizontal bending stresses f which are caused by bending alone. Tension stresses of various inclinations and magnitudes, resulting from shear alone (at the neutral axis) or from the combined action of shear and bending, exist in all parts of a beam and can impair its integrity if not adequately provided for. It is for this reason that the inclined tension stresses, known as *diagonal tension*, must be carefully considered in reinforced-concrete design.

b. Reinforced-concrete beams without shear reinforcement　The discussion of shear in a homogeneous elastic beam applies very closely to a plain concrete beam without reinforcement. As the load is increased in such a beam, a tension crack will form where the tension stresses are largest and will immediately cause the beam to fail. Except for beams of very unusual proportions, the largest tension stresses are those caused at the outer fiber by bending alone, at the section of maximum bending moment. In this case, shear has little, if any, influence on the strength of a beam.

However, when tension reinforcement is provided, the situation is quite different. Even though tension cracks form in the concrete, the required flexural tension strength is furnished by the steel, and much higher loads can be carried. Shear stresses increase proportionally to the loads. In consequence, diagonal tension stresses of significant intensity are created in regions of high shear forces, chiefly close to the supports. The longitudinal tension reinforcement has been so calculated and placed that it is chiefly effective in resisting longitudinal tension near the tension face. It does not reinforce the tensionally weak concrete against those diagonal tension stresses which occur elsewhere, caused by shear alone or by the combined effect of shear and flexure. Eventually, these stresses attain magnitudes sufficient to open additional tension cracks in a direction perpendicular to the local tension stress. These are known as *diagonal* cracks, in distinction to the vertical flexural cracks. The latter occur in regions of large moments, the former in regions in which the shear forces are high. Both types of cracks are shown in Fig. 2.16. In beams in which no reinforcement is provided to counteract the formation of large diagonal tension cracks, their appearance has far-reaching and generally detrimental effects. For this reason methods of predicting the loads at which these cracks will form are desired.

CRITERIA FOR FORMATION OF DIAGONAL CRACKS

It is seen from Eq. (2.9) that the diagonal tension stresses t represent the combined effect of the shear stresses v and the bending stresses f. These in turn are, respectively, proportional to the shear force V and the bending moment M at the particular location in the beam [Eqs. (2.10) and (2.12)]. Depending on configuration, support conditions, and load distribution, at a given location in a beam one may find a large moment combining with a small shear force, or the reverse, or large or small values for both shear and moment. Evidently, the relative values of M and V will affect the magnitude as well as the direction of the diagonal tension stresses. Figure 2.15 shows a few typical beams and their moment and shear diagrams and draws attention to locations at which various combinations of high or low V and M occur.

At a location of large shear force V and small bending moment M there will be little flexural cracking, if any, prior to the development of a diagonal tension crack. Consequently, the average shear stress prior to crack formation is

$$v = \frac{V}{bd} \tag{2.32}$$

The exact distribution of these shear stresses over the depth of the cross section is not known. It cannot be computed from Eq. (2.12) because this equation does not account for the influence of the reinforcement, and because concrete is not an elastic homogeneous material. The value computed from Eq. (2.32) must therefore be regarded merely as a measure of the average intensity of shear stresses in the section. The maximum value, which occurs at the neutral axis, will exceed this average by an unknown but moderate amount.

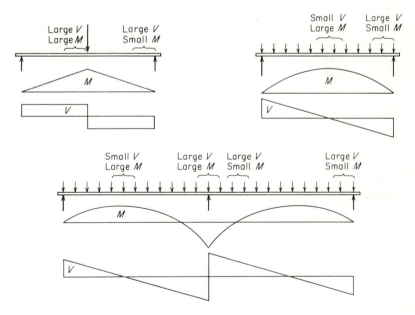

Fig. 2.15 Typical locations of critical combinations of shear and moment.

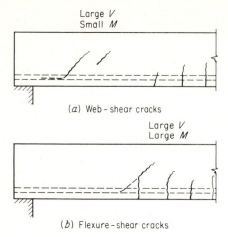

(a) Web - shear cracks

(b) Flexure - shear cracks

Fig. 2.16   Flexural and diagonal tension cracks.

If flexural stresses are negligibly small at the particular location, the diagonal tension stresses, as in Fig. 2.14*b* and *c*, are inclined at about 45° and are numerically equal to the shear stresses, with a maximum at the neutral axis. Consequently, diagonal cracks form mostly at or near the neutral axis and propagate from that location, as shown in Fig. 2.16*a*. These so-called *web-shear* cracks can be expected to form when the diagonal tension stress in the vicinity of the neutral axis becomes equal to the tension strength of the concrete. The former, as was indicated, is of the order of and somewhat larger than $v = V/bd$; the latter, as discussed in Art. 1.8, varies from about $3\sqrt{f_c'}$ to about $5\sqrt{f_c'}$. An evaluation of a very large number of beam tests is in fair agreement with this reasoning (Ref. 2.5). It was found that in regions with large shear and small moment, diagonal tension cracks form at an average or nominal shear stress v_{cr} of about

$$v_{cr} = \frac{V_{cr}}{bd} = 3.5\sqrt{f_c'} \tag{2.33a}$$

where V_{cr} is that shear force at which the formation of the crack was observed.[1] Web-shear cracking is relatively rare and occurs chiefly near supports of deep, thin-webbed beams or at inflection points of continuous beams.

The situation is different where both the shear force and the bending moment have large values. At such locations, in a well-proportioned and rein-forced beam, flexural tension cracks form first. Their width and length are well controlled and kept small by the presence of longitudinal reinforcement.

---

[1] Actually, diagonal tension cracks form at places where a compression stress acts in addition to and perpendicular to the diagonal tension stress, as shown in Fig. 2.14*c* and *e*. The crack therefore occurs at a location of biaxial stress, rather than uniaxial tension. However, the effect of this simultaneous compression stress on the cracking strength appears to be small, in agreement with the information in Fig. 1.9.

However, when the diagonal tension stress at the upper end of one or more of these cracks exceeds the tensile strength of the concrete, the crack bends in a diagonal direction and continues to grow in length and width (see Fig. 2.16b). These cracks are known as *flexure-shear* cracks and are more common than web-shear cracks.

It is evident that at the instant at which a diagonal tension crack of this type develops, the average shear stress is larger than that given by Eq. (2.32). This is so because the preexisting tension crack has reduced the area of uncracked concrete which is available to resist shear to a value smaller than that of the uncracked area bd used in Eq. (2.32). The amount of this reduction will vary, depending on the unpredictable length of the preexisting flexural tension crack. Furthermore, the simultaneous bending stress f combines with the shear stress v to further increase the diagonal tension stress t [see Eq. (2.9)]. No way has been found to calculate reliable values of the diagonal tension stress under these conditions, and recourse must be had to test results.

A large number of beam tests have been evaluated for this purpose (Ref. 2.5). They show that in the presence of large moments (for which adequate longitudinal reinforcement has been provided) the nominal shear stress at which diagonal tension cracks form and propagate is conservatively given by

$$v_{cr} = \frac{V_{cr}}{bd} = 1.9 \sqrt{f_c'} \qquad (2.33b)$$

Comparison with Eq. (2.33a) shows that large bending moments can reduce the shear force at which diagonal cracks form to roughly one-half the value at which they would form if the moment were zero or nearly so. This is in qualitative agreement with the discussion just given.

It is evident, then, that the shear at which diagonal cracks develop depends on the ratio of shear force to bending moment, or more precisely, on the ratio of shear stress v to bending stress f at the top of the flexural crack. Neither of these can be accurately calculated. It is clear, though, that $v = K_1(V/bd)$, where, by comparison with Eq. (2.32), K_1 depends chiefly on the depth of penetration of the flexural crack. On the other hand [see Eq. (2.18)], $f = K_2(M/bd^2)$, where K_2 also depends on crack configuration. Hence, the ratio

$$v/f = (K_1/K_2)(Vd/M)$$

must be expected to affect that load at which flexural cracks develop into flexure-shear cracks, the unknown quality K_1/K_2 to be explored by tests. Equation (2.33a) gives the cracking shear for very large values of Vd/M, and Eq. (2.33b) for very small values. Moderate values of Vd/M result in magnitudes of v_{cr} intermediate between these extremes. Again, from evaluations of large numbers of tests (Ref. 2.5), it has been found that the nominal shear stress at which diagonal flexure-shear cracking develops is conservatively predicted by

$$v_{cr} = \frac{V_{cr}}{bd} = 1.9 \sqrt{f_c'} + 2500 \frac{\rho V d}{M} \leq 3.5 \sqrt{f_c'} \qquad (2.33)$$

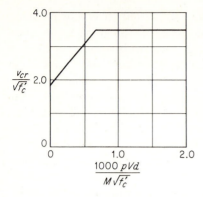

Fig. 2.17

where $\rho = A_s/bd$ as before, and 2500 is an empirical constant in psi units. A graph of this relation is given in Fig. 2.17.

Apart from the influence of Vd/M, it is seen from Eq. (2.33) that increasing amounts of tension reinforcement, i.e., increasing values of the steel ratio ρ, have a beneficial effect in that they increase the shear at which diagonal cracks develop. This is so because larger amounts of longitudinal steel result in smaller and narrower flexural tension cracks prior to the formation of diagonal cracking, leaving a larger area of uncracked concrete available to resist shear. [For more details on the development of Eq. (2.33), see Ref. 2.5.]

BEHAVIOR OF DIAGONALLY CRACKED BEAMS

In regard to flexural, as distinct from diagonal, tension, it was explained in Art. 2.4 that cracks on the tension side of a beam are permitted to occur and are in no way detrimental to the strength of the member. One might expect a similar situation in regard to diagonal cracking caused chiefly by shear. The analogy, however, is not that simple. Flexural tension cracks are harmless only because adequate longitudinal reinforcement has been provided to resist those flexural tension stresses which the cracked concrete is no longer able to transmit. In contrast, the beams now being discussed, although furnished with the usual longitudinal reinforcement, are not equipped with any other reinforcement to offset the effects of diagonal cracking. This makes the diagonal cracks much more decisive in subsequent performance and strength of the beam than the flexural cracks.

Two types of behavior have been observed in the many tests on which present knowledge is based:

1. The diagonal crack, once formed, spreads either immediately or at only slightly higher load, traversing the entire beam from the tension reinforcement to the compression face, splitting it in two and failing the beam. This process is sudden and without warning and occurs chiefly in the shallower beams, i.e., beams with span-depth ratios of about 8 or more.

Beams in this range of dimensions are very frequent. Complete absence of shear reinforcement would make them very vulnerable to accidental large overloads, which would result in catastrophic failures without warning. For this reason it is good practice to provide a minimum amount of shear reinforcement even if calculation does not require it, because such reinforcement restrains growth of diagonal cracks, thereby increasing ductility and providing warning in advance of actual failure. Only in situations where an unusually large safety factor against inclined cracking is provided, i.e., where actual shear stresses are very small as compared with v_{cr}, as in most slabs and footings, is it advisable to omit shear reinforcement.

2. Alternatively, the diagonal crack, once formed, spreads toward and partially into the compression zone but stops short of penetrating to the compression face. In this case no sudden collapse occurs, and the failure load is usually significantly higher than that at which the diagonal crack first formed. This behavior is chiefly observed in the deeper beams with smaller span-depth ratios and will be analyzed now.

Figure 2.18 shows a portion of a beam, arbitrarily loaded, in which a diagonal tension crack has formed. Consider the part of the beam to the left of

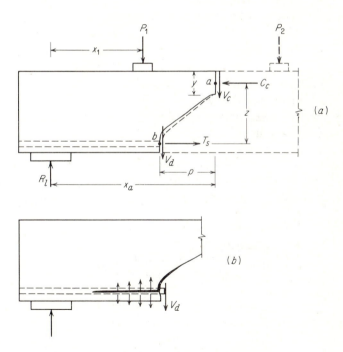

Fig. 2.18 Forces at diagonal crack in beam without web reinforcement.

the crack, shown in solid lines. There is an external upward shear force V_{ext} acting on this portion, which, for the particular loading shown, happens to be

$$V_{ext} = R_l - P_1$$

Once a crack is formed, no tension force perpendicular to the crack can be transmitted across it. However, as long as the crack is narrow, it can still transmit forces in its own plane through interlocking of the surface roughnesses. Sizable interlock forces of this kind have in fact been measured, amounting to one-third and more of the total shear force. However, at this writing, reliable values for it have not been determined over a sufficient range of variables to permit useful conclusions. Therefore, discounting the interlock force, the only internal vertical forces are those acting in the uncracked portion of the concrete, V_c, and in or across the longitudinal steel, V_d. Thus, the internal shear force is

$$V_{int} = V_c + V_d$$

Equilibrium requires $V_{int} = V_{ext}$, so that the part of the shear resisted by the uncracked concrete is

$$V_c = V_{ext} - V_d \qquad (2.34)$$

In a beam provided with longitudinal reinforcement only, the portion of the shear force resisted by the steel, in dowel action as it were, is usually quite small, In fact, the reinforcing rods on which the dowel force V_d acts are supported against vertical displacement chiefly by the thin concrete layer below. The bearing pressure caused by V_d creates, in this concrete, vertical tension stresses as shown in Fig. 2.18b. Because of these stresses, diagonal cracks often result in splitting of the concrete along the tension reinforcement, as shown. Correspondingly, it is likely that most of the shear is resisted by the uncracked portion of concrete at the head of the diagonal crack, or

$$V_c \cong V_{ext} \qquad (2.34a)$$

Next, consider moments about point a at the intersection of V_c and C_c; the external moment for the loading shown happens to be

$$(M_{ext})_a = R_l x_a - P_1(x_a - x_1)$$

and the internal moment is

$$(M_{int})_a = (T_s)_b z + (V_d)_b p$$

Here p is the horizontal projection of the diagonal crack, and the designations $(T_s)_b$ and $(V_d)_b$ are meant to emphasize that these two forces in the steel act at point b rather than vertically below point a. Again, equilibrium requires that $(M_{int})_a = (M_{ext})_a$, so that the longitudinal tension in the reinforcement at b is

$$(T_s)_b = \frac{(M_{ext})_a}{z} - \frac{(V_d)_b p}{z} \qquad (2.35)$$

Neglecting the dowel force V_d as being only a small fraction of the tension force T_s, one has, with very little error,

$$(T_s)_b \cong \frac{(M_{ext})_a}{z} \tag{2.35a}$$

The formation of the diagonal crack, then, is seen to produce the following redistribution of internal forces and stresses:

1. In the vertical section through point a the average shear stress before crack formation was V_{ext}/bd [see Eq. (2.32)]. After crack formation, that same shear force is almost entirely resisted by the much smaller area by of the remaining uncracked concrete [see Eq. (2.34a) and Fig. 2.18a]. Hence, the average shear stress in the concrete has now increased to V_{ext}/by.

2. The diagonal crack, as was described previously, usually rises above the the neutral axis and traverses some part of the compression zone before it is arrested by the compression stresses. Consequently, the compression force C_c also acts on an area by smaller than that on which it acted before the crack was formed. Correspondingly, formation of the crack has increased the compression stresses in the remaining uncracked concrete.

3. Prior to diagonal cracking, the tension force in the steel at point b was caused by and was proportional to the bending moment in a vertical section through the same point b. As a consequence of the diagonal crack, however, Eq. (2.35a) shows that the tension in the steel at b is now caused by and is proportional to the bending moment at a. Since the moment at a is evidently larger than that at b, formation of the crack has caused a sudden increase in the steel stress at b.

4. The previously described dowel action causes bearing and tension stresses which contribute to the tendency of the concrete to split along the bar toward the nearby support.

If the two materials are capable of resisting these increased stresses, equilibrium will establish itself after internal redistribution, and further load can be applied before failure occurs. Such failure can then develop in various ways. For one, if only enough steel has been provided at b to resist the moment at that section, then the increase of the steel force, in item 3 above, will cause the steel to yield because of the larger moment at a, thus failing the beam. If the beam is properly designed to prevent this occurrence, it is usually the concrete at the head of the crack which will eventually crush. This concrete is subject simultaneously to large compression and shear stresses, and this biaxial stress combination is conducive to earlier failure than would take place if either of these stresses were acting alone (see Art. 1.9, also Ref. 2.6). Finally, if there is splitting along the reinforcement, it will cause the bond between steel and concrete to weaken to such a degree that the reinforcement may pull loose. This may either be the cause of failure of the beam or may occur simultaneously with crushing of the remaining uncracked concrete.

It was noted earlier that relatively deep beams will usually show continued and increasing resistance after formation of a critical diagonal tension crack, but relatively shallow beams will fail almost immediately upon formation of the crack. The amount of reserve strength, if any, was found to be erratic. In fact, in several test series in which two specimens as identical as one can make them were tested, one failed immediately upon formation of a diagonal crack, while the other reached equilibrium under the described redistribution and failed at a higher load.

For this reason, this reserve strength is discounted in modern design procedures. As previously mentioned, most beams are furnished with at least a minimum of web reinforcement. For those flexural members which are not, such as slabs, footings, and others, design is based on that shear force V_{cr} or shear stress v_{cr} at which formation of inclined cracks must be expected. Thus, Eq. (2.33), or some simplification of it, has become the design criterion for such members.

PROBLEM

In the problem at the end of Art. 2.4 it was found that a rectangular beam with $b = 10$ in., $h = 25$ in., $d = 23$ in., three No. 9 bars, $f_c' = 4000$ psi, $f_y = 60,000$ psi will carry a uniform design load of 2450 lb per ft, plus its own weight with a safety factor slightly larger than the value 1.85 stipulated in that problem. Check whether this beam has a safety factor of at least 2.3 against the formation of diagonal cracks at the indicated load. Decide whether or not shear reinforcement is required. *Explanatory comments:* (*a*) It has been explained that if a beam without web reinforcement fails in shear, collapse is likely to be sudden and complete. On the other hand, if a beam were to be overloaded and to fail in bending, collapse would be gradual and preceded by excessive cracking and deflections which serve as warning of distress. For this reason a larger safety margin is desirable against shear than against flexural failure, as is here stipulated. The detailed manner in which these differential margins are provided in the ACI Code is explained in Chap. 3. (*b*) Tests have shown that diagonal cracks rarely form at a distance from the nearby support smaller than the depth d of the beam, because of the local stress conditions in the immediate vicinity of a support (see also Art. 3.5). For this reason it is now accepted practice that the maximum shear force for which the adequacy of a member must be checked is that acting at a distance d from the face of the support, rather than the larger shear force directly at that face.

c. Reinforced-concrete beams with web reinforcement   Economy of design demands, in most cases, that a flexural member be capable of developing its full moment capacity rather than having its strength limited by premature shear failure. This is also desirable because structures, if overloaded, should not fail in the sudden and explosive manner characteristic of many shear failures, but should show adequate ductility and warning of impending distress. The latter, as was pointed out, obtains for flexural failure caused by yielding of the longitudinal steel, which is preceded by gradual, excessive deflections

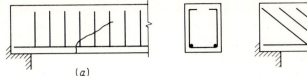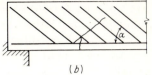

Fig. 2.19

and noticeable enlargement of cracks. If, then, a fairly large safety margin relative to the available shear strength as given by Eq. (2.33) does not exist, special shear reinforcement, known as *web reinforcement,* is used to increase this strength.

Web reinforcement may be in either of the two forms shown in Fig. 2.19, or a combination of the two. Figure 2.19a shows web reinforcement consisting of vertical stirrups; Fig. 2.19b shows inclined bars. The latter can consist of individual stirrups placed at an angle or part of the longitudinal flexural reinforcement bent up where it is no longer needed to furnish moment resistance. A diagonal crack is shown traversed by a number of web bars.

Web reinforcement has no noticeable effect prior to the formation of diagonal cracks. In fact, measurements show that the web steel is practically free of stress prior to crack formation. After diagonal cracks have developed, web reinforcement augments the shear resistance of a beam in three separate ways: (1) Part of the shear force is resisted by the bars which traverse a particular crack. The mechanism of this added resistance is discussed below. (2) The presence of these same bars restricts the growth of diagonal cracks and reduces their penetration into the compression zone. This leaves more uncracked concrete available at the head of the crack for resisting the combined action of shear and compression, already discussed. (3) As seen in the cross section of Fig. 2.19, the stirrups are so arranged that they tie the longitudinal reinforcement into the main bulk of the concrete. This provides some measure of restraint against the splitting of concrete along the longitudinal reinforcement, shown in Fig. 2.18b, and increases the share of the shear force resisted in dowel action.

It will be realized from this description that the behavior, once a crack is formed, is quite complex and dependent in its details on the particulars of crack configuration (length, inclination, and location of the main or critical crack). The latter, in turn, is quite erratic and has so far defied purely analytical prediction. For this reason, the concepts which underlie present design practice are not wholly rational. They are based partly on rational analysis, partly on test evidence, and partly on successful longtime experience with structures in which certain procedures for designing web reinforcement have resulted in satisfactory performance. Research in this field, both on plain and web-reinforced members, performed mostly since 1945, has furnished important experimental data for recent improvements in shear design and analysis. It has not yet resulted in a completely consistent rational analysis of shear behavior.

BEAMS WITH VERTICAL STIRRUPS

The web reinforcement being ineffective in the uncracked beam, the magnitude of the shear force or stress which causes cracking to occur is the same as in a beam without web reinforcement and is given by Eq. (2.33). Most frequently, web reinforcement consists of *vertical stirrups;* the forces acting on the portion of such a beam between the crack and the nearby support are shown in Fig. 2.20. They are the same as those of Fig. 2.18, except that each stirrup traversing the crack exerts a force $A_v f_v$ on the given portion of the beam. Here A_v is the cross-sectional area of the stirrup (in the case of the U-shaped stirrup of Fig. 2.19, it is twice the area of one bar), and f_v is the tension stress in the stirrup. Equilibrium in the vertical direction requires

$$V_c + \sum_n A_v f_v + V_d = V_{ext}$$

where n is the number of stirrups traversing the crack. If s is the stirrup spacing, and p the horizontal projection of the crack, as shown, then

$$n = \frac{p}{s}$$

Prior to failure, part of the load, as before, is carried by the uncracked section at the head of the crack. The dowel force V_d and the interlocking force along the crack will be disregarded since at present they are of uncertain value. Further opening of the crack is counteracted by the stirrups until the stress in them has reached the yield point f_y. The large stirrup extensions which occur at that stage permit the crack to open and extend, causing the remaining concrete to fail. Consequently, the stirrup stress being f_y when failure occurs, the ultimate shear force V'_u is

$$V'_u = V_c + n A_v f_y$$

While the share of the total shear carried by the stirrups is known, the shear V_c carried by the uncracked concrete at the head of the crack has yet to

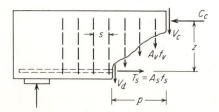

Fig. 2.20 Forces at diagonal crack in beam with vertical stirrups.

be defined. It is assumed in present-day methods that this shear, at failure, is the same as that which caused the diagonal crack to form; i.e.,

$$V_c = V_{cr}$$

[V_{cr} is given by Eq. (2.33).] This assumption seems reasonable and conservative. In fact, it has been pointed out that beams without web reinforcement generally fail at loads above that which causes the crack to form, the difference often being insignificant, but under certain conditions sizable. Also, since the presence of stirrups restricts crack growth, the shear resistance of the larger uncracked concrete area in web-reinforced beams should, if anything, exceed that of identical beams without such reinforcement. In addition, as has been pointed out, the dowel force V_d and the roughness interlock force are likely to be larger in web-reinforced than in plain-web beams. All this points to the fact that in a web-reinforced beam the part of the shear which is not resisted by the stirrups should be at least equal to, and is most likely larger than, the shear force which will cause an identical plain-web beam to crack.

The number of stirrups n spaced a distance s apart was seen to depend on the length p of the horizontal projection of the diagonal crack. If this length is assumed equal to the effective depth of the beam ($p = d$), implying a crack somewhat flatter than 45°, the above equation for the ultimate shear force V'_u becomes

$$V'_u = V_{cr} + \frac{A_v f_v d}{s} \tag{2.36}$$

Dividing both sides by bd, one obtains the same relation expressed in terms of the nominal ultimate shear stress:

$$v'_u = \frac{V'_u}{bd} = v_{cr} + \frac{A_v f_v}{bs} \tag{2.36a}$$

In Ref. 2.5 the results of 166 beam tests are compared with Eq. (2.36a). It is shown that the equation predicts the actual shear strength quite conservatively, the observed strength being on the average 45 per cent larger than predicted; a very few of the individual test beams developed strength just slightly below that of Eq. (2.36a). At the same time the very considerable scatter, even though almost entirely on the conservative side, indicates that a deeper and more precise understanding of shear strength has yet to be developed.

BEAMS WITH INCLINED BARS

The function of *inclined web reinforcement* (Fig. 2.19b) can be discussed in very similar terms. Figure 2.21 again indicates the forces which act on the portion of the beam to one side of that diagonal crack which results in eventual failure. The dowel force V_d has been omitted for clarity. The crack with horizontal

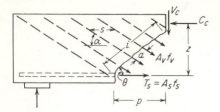

Fig. 2.21 Forces at diagonal crack in beam with inclined web reinforcement.

projection p and inclined length $i = p/\cos\theta$ is crossed by inclined bars horizontally spaced a distance s apart. The inclination of the bars is α, and that of the crack θ, as shown. The distance between bars measured parallel to the direction of the crack is seen from the irregular triangle to be

$$a = \frac{s}{\sin\theta(\cot\theta + \cot\alpha)}$$

The number of bars crossing the crack, $n = i/a$, after some transformation, is obtained as

$$n = \frac{p}{s}(1 + \cot\alpha\tan\theta)$$

The vertical component of the force in one bar or stirrup is $A_v f_v \sin\alpha$, so that the total vertical component of the forces in all bars which cross the crack is

$$V_s = nA_v f_v \sin\alpha = A_v f_v \frac{p}{s}(\sin\alpha + \cos\alpha\tan\theta)$$

As in the case of vertical stirrups, shear failure occurs when the stress in the web reinforcement reaches the yield point. Also, the same assumptions are made as in the case of stirrups, namely, that the horizontal projection of the diagonal crack is equal to the effective depth d, and that the concrete, at failure, resists the same force V_{cr} which caused the formation of the diagonal crack. Lastly, the inclination θ of the diagonal crack, which varies somewhat depending on various influences, is generally assumed to be 45°. On this basis, the ultimate strength when failure is caused by shear is obtained as

$$V'_u = V_{cr} + \frac{A_v f_y d(\sin\alpha + \cos\alpha)}{s} \tag{2.37}$$

or dividing both sides by bd,

$$v'_u = v_{cr} + \frac{A_v f_y(\sin\alpha + \cos\alpha)}{bs} \tag{2.37a}$$

It is seen that Eqs. (2.36) and (2.36a), developed for vertical stirrups, are only special cases, for $\alpha = 90°$, of the more general expressions (2.37) and (2.37a).

It should be noted that Eqs. (2.36) and (2.37) apply only if web reinforcement is so spaced that any conceivable diagonal crack is traversed by at least one stirrup or inclined bar. Otherwise web reinforcement would not contribute to the shear strength of the beam, because diagonal cracks which could form between widely spaced web reinforcement would fail the beam at the load at which it would fail if no web reinforcement were present. This imposes upper limits on the permissible spacing s to ensure that the web reinforcement is actually effective as calculated.

The inconsistencies in the preceding derivations are fairly evident. For instance, in the case of inclined bars, the assumptions that $p = d$ and $\theta = 45°$ actually mean, by simple geometry, that the diagonal crack has spread all the way to the compression face of the beam. If this were so, no uncracked compression zone would remain available to resist the share V_{cr} of the total shear force. However, this assumption, which is evidently on the unconservative side, appears to be compensated for by other conservative approximations. For one, the share of the shear which is resisted by dowel action and roughness interlock has been neglected; for another, as has been pointed out, the share which the uncracked concrete is able to carry has probably been underestimated.

One must conclude that at this time the nature and mechanism of diagonal tension failure are rather clearly understood qualitatively, that some of the quantitative assumptions which have been made cannot be justified by rational analysis, that the calculated results are in acceptable and generally conservative agreement with a very large body of empirical data, and that structures designed on this basis have proved satisfactory.

PROBLEM

The beam discussed in the problem at the end of Art. 2.5b ($b = 10$ in., $h = 25$ in., $d = 23$ in., $f'_c = 4000$ psi, span = 20 ft) has been more heavily reinforced with three No. 10 bars, $f_y = 60$ ksi, in order to enable it to carry in flexure a uniform design load of 4000 lb per ft in addition to its own weight, with a minimum safety factor of 1.7. For the section of the beam at a distance d from the support, determine whether web reinforcement is required if a safety factor of at least 1.9 is stipulated against failure by shear. If so, use vertical U-shaped stirrups as shown on Fig. 2.19a, made of No. 2 bars with $f_y = 60$ ksi, and determine the required spacing s at two locations: (1) a distance equal to $d = 23$ in. and (2) a distance of 36 in. from the support.

2.6 BOND AND ANCHORAGE

In a short piece of a beam of length dx, shown in Fig. 2.22, the moment at one end will generally differ from that at the other end by a small amount dM. If this piece is isolated, and if one assumes that after cracking the concrete does not resist any tension stresses, the internal forces are those shown in Fig. 2.22b. The change in bending moment dM produces a change in the bar force

$$dT = \frac{dM}{z}$$

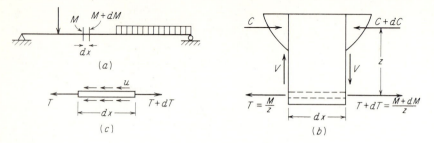

Fig. 2.22

Since the bar or bars must be in equilibrium, this change in bar force is resisted at the contact surface between steel and concrete by an equal and opposite force produced by the bond between steel and concrete. If U is the magnitude of this bond force per unit length of bar, then

$$(a) \quad U \, dx = (T + dT) - T = dT$$

from which

$$(b) \quad U = \frac{dT}{dx} = \frac{1}{z} \frac{dM}{dx} = \frac{V}{z}$$

Assuming that this bond force per unit length is the resultant of shear-type bond stresses u (see Fig. 2.22c) uniformly distributed over the contact area, the nominal bond stress is $u = U/\Sigma_0$, or

$$u = \frac{V}{\Sigma_0 z} \tag{2.38}$$

where Σ_0 is the sum of the perimeters of all the bars. This bond stress is caused by the change dM in bending moment and, for this reason, is known as *flexural bond*.

When plain bars without surface deformations were used, initial bond strength was provided only by the relatively weak chemical adhesion and mechanical friction between steel and concrete. Once adhesion and static friction were overcome at larger loads, small amounts of slip led to interlocking of the natural roughness of the bar with the concrete. However, this natural bond strength is so low that in beams reinforced with plain bars the bond between steel and concrete is frequently broken. Such a beam will collapse as the bar is pulled through the concrete. To prevent this occurrence, end anchorage is provided, chiefly in the form of hooks, as in Fig. 2.23. If the anchorage is adequate, such a beam will not collapse even if the bond is broken over the entire length between anchorages. This is so because the member acts as a tied arch, as shown in Fig. 2.23, the uncracked concrete representing the arch (shaded), and the anchored bars the tie rod. In this case, over the length in which the bond is broken, $u = 0$, and therefore $dT = 0$. This means that over the entire unbonded length the force in the steel is con-

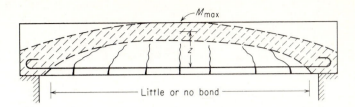

Fig. 2.23 Tied-arch action in beam with little or no bond.

stant and equal to $T = M_{max}/z$. In consequence, the total steel elongation in such beams is larger than in those in which bond is preserved, resulting in larger deflections and larger widths of cracks.

To improve this situation, deformed bars are now universally used in the United States (see Art. 1.12). With such bars, the shoulders of the projecting ribs bear on the surrounding concrete and result in greatly increased bond strength. It is then possible in most cases to dispense with special anchorage devices such as hooks and anchor plates. In addition, crack widths, as well as deflections, are reduced.

The actual distribution of bond stresses along deformed reinforcing bars is more complex than that represented by Eq. (2.38). In actuality, concrete fails to resist any tension stresses only where a crack is located. Between cracks, the concrete does resist moderate amounts of tension stress; this reduces the tension force T in the steel, as shown in Fig. 2.24 for a beam portion in pure bending. Since, from Eq. (*b*),

$$u = \frac{dT/dx}{\Sigma_0}$$

local bond stresses, being proportional to the rate of change of the bar force, act in the immediate vicinity of tension cracks and are distributed as shown in

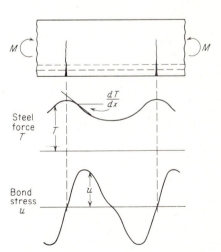

Fig. 2.24 Variation of steel force and bond stress at flexural cracks.

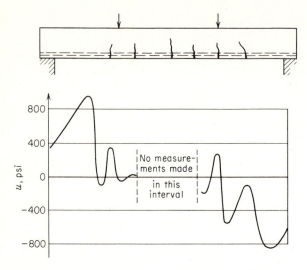

Fig. 2.25   Measured bond-stress distribution. (*From Ref. 2.7.*)

Fig. 2.24. Very high values of such bond stresses adjacent to cracks have been measured at Cornell University (Ref. 2.7). They are so high, particularly when high-strength steels are used, that a certain amount of localized slip between steel and concrete occurs at each crack.

If bond stresses become excessive, bond failure ensues. Equation (2.38) shows that this will generally occur where shear forces are large, i.e., in the vicinity of supports. In these regions, bond failure starts at cracks, because there the described localized bond stresses superpose on the flexural bond. Thus, bond stresses at loads close to those which cause bond failure have been measured to be distributed in the general manner shown in Fig. 2.25 (Ref. 2.7).

When bond failure ensues, it results generally in splitting of the concrete along the bars, either in vertical planes as in Fig. 2.26a or in horizontal planes as in Fig. 2.26b. Such splitting comes largely from wedging action when the

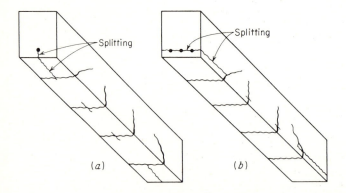

Fig. 2.26   Splitting along reinforcement (schematic).

ribs of the deformed bars bear against the concrete. The horizontal type of splitting of Fig. 2.26b frequently begins at a diagonal crack. In this case, as discussed in connection with Fig. 2.18b, the dowel action increases the tendency toward splitting. This indicates that shear and bond failures are often intricately interrelated.

When splitting has spread all the way to the end of an unanchored bar, complete bond failure occurs. That is, sliding of the steel relative to the concrete leads to immediate collapse of the beam.

For modern deformed bars, recent tests at the University of Texas and at the United States Bureau of Standards (Refs. 2.8, 2.9, and 2.19) seem to indicate that splitting occurs when the total bond force U per inch of length of bar, which is transmitted from steel to concrete, reaches a critical value. This ultimate bond force, in pounds per linear inch of bar, is largely independent of bar size or perimeter. The concept of a wedging action is in reasonable conformity with this finding, since the effects of a wedge of given shape depend more on the force with which it is driven than on its size.

It was found that the ultimate average bond force per inch of length of bar is, approximately,

$$U'_u = 35 \sqrt{f'_c} \tag{2.39}$$

Bond forces of this magnitude were observed to result in failure by splitting, in large and excessive slip, or in other prohibitive deformations. These tests were made mostly on beams of the type in Fig. 2.26a, a single bar causing vertical splitting. For several bars in one layer, spaced laterally 6 in. or less from each other (Fig. 2.26b), the ultimate bond force is about 80 per cent of that of Eq. (2.39).

This information can be expressed in terms of the conventional nominal average bond stress per square inch of contact area by dividing U'_u by $\Sigma_0 = \pi d_b$, with d_b the bar diameter. This gives, for the ultimate bond stress, approximately

$$u'_u = 11 \frac{\sqrt{f'_c}}{d_b} \qquad \text{psi} \tag{2.40}$$

The fact that bond resistance was found to correlate better with $\sqrt{f'_c}$ than with f'_c agrees with the concept that the resistance of concrete to splitting depends chiefly on its tensile, rather than compressive, strength.

If one considers the large local variations of bond stress caused by flexural and diagonal cracks (see Figs. 2.24 and 2.25), it becomes clear that local bond failures immediately adjacent to cracks will often occur at loads considerably below the failure load of the beam. These result in small local slips and some widening of cracks and increase of deflections, but will be harmless as long as failure does not propagate all along the bar, with resultant total slip. In fact, as was discussed in connection with Fig. 2.23, in those cases in which end anchorage is reliable, bond can be severed along the entire length of bar, excluding the anchorages, without endangering the carrying capacity of the beam.

In the beam of Fig. 2.27, the moment, and therefore the steel stress, is evidently zero at the supports and maximal at point a (neglecting the weight of the beam). If one designates the steel stress at a as f_s, the total tension force at that point in a bar of area A_b is $T_s = A_b f_s$, while at the end of the bar it is zero. Evidently, this force has been transferred from the concrete into the bar over the length l by bond stresses on the surface. Therefore, the average bond force per unit length over the length l is

$$U = \frac{A_b f_s}{l}$$

If this bond force U is smaller than the ultimate value U'_u of Eq. (2.39), no splitting or other complete failure will occur over the length L.

To put it differently, the minimum length which is necessary to develop, by bond, a given bar force $A_b f_s$ is

$$l_d = \frac{A_b f_s}{U'_u} \tag{2.41}$$

This length l_d is known as the *development length* of the bar. In particular, in order to ensure that a bar is securely anchored by bond to develop its maximum usable strength (the yield stress), this development length must be, approximately,

$$l_{du} = \frac{A_b f_y}{U'_u} = \frac{0.028 A_b f_y}{\sqrt{f'_c}} \tag{2.41a}$$

for bars spaced at least 6 in. from each other. For bars of closer lateral spacing, dividing by 0.8, one has

$$l_{du} = \frac{0.035 A_b f_y}{\sqrt{f'_c}} \tag{2.41b}$$

Therefore if, in the beam of Fig. 2.27, the actual length l is equal to or larger

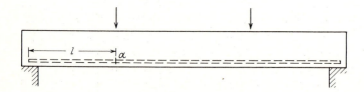

Fig. 2.27 Development length.

than the development length l_{du}, no premature bond failure will occur. That is, the beam will fail in bending or shear, rather than by bond failure. This will be so even if, in the immediate vicinity of cracks, local slips may have occurred over small regions along the beam.

From this discussion it is seen that the main requirement for safety against bond failure is this: The length of the bar, from any point of given steel stress (f_s, or at most f_y) to its nearby free end must be at least equal to its development length, as given by Eq. (2.41), (2.41a), or (2.41b). If this requirement is satisfied, the magnitude of the nominal flexural bond stress along the beam, as given by Eq. (2.38), is of only secondary importance, since the integrity of the members is assured even in the face of possible minor local bond failures. However, if the actual available length is inadequate for full development, special anchorage, such as by hooks, must be provided to ensure adequate strength.

The actual local intensity of the bond force along the embedded length of a bar varies, depending on the crack pattern, on the distance from the section of maximum steel stress, and on other factors. This was illustrated by Figs. 2.24 and 2.25. The ultimate value of Eq. (2.39) was obtained, from tests with a variety of embedded lengths l, by dividing the bar force by this length; that is, $U'_u = A_b f_s / l$. This quantity, therefore, is merely the average value over the embedded length at bond failure, regardless of local variations.

It is seen that U'_u is largely independent of the diameter of the bar. This means that the bond force per inch is substantially the same for large as for small bars. On the other hand, for a given steel stress f_s, the bar force $A_b f_s = f_s \pi d_b^2 / 4$ is proportional to the square of the diameter. If this is substituted in Eq. (2.41), the result is $l_d = f_s \pi d_b^2 / 4 U'_u$, showing that the necessary development length increases with the square of the bar diameter. Therefore, a beam reinforced with a larger number of small bars requires a smaller development length than a beam reinforced with a small number of large bars of the same total area. This demonstrates the superiority of small bars in developing bond strength.

PROBLEM

Verify whether the beam discussed at the end of Art. 2.5 is adequately designed to provide a safety factor of at least 2.0 against failure in bond. *Notes:* (a) For identifying the development length, assume that the beam has a clear span of 20 ft and that the reinforcement is carried 6 in. beyond the face of the support, as shown schematically in Fig. 2.27. (b) The maximum steel stress evidently occurs at the point of maximum moment, and development length must be checked for that point. However, in a uniformly loaded beam the bending moment, and thereby the steel stress, at first decrease quite slowly with increasing distance from midspan. For this reason, adequacy of development length should also be checked at one or more additional locations, closer to the supports; it is suggested that this be done for the quarter-points of the span. (c) For bar diameters and areas see Tables 1 and 2 of the Appendix. (d) If the development length is insufficient, bond strength can be augmented by providing end anchorage in the form of hooks (see Art. 3.6).

2.7 TORSION, AND TORSION PLUS SHEAR

Reinforced-concrete members are rarely designed to resist torsion alone. However, in many situations beams and other members are subject to torsion in addition to shear and bending, and in some cases, to axial compression or tension. Typical situations are spandrel beams or any edge beams of slabs, where the torsional rigidity of the beam provides rotational edge restraint to the slab. This causes corresponding flexural restraining moments in the slab, which in turn are balanced by torsional moments in the beam. Another frequent example is edge members of shells. To understand the behavior of members subject to such combinations of load effects, it is first necessary to discuss the effects of simple torsion on reinforced-concrete members. For simplicity the discussion will be restricted to rectangular members.

a. Torsion in plain concrete members   Figure 2.28 shows a portion of a prismatic member subject to equal and opposite torques T at both ends. If the material is elastic, St. Venant's classical torsion theory indicates that torsional shear stresses are distributed over the cross section, as shown in Fig. 2.28b. The largest shear stresses occur at the middle of the wide faces and are equal to

$$\tau_{\max} = \eta \, \frac{T}{x^2 y} \tag{2.42}$$

where η is a shape factor, and x and y are, respectively, the shorter and longer sides of the rectangle, as shown. If the material is inelastic, the stress distribution is similar, as shown by dashed lines, and the maximum shear stress is still given by Eq. (2.42), except that η assumes a different value.

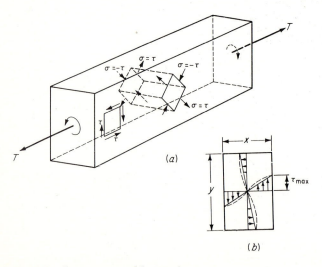

Fig. 2.28   Stresses caused by torsion.

Shear stresses in pairs act on an element at or near the wide surface, as shown in Fig. 2.28a. As is shown in strength-of-materials (see also Fig. 2.14), this state of stress corresponds to equal tension and compression stresses on the faces of an element at 45° to the direction of shear. These inclined tension stresses are of the same kind as those caused by transverse shear, and discussed in Art. 2.5a. However, in the case of torsion, since the torsional shear stresses are of opposite signs in the two halves of the member (Fig. 2.28b), the corresponding diagonal tension stresses in the two halves are at right angles to each other (Fig. 2.28a).

When the diagonal tension stresses exceed the tension resistance of the concrete, a crack forms at some accidentally weaker location and spreads immediately across the beam, as shown in Fig. 2.29. Observation shows that the tension crack (on the near face on Fig. 2.29) forms at practically 45°, that is, perpendicular to the diagonal tension stresses. The cracks on the two narrow faces, where diagonal tension stresses are smaller, are of more indefinite inclination, as shown, and the fracture line on the far face connects the cracks

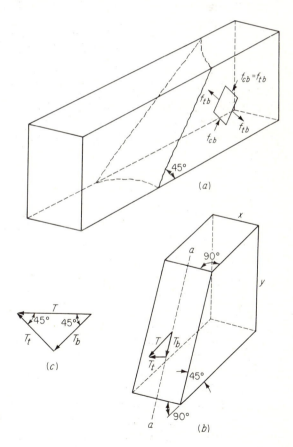

Fig. 2.29 Torsional crack in plain concrete member.

at the short faces. This completes the formation of an entire fracture surface across the beam which fails the member.

For purposes of analysis this somewhat warped fracture surface can be replaced by a plane section inclined at 45° to the axis, as in Fig. 2.29b. Test observation shows (Ref. 2.13) that on such a plane, failure is more nearly by bending than by twisting. As shown in Fig. 2.29b and c, the applied torque T can be resolved into a component T_b which causes bending about the axis a-a of the failure plane and a component T_t which causes twisting. It is seen that

$$T_b = T \cos 45°$$

The section modulus of the failure plane about a-a is

$$Z = \frac{x^2 y \csc 45°}{6}$$

Then the maximum bending (tension) stress in the concrete

$$f_{tb} = \frac{T_b}{Z} = T \sin 45° \cos 45° \frac{6}{x^2 y}$$

or

$$f_{tb} = \frac{3T}{x^2 y} \tag{2.43}$$

It is seen that the tension stress so calculated is identical with the St. Venant shear stress τ_{max} (Eq. 2.42) or with the corresponding diagonal tension stress σ (Fig. 2.28) for $\eta = 3$.

If f_{tb} were the only stress acting, cracking should occur when $f_{tb} = f_r$, the modulus of rupture of concrete, which can be taken as $f_r = 7.5 \sqrt{f_c'}$ (see Art. 1.8). However, at right angles to the tension stress f_{tb} there exists a compression stress f_{cb} of equal magnitude (see Figs. 2.28a and 2.29a). For this state of biaxial stress, tests show that the presence of equal perpendicular compression reduces the tension strength of concrete by about 15 per cent (see Fig. 1.9). Consequently, a crack forms and the member fails approximately when $f_{tb} = 0.85 f_r = 6 \sqrt{f_c'}$. Let this value of f_{tb} be designated as the cracking stress

$$f_{cr} = \tau_{cr} = 6 \sqrt{f_c'} \tag{2.44}$$

Then, upon substitution of f_{cr} for f_{tb} in Eq. (2.43), one obtains the magnitude of the torque which will crack and fail a plain rectangular concrete member.

$$T_{cr} = 6 \sqrt{f_c'} \frac{x^2 y}{3} \tag{2.45}$$

b. Torsion in reinforced-concrete members To resist torsion, reinforcement must consist of closely spaced closed stirrups and of longitudinal bars. Tests have shown that longitudinal bars alone hardly increase the torsional strength, test results showing an improvement of at most 15 per cent (Ref. 2.14). This is understandable because the only way in which longitudinal steel alone can contribute to torsional strength is by dowel action, which is particularly weak and unreliable if longitudinal splitting along bars is not restrained by transverse reinforcement. Thus, the torsional strength of members reinforced only with longitudinal steel is satisfactorily, and somewhat conservatively, predicted by Eqs. (2.44) and (2.45).

When members are adequately reinforced as in Fig. 2.30a, the concrete cracks at a torque equal to or only somewhat larger than in an unreinforced member [Eq. (2.45)]. The cracks form a spiral pattern as shown, for one single crack, in Fig. 2.30b. In actuality, a great number of such spiral cracks develop at close spacing. Upon cracking, the torsional resistance of the concrete drops to about half of that of the uncracked member, the remainder being now resisted by reinforcement. This redistribution of internal resistance is reflected in the torque-twist curve (Fig. 2.31), which at the cracking torque shows continued twist at constant torque until the reinforcement has picked up the portion of the torque no longer carried by concrete. Any further increase of applied torque must then be carried by the reinforcement. Failure occurs when somewhere along the member the concrete crushes along a line such as a-d in Fig. 2.30. In a well-designed member such crushing occurs only after the stirrups have started to yield.

The torsional strength can be analyzed by considering the equilibrium of the internal forces which are transmitted across the potential failure surface,

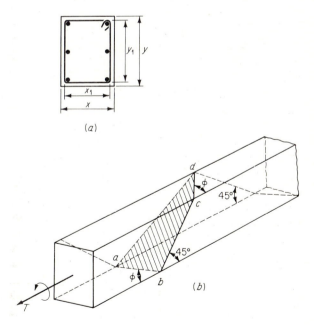

Fig. 2.30 Torsional crack in reinforced-concrete member.

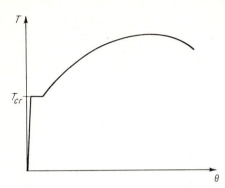

Fig. 2.31 Torque-twist curve in reinforced concrete.

shown shaded in Fig. 2.30. This surface is seen to be bounded by a 45° tension crack across one wider face, two cracks across the narrower faces of inclination ϕ, an angle generally between 45° and 90°, and the zone of concrete crushing along line a-d. The failure is basically flexural, as for plain beams, with a concrete compression zone developing adjacent to a-d (Ref. 2.15).

Figure 2.32 shows the partially cracked failure surface, including the compression zone of concrete (shaded) and the horizontal and vertical stirrup forces S_h and S_v of all the stirrups intersecting the failure surface, except for those located in the compression zone. The number of horizontal stirrup legs, top or bottom, crossing the surface is seen to be $n_h = (x_1 \cot \phi)/s$, and the number of vertical legs opposite the compression zone, $n_v = y_1/s$. It is known

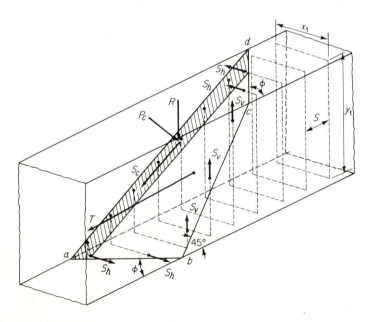

Fig. 2.32

from tests that at failure the vertical stirrup legs yield while the horizontal legs are generally not stressed to yielding. Correspondingly, the twisting couple produced by the horizontal stirrup forces is

$$(a) \quad T_h = n_h S_h y_1 = \left(\frac{x_1 \cot \phi}{s}\right) A_t f_{sh} y_1 = \kappa_h \frac{x_1 y_1}{s} A_t f_y$$

where A_t = area of one stirrup leg
$\quad f_{sh}$ = tension stress in horizontal stirrup leg
$\quad f_y$ = yield point
$\quad \kappa_h = \cot \phi (f_{sh}/f_y)$

To analyze the torque produced by the vertical stirrup forces near the front face, one must note, first, that the equilibrating forces near the rear face, in the compression zone, are fairly indeterminate. They consist, at the least, of a shear force S_c and a compression force P_c in the concrete and forces in the stirrup legs located in that zone. It is clear, however, that because of equilibrium, all these forces must have a resultant R equal and opposite to the sum of the vertical stirrup forces S_v. Correspondingly, the torque produced by the vertical stirrup forces can be written

$$(b) \quad T_v = n_v S_v x_v = \frac{y_1}{s} A_t f_y \kappa_v x_1 = \kappa_v \frac{x_1 y_1}{s} A_t f_y$$

where x_v is the lever arm of the internal forces S_v and R, and $\kappa_v = x_v/x_1$.

It is seen that Eqs. (a) and (b) are identical except for κ_h and κ_v. At the present stage of knowledge, neither of these constants can be determined analytically and recourse must be had to extensive tests. Designating $\alpha_t = \kappa_h + \kappa_v$, the total torque contributed by the stirrups, $T_s = T_v + T_h$, is seen to be

$$T_s = \alpha_t \frac{x_1 y_1}{s} A_t f_y \tag{2.46}$$

Tests (Ref. 2.15) have shown that α_t depends primarily on the ratio of cross-sectional dimensions and can be taken as

$$\alpha_t = 0.66 + 0.33 \frac{k_1}{y_1} \leq 1.5 \tag{2.47}$$

It was mentioned previously that after cracking, the torque T_0 contributed by the concrete compression zone is about half of the cracking torque T_{cr} [Eq. (2.45)]. Taking the fraction conservatively to be 40 per cent,

$$T_0 = 2.4 \sqrt{f_c'} \frac{x^2 y}{3} \tag{2.48}$$

corresponding to a nominal torsional shear stress resisted by the concrete of

$$\tau_0 = 2.4 \sqrt{f_c'} \tag{2.49}$$

The total ultimate torque is then $T'_u = T_0 + T_s$, or

$$T'_u = 2.4 \sqrt{f'_c} \frac{x^2 y}{3} + \alpha_t A_t \frac{x_1 y_1}{s} f_v \tag{2.50}$$

From the derivation of T'_u it is evident that this ultimate torsional strength will be developed only if the stirrups are sufficiently closely spaced so that any failure surface will intersect an adequate number of stirrups. For this reason the maximum spacing of stirrups is $s_{max} = y_1/2$ and also should not exceed x_1.

The role of the longitudinal reinforcement in providing torsional strength is not yet clearly understood, but it is known that T'_u can be developed only if adequate longitudinal reinforcement is provided. Its chief functions are: (1) It anchors the stirrups, particularly at the corners, which enables them to develop their full yield strength. (2) It provides at least some resisting torque because of the dowel forces which develop where the bars cross torsional cracks. (3) It has been observed that, after cracking, members subject to torsion tend to lengthen as the spiral cracks widen and become more pronounced. Longitudinal reinforcement counteracts this tendency and controls crack width. Tests indicate that for Eq. (2.50) to be valid the total volume of longitudinal steel in a unit length of the member should be between 0.7 and 1.5 times the total volume of stirrups in that same length. It is customary to design torsional members so that these two volumes are equal. It is easily verified that this is so if the total area of longitudinal reinforcement is

$$A_l = 2A_t \frac{x_1 + y_1}{s} \tag{2.51}$$

Lastly, just as for flexural members, it is desirable that torsional members show ductile rather than brittle behavior. This means that the stirrups should yield before the compression zone of the failure surface crushes. For this to be true (Ref. 2.14) the amount of reinforcement must be limited to that which, according to Eq. (2.50), will produce a calculated ultimate torque no larger than

$$T'_{u,max} = 12 \sqrt{f'_c} \frac{x^2 y}{3} \tag{2.52}$$

c. Torsion plus shear, members without stirrups It was mentioned earlier that reinforced-concrete members designed to carry torsion alone are quite unusual. The prevalent situation is that of a beam subject to the usual flexural moments and shear forces, which, in addition, must also resist torsional moments. In an uncracked member, shear forces as well as torques produce shear stresses. It must be expected, therefore, that simultaneously applied flexural shear forces and torques interact in a manner which will reduce the strength of the member compared with what it would be if shear or torsion were acting alone.

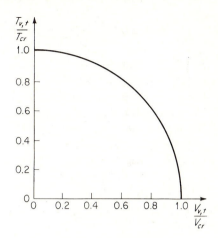

Fig. 2.33 Interaction curve for combined torsion plus flexural shear.

No satisfactory theories of this complex interaction have yet been developed, so that reliance must be placed on the extensive experimental investigations of this situation (Ref. 2.16). These test results are best represented in terms of an interaction equation or curve. Let V_{cr} and T_{cr}, according to Eqs. (2.33) and (2.45), be the cracking shear and torque of the member when subject, respectively, to flexural shear or to torsion alone. It will be recalled that for all practical purposes failure occurs at these same values almost immediately following cracking, so that, for members without web reinforcement, V_{cr} and T_{cr} adequately represent their ultimate strengths in the two modes. Further, let $V_{v,t}$ and $T_{v,t}$, respectively, represent the shear capacity and the torsion capacity under combined loading, that is, when the member is subject to simultaneous flexural shear and torsion. Then the numerous test results are well, and somewhat conservatively, represented by the circular-interaction equation

$$\left(\frac{V_{v,t}}{V_{cr}}\right)^2 + \left(\frac{T_{v,t}}{T_{cr}}\right)^2 = 1 \tag{2.53}$$

A graphical representation of this equation is shown in Fig. 2.33. It is seen that this interaction curve is quite favorable; i.e., the two modes do not very strongly interfere with each other. For instance, if a member carries a torque $T_{cr}/2$, that is, one half of its pure torsion capacity, the curve shows that it can carry simultaneously about $0.85V_{cr}$, that is, only 15 per cent less than it could carry if no torsion were present at all.

Equation (2.53) can also be written in terms of the nominal shear stresses due to flexural shear and torsion, i.e.,

$$\left(\frac{v_{v,t}}{v_{cr}}\right)^2 + \left(\frac{\tau_{v,t}}{\tau_{cr}}\right)^2 = 1 \tag{2.54}$$

which, after some transformation, results in

$$\tau_{v,t} = \frac{\tau_{cr}}{\sqrt{1 + (\tau_{cr}/v_{cr})^2 (v_{v,t}/\tau_{v,t})^2}} \tag{2.54a}$$

But according to Eq. (2.33b), $v_{cr} = 1.9 \sqrt{f_c'}$, while according to Eq. (2.44), $\tau_{cr} = 6 \sqrt{f_c'}$, so that, with slight rounding off, $\tau_{cr}/v_{cr} = 3$. Substituting this value in Eq. (2.54a), one obtains that amount of torsional shear stress $\tau_{v,t}$ which a member can carry in the presence of a simultaneous flexural shear stress $v_{v,t}$.

$$\tau_{v,t} = \frac{\tau_{cr}}{\sqrt{1 + (3v_{v,t}/\tau_{v,t})^2}} \tag{2.55}$$

In Eq. (2.55) the reduction in torsional strength caused by the simultaneous presence of flexural shear is expressed by the denominator and is seen to depend on the ratio of the flexural to torsional nominal shear stress.

d. Torsion plus shear, members with stirrups In section b above, it has been shown that in members subject only to torsion and reinforced with stirrups and longitudinal bars, the ultimate torque T_u' is furnished in part by the torsional strength of the concrete compression zone T_c, and in part by the torsional resistance of the steel, T_s [see Eqs. (2.46), (2.48), and (2.50)].

Experimental evidence on the strength of such reinforced members when subject to simultaneous torsion and shear at this time is fairly sparse (Ref. 2.17). Pending more extensive research, both experimental and analytical, the following approach appears reasonable and correlates satisfactorily and conservatively with the limited test evidence (Refs. 2.14 and 2.17):

1. In members with stirrups the portion of the total torsion carried by the concrete is determined by the same type of interaction equation [Eq. (2.54)] as in members without stirrups.
2. To carry the excess torque, over and above that resisted by the concrete, the same amount of reinforcement must be provided in members subject to torsion plus shear as would be required for purely torsional members. This torsional reinforcement is to be added to that required in the same member for carrying bending moments and flexural shears.

To determine the portions of the total shear and torsion carried by the concrete according to point 1 above:

Let v_c = nominal shear stress carried by concrete in member subject only to flexural shear

τ_c = nominal shear stress carried by concrete in member subject only to torsion

$v_{cv,t}$ = nominal shear stress carried by concrete, caused by flexural shear in member subject to both shear and torsion

$\tau_{cv,t}$ = nominal shear stress carried by concrete, caused by torsion in member subject to both shear and torsion

Then, in analogy with Eq. (2.54) for members without stirrups, one has for members with stirrups

(a) $\quad \left(\dfrac{v_{cv,t}}{v_c}\right)^2 + \left(\dfrac{\tau_{cv,t}}{\tau_c}\right)^2 = 1$

From this [see also Eq. (2.54a)], assuming $v_{v,t}/\tau_{v,t} = v_{cv,t}/\tau_{cv,t}$, one obtains

(b) $\quad \tau_{cv,t} = \dfrac{\tau_c}{\sqrt{1 + (\tau_c/v_c)^2(v_{v,t}/\tau_{v,t})^2}}$

(c) $\quad v_{cv,t} = \dfrac{v_c}{\sqrt{1 + (v_c/\tau_c)^2(\tau_{v,t}/v_{v,t})^2}}$

where, as before, $v_{v,t}$ and $\tau_{v,t}$ are the total nominal shear stresses caused in the member by the simultaneously acting shear force and torque, respectively.

In regard to flexural shear alone it has been shown in Art. 2.5 [Eqs. (2.36) and (2.36a)] that in members with stirrups the concrete after cracking continues to carry the same amount of shear which will cause cracking in the same member without stirrups. Hence, $v_c = 2\sqrt{f_c'}$, according to Eq. (2.33b) slightly rounded off. On the other hand, in regard to torsion, it was mentioned in section c above that in members with stirrups, the concrete after cracking continues to carry only about 40 per cent of the torsion which would cause cracking in the same member without stirrups. Therefore, according to Eq. (2.49), $\tau_c = 2.4\sqrt{f_c'}$. Hence, for use in the denominators of Eqs. (b) and (c), $\tau_c/v_c = 1.2$. Substituting this value in Eq. (b), one obtains for the torsional shear stress carried by the concrete

$$\tau_{cv,t} = \frac{2.4\sqrt{f_c'}}{\sqrt{1 + (1.2v_{v,t}/\tau_{v,t})^2}} \tag{2.56}$$

or for the torque carried by the concrete [cf. Eq. (2.48)]

$$T_{cv,t} = \frac{2.4\sqrt{f_c'}}{\sqrt{1 + (1.2v_{v,t}/\tau_{v,t})^2}}\frac{x^2y}{3} \tag{2.56a}$$

If the torque T_u' actually applied to the member is larger than that carried by the concrete, reinforcement must be provided to carry the excess torque

$$T_s = T_u' - T_{cv,t}$$

Then from Eq. (2.46) the required area of one stirrup leg at spacing s becomes

$$A_t = \frac{T_s s}{\alpha_t f_y x_1 y_1} = \frac{(\tau_{v,t} - \tau_{cv,t})s x^2 y}{3\alpha_t x_1 y_1 f_y} \tag{2.57}$$

This amount of stirrup reinforcement, plus the corresponding amount of longitudinal steel calculated from Eq. (2.51), must be provided in addition to that required in the same member for flexural shears and moments.

For determining the latter, i.e., the amount of stirrup reinforcement necessary to carry the flexural shear in the presence of simultaneous torsion, Eqs. (2.36a) and (2.37a) are used with the following modification: In these equations, v_{cr} is the nominal shear stress carried by the concrete when no torsion is present. In the presence of torsion, Eq. (c), above, gives the flexural shear stress carried by concrete, and hence, in Eqs. (2.36a) and (2.37a), v_{cr} must be replaced by $v_{cv,t}$. For determining the latter, $v_c = v_{cr} = 1.9 \sqrt{f_c'}$ is again rounded off to $v_c = 2.0 \sqrt{f_c'}$ and, as before, $\tau_c/v_c = 1.2$. Substituting these values in Eq. (c), one has

$$v_{cv,t} = \frac{2.0 \sqrt{f_c'}}{\sqrt{1 + (\tau_{v,t}/1.2v_{v,t})^2}} \tag{2.58}$$

for use in Eqs. (2.36a) and (2.37a), in lieu of v_{cr}.

This presentation of torsion in reinforced concrete demonstrates the basis of present Code provisions, omitting some minor details. The notation for torsional shear stresses differs from that used in the Code, but makes it easier to distinguish shear stresses caused by torsion from those caused by flexural shear.

2.8 DEFLECTIONS AT SERVICE LOADS

In order to serve its intended purpose, a structure must be (1) safe and (2) serviceable. A structure is safe if it is able to resist, without distress and with some margin to spare, all forces which foreseeably will act on it during its lifetime. Serviceability implies, among other things, that deflections and other distortions under load shall be unobjectionably small. For example, excessive beam and slab deflections can lead to objectionable cracking of partitions, ill-fitting doors and windows, poor drainage, misalignment of sensitive machinery or other equipment, excessive vibrations, etc. It becomes important, therefore, to be able to predict deflections with reasonable accuracy, so that members can be dimensioned to ensure both adequate strength and appropriately small deflections.

The deflections of interest are those which occur under normal service conditions. In service, a structure sustains the full dead load plus some fraction or all of the maximum design live load. The usual safety factors provide that under service loading the steel and concrete are not stressed beyond their respective elastic ranges. For this reason the deflections which occur immediately upon application of the load, the so-called *instantaneous deflections*, can be calculated by methods based on the elastic behavior of flexural members, as analyzed in Arts. 2.4c and 2.4d.

It was pointed out in Arts. 1.7 and 1.10 that, in addition to those concrete deformations which occur immediately upon load application, other deformations and volume changes take place gradually and over long intervals of time. These are chiefly creep deformations and shrinkage. In consequence, deflections of reinforced-concrete members continue to increase for some time after load application, at a decreasing rate, and the *longtime deflections* may exceed the instantaneous deflections by a large amount. For this reason, methods for estimating both the instantaneous and the longtime deflections are essential.

a. Instantaneous deflections   Elastic deflections can be expressed in the general form

$$\delta = \frac{F \text{ (loads, spans)}}{EI}$$

where EI is the flexural rigidity, and F (loads, spans) is a function of the particular load-and-span arrangement. For instance, the deflection of a uniformly loaded simple beam is $5wL^4/384EI$, so that $F = 5wL^4/384$. Similar deflection equations have been tabulated, or can easily be computed, for many other loadings and span arrangements, simple, fixed, or continuous, and the corresponding functions F may be determined. The particular problem in reinforced-concrete structures is therefore the determination of the appropriate flexural rigidity EI for a member consisting of two materials with properties and behavior as widely different as steel and concrete.

If the maximum moment in a flexural member is sufficiently small so that the tension stress in the concrete does not exceed the flexural tensile strength (modulus of rupture, which can be taken as about $7.5 \sqrt{f_c'}$), no flexural tension cracks will occur. The full, uncracked section is then available for resisting stress and providing rigidity. This stage of loading has been analyzed in Art. 2.4c. In agreement with this analysis, the effective moment of inertia for this range of low loads is that of the uncracked, transformed section I_{ut}, and E is the modulus of concrete E_c as given in Art. 1.7. Correspondingly, for this load range,

$$\delta_{iu} = \frac{F}{E_c I_{ut}} \tag{2.59}$$

At higher loads flexural tension cracks are formed; in addition, if shear stresses exceed v_{cr} [see (Eq. 2.33)], and web reinforcement is employed to resist them, diagonal tension cracks can exist at service loads. In the region of flexural cracks the position of the neutral axis varies: directly at each crack it is located at the level calculated for the cracked, transformed section (see Art. 2.4d); midway between cracks it dips to a location closer to that calculated for the uncracked, transformed section (Art. 2.4c). Correspondingly, flexural tension cracking causes the effective moment of inertia to be that of the cracked, transformed section in the immediate neighborhood of flexural tension cracks, and closer to that of the uncracked, transformed section midway between cracks, with a gradual transition between these extremes.

It is seen that the value of the local moment of inertia varies in those portions of the beam in which the bending moment exceeds the cracking moment of the section.

$$M_{cr} = f_r I_{ut}/y_t \qquad (2.60)$$

where y_t is the distance from the neutral axis to the tension fiber, and f_r is the modulus of rupture $= 7.5\sqrt{f_c'}$. The exact variation of I depends on the shape of the moment diagram and on crack pattern and is difficult to determine. This makes an exact deflection calculation tedious, if not impossible. It was found, however, that the deflections δ_{ic} occurring in a beam after the maximum M_{max} moment has reached and exceeded the cracking moment M_{cr} can be calculated by utilizing an effective moment of inertia I_{eff}; that is,

$$\delta_{ic} = \frac{F}{E_c I_{eff}} \qquad (2.61)$$

where

$$I_{eff} = \left(\frac{M_{cr}}{M_{max}}\right)^3 I_{ut} + \left[1 - \left(\frac{M_{cr}}{M_{max}}\right)^3\right] I_{ct} \qquad (2.62)$$

where I_{ct} is the moment of inertia of the cracked, transformed section. It is seen that $I_{ut} > I_{eff} > I_{ct}$ and that I_{eff} approaches closer to I_{ct} the more M_{max} exceeds M_{cr}.

Figure 2.34 shows growth and illustrates calculation of instantaneous deflections. For moment values no larger than M_{cr}, deflections are practically proportional to loads (moments) and the deflection d_{iu} at which cracking begins is obtained from Eq. (2.59) with $M = M_{cr}$. At larger loads (moments), the effective moment of inertia I_{eff} becomes progressively smaller [see Eq. (2.62)]. Correspondingly, a curved moment-deflection relation is obtained, as shown. The figure indicates the deflections δ_{ic} which obtain at two different values of M_{max}.

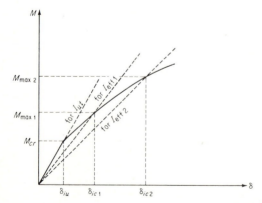

Fig. 2.34 Deflections of reinforced-concrete beams (schematic).

b. Longtime deflections Longtime deflections are caused by shrinkage and by creep (chiefly the latter). It was pointed out in Art. 1.7 that creep deformations of concrete are directly proportional to the compression stress. They increase asymptotically with time and, for the same stress, are larger for low- than for high-strength concretes. Correspondingly longtime deflections caused by sustained loads can be estimated by replacing E_c with an equivalent sustained modulus. This modulus must also be used in determining n when calculating I_{ut} or I_{ct}. One such method, including data on the magnitude of this equivalent sustained modulus, is given in Ref. 2.10.

In this same investigation at Cornell University (Ref. 2.10), it was shown from many tests that the following simplified method leads to an adequate estimate of longtime deflections, including the effect of normal shrinkage: (1) Calculate the instantaneous deflection due to the sustained load. (2) Multiply this short-time deflection by a coefficient λ; the resulting value is the additional longtime deflection caused by creep and normal shrinkage. That is, (*additional longtime deflection*) = $\lambda \times$ (*instantaneous deflection*).

The coefficient λ depends on the duration of the sustained load. It also depends on whether the flexural member carries only reinforcement of amount A_s on the tension side, or whether additional longitudinal reinforcement A_s' is provided on the compression side. In the latter case, the longtime deflections are much reduced. This is so because, when no compression reinforcement is provided, the compression concrete is subject to unrestrained creep and shrinkage. On the other hand, since steel is not subject to creep, if additional bars are located close to the compression face, they will resist, and thereby reduce, the amount of creep and shrinkage and the corresponding deflection. The appropriate coefficient can be read from Table 2.1.

On the basis of experimental information such as in Table 2.1, the ACI Code gives a curve-fitting expression for the multiplier for longtime deflections as follows:

$$\lambda = 2 - 1.2 \frac{A_s'}{A_s} \geq 0.6 \tag{2.63}$$

which gives values very close to those in the last line of Table 2.1.

Table 2.1   Multiplier λ for additional longtime deflections (Ref. 2.10)

Duration of loading	A_s only	$A_s' = A_s/2$	$A_s' = A_s$
1 month	0.6	0.4	0.3
6 months	1.2	1.0	0.7
1 year	1.4	1.1	0.8
5 years and more	2.0	1.2	0.8

If a beam carries a certain sustained load P_{sus} (e.g., the dead load plus the average traffic load on a bridge) and is subjected to a short-time peak load P_{sh} (e.g., the weight of an unusually heavy-vehicle combination), the maximum deflection under this short-time load is obtained as follows:

1. Calculate the instantaneous deflection $\delta_{i,sus}$ caused by P_{sus} by methods given in Art. 2.8a.
2. Calculate the additional longtime deflection caused by P_{sus}, $\delta_{t,sus} = \lambda \delta_{i,sus}$.
3. Then the total deflection caused by the sustained part of the load is

$$\delta_{sus} = \delta_{i,sus} + \delta_{t,sus}$$

4. In calculating the additional instantaneous deflection caused by P_{sh}, account must be taken of the fact that the load-deflection relation after cracking is nonlinear (see Fig. 2.33). Hence,

$$\delta_{i,sh} = \delta_{i,sus+sh} - \delta_{i,sus} \qquad (2.64)$$

where $\delta_{i,sus+sh}$ is the total instantaneous deflection which would obtain if $P_{sus} + P_{sh}$ were applied simultaneously, calculated by using I_{eff} determined for the moment caused by $P_{sus} + P_{sh}$.
5. Then, the total deflection under the heavy, short-time load is

$$\delta_{tot} = \delta_{sus} + \delta_{i,sh}$$

It will be realized from this discussion that the magnitude of the deflections in reinforced-concrete structures depends on so many influences that no precise calculation is possible. On the other hand, great precision is generally not necessary. From extensive experimental information it can be said that the methods presented here permit deflections to be estimated with an accuracy of about ± 20 per cent, which is adequate for most purposes of design.

The ACI Code contains a somewhat simplified version of the procedure just described. On the other hand, more accurate but also more elaborate procedures are available in the literature, such as those given in Ref. 2.18.

PROBLEM

For the beam analyzed in the problem at the end of Art. 2.4, estimate the longtime deflection at full service load. Assume that in the particular structure that part of the total service load which acts continually and constitutes the longtime loading consists of the weight of the beam plus 1400 lb per ft. This longtime loading consists of the weight of the floor structure and of other permanent portions of the building supported by the given beam, plus that part of the live load which acts more or less continuously, such as furniture, equipment, the time average of the weight of occupants, etc. The remaining 1050 lb per ft are short-duration peak loads, such as the brief peak loading of corridors in an office building at the end of the working day. *Note:* To avoid repetitive calculations, utilize those cross-sectional properties which have been calculated in connection with the problem in Art. 2.4.

2.9 COMPRESSION PLUS BENDING

The behavior and strength of members which are axially, i.e., concentrically, compressed have been discussed in Art. 2.2. Such members occur rarely, if ever, in buildings and other structures. Components such as columns and arches chiefly carry loads in compression, but simultaneous bending is almost always present. Bending moments are caused by continuity, i.e., by the fact that building columns are parts of monolithic frames in which the support moments of the girders are partly resisted by the abutting columns, by transverse loads such as wind forces, by loads carried eccentrically on column brackets, or in arches when the arch axis does not coincide with the pressure line. Even when design calculations show a member to be loaded purely axially, inevitable imperfections of construction will introduce eccentricities and consequent bending in the member as built. For this reason members which must be designed for simultaneous compression and bending are very frequent in almost all types of concrete structures. Their behavior is easily understood by combining the information and principles of Arts. 2.2 and 2.4.

a. Elastic behavior Up to loads of the order of one-half the ultimate, eccentrically compressed members behave substantially elastically in the manner, and for the reasons, discussed for axially compressed members and for beams. In elastic members made of materials with substantial tension strength, the stresses can be calculated by a simple addition of those caused by axial compression and those caused by bending. A similar situation prevails in eccentrically compressed reinforced-concrete members, provided that no tension cracking has occurred to reduce the effective concrete section. This will be the case if the eccentricity is small enough so that tension stresses in the concrete will not exist, or if present, will be smaller than the bending tension strength f_r (modulus of rupture). This case of small eccentricity will be analyzed first.

(1) SMALL ECCENTRICITY, SECTION UNCRACKED

Figure 2.35a shows the cross section of a rectangular eccentrically loaded column, (b) its transformed section, and (c) the load P acting at a distance (eccentricity) e measured from the centroid of the transformed section. If $A'_s = A_s$, as is most frequently the case and is assumed here, then the eccentricity is measured from the center line of the column. The eccentric force P can be replaced by an axial force of equal magnitude and a moment $M = Pe$, as shown.

The axial force causes uniform concrete compression stresses

$$f_a = \frac{P}{A_{ut}}$$

and the moment causes maximum concrete bending stresses

$$f_b = \pm \frac{Mc}{I_{ut}} = \pm \frac{Mh/2}{I_{ut}}$$

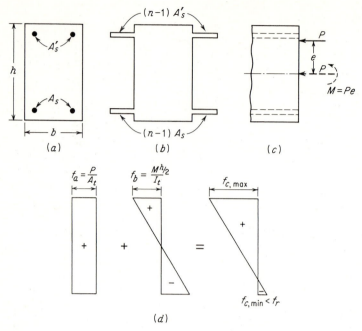

Fig. 2.35

the first of the two expressions for f_b being general, and the second applying when $A'_s = A_s$. Here A_{ut} and I_{ut} are, respectively, the area and moment of inertia of the transformed uncracked section. Substituting $Pe = M$, the outer fiber stresses in the concrete, by superposition, become

$$f_{c,\substack{\max \\ \min}} = \frac{P}{A_{ut}} \pm \frac{Mc}{I_{ut}} = \frac{P}{A_{ut}} \pm \frac{Mh/2}{I_{ut}} \tag{2.65}$$

and are shown in Fig. 2.35d. Depending on the relative magnitudes of the two terms, $f_{\min}$ can be either positive (compression), in which case the entire cross section is under compression of varying intensity, or negative, in which case a portion of the section is in tension. In the latter case, which is shown in Fig. 2.35d, it is assumed that the section remains uncracked as long as the numerically largest tension stress $f_{c,\min}$ does not exceed the modulus of rupture f_r. Then the transformed section of Fig. 2.35b continues to be correct, and the above analysis applies. Otherwise, if $f_{c,\min}$ exceeds f_r in absolute value, tension cracks develop, and the transformed section must be modified accordingly.

It is evident from Eq. (2.65) that, for a given cross section and a given load P, there is a limiting value for the eccentricity e beyond which $f_{\min}$ will numerically exceed f_r, causing tension cracks to appear. This limiting eccentricity can be calculated from Eq. (2.65) by substituting $-f_r$ for $f_{\min}$ and Pe for M and solving for e.

(2) LARGE ECCENTRICITY, SECTION CRACKED

If the eccentricity exceeds the value just defined, tension cracks will make part of the concrete area ineffective. Then it is usually assumed, just as in the case of beams, that the concrete is not capable of resisting any tension whatever. The shape of the transformed cracked section is then that shown in Fig. 2.36b, and the elastic stresses are distributed as in (c).

In this case the depth kh of that part of the section which is in compression becomes an additional unknown. It can be found from the equilibrium requirement that the moment of all internal forces about the action line of the external force P must be zero. The internal forces are the stresses shown in Fig. 2.36c acting on the transformed area of (b). The mentioned condition

$$M_P = 0$$

results in a cubic equation for kh, which can be solved either explicitly or by successive approximation. Once kh is known, the stresses are calculated from the first (general) form of Eq. (2.65). Thus, the maximum concrete stress becomes

$$f_{c,\max} = \frac{P}{A_t} + \frac{Pec_c}{I_t} \qquad\qquad (2.65a)$$

and the stress in the tension steel

$$f_s = n\left(\frac{P}{A_t} - \frac{Pec_s}{I_t}\right) \qquad\qquad (2.65b)$$

Here A_t and I_t refer, of course, to the transformed cracked section. The eccentricity e and the fiber distances c_c and c_s for concrete and steel, respectively, are measured from the centroidal axis of the transformed section; compression is designated as positive, tension negative.

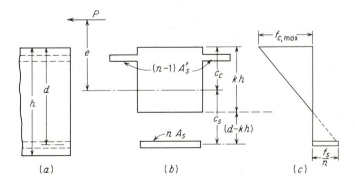

Fig. 2.36

b. Ultimate strength   In eccentrically compressed members, at loads approaching to the ultimate, a nonelastic redistribution of stresses takes place which is similar to that discussed for axially compressed members in Art. 2.2 and for flexural members in Art. 2.4. In compression members, even more than in beams and girders, the designer is chiefly interested in the carrying capacity, i.e., the ultimate strength. Stresses and rigidity at service loads, which govern deflections of flexural members (see Art. 2.8), are of little interest for compression members, whose deflections or other service behavior need rarely be computed. In earlier days, the elastic stresses under service loads, calculated as just described, were taken as a measure of carrying capacity by limiting them to appropriate fractions of f_c' or f_y. Experimental evidence showed, however, that satisfactory correlation of actual strength with elastically computed stresses could not be obtained. This is not surprising if one considers that, as loads become sufficiently large to cause inelastic behavior, not only the magnitude, but also the manner of distribution of the stresses over the cross section, changes decisively. Ways of calculating ultimate strength on the basis of inelastic behavior are now available for eccentrically compressed members as for other members. Their results are in satisfactory agreement with extensive test evidence.

If a rectangular column with cross section as in Fig. 2.37a is compressed by an eccentric load P applied as shown, the steel and concrete stresses just prior to failure, i.e., at ultimate load, are distributed in the manner of Fig. 2.37b. That is, with plane sections assumed to remain plane, the strains are those shown in Fig. 2.37c, which, for the reasons discussed in connection with Fig. 2.9, result in the shown stress distribution. Just as for simple bending, this actual distribution can be replaced, for purposes of calculation, by the equivalent rectangular stress distribution (Fig. 2.37d), discussed in Art. 2.4f and shown in Fig. 2.11. A large number of tests on columns of a variety of shapes have shown that the ultimate strengths computed on this basis are in satisfactory agreement with test results (Ref. 2.4).

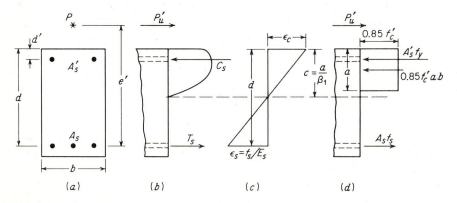

Fig. 2.37   Eccentric compression, ultimate strength.

It is assumed in Fig. 2.37d that at ultimate load, when the concrete fails in compression, i.e., when $\epsilon_c = \epsilon_u = 0.003$, the compression steel is yielding, so that $f_s' = f_y'$ as shown. For ordinary conditions this is usually the case; it can easily be checked, as will be demonstrated in a numerical example.

Equilibrium between external and internal axial forces requires that

$$P_u' = 0.85f_c'ab + A_s'f_y' - A_sf_s \qquad (2.66)$$

Also, the moment about the centroid of the tension steel A_s of the internal stresses must be equal and opposite to the moment of the external force P_u', so that

$$P_u'e' = 0.85f_c'ab\left(d - \frac{a}{2}\right) + A_s'f_y'(d - d') \qquad (2.67)$$

These are the two basic equilibrium relations for rectangular, eccentrically compressed members. The fact that the presence of the compression reinforcement A_s' has displaced a corresponding amount of concrete, of area A_s', is neglected in Eqs. (2.66) and (2.67). If necessary, particularly for large steel ratios, one can account for this very simply. Evidently, in the above equations a nonexistent concrete compression force of amount $A_s'(0.85f_c')$ has been included as acting in the displaced concrete at the level of the compression steel. This excess force can be removed in both equations by multiplying A_s' by $(f_y' - 0.85f_c')$ rather than by f_y'.

For simplicity, the following presentation will be limited to the most frequent case of *symmetrical cross section*, i.e., where

$$A_s' = A_s \qquad \rho = \frac{A_s}{bd} = \frac{A_s'}{bd} \qquad \text{and} \qquad f_y' = f_y \qquad (2.68)$$

Other conditions, dealt with by the same type of analysis, merely result in lengthier expressions and calculations.

For large eccentricities, failure will be initiated by yielding of the tension steel, followed by a shift of the neutral axis toward the compression side until crushing of the concrete causes a secondary compression failure. Conversely, for small eccentricities, the concrete may crush while the tension steel may be far from yielding. For any cross section of given dimensions and material strength values, there is one specific eccentricity e' such that a force applied at that distance will cause failure by simultaneous yielding of the tension steel and crushing of the concrete. For this so-called *balanced condition*, let the particular eccentricity be e_b', and the corresponding ultimate load P_b.

For the *balanced condition*, the concrete strain is $\epsilon_c = \epsilon_u = 0.003$, and simultaneously the strain in the tension steel at the instant at which yielding commences is $\epsilon_s = \epsilon_y = f_y/E_s$. With these values, from the geometry of Fig. 2.37c, the distance to the neutral axis is

$$c_b = \frac{\epsilon_u}{\epsilon_y + \epsilon_u}d = \frac{0.003}{f_y/E_s + 0.003}d \qquad (2.69)$$

On the other hand, with Eq. (2.68) for symmetrical sections, the depth of the rectangular stress block, from Eq. (2.66) becomes

$$a = \frac{P'_u}{0.85f'_c b} \tag{2.70}$$

Also, by definition of the rectangular stress block [see Eq. (2.31)],

$$a_b = \beta_1 c_b \tag{2.71}$$

Equating the right sides of Eqs. (2.70) and (2.71), and using c_b from Eq. (2.69), one has, for the ultimate load under balanced conditions,

$$P'_b = 0.85\beta_1 f'_c bd \frac{0.003}{f_y/E_s + 0.003} \tag{2.72}$$

In particular, for concretes with f'_c not exceeding 4000 psi, from Eq. (2.31), $\beta_1 = 0.85$, so that in this case

$$P'_b = 0.72f'_c bd \frac{0.003}{f_y/E_s + 0.003} \tag{2.72a}$$

The eccentricity e'_b which results in balanced conditions can then be obtained from Eq. (2.67) by substituting P'_b for P'_u and a_b for a and solving for e'. Loads with eccentricities smaller than e'_b result in primary compression failures at ultimate values larger than P_b; loads with eccentricities larger than e'_b result in primary tension failures at loads smaller then P_b.

(1) FAILURE BY YIELDING OF TENSION STEEL

When $e' > e'_b$, failure is initiated by yielding of the tension steel, followed by lengthening of cracks, shift of the neutral axis, and secondary compression failure by crushing of the concrete. In this case, $f_s = f_y$ and, for a symmetrical section, from Eq. (2.68),

$$A_s f_y = A'_s f'_y$$

Therefore, Eqs. (2.66) and (2.67) become

$$P'_u = 0.85f'_c ab \tag{2.66a}$$

$$P'_u e' = P'_u \left(d - \frac{a}{2}\right) + A'_s f'_y (d - d') \tag{2.67a}$$

Solving Eq. (2.66a) for the depth of the rectangular stress block, one has

$$a = \frac{P'_u}{0.85f'_c b} \tag{2.73}$$

If this value is substituted in Eq. (2.67), one obtains the following quadratic equation:

$$\frac{P_u'^2}{2 \times 0.85 f_c' b} + (e' - d)P_u' - A_s' f_y (d - d') = 0$$

Dividing through by d and solving, one obtains the ultimate load for large eccentricities:

$$P_u' = 0.85 f_c' bd \left[-\left(\frac{e'}{d} - 1\right) + \sqrt{\left(\frac{e'}{d} - 1\right)^2 + 2\rho\mu\left(1 - \frac{d'}{d}\right)} \right] \qquad (2.74)$$

where the parameter

$$\mu = \frac{f_y}{0.85 f_c'} \qquad (2.75)$$

is the ratio of the effective strengths of the two materials in connection with the rectangular stress distribution and is useful in this and other connections.

(2) FAILURE BY COMPRESSION OF CONCRETE

When $e' < e_b'$, failure is initiated by crushing of the concrete. That is, when the concrete reaches its ultimate strain $\epsilon_u = 0.003$, the tension steel has not yet reached its yield point, so that $f_s < f_y$. From the geometry of the strain distribution of Fig. 2.37c with $\epsilon_c = \epsilon_u$, the depth to the neutral axis is

$$c = \frac{\epsilon_u}{f_s / E_s + \epsilon_u} d \qquad (2.76)$$

from which the tension steel stress becomes

$$f_s = E_s \epsilon_u \frac{d - c}{c} = E_s \epsilon_u \frac{d - a/\beta_1}{a/\beta_1} \qquad (2.76a)$$

If f_s is substituted in Eq. (2.66), and the depth of the stress block a is eliminated between Eqs. (2.66) and (2.67), one obtains a cubic equation in P_u'. From this, the ultimate load for small eccentricities can be determined by solving for P_u' either explicitly or, often more quickly, by successive approximation.

Examples The following two examples (in one of which tension governs, and in the other compression) are directly solved numerically, without reference to the previously derived equations. The simplicity of the equivalent rectangular stress block enables such direct numerical analyses, without reference to elaborate formulas, to be carried out for most situations.

   The cross section of the column is shown in Fig. 2.38a. Four No. 9 bars are used, of 1.0 in.² area each. The cylinder strength is $f_c' = 3500$ psi, and the yield point is $f_y = 50,000$ psi.

   (1) *Eccentricity e = 15 in. from center line of column.* The compression force is applied outside the boundaries of the cross section; it is therefore sensible to assume

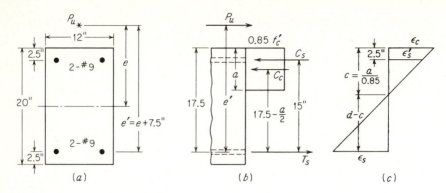

Fig. 2.38

that failure will be initiated by yielding of the tension steel. Assuming the compression steel to yield as well (to be checked later), the three internal forces of Fig. 2.38b are

$T_s = -2.0 \times 50,000 = -100,000$ lb
$C_s = 2.0 \times 50,000 = 100,000$ lb
$C_c = 0.85 \times 3500 \times 12a = 35,700a$

Since T_s and C_s are equal and opposite, the total internal force being equal and opposite to the external load, P'_u is

$$P'_u = C_c = 35,700a$$

from which the depth of the rectangular stress block is

$$a = \frac{P'_u}{35,700}$$

Taking moments about the tension steel,

$$P'_u \times 22.5 - 100,000 \times 15 - C_c \left(17.5 - \frac{a}{2}\right) = 0$$

Substituting P'_u for C_c and $P'_u/35,700$ for a, as previously calculated, one obtains the quadratic equation

$$\frac{P'^2_u}{71,400} + 5P'_u - 1,500,000 = 0$$

whose solution gives the ultimate load

$$P'_u = 194,000 \text{ lb}$$

It was assumed that the compression steel yields when, after failure is initiated by yielding of the tension steel, the concrete reaches its crushing strain $\epsilon_u = 0.003$. For this to be true, the strain of the compression steel, ϵ'_s of Fig. 2.38c, must be at least equal to that strain at which yielding starts, that is, $\epsilon_y = f'_y/E_s = 50,000/29,000,000 = 0.00172$. To check this, use is made of the strain distribution of Fig. 2.38c.

The depth of the rectangular stress block is

$$a = \frac{P'_u}{35,700} = \frac{194,000}{35,700} = 5.44 \text{ in.}$$

and the depth to the neutral axis is

$$c = \frac{a}{\beta_1} = \frac{5.44}{0.85} = 6.40 \text{ in.}$$

From similar triangles it is seen that

$$\epsilon'_s = \epsilon_c \frac{c - d'}{c} = 0.003 \frac{6.4 - 2.5}{6.4} = 0.00183 > 0.00172$$

Hence, at the instant at which the ultimate concrete strain is reached, the compression steel strain exceeds the yield strain, verifying the assumption that at ultimate load the compression steel is, in fact, yielding.

(2) *Eccentricity e = 6 in. from center line of column.* For this small eccentricity it is sensible to assume that failure will be initiated by crushing of the concrete. Proceeding by successive approximations, assume that for this small eccentricity the major part of the cross section is in compression. Thus, try

$$a = 11.0 \text{ in.}$$

Hence,

$$c = \frac{a}{\beta_1} = \frac{11}{0.85} = 12.95 \text{ in.}$$

From the similar triangles of the strain distribution of Fig. 2.38c,

$$\epsilon_s = \frac{f_s}{E_s} = \epsilon_c \frac{d - c}{c}$$

from which the steel stress in the tension reinforcement is

$$f_s = 0.003 \times 29{,}000{,}000 \times \frac{17.5 - 12.95}{12.95} = 30{,}600 \text{ psi}$$

Taking moments about the centroid of the tension steel,

$$P'_u \times 13.5 = 0.85 \times 3500 \times 11 \times 12(17.5 - 5.5) + 100{,}000 \times 15$$

from which, as a first approximation,

$$P'_u = 460{,}000 \text{ lb}$$

Taking the sum of the axial forces,

$$460{,}000 = 0.85 \times 3500 \times 12a + 2.0 \times 50{,}000 - 2.0 \times 30{,}600$$

from which

$$a = 11.8 \text{ in.}$$

which differs from the assumed value of $a = 11.0$ in.

If this simple calculation is repeated assuming $a = 11.8$ in., one obtains $P'_u = 473{,}000$ lb and $a = 11.7$ in. The latter being satisfactorily close to the assumed value for a, the second approximation,

$$P'_u = 473{,}000 \text{ lb}$$

is taken as the final answer.

Finally, if one computes the ultimate load for balanced conditions from Eq. (2.72a), one finds $P'_b = 337,000$ lb. For case 1, $P'_u = 194,000$ lb $< P'_b$, indicating that failure is, in fact, initiated by yielding of the tension steel; for case 2, $P'_u = 473,000$ lb $> P'_b$, indicating that failure is, in fact, caused by crushing of the concrete, as was assumed.

PROBLEM

If the column of the preceding examples is made of high-strength concrete with $f'_c = 5000$ psi and reinforced with high-strength steel with $f_y = 75,000$ psi, calculate the ultimate loads. Check whether the substitution of high-strength materials has led to the same percentage increase in carrying capacity for the two conditions of loading. If not, why not? *Note:* As is seen from Table 1.2, steel with $f_y = 75$ ksi is the highest-strength reinforcing steel available at this time under ASTM standards. The stress-strain curve for such steels frequently has the shape shown in Fig. 1.13. For simplicity of calculation, in the given problem assume that the shape of the curve is similar to that for the lower-strength steels, i.e., that it exhibits a horizontal yield plateau before entering the strain-hardening range.

2.10 PRESTRESSED CONCRETE

a. Prestressing The successful combination of two materials as unlike in almost every respect as are concrete and steel enables reinforced concrete to occupy an important place in modern construction. Yet there are weaknesses inherent in this combination. They result mainly from the presence of tension cracks which, wherever substantial flexure or shear is present, affect the value of the ultimate load; cracks also occur under service conditions, i.e., at design loads. This has the following disadvantages:

1. In bending, all the concrete on the tension side of the neutral axis, two-thirds or more in flexural members, is lost as a means of resisting stress; its chief structural function is merely to hold the reinforcing bars in place and protect them from corrosion.
2. In shear, as was discussed, the formation of diagonal tension cracks is decisive in limiting the available strength.
3. Deflections being inversely proportional to the moment of inertia of the transformed section, cracked or uncracked as the case may be, beams in which tension cracks have formed show deflections sizably larger than would be the case if cracking could have been prevented.
4. High-strength steels with yield points in excess of 200,000 psi and ultimate strengths of the order of 250,000 psi, 3 to 5 times the strength of ordinary reinforcing steels, are available for use in concrete structures. However, the amount of cracking (width and number of cracks) is roughly proportional to the steel strain, i.e., to the steel stress. Therefore, the high stresses which these steels could sustain cannot be utilized with conventional reinforcement because they would cause an amount of cracking prohibitive with respect to both appearance and corrosion protection and would result in large and frequently prohibitive deflections.

These limitations have been largely overcome by the development of *prestressed concrete*. The essence of prestressing consists in intentionally applying forces to a structure or structural member in a manner which will counteract the effects of the subsequent loads. Specifically, forces are introduced in advance of loading in such a way that they cause internal stresses opposite in sign and distribution to those subsequently produced by the external loads. The resulting total stresses being the sums of those caused by prestressing and those caused by useful load, the performance of the member is improved. The prestressing steels used for this purpose are discussed in Art. 1.12.

Suppose a beam is fabricated in the following manner: First, high-strength steel strands are placed in the form and passed through holes in two anchor plates, one at each end. The strands are then pretensioned by applying the *initial prestress force* P_i by a suitable means, such as jacks bearing against abutments (Fig. 2.39a). Then the concrete, poured so that the anchor plates bear against the ends of the beam, is allowed to harden. Next the steel strands are secured to the anchor plates, and the jacks released. The steel, tending to contract to its original length, is prevented from doing so by the concrete interposed between the anchor plates. Hence, it remains under tension, and this tension is equilibrated by a corresponding compression in the concrete. The effect of the prestressing force, then, is the same as that of an external compression force P_e applied to the concrete as shown in Fig. 2.39b. This stage is known as *transfer;* i.e., the force which equilibrates the steel tension has been transferred from the jacks to the concrete.

After some time has elapsed, the initial prestress P_i will have decreased somewhat, for various reasons which include the elastic compression of the con-

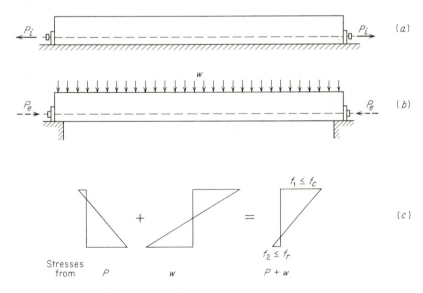

Fig. 2.39 Principle of prestressing.

crete when stressed by the prestressing force, concrete creep and shrinkage, and relaxation of the steel. With current prestressing methods, these losses amount to about 12 to 20 per cent, so that the *final stress* P_e is smaller by this amount than the initial prestress P_i.

b. Elastic behavior at service loads The principal advantage of a prestressed member is its behavior at service loads. At such loads, stresses in both steel and concrete are kept within the elastic range. The magnitude and location of the prestressing force generally are so arranged that tension cracks will not occur under working loading, dead plus live. In most cases, tension cracking is also prevented, or at least minimized, under temporary conditions, such as when a girder, having been fabricated in a prestressing yard, is transported to the site and placed in position, during which time the only load it carries is its own weight. Cracking is prevented by limiting the elastically computed tension stress in the concrete to some fraction of the modulus of rupture f_r. The latter is of the order of 7.5 $\sqrt{f_c'}$. When a crackfree member is desired, but the accidental formation of occasional small cracks would be of no consequence, an allowable maximum tension stress of 6 $\sqrt{f_c'}$ is considered adequate.

Corrosion protection is of more critical importance in prestressed than in reinforced concrete. This is so for two reasons: (1) Other things being equal, smaller amounts of high-strength prestressing tendons than of ordinary reinforcing rods are employed to carry the same loads. If corrosion were to penetrate to a certain depth in the reinforcement and thereby to remove a certain cross-sectional area of the steel, this area would constitute a larger percentage of the light prestressing steel than of the heavier conventional reinforcement; it would therefore constitute a greater danger to the integrity of the member. (2) More important, steel under permanent high stress, as used in prestressing, is more easily subject to corrosion than steel under no or low stress. This phenomenon is known as *stress corrosion*. Mildly corrosive conditions which may be harmless to reinforced concrete (such as those connected with the usual hairline cracks) can be damaging to prestressing tendons. Thus, it has been observed that some antifreeze compounds and some high-alumina cements, used chiefly abroad as high-early-strength cements, have led to corrosion difficulties in prestressed concrete, but not in conventionally reinforced concrete. For reasons such as this, it is customary to permit no tensile stresses to occur under service conditions in outdoor structures or in other corrosive environments.

The following discussion of behavior, stresses, and strength of prestressed beams is limited to the situation in the midportions of simple beams, for the sake of simplicity.

In the cross section of a prestressed beam, the prestress force P_e (Fig. 2.40a), applied with an eccentricity e below the centroidal axis, gives rise to the compression stress $-P_e/A_t$ and the bending stress $\pm P_e ec/I_t$ as shown. (In this discussion, compression stresses are designated as negative, tension

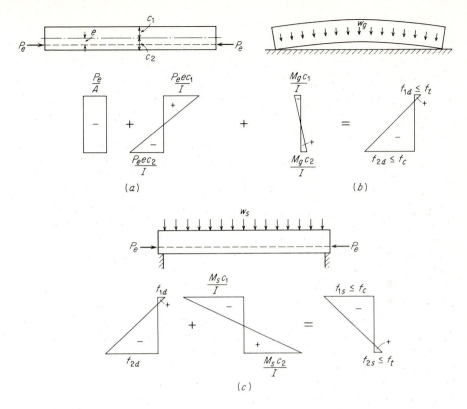

Fig. 2.40

as positive.) The bending stress produces a shortening of the bottom fibers and a lengthening of the top fibers, which results in an upward deflection of the beam. In other words, when the prestress force is transferred to the concrete (i.e., in the above described procedure, when the jacks are released), the beam lifts off its soffit, as shown in Fig. 2.40b. Then the weight of the girder w_g becomes an active load which, for a prismatic girder, results at midspan in the moment

$$M_g = \frac{w_g L^2}{8} \tag{2.77}$$

This moment causes additional bending stresses $\pm M_g c/I_t$ of sign opposite those due to prestress, as shown. These three types of stresses are caused by prestressing and the consequent dead-load stresses resulting from upward deflection; their sums are designated here as f_{1d} and f_{2d} in top and bottom fibers, respectively. They are distributed as shown. In order to prevent cracking, the maximum concrete tension stress, which at this stage occurs at the top fiber, must not exceed the safe value f_t, stipulated as discussed before.

Hence, at this stage, the outer fiber stresses are

$$f_{1d} = -\frac{P_e}{A_t} + \frac{P_e c_1}{I_t} - \frac{M_g c_1}{I_t} \qquad \text{and} \qquad |f_{1d}| \le |f_t| \qquad\qquad (2.78a)$$

$$f_{2d} = -\frac{P_e}{A_t} - \frac{P_e c_2}{I_t} + \frac{M_g c_2}{I_t} \qquad \text{and} \qquad |f_{2d}| \le |f_c| \qquad\qquad (2.78b)$$

In other words, the location and magnitude of the prestress force must be so arranged that at this stage the tension in the top fiber does not exceed the stipulated maximum f_t. Likewise, the compression at the bottom fiber must not exceed, in absolute value, the maximum compression stress f_c permitted to prevent compression failure and ensure substantially elastic behavior. This f_c is of the order of one-half the cylinder strength f_c'.

Next, when the girder is installed, it is subjected to the service load, which consists of additional dead load, such as a floor slab which may be supported on the girder, plus the design live load. This superimposed load w_s causes the additional moment M_s and the consequent bending stresses $M_s c/I$ as shown in Fig. 2.40c. These bending stresses superpose on those caused by prestress and girder weight. Hence, under design load, the outer fiber stresses are

$$f_{1s} = f_{1d} - \frac{M_s c_1}{I_t} \qquad \text{and} \qquad |f_{1s}| \le |f_c| \qquad\qquad (2.79a)$$

$$f_{2s} = f_{2d} + \frac{M_s c_2}{I_t} \qquad \text{and} \qquad |f_{2s}| \le |f_t| \qquad\qquad (2.79b)$$

In other words, the location and magnitude of the prestress force must also be arranged so that, at this stage, the maximum tension stress, which now acts in the bottom fiber, and the maximum compression stress, in the top fiber, do not exceed their respective stipulated limits.

Since tension cracking has been prevented in both stages, in these equations A_t and I_t are the area and moment of inertia of the uncracked transformed section, and c_1 and c_2 are the outer fiber distances of that section.

It will be observed that the stresses caused by prestressing, f_{1d} and f_{2d}, are of opposite sign to those caused by the design load. It is this arrangement which prevents cracking at design loads, so that the entire cross section of the concrete is available to resist stress. In combination with the fact that high-strength steel is used for reinforcement and that high-strength concrete can be used more effectively in prestressed than in reinforced concrete, this leads to sizable reductions in the required quantities of steel and concrete, to corresponding reductions of dead load, to smaller deflections (owing to the fact that the entire uncracked transformed section is available to provide flexural rigidity), and to improved durability and reduced maintenance because of the absence of hairline cracks.

c. Ultimate strength If the design load is significantly exceeded, the maximum tension stress in the concrete will eventually exceed the modulus of rupture. Tension cracks will form, and the beam will behave essentially like an ordinary cracked reinforced-concrete beam, except for the following: (1) In reinforced concrete, under zero load the strain in the reinforcement is, likewise, zero. In prestressed concrete, in contrast, the strain in the tendons at zero load is not zero, but corresponds to the effective prestress after losses; i.e.,

$$\epsilon_{se} = \frac{f_{se}}{E_s} = \frac{P_e}{A_{ps}E_s}$$

where P and f_{se} are, respectively, the effective prestress force and corresponding stress. Any further steel strain caused by the applied loads adds to this pre-existing strain. (2) The stress-strain characteristics of prestressing steel are quite different from those of reinforcing bars. Figure 1.14 shows typical stress-strain curves for prestressing wire and high-tensile bars. It is seen that in contrast to ordinary reinforcing bars these steels do not show a definite yield plateau. Yielding develops more gradually, and in the inelastic range the stress-strain curve continues to rise smoothly until the tensile strength is reached. Also, the spread between tensile strength f_{pu} and yield strength f_{py} is much smaller in prestressing steels than in reinforcing steels, and the total elongation at rupture ϵ_s' is much smaller (see Table 1.3).

For relatively small steel ratios, failure occurs by rupture of the steel when it has reached its tensile strength f_{pu}. In this case the beam is said to be *underreinforced*. The ultimate-strength theory, utilizing the equivalent rectangular stress block (Art. 2.4*f*), predicts such failure with adequate accuracy. With reference to Fig. 2.41*a*, the tension force at failure, in this case, is

$$T = A_{ps}f_{pu}$$

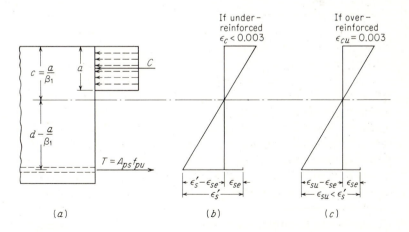

(a)   (b)   (c)

Fig. 2.41

while, for a rectangular section, the compression force is

$$C = 0.85 f'_c ab$$

Equating the two, $T = C$, gives the depth of the rectangular stress block at failure:

$$a = \frac{A_{ps} f_{pu}}{0.85 f'_c b} \tag{2.80}$$

The ultimate moment, then, is

$$M'_u = T\left(d - \frac{a}{2}\right) = A_{ps} f_{pu}\left(d - \frac{A_{ps} f_{pu}}{1.7 f'_c b}\right) \tag{2.81}$$

Introducing the steel ratio $\rho = A_{ps}/bd$, this can be written in the alternative form

$$M'_u = A_{ps} f_{pu} d\left(1 - 0.59 \rho \frac{f_{pu}}{f'_c}\right) \tag{2.81a}$$

which is seen to be identical with Eq. (2.27a), except that the breaking strength of the strand f_{pu}, rather than the yield point of the bars f_y, is seen to govern.

To determine whether the beam actually is underreinforced, one ascertains whether, at the instant of steel failure, i.e., when the steel has reached its ultimate breaking strain ϵ'_s, the concrete strain ϵ_c is indeed smaller than the crushing strain, assumed to be equal to 0.003. Because the steel fractures in the inelastic range, the ultimate strain cannot be obtained by use of Young's modulus; it must be measured on the stress-strain diagram or obtained from pertinent specifications (see Table 1.3). From the strain distribution of Fig. 2.41b it is seen that

$$\epsilon_c = (\epsilon'_s - \epsilon_{se}) \frac{a/\beta_1}{d - a/\beta_1} \tag{2.82}$$

the depth a having been found from Eq. (2.80). If ϵ_c does not exceed the crushing strain, the beam is underreinforced, and Eqs. (2.81) and (2.81a) apply. If ϵ_c exceeds this value, the beam is *overreinforced*. That is, crushing of the concrete will occur before the steel has reached its tensile strength.

In the case of *overreinforced beams*, an explicit calculation of the ultimate moment is not possible for the following reason: The stress in the steel at the moment of concrete failure f_{ps}, being smaller than f_{pu}, is not known at the outset. Hence, the depth of the stress block cannot be computed [cf. Eq. (2.80)], making it impossible to calculate the moment of internal forces [cf. Eq .(2.81)]. In this case, successive approximation can be employed as follows: Assume a steel stress f_{ps}, read the corresponding strain ϵ_{su} from the stress-strain diagram for the strand, and compute $T = A_{ps}f_{ps}$. Then the depth of the stress block can be calculated from

$$a = \frac{A_{ps}f_{ps}}{0.85f'_cb} \tag{2.83}$$

and from the strain relations of Fig. 2.41c, the steel strain is

$$\epsilon_{su} = 0.003 \frac{d - a/\beta_1}{a/\beta_1} + \epsilon_{se} \tag{2.84}$$

If this value differs from that which corresponds to the assumed steel stress f_{ps}, the assumption was incorrect. A new value of ϵ_{su} is assumed on the basis of this first trial, the corresponding next approximation for f_{ps} read from the stress-strain curve, and the procedure repeated until adequate convergence is obtained, usually upon the second, or at most third, trial.

Once the steel stress f_{ps} and the corresponding depth of the stress block a are known, the ultimate moment is

$$M'_u = A_{ps}f_{ps}\left(d - \frac{a}{2}\right) \tag{2.85}$$

The strain distributions of Fig. 2.41b and c and the calculations based on them neglect the small additional elastic compression strain caused in the concrete at the level of the steel by the prestressing force prior to load application. This strain is of the order of 5 per cent or less of the steel strain at failure, and neglecting it will cause no significant error in the analysis.

When a prestressed beam is underreinforced, failure is preceded by adequate warning, because the large inelastic steel strains cause visible cracking and sagging of the member. The failure of overreinforced beams can occur suddenly and without warning, for which reason this situation is generally avoided, except where special safeguards are used.

A particular, and undesirable, situation may obtain for beams with unusually small steel ratios. In this case, upon formation of tension cracks when the steel has to carry the entire tension, its total breaking strength $A_{ps}f_{pu}$ may be insufficient to resist that moment at which the tension cracks form. Formation of the first tension crack is then followed by immediate rupture of the steel and sudden collapse of the beam. To guard against this eventuality, the cracking moment can be calculated from Eq. (2.79b), with the tension fiber stress f_{2d} taken as the modulus of rupture of the particular concrete. If this cracking moment exceeds the ultimate moment of the cracked section as calculated with Eq. (2.81), the described situation would obtain, and the beam should be redesigned.

These methods of analyzing the ultimate strengths of prestressed rectangular beams assume perfect bonding between steel and concrete. Only in this case is the steel strain equal to the strain in the immediately adjacent concrete so that the strain relationships of Fig. 2.41b and c hold. For so-called *bonded* members, i.e., where the concrete is poured around and in immediate contact with the strands, nearly perfect bond can be assumed. In some methods of prestressing, unbonded steel is employed. Corrections have been derived (Ref. 2.12) to account for possible slight bond deficiencies in bonded members and for the markedly reduced steel strain in unbonded members. Their design use is discussed in Chap. 11.

Example   The beam of Fig. 2.42 is reinforced with prestressing strands whose stress-strain curve is that of Fig. 1.14, with $f_{py} = 224,000$ psi, $f_{pu} = 256,500$, $E_s = 27 \times 10^6$ psi; the concrete cylinder strength is $f'_c = 5000$ psi. The span is 36 ft. Determine (1) the available bending moment for superimposed (useful) loads M_s and (2) the ultimate strength of the beam. Permissible stresses at service load, to avoid cracking and inelastic deformations, are $f_c = 0.45 f'_c = 2250$ psi and $f_t = 6\sqrt{f'_c} = 425$ psi. Prestress after losses is $f_{se} = 140,000$ psi.

The contribution of the small steel area to the properties of the transformed uncracked section being negligible, those of the gross section will be used. Then $I = 20^3 \times 1\%_{12} = 6670$ in.[4]; $A = 10 \times 20 = 200$ in.[2] The prestressing force $P_e = 140,000 \times 1.3 = 182,000$ lb. To calculate the maximum moment of the girder due to its own weight, one has weight per foot $= {}^{200}\!\!/_{144} \times 150 = 208$ lb per ft, and $M_g = (208 \times 36^2)/8 = 33,700$ ft-lb.

Checking the *outer fiber stresses after transfer* of prestress, the girder having deflected away from its soffit [cf. Eqs. (2.78)],

$$
f_{\substack{1d\\2d}} = -\frac{182,000}{200} \pm \frac{182,000 \times 7 \times 10}{6670} \mp \frac{33,700 \times 12 \times 10}{6670} = \begin{array}{l} 405 < \quad 425 \text{ psi} \\ -2215 < -2250 \text{ psi} \end{array}
$$

The moment M_s of the *superimposed loads* is limited to a value which will keep the maximum compression and tension stresses within the stipulated limits. Hence, the two values of M_s are computed from Eqs. (2.79), one as governed by the tension stress f_t, and the other by the compression stress f_c. The smaller of the two, evidently, is the maximum admissible moment under design loading. Hence, equating f_{1d} to f_c in Eq. (2.79a),

$$
-2250 - 405 = -\frac{M_{s1} \times 10}{6670} \qquad M_{s1} = 1,775,000 \text{ in.-lb}
$$

Equating f_{2s} to f_t in Eq. (2.79b),

$$
425 + 2215 = \frac{M_{s2} \times 10}{6670} \qquad M_{s2} = 1,760,000 \text{ in.-lb}
$$

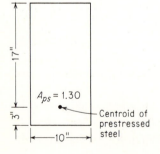

17"

3"

$A_{ps} = 1.30$

Centroid of prestressed steel

←—10"—→

Fig. 2.42

M_{s2}, being smaller than M_{s1}, is the limiting moment of the superimposed loads under service conditions; i.e.,

$$M_s = 1,760,000 \text{ in.-lb} = 146,700 \text{ ft-lb}$$

To calculate the *ultimate strength*, one must first check whether the beam is underreinforced. If so, it would fail by rupture of the strands at their tension strength of $f_{pu} = 256,500$ psi. Hence, the depth of the rectangular stress block [Eq. (2.80)] is

$$a = \frac{1.30 \times 256,500}{0.85 \times 5000 \times 10} = 7.85 \text{ in.}$$

For $f'_c = 5000$, from Eq. (2.36), $\beta_1 = 0.80$, so that the distance to the neutral axis is $a/\beta_1 = 7.85/0.80 = 9.8$ in. From Fig. 2.42 it is seen that the ultimate strain of the steel is $\epsilon'_s = 0.056$; the strain due to prestress after losses, $f_{se} = 140,000$ psi being in the elastic range of the steel (see Fig. 2.42), is $\epsilon_{se} = 140,000/27 \times 10^6 = 0.0052$. Then the maximum compression strain in the concrete [cf. Eq. (2.82)] would be

$$\epsilon_c = \frac{(0.056 - 0.0052)9.8}{17.0 - 9.8} = 0.069 > 0.003$$

This strain is seen to exceed by far the assumed ultimate strain of the concrete, which indicates that the concrete would crush before the steel ruptures. In this sense, then, the beam is overreinforced. Its ultimate strength will therefore be calculated by successive approximation.

The steel stress at the moment at which the concrete fails by crushing may be assumed to be $f_{ps} = 225,000$ psi; it is seen from Fig. 1.14 that this stress is in the nonlinear, i.e., inelastic, portion of the stress-strain curve. The corresponding steel strain is read from the curve to be $\epsilon_{su} = 0.010$. Then, from Eq. (2.83), $a = 6.9$ in., and from Eq. (2.84), the corresponding steel strain is $\epsilon_{su} = 0.0081$. The strain read for the assumed f_{ps} from the stress-strain curve (0.010) and that calculated for the same f_{ps} from the strain relationships (0.081) differ by about 20 per cent. Hence, a second approximation is in order.

It will now be assumed that ϵ_{su} is about midway between the two values of the first trial, that is, $\epsilon_{su} = 0.0090$. The corresponding steel stress is read from Fig. 1.14 to be $f_{su} = 220,000$ psi. Repeating the steps of the previous calculations, one finds that for this stress $a = 6.75$ in. and, according to the assumed strain relations, $\epsilon_{su} = 0.0086$. It is seen that the two values of the steel strain for this approximation differ by less than 5 per cent. The accuracies of the various data and assumptions involved in this calculation being, at best, of the same order, no further approximation is indicated, and the steel stress at failure can be taken as $f_{ps} = 220,000$ psi.

Then, with $a = 6.75$ in., the ultimate moment is [Eq. (2.85)] $M'_u = 1.3 \times 220,000 \times 13.13 = 3,900,000$ in.-lb $= 325,000$ ft-lb. The service moment, consisting of that from the weight of the girder M_g and that from the superimposed loads M_s, is

$$M = 33,700 + 146,700 = 180,400 \text{ ft-lb}$$

The ultimate moment is $325/180.4 = 1.8$ times the design or service moment.

It is seen that this method of determining the ultimate strength of prestressed beams, which reflects the actual physical behavior of the member, presupposes a knowledge of the entire stress-strain curve of the particular steel.

Even though minimum properties of prestressing steels are standardized in appropriate specifications, details of the shapes of stress-strain curves vary within limits. For this and other reasons, for routine design calculations, simpler approximate formulas have been developed on the basis of experimental information and calculations such as those just presented. These are discussed in Chap. 11.

2.11 STRUCTURAL SAFETY

a. Safety and serviceability   To serve its purpose, a structure must be safe against collapse and serviceable in use. Serviceability requires that deflections be adequately small, that cracks, if any, be kept to tolerable limits, that vibrations be minimized, etc. Safety requires that the strength of the structure be adequate for all loads which may foreseeably act on it. If the strength of a structure, built as designed, could be predicted with precision, and if the loads and their internal effects (moments, shears, axial forces) were known with equal precision, then safety could be assured by providing a carrying capacity just barely in excess of the known loads. However, there are a number of sources of uncertainty in the analysis, design, and construction of reinforced-concrete structures. These sources of uncertainty, which require a definite margin of safety, may be listed as follows:

1. Actual loads may differ from those assumed in the design.
2. Actual loads may be distributed in a manner different from that assumed in the design.
3. The assumptions and simplifications inherent in any analysis may result in calculated load effects—moments, shears, etc.—different from those which a more rigorous analysis would furnish.
4. The actual structural behavior may differ from that assumed, owing to imperfect knowledge.
5. Actual member dimensions may differ from those specified by the designer.
6. Reinforcement may not be in its proper position.
7. Actual material strength may be different from that specified by the designer.

   In addition, in establishing a safety specification, consideration must be given to the consequences of failure. In some cases, a failure would merely be an inconvenience. In other cases, loss of life and significant loss of property may be involved. A further consideration should be the nature of the failure, should it occur. A gradual failure, with ample warning permitting remedial measures, is preferable to a sudden, unexpected collapse.

   It is evident that the selection of an appropriate margin of safety is not a simple matter. Although progress has been made toward rational safety provisions in design codes, the ideal safety specification has not yet been written.

b. Loads The loads which act on structures can be divided into two broad categories: *dead loads* and *live loads*. Dead loads are those that are constant in magnitude and fixed in location throughout the lifetime of the structure. Usually, the major part of the dead load is the weight of the structure itself. Dead load may generally be calculated with good precision from the design dimension of the structure. On the other hand, live loads, such as occupancy loads, snow loads, etc., may be either fully or partially in place or not present at all, and may also change in location. Their magnitude and distribution at any given time is uncertain, and even their maximum intensities throughout the lifetime of the structure are not known with precision. Design live loads for buildings are usually established by local or regional building codes (see Art. 8.2) and include some margin against overload and often some allowance for dynamic effects. Some live loads, such as those resulting from occupancy by persons or storage of materials and goods, may be controlled by the owner. Others, resulting from environmental agents, such as wind, snow, or earthquake loads, cannot be controlled.

Since the maximum load which will occur during the life of a structure is uncertain, it can be considered a random variable. In spite of this uncertainty the engineer must provide an adequate structure. A probability model for the maximum load can be devised by means of a probability density function for loads as shown in Fig. 2.43a. The exact form of this distribution curve cannot be known because of necessarily limited information on complex combinations of loads of various types and with different degrees of control and uncertainty. In repetitive structures, such as highway overpasses, a frequency curve constructed from actual measured load data would have the same general form and meaning as this probability density function. In such a function, the area under the curve between two abscissas, such as the loads L_1 and L_2, represents the probability of occurrence of loads of magnitude $L_1 < L < L_2$. Correspondingly, the probability of occurrence of loads larger than L_d is given by the shaded area to the right of L_d.

c. Strength The strength of a structure depends primarily on the strength of the materials from which it is made. For this purpose, minimum material's strengths are specified in standardized ways. Actual material's strengths cannot be precisely known and therefore also constitute random variables (see Art. 1.6). Structural strength depends, furthermore, on the care with which a structure is built, which in turn reflects the quality of supervision and inspection. Member sizes may differ from specified dimensions, reinforcement may be out of position, poorly placed concrete may show voids, etc.

Strength of the entire structure, or of a population of repetitive structures, such as the already mentioned overpasses, can also be considered a random variable with a probability density function of the type shown on Fig. 2.43b. As in the case of loads, the exact form of this function cannot be known, but can be approximated from known data, such as statistics of actual, mea-

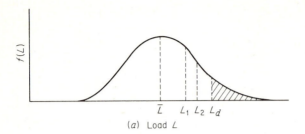

(a) Load L

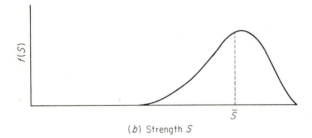

(b) Strength S

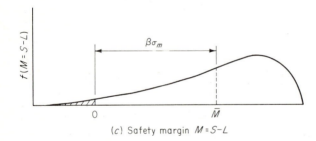

(c) Safety margin $M = S - L$

Fig. 2.43 Frequency curves for (a) loads L, (b) strengths S, (c) safety margin M.

sured materials and member strengths and similar information.

d. Structural safety   A given structure has a *safety margin M* if

$$M = S - L > 0 \tag{2.86}$$

that is, if the strength of the structure is larger than the load acting on it. Since S and L are random variables, the safety margin $M = S - L$ is also a random variable. A plot of the probability function of M may appear as in Fig. 2.43c. Failure occurs when M is less than zero. Thus, the probability of failure is represented by the shaded area in the figure.

Even though the precise form of the probability density functions for S and L, and therefore for M, is not known, much can be achieved in the way of a rational approach to structural safety. One such approach is to require that the mean safety margin $\bar{M}$ be a specified number β of standard deviations

σ_m above zero. It can be demonstrated (Ref. 2.20) that this results in the requirement that

$$a\bar{L} \leq b\bar{S} \tag{2.87}$$

where a is a partial safety coefficient larger than 1 applied to the mean load $\bar{L}$, and b a partial safety coefficient smaller than 1 applied to the mean strength $\bar{S}$. The magnitude of each partial safety coefficient depends on the variance of the quantity to which it applies, L or S, and on the chosen value of β, the safety index of the structure.

The safety provisions of the ACI Code are given in the form of Eq. (2.87); i.e., they are formulated in terms of load factors (here designated by a) and capacity reduction factors, b. The Code relates these to design values rather than mean values of S and L. It assigns different values of a to different kinds of loads, and different values of b to different kinds of strength, based to some extent on statistical information, but to a much larger extent on experience, intuition, and compromise.

As more information becomes available on the factors which influence structural safety, safety provisions may develop that are similar in form to those now in the Code, but more explicitly based on probabilistic methods of the type just outlined. In this event the safety index β may be varied, depending on the consequences of failure; these may range all the way from a mere nuisance to major loss of life and property. In this and other respects, safety provisions in design codes will always be a blend of rational methods modified and supplemented by experience and judgment. Probable future trends in this important area are discussed in realistic terms in Refs. 2.21 and 2.22, which also contain extensive bibliographies.

REFERENCES

2.1. F. E. Richart and R. L. Brown: An Investigation of Reinforced Concrete Columns, *Univ. Illinois Eng. Exp. Sta. Bull.* 267, 1934.

2.2. H. Rüsch: Researches toward a General Flexural Theory of Structural Concrete, *J. ACI*, vol. 57, p. 1, July, 1960.

2.3. L. B. Kriz: Ultimate Strength Criteria for Reinforced Concrete, *Proc. ASCE*, vol. 85, no. EM3, 1959; L. B. Kriz and S. L. Lee, Ultimate Strength of Overreinforced Beams, *Proc. ASCE*, vol. 86, no. EM3, 1960.

2.4. A. H. Mattock, L. B. Kriz, and E. Hognestad: Rectangular Concrete Stress Distribution in Ultimate Strength Design, *J. ACI*, vol. 57, p. 875, February, 1961.

2.5. Shear and Diagonal Tension, part 2, Report of ACI-ASCE Committee 326, *J. ACI*, vol. 59, p. 277, 1962 (with extensive bibliography).

2.6. S. A. Guralnick: Shear Strength of Reinforced Concrete Beams, *Trans. ASCE*, vol. 125, p. 603, 1960.

2.7. R. M. Mains: Measurement of the Distribution of Tensile and Bond Stresses along Reinforcing Bars, *J. ACI*, vol. 48, p. 225, 1951.

2.8. P. M. Ferguson and J. N. Thompson: Development Length of High Strength Reinforcing Bars, *J. ACI*, vol. 49, p. 887, 1962.

2.9. R. G. Mathey and D. Watstein: Investigation of Bond in Beam and Pullout Specimens with High-yield-strength Deformed Bars, *J. ACI*, vol. 57, p. 1071, 1961.
2.10. W. W. Yu and G. Winter: Instantaneous and Long-time Deflections of Reinforced Concrete Beams under Working Loads, *J. ACI*, vol. 57, p. 29, 1960.
2.11. S. A. Guralnick: High-strength Deformed Steel Bars for Concrete Reinforcement, *J. ACI*, vol. 57, p. 241, 1960.
2.12. J. Warwaruk, M. S. Sozen, and C. P. Siess: Strength and Behavior in Flexure of Prestressed Concrete Beams, *Univ. Illinois Eng. Exp. Sta. Bull.* 464, 1962.
2.13. T. T. C. Hsu: Torsion of Structural Concrete: Plain Concrete Rectangular Sections, in "Torsion of Structural Concrete," ACI Publ. SP-18, p. 203, 1968.
2.14. Tentative Recommendations for the Design of Reinforced Concrete Members to Resist Torsion, Report of ACI Committee 438; T. T. C. Hsu and E. L. Kemp, Background and Practical Application of Tentative Design Criteria for Torsion, *J. ACI*, vol. 66, pp. 1 and 12, January, 1969.
2.15. T. T. C. Hsu: Torsion of Structural Concrete: Behavior of Reinforced Concrete Rectangular Members, in "Torsion of Structural Concrete," ACI Publ. SP-18, p. 261, 1968.
2.16. U. Ersoy and P. M. Ferguson: Concrete Beams Subjected to Combined Torsion and Shear: Experimental Trends, in "Torsion of Structural Concrete," ACI Publ. SP-18, p. 441, 1968.
2.17. D. L. Osburn, B. Mayoglou, and A. H. Mattock: Strength of Reinforced Concrete Beams with Web Reinforcement in Combined Torsion, Shear and Bending, *J. ACI*, vol. 66, p. 31, January, 1969.
2.18. D. E. Branson: Design Procedures for Computing Deflections, *J. ACI*, vol. 65, p. 730, 1968 (with extensive bibliography).
2.19. Bond Stress: The State of the Art, Report of ACI Committee 408, *J. ACI*, vol. 63, p. 1161, 1966.
2.20. N. C. Lind: Consistent Partial Safety Factors, *Proc. ASCE*, vol. 97, no. ST6, p. 1651, 1971.
2.21. C. A. Cornell: A Probability-based Structural Code, *J. ACI*, vol. 66, p. 974, 1969.
2.22. R. G. Sexsmith and M. F. Nelson: Limitations in Application of Probabilistic Concepts, *J. ACI*, vol. 66, p. 823, 1969.

3
Beams

3.1 DESIGN BASIS

In Art. 2.4, the behavior of reinforced-concrete beams was traced through the entire range of loads, from zero to failure. In designing beams, i.e., in determining the concrete proportions and the reinforcement to resist given loads, one or another of these loading stages may serve as the starting point.

Probably the most important single characteristic of a beam is its strength. Accordingly, a logical design basis is that the beam be proportioned so that its calculated strength is fully utilized only if the expected loads are increased by an overload factor substantially larger than unity. The possibility of deficient strength due to poor materials or careless workmanship must be recognized. It is clear from the discussion of Art. 2.4e that if this approach is followed, the designer must allow for the nonlinear stress-strain behavior of concrete at high stresses.

A member designed on a strength basis must also perform well under service loads, particularly with regard to deflection and cracking. Methods have been developed recently which permit accurate prediction of deflection and crack widths at service loads. Thus, by the *strength method* of design, the proportions of a beam are selected so that it will have the proper strength when overloaded. With dimensions known, the member is then checked for adequacy at service load. The dimensions may be revised as necessary.

Alternatively, a beam may be proportioned so that stresses in the concrete and steel under service loads are within acceptable limits. These allowable stresses are a fractional part of the stress capacities of the materials. From the concrete stress-strain curves of Fig. 1.4, it is clear that concrete responds elastically as long as the maximum compressive stress is not more than about $1/2 f'_c$. From Fig. 1.13 it is apparent that steel is elastic up to the yield stress. A beam may be designed on an elastic basis, as long as stresses remain within these limits. They generally do, up to service-load level.

An adequate margin of safety against failure for a member designed in this way is assured indirectly by requiring that the maximum concrete and steel stresses at service load be a sufficiently small fraction of the compressive strength and yield stress, respectively. Because of the nonlinear response of the concrete as load is increased above service load, no direct evaluation of the margin of safety is possible when designing by this means. The implied safety factors (strength divided by allowable stress) are conservatively lower than the actual safety factors.

It is common practice, when using this method, to provide crack and deflection control indirectly also, by setting limits on stresses which have produced satisfactory members in the past.

The first method of designing described above is known as *strength design* or *ultimate-strength design*. The second is known as *service-load design* or *working-stress design*. Because it permits an accurate appraisal of the strength of a reinforced-concrete member, generally the most important consideration, strength design has gained favor rapidly over the past decade in the United States. It is the method which will be emphasized throughout this text. A brief summary of the essential features of service-load design will be found in Art. 3.14.

3.2 HISTORICAL SUMMARY

Historically, it is interesting to observe that early proposals for the design of reinforced-concrete members included both strength design and service-load design (Ref. 3.1). The theories advanced by Koenen (1886), Neumann (1890), and Coignet and Tedesco (1894) were elastic theories which assumed a linear variation of concrete stress with distance from the neutral axis. The theories of Thullie (1897) and Ritter (1899) were based on a prediction of the ultimate strengths of beams and assumed nonlinear stress distributions.

Because of its relative simplicity, the "straight-line" theory proposed by Coignet and Tedesco gained general acceptance about 1900. It was adopted by code-writing organizations such as the Joint Committee on Standard Specifications for Concrete and Reinforced Concrete (1909), which introduced the concepts of working loads and working stresses. Although modifications were made over the ensuing years, the basic concepts of the service-load method became firmly established, to the extent that its limitations were generally overlooked.

In the 1930's, however, extensive testing of columns initiated by the American Concrete Institute indicated that measured steel stresses were considerably higher than those predicted by elastic theory. Work by Lyse, Slater, and Richart led to design formulas for columns which could only be described as strength theories in concept, although they were included in modified form in the service-load design specifications of that time. A 1931 paper by Emperger generated intensive investigation into the ultimate strengths of beams, with significant work done in the United States by Jensen and Whitney.

In the early 1950's, a joint ASCE-ACI committee was organized in the United States to study current knowledge of strength design methods and to coordinate research activities. In 1956 that committee published its Report on Ultimate Strength Design, which contained specific design recommendations. Its proposed formulas were incorporated on an optional basis in the 1956 ACI Building Code (ACI 318-56),[1] while strength design was included in parallel with service-load design, as an equally acceptable alternative, in the 1963 Code (ACI 318-63). The current ACI Building Code (ACI 318-71) is based entirely upon the strength approach to design, except that service-load or allowable-stress design is mentioned as an acceptable alternative.

3.3 SAFETY PROVISIONS OF THE ACI CODE

Important aspects of structural safety have been discussed in Art. 2.11. It was shown that the general requirement for structural safety can be expressed in the form

$$\gamma_L \bar{L} = \gamma_S \bar{S} \tag{2.86}$$

where γ_L is a partial safety coefficient larger than 1, applied to the mean load

---

[1] Throughout this text, repeated reference will be made to the American Concrete Institute and its recommendations. This organization of more than 13,000 members is dedicated to the advancement of concrete technology and of the analysis, design, and construction of reinforced-concrete buildings. As one part of its activity, the American Concrete Institute has published the widely recognized Building Code Requirements for Reinforced Concrete (ACI 318-71), which serves as a guide in the design and construction of reinforced-concrete buildings. The Code has no official status in itself. However, it is generally regarded as an authoritative statement of current good practice in the field of reinforced concrete. As a result, it has been incorporated by law into countless municipal and regional building codes which do have legal status. Its provisions thereby attain, in effect, legal significance. Most reinforced-concrete structures in the United States and in many other countries are designed in accordance with the current ACI Building Code. A second ACI publication, Commentary on Building Code Requirements for Reinforced Concrete (ACI 318-71), provides background material and rationale for the Code provisions.

$\bar{L}$, and γ_S a partial safety coefficient smaller than 1, applied to the mean strength $\bar{S}$. It was pointed out that the magnitude of each partial safety coefficient should depend upon the variance of the quantity to which it applies, $\bar{L}$ or $\bar{S}$, and on the chosen safety index of the structure.

The approach to structural safety incorporated in the ACI Code follows these concepts, using load factors and capacity reduction factors. The load factors larger than 1 are given different values, depending upon the type and combination of loads considered, and are related to calculated dead load and legally specified live load. The capacity reduction factors, denoted by ϕ and taking values less than 1, are given different values, depending upon the type of strength to be calculated. The variation of ϕ factors reflects the state of knowledge on which the strength calculations are based (higher values are permitted for moment than for shear strength), the probable importance of the member in the structure (a lower, more conservative value is used for columns than for beams), and the quality of construction control probably attainable (higher values used for beams than for columns).

According to the ACI Code, the required strength U provided to resist the calculated dead load D and the calculated or legally specified live load L shall be at least equal to

$$U = 1.4D + 1.7L \tag{3.1a}$$

If resistance to the structural effects of a specified wind load W must be included in the design, the following combinations of D, L, and W shall be investigated in determining the greatest required strength U:

$$U = 0.75(1.4D + 1.7L + 1.7W) \tag{3.1b}$$

(where the cases of L having its full value or being completely absent shall both be checked to determine the most severe condition) and

$$U = 0.9D + 1.3W \tag{3.1c}$$

but in any case the strength of the member or structure shall not be less than required by Eq. (3.1a).

If resistance to specified earthquake loads or forces E must be considered, then $1.1E$ shall be substituted for W in Eqs. (3.1b) and (3.1c).

If lateral earth pressure H must be included, the strength U shall be at least equal to

$$U = 1.4D + 1.7L + 1.7H \tag{3.1d}$$

but where D or L reduces the effect of H, the corresponding load factors shall be taken as 0.9 for D and zero for L. For lateral pressures from liquids F, Eq. (3.1d) and its restrictions shall apply, except that $1.4F$ is substituted for $1.7H$. The vertical pressure of liquids is considered as dead load.

Table 3.1   Capacity reduction factors

Kind of stress	Capacity reduction factor ϕ
Bending, with or without axial tension, and axial tension	0.90
Axial compression or axial compression combined with bending:	
Spirally reinforced members	0.75*
Other members	0.70*
Shear and torsion	0.85
Bearing on concrete	0.70
Bending in plain concrete	0.65

* These values may be increased linearly to 0.90 as P_u decreases from $0.10f'_cA_g$ to zero for sections with symmetrical reinforcement and $(h - d' - d_s)/h$ not less than 0.70. For sections with small axial compression not satisfying this restriction, these values may be increased linearly to 0.90 as P_u decreases from $0.10f'_cA_g$ or P_b, whichever is smaller, to zero.

In all the above, impact effects, if they exist, are included with the live load L. Where the structural effects of differential settlement, creep, shrinkage, or temperature change may be significant, they are to be included with the dead load D, and the strength shall be at least equal to

$$U = 0.75(1.4D + 1.7L) \tag{3.1e}$$

The calculated strength of members is to be reduced, according to the ACI Code, by capacity reduction factors ϕ to obtain a conservative estimate of strength provided. These capacity reduction factors are given in Table 3.1.

3.4   DESIGN FOR FLEXURE

a. Singly reinforced rectangular beams   It was shown in Art. 2.4e that singly reinforced concrete beams (i.e., beams reinforced with tensile steel only) can fail in two principal ways. If the steel ratio ρ is relatively low, then at some value of the load the steel will commence yielding. As it yields, the neutral axis migrates upward, the area available to resist compression is reduced, and a secondary compression failure occurs. If, on the other hand, the steel ratio is high, the concrete will reach its ultimate capacity before the steel yields. While failure due to yielding is a gradual one, with adequate warning of collapse, failure due to crushing of the concrete is sudden and without warning. The first type of failure is therefore preferred. It can be ensured by setting an upper limit on the ratio of tensile reinforcement.[1]

[1] A third, undesirable, mode of failure may occur in very lightly reinforced beams. If the ultimate resisting moment of the cracked section is less than the moment which produces cracking of the transformed, uncracked section, the beam will fail immediately upon formation of the crack. To avoid this type of failure, which would occur without warning of collapse, the ACI Code establishes a minimum tensile steel ratio equal to $200/f_y$.

It was shown in Art. 2.4e that the balanced steel ratio, such that the steel starts to yield and the concrete reaches its ultimate strain capacity at precisely the same load, is given by the expression

$$\rho_b = \alpha \frac{0.003}{f_y/E_s + 0.003} \frac{f_c'}{f_y} \tag{2.29}$$

Adopting the rectangular stress block, for which $\alpha = 0.85\beta_1$ (see Art. 2.4f) and substituting $E_s = 29,000,000$ psi, one obtains the following equation for balanced steel ratio:

$$\rho_b = 0.85\beta_1 \frac{f_c'}{f_y} \frac{87,000}{87,000 + f_y} \tag{3.2}$$

where β_1 relates the depth of the equivalent rectangular stress block to the depth of the actual neutral axis. It equals 0.85 for $f_c' \leq 4000$ psi, and decreases by 0.05 for every 1000 psi above 4000 psi. The Code specifies that the steel ratio shall not exceed $0.75\rho_b$, in order that yielding-type failure be ensured. Thus,

$$\rho_{max} = 0.75\rho_b \tag{3.3}$$

and for all sections designed according to this specification, $f_s = f_y$ at failure. From equilibrium of horizontal forces shown in Fig. 3.1,

$$a = \frac{A_s f_y}{0.85 f_c' b} = \frac{\rho f_y d}{0.85 f_c'} \tag{3.4}$$

and

$$M_u' = A_s f_y \left(d - \frac{a}{2}\right) \tag{3.5a}$$

By substituting the value of a obtained from Eq. (3.4), and by substituting ρbd for A_s into Eq. (3.5a), one obtains the alternative expression

$$M_u' = \rho f_y bd^2 \left(1 - 0.59\rho \frac{f_y}{f_c'}\right) \tag{3.5b}$$

which is the same as Eq. (2.27a). In accordance with the safety provisions of the Code, this ultimate-moment capacity M_u' is reduced by the factor ϕ (equal

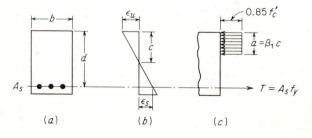

Fig. 3.1 Singly reinforced rectangular beam.

to 0.90 for bending) to obtain the ultimate moment M_u to be used in design, so that

$$M_u = \phi A_s f_y \left(d - \frac{a}{2} \right) \tag{3.6a}$$

$$M_u = \phi \rho f_y b d^2 \left(1 - 0.59 \rho \frac{f_y}{f_c'} \right) \tag{3.6b}$$

Example 1: ultimate-moment capacity of a given section   A rectangular beam has a width of 12 in. and an effective depth to the centroid of the reinforcing steel of 17.5 in. It is reinforced with four No. 9 bars[1] in one row. If $f_y = 60$ ksi and $f_c' = 4$ ksi what is the moment capacity of the beam to be used in design?

The actual steel ratio, $\rho = 4.00/(12 \times 17.5) = 0.0190$, is found to be less than

$$\rho_{max} = 0.75 \times 0.85 \beta_1 \frac{f_c'}{f_y} \frac{87,000}{87,000 + f_y}$$

$$= 0.75 \times 0.85^2 + \tfrac{4}{60} \times \tfrac{87}{147} = 0.0214$$

Consequently, failure by yielding is assured. Then

$$a = \frac{A_s f_y}{0.85 f_c' b} = \frac{4.0 \times 60}{0.85 \times 4 \times 12} = 5.89 \text{ in.}$$

$$M_u = \phi A_s f_y \left(d - \frac{a}{2} \right) = 0.90 \times 4.0 \times 60(17.5 - 5.89/2) = 3140 \text{ in.-kips}$$

Example 2: proportioning a section to resist a given moment   Determine the cross section of concrete and area of steel required for a simply supported rectangular beam with a span of 15 ft which is to carry a computed dead load of 1.27 kips per ft and a working live load of 2.44 kips per ft. A 3-ksi concrete is to be used, and the specified steel yield stress is 40 ksi.

Overload factors are first applied to the given service loads to obtain the load for which the beam is to be designed:

$$w_t = 1.4 \times 1.27 + 1.7 \times 2.44 = 5.92 \text{ kips per ft}$$
$$M_u = \tfrac{1}{8} = 5.92 \times 15^2 \times 12 = 2000 \text{ in.-kips}$$

The dimensions of the concrete will be directly influenced by the designer's choice of steel ratio. Selecting the maximum steel ratio $\rho = 0.75 \, \rho_b$, which will result in the minimum concrete section,

$$\rho = 0.75 \times 0.85^2 \frac{f_c'}{f_y} \frac{87}{127} = 0.371 \frac{f_c'}{f_y}$$

$$M_u = \phi \rho f_y b d^2 \left(1 - 0.59 \rho \frac{f_y}{f_c'} \right)$$

$$M_u = 0.90 \times 0.371 f_c' b d^2 (1 - 0.59 \times 0.371) = 0.260 f_c' b d^2$$

$$b d^2 = \frac{M_u}{0.260 f_c'} = \frac{2000}{0.260 \times 3} = 2560 \text{ in.}^3$$

[1] See Table 2 in the Appendix for bar areas.

A concrete section of 10-in. width and 16-in. effective depth will be used. For those dimensions,

$$A_s = \rho bd = 0.371 \times \tfrac{3}{40} \times 10 \times 16 = 4.44 \text{ in.}^2$$

Three No. 11 bars will be used, providing an area of 4.68 in.^{2}

A somewhat larger beam cross section using less steel may be more economical. In addition, the deflections associated with beams of minimum cross section may be undesirably large. As an alternative solution, the beam will be redesigned with a lower reinforcement ratio, say $\rho = 0.18f'_c/f_y$. In this case,

$$M_u = 0.90 \times 0.18f'_c bd^2(1 - 0.59 \times 0.18)$$
$$M_u = 0.145f'_c bd^2$$

$$bd^2 = \frac{2000}{0.145 \times 3} = 4600 \text{ in.}^3$$

A beam 11.5 in. wide with an effective depth of 20 in. will be used. For this case,

$$A_s = 0.18 \times \tfrac{3}{40} \times 11.5 \times 20 = 3.11 \text{ in.}^2$$

Four No. 8 bars will be used, providing an area of 3.14 in.^{2}

It is apparent that an infinite number of solutions to the stated problem are possible, depending upon the steel ratio selected. That ratio may vary, according to the Code, from an upper limit of $0.75\rho_b$ to a lower limit of $200/f_y$ for beams, where f_y is given in psi units.

Certain design aids suggest themselves. For example, it would be convenient in practice to graph the relation between the steel ratio ρ and the value $M_u/\phi bd^2$ for frequently used values of f'_c and f_y. A typical set of such design curves is given by Graph 1 of the Appendix.[1] Other ultimate-strength design aids are given in Ref. 3.2.

Rectangular-beam problems can generally be classified as review problems and design problems. The first example illustrated a review problem, in which the section dimensions and material strength capacities were given and it was required to find the moment capacity. The second example illustrated a design problem in which the design moment was known, as were the material strengths, and it was required to find the proportions of the cross section. There is a certain type of problem which falls into neither category but which occurs

---

[1] To solve the preceding example by means of Graph 1, for $f_y = 40$ ksi and $f'_c = 3$ ksi one reads $\rho_{max} = 0.028$ (at the end of the given curve). The corresponding value of $M_u/\phi bd^2$, from the graph, is 870 psi. Then

$$bd^2 = \frac{M_u}{870 \times 0.9} = \frac{2000 \times 1000}{785} = 2560 \text{ in.}^3$$

If $b = 10$ in. and $d = 16$ in. are selected as before, then

$$A_s = \rho bd = 0.028 \times 10 \times 16 = 4.48 \text{ in.}^2$$

The alternative solution used a steel ratio $\rho = 0.18f'_c/f_y = 0.18(3/40) = 0.0135$. From the graph, $M_u/\phi bd^2 = 483$, and $bd^2 = 2000 \times 1000/(483 \times 0.9) = 4600$ in.^{3} With $b = 11.5$ in. and $d = 20$ in., $A_s = \rho db = 0.0135 \times 11.5 \times 20 = 3.11$ in.^{2}

frequently. In this case, the concrete dimensions are given and are known to be adequate to carry the specified moment. It is required only to find the necessary steel area. Typically, this situation obtains at critical design sections of continuous beams, in which the concrete dimensions are often kept constant although the steel reinforcement varies along the span. The concrete dimensions b and d are determined at the maximum moment section, usually at one of the supports. At other supports and at midspan locations, where moments are smaller, the concrete dimensions are known to be adequate, and only the tensile steel is to be found.

Example 3: determination of steel area for a given moment   Using the concrete dimensions of the preceding example ($b = 11.5$ in. and $d = 20$ in.) and the same specified material strengths, find the steel area required to resist a design moment of 1600 in.-kips.

It is possible to obtain a direct solution to this problem by solving Eq. (3.6b) for ρ, which would appear in quadratic form. A somewhat quicker solution may usually be obtained by means of Eq. (3.6a). A reasonable value for a is assumed, and A_s is found. By use of Eq. (3.4), a revised estimate of a is obtained, and a modified A_s is obtained from Eq. (3.6a). This method converges very rapidly to the final answer. For example:

Assume $a = 4$ in.

$$A_s = \frac{M_u}{\phi f_y(d - a/2)} = \frac{1600}{0.90 \times 40(20 - 2)} = 2.47 \text{ in.}^2$$

Check:

$$a = \frac{A_s f_y}{0.85 f'_c b} = \frac{2.47 \times 40}{0.85 \times 3 \times 11.5} = 3.38 \text{ in.}$$

Assume $a = 3.30$ in.

$$A_s = \frac{1600}{0.90 \times 40 \times (20 - 1.65)} = 2.42 \text{ in.}^2$$

Check:

$$a = 3.38 \times 2.42/2.47 = 3.30 \text{ in.}$$

as assumed. Use $A_s = 2.42$ in.² Four No. 7 bars will be provided.

Problems such as this can readily be solved using Graph 1. The required steel ratio ρ can be found for the known $M_u/\phi bd^2$. $A_s = \rho bd$ is then easily determined.

In the design of reinforced-concrete structures, the decision to use either a relatively high or low strength concrete or steel depends upon economics, the availability of materials, the importance of special requirements such as minimum member size, and concern for such factors as deflection and crack widths. In general, higher-strength concrete is attained by increasing the amount of cement in the mix. Since cement is the most expensive of the several ingredients, the cost of the concrete increases substantially with increasing strength. On the other hand, high-strength steels, produced either metallurgically or by cold-working lower-strength bars, are available at only slightly increased cost. Consequently, the present trend is toward use of reinforcement having f_y of 60 ksi, while f_y of 40 ksi was almost standard a decade ago. Typical concrete strength, on the other hand, is not likely to change much from the present range of 3 to 6 ksi for job-mixed material. The reader may also wish to study Graph 1 of the Appendix with regard to the effect of changes in f_y and f_c' at various steel ratios.

PROBLEMS

1. A tensile-reinforced beam has a width of 10 in. and an effective depth of 20 in. to the centroid of the bars, which are placed all in one row. If steel with a yield stress of 60 ksi is used with concrete having $f_c' = 4$ ksi, find the resisting moment M_u for the following cases: (a) $A_s = 2$ No. 8 bars, (b) $A_s = 2$ No. 10 bars, (c) $A_s = 3$ No. 10 bars.

2. A singly reinforced rectangular beam is to be designed, with effective depth 1.5 times the width, to carry a service live load of 1500 plf, in addition to its own weight, on a 24-ft simple span. With $f_y = 40$ ksi and $f_c' = 4$ ksi, determine the required concrete dimensions and steel reinforcement (a) for $\rho = 0.18f_c'/f_y$ and (b) for $\rho = \rho_{max}$. Comment on your results, and include a sketch of each cross section drawn to scale.

3. A rectangular beam is to span 20 ft between simple supports and must carry a parallel partition weighing 800 plf of beam, in addition to a concentrated live load of 15,000 lb which acts at midspan. Design the beam, using $f_y = 50$ ksi and $f_c' = 5$ ksi.

4. A four-span continuous beam of constant rectangular section is supported at A, B, C, D. and E. Factored moments resulting from analysis are:

At supports, ft-kips	At midspan, ft-kips
$M_a = 92$	$M_{ab} = 105$
$M_b = 147$	$M_{bc} = 92$
$M_c = 134$	$M_{cd} = 92$
$M_d = 147$	$M_{de} = 105$
$M_e = 92$	

Determine the required concrete dimensions for this beam, using $d = 1.75b$, and find the required reinforcement for all critical moment sections. Use a maximum steel ratio of $\rho = 0.50\rho_b$, $f_y = 60$ ksi, and $f_c' = 5$ ksi.

b. Doubly reinforced rectangular beams

If a beam is limited in cross section, it may happen that the concrete cannot develop the compression force required to resist the given bending moment. In this case reinforcing is added in the compression zone, resulting in a so-called *doubly reinforced* beam, i.e., one with compression as well as tension reinforcement. The use of compression reinforcement has decreased markedly with the introduction and wider use of strength design methods, which account for the full strength potential of the concrete on the compressive side of the neutral axis. However, there are

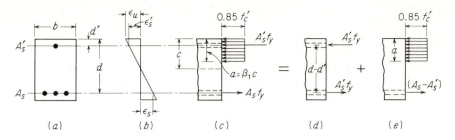

Fig. 3.2 Doubly reinforced rectangular beam.

situations in which compressive reinforcement is used for reasons other than strength. It has been found that the inclusion of some compression steel will reduce the long-term deflections of members (see Art. 2.8). In addition, in some cases, bars will be placed in the compression zone for minimum-moment loading (see Art. 7.3) or as stirrup support bars continuous throughout the beam span (see Art. 3.5). It is often desirable to account for the presence of such reinforcement in flexural design.

If, in a doubly reinforced beam, the tensile-steel ratio ρ is equal to or less than $\rho_{max} = 0.75\rho_b$, the strength of the beam may be approximated within acceptable limits by disregarding the compression bars. The strength of such a beam will be controlled by tensile yielding, and the lever arm of the resisting moment will ordinarily be but little affected by the presence of the compression bars.

If the tensile steel ratio is larger than $0.75\rho_b$, a somewhat more elaborate analysis is indicated. In Fig. 3.2a, a rectangular beam cross section is shown with compression steel A_s' placed a distance d' from the compression face, and with tensile steel A_s at effective depth d. It is assumed for the moment that both A_s' and A_s are stressed to f_y at failure. The total resisting moment can be thought of as the sum of two parts. The first part, M_1', is provided by the couple consisting of the force in the compression steel A_s' and the force in an equal area of tension steel:

$$M_1' = A_s'f_y(d - d') \tag{3.7a}$$

as shown in Fig. 3.2d. The second part, M_2', is the contribution of the remaining tension steel $(A_s - A_s')$ acting with the compression concrete:

$$M_2' = (A_s - A_s')f_y \left(d - \frac{a}{2}\right) \tag{3.7b}$$

as shown in Fig. 3.2e, where the depth of the stress block is

$$a = \frac{(A_s - A_s')f_y}{0.85f_c'b} \tag{3.8}$$

Defining $\rho = A_s/bd$ and $\rho' = A_s'/bd$, this can be written

$$a = \frac{(\rho - \rho')f_y d}{0.85f_c'} \tag{3.8a}$$

The total resisting moment is then

$$M_u' = M_2' + M_1' = (A_s - A_s')f_y\left(d - \frac{a}{2}\right) + A_s'f_y(d - d') \tag{3.9}$$

In accordance with the safety provisions of the ACI Code, this moment capacity is reduced by the factor ϕ to obtain the ultimate moment M_u to be used in design:

$$M_u = \phi M_u' = \phi\left[(A_s - A_s')f_y\left(d - \frac{a}{2}\right) + A_s'f_y(d - d')\right] \tag{3.10}$$

It is highly desirable, for reasons given earlier, that failure be precipitated by tensile yielding rather than by crushing of the concrete. This can be assured by setting an upper limit on the tensile-steel ratio. Setting the tensile-steel strain in Fig. 3.2b equal to ϵ_y to establish the location of the neutral axis for the balanced-failure condition, then summing horizontal forces shown in Fig. 3.2c (still assuming the compressive steel to be at the yield stress at failure), it is easily shown that the balanced steel ratio $\bar{\rho}_b$ for a doubly reinforced beam is

$$\bar{\rho}_b = \rho_b + \rho' \tag{3.11}$$

where ρ_b is the balanced steel ratio for the corresponding singly reinforced beam and is calculated from Eq. (3.2). To ensure the same margin against brittle concrete failure in a doubly reinforced beam as in a singly reinforced beam, according to the Code,

$$\bar{\rho}_{\max} = 0.75(\rho_b + \rho') \tag{3.11a}$$

Whether or not the compression steel will have yielded at failure can be determined as follows: Referring to Fig. 3.2b, and taking as the limiting case $\epsilon_s' = \epsilon_y$, one obtains, from geometry,

$$\frac{c}{d'} = \frac{\epsilon_u}{\epsilon_u - \epsilon_y} \quad \text{or} \quad c = \frac{\epsilon_u}{\epsilon_u - \epsilon_y}d'$$

Summing forces in the horizontal direction (Fig. 3.2c) gives

$$\rho f_y bd = 0.85\beta_1 f_c' bc + \rho' f_y bd$$

$$\rho = 0.85\beta_1 \frac{f_c'}{f_y}\frac{c}{d} + \rho'$$

Thus, the minimum tensile-steel ratio $\bar{\rho}_{\min}$ which will ensure yielding of the compressive steel at failure is

$$\bar{\rho}_{\min} = 0.85\beta_1 \frac{f_c'}{f_y} \frac{d'}{d} \frac{\epsilon_u}{\epsilon_u - \epsilon_y} + \rho' \tag{3.12}$$

Taking $\epsilon_u = 0.003$ as usual and $\epsilon_y = f_y/E_s$ with $E_s = 29,000,000$ psi, one obtains

$$\bar{\rho}_{\min} = 0.85\beta_1 \frac{f_c'}{f_y} \frac{d'}{d} \frac{87,000}{87,000 - f_y} + \rho' \tag{3.12a}$$

If the tensile-steel ratio is less than this limiting value, the compressive-steel stress at failure is less than the yield stress. This situation, which is common for shallow beams using relative high strength steel, requires slight modifications in the equations previously developed. In this case, it can easily be shown on the basis of Fig. 3.2b and c that the balanced steel ratio is

$$\bar{\rho}_b = \rho_b + \rho' \frac{f_s'}{f_y} \tag{3.13}$$

where

$$f_s' = E_s\epsilon_s' = E_s \left[\epsilon_u - \frac{d'}{d} (\epsilon_u + \epsilon_y) \right] \leq f_y$$

Hence, the maximum steel ratio permitted by the Code is

$$\bar{\rho}_{\max} = 0.75 \left(\rho_b + \rho' \frac{f_s'}{f_y} \right) \tag{3.13a}$$

If the tensile-steel ratio is less than $\bar{\rho}_{\min}$ and less than $\bar{\rho}_b$ as given by Eq. (3.13), the tensile steel is at the yield stress at failure but the compressive steel is not. In such a case, the compressive-steel stress can be written in terms of the unknown neutral axis depth:

$$f_s' = \epsilon_u E_s \frac{c - d'}{c}$$

after which consideration of horizontal force equilibrium gives

$$A_s f_y = 0.85\beta_1 f_c' bc + A_s' \epsilon_u E_s \frac{c - d'}{c}$$

The last equation is solved for c, the only unknown, after which the ultimate resisting moment can be found from the expression

$$M_u' = 0.85 f_c' ab \left(d - \frac{a}{2} \right) + A_s' f_s'(d - d') \tag{3.14}$$

This ultimate-moment capacity M'_u is reduced by the capacity reduction factor ϕ to obtain the design moment capacity M_u:

$$M_u = \phi \left[0.85 f'_c ab \left(d - \frac{a}{2} \right) + A'_s f'_s (d - d') \right] \tag{3.15}$$

If compression bars are used in a flexural member, precautions must be taken to ensure that these bars will not buckle outward under load, spalling off the outer concrete. The ACI Code includes the requirement that such bars be anchored in the same way that compressive bars in columns are anchored by lateral ties (see Art. 5.2). Such ties must be used throughout the distance where the compression reinforcement is required.

Example 1: moment capacity of a given section. A rectangular beam has a width of 12 in. and an effective depth to the centroid of the tension reinforcement of 18 in. The tension reinforcement consists of six No. 10 bars in two rows. Compression reinforcement consisting of two No. 9 bars is placed $2\frac{1}{2}$ in. from the compression face of the beam. If $f_y = 50$ ksi and $f'_c = 5$ ksi, what is the ultimate-moment capacity of the beam to be used in design?

$$A_s = 7.59 \qquad \rho = \frac{A_s}{bd} = \frac{7.59}{12 \times 18} = 0.0352$$

$$A'_s = 2.00 \qquad \rho' = \frac{A'_s}{bd} = \frac{2.00}{12 \times 18} = 0.0093$$

Check:

$$\rho_b = 0.85 \beta_1 \frac{f'_c}{f_y} \frac{87,000}{87,000 + f_y}$$
$$= 0.85 \times 0.80 \times \tfrac{5}{50} \times \tfrac{8.7}{13.7} = 0.0432$$
$$\rho_{max} = 0.75 \times 0.0432 = 0.0324$$

Since $\rho = 0.0352$ is larger than ρ_{max}, the beam must be analyzed as doubly reinforced. Checking limits:

$$\bar\rho_{max} = 0.75(\rho_b + \rho')$$
$$= 0.75(0.0432 + 0.0093) = 0.0394$$
$$\bar\rho_{min} = 0.85 \beta_1 \frac{f'_c}{f_y} \frac{d'}{d} \frac{87,000}{87,000 - f_y} + \rho'$$
$$= 0.85 \times 0.80 \times \frac{5}{50} \times \frac{2.5}{18} \times \frac{87}{37} + 0.0093 = 0.0315$$

Since the actual $\rho = 0.0352$ is between these limits, failure will be initiated by tensile yielding, and the compression steel will have yielded at failure. Then

$$M_1 = \phi A'_s f_y (d - d')$$
$$= 0.90 \times 2.00 \times 50(15.5) = 1400 \text{ in.-kips}$$
$$a = \frac{(A_s - A'_s)f_y}{0.85 f'_c b} = \frac{5.59 \times 50}{0.85 \times 5 \times 12} = 5.48 \text{ in.}$$
$$M_2 = \phi(A_s - A'_s)f_y \left(d - \frac{a}{2} \right)$$
$$= 0.90(5.59)50 \times 15.26 = 3840 \text{ in.-kips}$$
$$M_u = M_1 + M_2 = 1400 + 3840 = 5240 \text{ in.-kips}$$

Example 2: determination of steel areas for a known moment A rectangular beam which must carry a working live load of 2.47 kips per ft and a calculated dead load of 1.05 kips per ft on an 18-ft simple span is limited in cross section for architectural reasons to 10 in. width and 20 in. total depth. If $f_y = 40$ ksi and $f'_c = 3$ ksi, what steel area(s) must be provided?

The working loads are first increased by load factors to obtain the design load of $1.4 \times 1.05 + 1.7 \times 2.47 = 5.66$ kips per ft. Then $M_u = \frac{1}{8} \times 5.66 \times 18^2 = 229$ ft-kips $= 2750$ in.-kips. To satisfy spacing and cover requirements (see Art. 3.7), it will be assumed that the tension-steel centroid will be 4 in. above the bottom face of the beam, and that compression steel, if required, will be placed $2\frac{1}{2}$ in. below the beam's top surface. Then $d = 16$ in., and $d' = 2\frac{1}{2}$ in.

First, checking the capacity of the section if singly reinforced, using $\rho = \rho_{\max}$,

$$\rho_b = 0.85^2 \times \tfrac{3}{40} \times \tfrac{87}{127} = 0.0371$$
$$\rho_{\max} = 0.0278$$
$$A_s = 0.0278 \times 10 \times 16 = 4.44 \text{ in.}^2$$
$$a = \frac{A_s f_y}{0.85 f'_c b} = \frac{4.44 \times 40}{0.85 \times 3 \times 10} = 6.96 \text{ in.}$$
$$M_{\max} = \phi A_s f_y \left(d - \frac{a}{2} \right) = 0.90 \times 4.44 \times 40 \times 12.52 = 2000 \text{ in.-kips*}$$

Since the moment capacity of the singly reinforced beam, 2000 in.-kips, is less than the required capacity, 2750 in.-kips, compression steel is needed. Assuming that $f'_s = f_y$ at failure, then

$$M_1 = 2750 - 2000 = 750 \text{ in.-kips}$$
$$A'_s = \frac{M_1}{\phi f_y (d - d')} = \frac{750}{0.90 \times 40 \times 13.5} = 1.54 \text{ in.}^2$$

giving the additional tensile-steel area required above that provided as the upper limit of the singly reinforced beam of the same concrete dimensions. While this is also the theoretical compressive-steel requirement, this area must be increased according to the Code [see Eqs. (3.11a) and (3.13a)] to ensure failure by tensile yielding. Accordingly, the compressive-steel area will be

$$A'_s = 1.54/0.75 = 2.05 \text{ in.}^2$$

while the tensile-steel area is

$$A_s = 4.44 + 1.54 = 5.98 \text{ in.}^2$$

The design should now be checked to confirm that the compressive bars would yield at failure as assumed. With $\rho' = 2.05/(10 \times 16) = 0.0128$, the limiting tensile-steel ratio for yielding of the compression bars is found from Eq. (3.12a):

$$\bar{\rho}_{\min} = 0.85 \times 0.85 \times \frac{3}{40} \times \frac{2.5}{16} \times \frac{87}{47} + 0.0128 = 0.0285$$

* Alternatively, the quantity $M_{\max}$ may easily be found from Graph 1 of the Appendix. For $f_y = 40$ ksi and $f'_c = 3$ ksi, the graph gives $\rho_{\max} = 0.028$, and the corresponding value of $M_u/\phi bd^2$ is 870 psi. Then, for $b = 10$ in. and $d = 16$ in.,

$$M_u = M_{\max} = \frac{0.9 \times 870 \times 10 \times 16^2}{1000} = 2000 \text{ in.-kips}$$

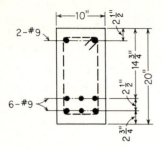

Fig. 3.3

The tentative tensile-steel ratio, $\rho = 5.98/(10 \times 16) = 0.0374$, is above the lower limit, thus ensuring that the compressive bars would yield as assumed.[1]

Two No. 9 bars will be used for compression reinforcement, six No. 9 bars providing the required tensile-steel area, as shown in Fig. 3.3. To place the tension bars within the 10-in. beam width, two rows of three bars each are used.

PROBLEMS

1. A rectangular concrete beam measures 12 in. in width and has an effective depth to the tensile-steel centroid of 18 in. Compression steel consisting of two No. 8 bars is located 2.5 in. from the compression face of the beam. If $f'_s = 4$ ksi and $f_y = 60$ ksi, what is the moment capacity of the beam, according to the ACI Code, for the following alternative tensile-steel areas: (a) $A_s = 3$ No. 10 bars, (b) $A_s = 4$ No. 10 bars, (c) $A_s = 6$ No. 10 bars?

2. For architectural reasons, a certain rectangular beam of 16-ft span is limited in cross section to a width of 8 in. and a total depth of 16 in. (effective depth = 12 in.). It must carry a total service live load of 1.40 kips per ft and a total dead load (including its own weight) of 1.10 kips per ft. If $f'_c = 4$ ksi and $f_y = 40$ ksi, what reinforcement is required?

c. T beams   With the exception of precast systems, reinforced-concrete floors, roofs, decks, etc., are almost always monolithic. Forms are built for beam soffits and sides and for the underside of slabs, and the entire construction is poured at once, from the bottom of the deepest beam to the top of the slab (see Fig. 1.1a and d). Beam stirrups and bent bars extend up into the slab. It is evident, therefore, that a part of the slab will act with the upper part of the beam to resist longitudinal compression. The resulting beam cross section is T-shaped rather than rectangular. The slab forms the beam flange, while the part of the beam projecting below the slab forms what is called the *web* or *stem*. The upper part of such a T beam is stressed laterally due to slab action in that direction. Although transverse compression may increase the longitudinal compressive strength by as much as 15 per cent (see Art. 1.9), this is not usually

---

[1] If the moment capacity of the resulting cross section is reviewed, it will be found to be slightly in excess of the required 2750 in.-kips. This is because of the Code requirement that 33 per cent more steel be added on the compression side of the member than on the tension side. This has the effect of reducing the depth of the concrete compressive-stress block and increasing the internal arm of the resisting moment M'_2. The difference is small, on the safe side, and can be neglected.

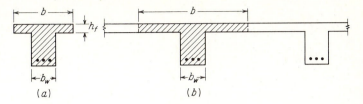

Fig. 3.4 Effective flange widths of T beams.

taken into account in design. Transverse tensile stresses are generally small and do not affect the longitudinal strength appreciably.

The next question to be resolved is that of effective width of flange. In Fig. 3.4a it is evident that, if the flange is but little wider than the stem width, the entire flange can be considered effective in resisting compression. For the floor system shown in Fig. 3.4b, however, it may be equally obvious that elements of the flange midway between the beam stems are less highly stressed in longitudinal compression than those elements directly over the stem. This is so because of shearing deformation of the flange, which relieves the more remote elements of some compressive stress.

While the actual longitudinal compression varies because of this effect, it is convenient in design to make use of an effective flange width which may be smaller than the actual flange width, but which is considered to be uniformly stressed. This effective width has been found to depend primarily on the beam span and on the relative thickness of the slab.

The recommendations for effective width given in the ACI Code are as follows:

1. For symmetrical T beams the effective width b shall not exceed one-fourth the span length of the beam. The overhanging width $(b - b_w)/2$ on either side of the beam web shall not exceed 8 times the thickness of the slab nor one-half the clear distance to the next beam.
2. For beams having a flange on one side only, the effective overhanging flange width shall not exceed one-twelfth the span length of the beam, nor 6 times the slab thickness, nor one-half the clear distance to the next beam.
3. For isolated beams in which the T form is used only for the purpose of providing additional compressive area, the flange thickness shall not be less than one-half the width of the web, and the total flange width shall not be more than 4 times the web width.

The neutral axis of a T beam may be either in the flange or in the web, depending upon the proportions of the cross section, the amount of tensile steel, and the strengths of the materials. If the calculated depth to the neutral axis is less than or equal to the slab thickness h_f, the beam can be analyzed as if it were a rectangular beam of width equal to b, the effective flange width.

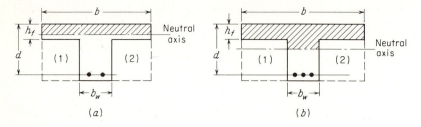

Fig. 3.5 Effective cross sections of T beams.

The reason for this is illustrated in Fig. 3.5a, which shows a T beam with neutral axis in the flange. The compressive area is indicated by the shaded portion of the figure. If the additional concrete indicated by areas 1 and 2 had been added when the beam was poured, the physical cross section would have been rectangular in shape, with a width b. No bending strength would have been added, because areas 1 and 2 are entirely in the tension zone, and tension concrete is disregarded in flexural calculations. The original T beam and the modified rectangular beam are equal in flexural strength, and rectangular beam analysis for flexure applies.

When the neutral axis is in the web, as in Fig. 3.5b, the above argument no longer is valid. In this case, methods must be developed which account for the actual T-shaped compressive zone.

In treating T beams, it is convenient to adopt the same equivalent stress distribution that is used for beams of rectangular cross section. The rectangular stress block, having a uniform compressive-stress intensity $0.85f_c'$, was devised originally on the basis of tests of rectangular beams (see Art. 2.4f), and its suitability for T beams may be questioned. However, extensive calculations based on actual stress-strain curves (reported in Ref. 3.3) indicate that its use for T beams, as well as for beams of circular or triangular cross section, introduces only minor error, and is fully justified.

Accordingly, a T beam may be treated as a rectangular beam if the depth of the equivalent stress block is equal to or less than the flange thickness. Figure 3.6 shows a tensile-reinforced T beam with effective flange width b, web

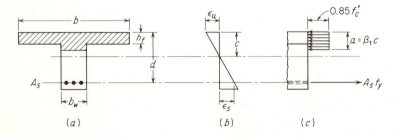

Fig. 3.6 Strain and equivalent stress distribution.

width b_w, effective depth to the steel centroid d, and flange thickness h_f. Assuming for trial purposes that the stress block is completely within the flange,

$$a = \frac{A_s f_y}{0.85 f'_c b} = \frac{\rho f_y d}{0.85 f'_c} \qquad (3.4)$$

where $\rho = A_s/bd$. If a is equal to or less than the flange thickness h_f, the member may be treated as a rectangular beam of width b and depth d. If a is greater than h_f, a T-beam analysis is required.

It will be assumed in the development that follows that the strength of the T beam is controlled by yielding of the tensile steel. This will usually be the case because of the large compressive concrete area provided by the flange. In addition, an upper limit can be established for the steel ratio to ensure that this is so, as will be shown.

As a computational device, it is convenient to divide the total tensile steel into two parts. The first part, A_{sf}, represents the steel area which, when stressed to f_y, is required to balance the longitudinal compressive force in the overhanging portions of the flange which are stressed uniformly at $0.85f'_c$. Thus,

$$A_{sf} = \frac{0.85 f'_c (b - b_w) h_f}{f_y} \qquad (3.16)$$

The force $A_{sf} f_y$ and the equal and opposite force $0.85 f'_c (b - b_w) h_f$ act with a lever arm $(d - h_f/2)$ to provide the resisting moment:

$$M'_1 = A_{sf} f_y \left(d - \frac{h_f}{2} \right) \qquad (3.17)$$

The remaining steel area $(A_s - A_{sf})$, at a stress f_y, is balanced by the compression in the rectangular portion of the beam. The depth of the equivalent rectangular stress block in this zone is found from horizontal equilibrium:

$$a = \frac{(A_s - A_{sf}) f_y}{0.85 f'_c b_w} \qquad (3.18)$$

An additional moment M'_2 is thus provided by the forces $(A_s - A_{sf}) f_y$ and $0.85 f'_c a b_w$, acting at the lever arm $(d - a/2)$,

$$M'_2 = (A_s - A_{sf}) f_y \left(d - \frac{a}{2} \right) \qquad (3.19)$$

and the total resisting moment is the sum of the moments M'_1 and M'_2:

$$M'_u = M'_1 + M'_2 = A_{sf} f_y \left(d - \frac{h_f}{2} \right) + (A_s - A_{sf}) f_y \left(d - \frac{a}{2} \right) \qquad (3.20)$$

This moment is reduced by the factor ϕ in accordance with the safety provisions of the ACI Code to obtain the design ultimate moment:

$$M_u = \phi M'_u = \phi \left[A_{sf} f_y \left(d - \frac{h_f}{2} \right) + (A_s - A_{sf}) f_y \left(d - \frac{a}{2} \right) \right] \qquad (3.21)$$

As for rectangular beams, it is desired to ensure that the tensile steel will yield prior to sudden crushing of the compression concrete, as assumed in the preceding development. For balanced failure the steel strain in Fig. 3.6b will reach ϵ_y at the same time that the concrete strain reaches its ultimate value ϵ_u. Then from geometry

$$\frac{c}{d} = \frac{\epsilon_u}{\epsilon_u + \epsilon_y}$$

Setting the sum of the horizontal forces shown in Fig. 3.6c equal to zero gives

$$A_s f_y = 0.85\beta_1 f'_c b_w c + 0.85 f'_c (b - b_w) h_f$$

or

$$A_s f_y = 0.85\beta_1 f'_c b_w c + A_{sf} f_y$$

Defining $\rho_w = A_s/b_w d$ and $\rho_f = A_{sf}/b_w d$ (i.e., expressing both total and partial steel areas in terms of the rectangular portion of the beam), the last equation can be rewritten

$$\rho_w = 0.85\beta_1 \frac{f'_c}{f_y} \frac{\epsilon_u}{\epsilon_u + \epsilon_y} + \rho_f$$

The first term on the right-hand side of the last equation is simply the balanced steel ratio ρ_b for the rectangular portion of the beam, as comparison with Eq. (3.2) will confirm. Thus the balanced steel ratio for a T beam is

$$\rho_{wb} = \rho_b + \rho_f \qquad (3.22)$$

where all ratios are expressed in terms of the rectangular portion of the beam. To provide a margin against brittle failure of T beams, the ACI Code establishes that the steel ratio used shall not exceed

$$\rho_{w,\max} = 0.75(\rho_b + \rho_f) \qquad (3.22a)$$

(The practical result of this restriction on tensile-steel area is that the stress block of T beams will almost always be within the flange, except for unusual geometry or combinations of material strength. Consequently, rectangular-beam equations may be applied in most cases.)

In designing T beams, as contrasted to reviewing the capacity of a given section, normally the slab dimensions and beam spacing will have been established by transverse flexural requirements. Consequently, the only additional section proportions that must be determined from flexural considerations are the width and depth of the web and the tensile-steel area.

If the stem dimensions were selected, as for rectangular beams, on the basis of concrete stress capacity in compression, they would be very small because of the large compression flange width furnished by the presence of the slab. Such a design would not represent the optimum solution because of the large tensile-steel requirement resulting from the small effective depth, because of the excessive web reinforcement that would be required for shear, and because of large deflections associated with such a shallow member. It is better practice to select the proportions of the web (1) so as to keep an arbitrarily low web steel ratio ρ_w or (2) so as to keep web-shear stress at desirably low limits or (3) for continuous T beams, on the basis of the flexural requirements at the supports, where the effective cross section is rectangular and of width b_w.

In addition to the main reinforcement calculated according to the requirements set forth above, it is necessary to ensure the integrity of the compressive flange of T beams by providing steel in the flange in the direction transverse to the main span. In typical construction the slab steel serves this purpose. In other cases separate bars must be added, to permit the overhanging flanges to carry, as cantilever beams, the loads directly applied. According to the Code, the spacing of such bars shall not exceed 5 times the thickness of the flange nor in any case 18 in.

Example 1: ultimate-moment capacity of a given section   An isolated T beam is composed of a flange 28 in. wide and 6 in. deep poured monolithically with a web of 10 in. width which extends 24 in. below the bottom surface of the flange to produce a beam of 30 in. total depth. Tensile reinforcement consists of six No. 10 bars placed in two horizontal rows. The centroid of the bar group is 26 in. from the top of the beam. It has been determined that the concrete has a strength of 3 ksi and that the yield stress of the steel is 60 ksi. What is the useful moment capacity of the beam?

It is easily confirmed that the flange dimensions are satisfactory according to the ACI Code for an isolated beam. The entire flange can be considered effective. For six No. 10 bars, $A_s = 7.59$ in.^{2} First checking the location of the neutral axis, on the assumption that rectangular-beam equations may be applied,

$$\rho = \frac{A_s}{bd} = \frac{7.59}{28 \times 26} = 0.0104$$

$$a = \frac{\rho f_y d}{0.85 f_c'} = \frac{0.0104 \times 60 \times 26}{0.85 \times 3} = 6.37 \text{ in.}$$

This exceeds the flange thickness, and so a T-beam analysis is required. From Eq. (3.16),

$$A_{sf} = 0.85 \times \tfrac{3}{60} \times 18 \times 6 = 4.59 \text{ in.}^2$$

Hence,

$$A_s - A_{sf} = 7.59 - 4.59 = 3.00 \text{ in.}^2$$

The steel ratios are

$$\rho_w = 7.59/260 = 0.0292$$
$$\rho_f = 4.59/260 = 0.0177$$

while from Eq. (3.2),

$\rho_b = 0.85 \times 0.85 \times \frac{3}{60} \times \frac{87}{147} = 0.0214$

According to the Code, the maximum permitted tensile-steel ratio is [Eq. (3.22a)]

$\rho_{w,\max} = 0.75(0.0214 + 0.0177) = 0.0294$

and comparison with ρ_w indicates that a ductile failure would be assured. Then, from Eq. (3.17),

$M_1' = 4.59 \times 60(26 - 3) = 6330$ in.-kips

while from Eqs. (3.18) and (3.19),

$a = \dfrac{3.00 \times 60}{0.85 \times 3 \times 10} = 7.07$ in.

$M_2' = 3.00 \times 60(26 - 3.53) = 4050$ in.-kips

Incorporating the ACI strength reduction factor, the useful capacity is

$M_u = 0.90(6330 + 4050) = 9350$ in.-kips

Example 2: determination of steel area for a given moment   A floor system consists of a 3-in. concrete slab supported by continuous T beams of 24-ft span, 47 in. on centers. Web dimensions, as determined by negative-moment requirements at the supports, are $b_w = 11$ in. and $d = 20$ in. What tensile-steel area is required at midspan to resist a design moment of 6400 in.-kips if $f_y = 60$ ksi and $f_c' = 3$ ksi?
   First determining the effective flange width,

$16h_f + b_w = 16 \times 3 + 11 = 59$ in.

$\dfrac{\text{Span}}{4} = 24 \times \dfrac{12}{4} = 72$ in.

Center-line beam spacing $= 47$ in.

The center-line T-beam spacing controls in this case, and $b = 47$ in. The concrete dimensions b_w and d are known to be adequate in this case, since they have been selected for the larger negative support moment applied to the effective rectangular section $b_w d$. The tensile steel at midspan is most conveniently found by trial. Assuming the stress-block depth equal to the flange thickness of 3 in.,

$d - \dfrac{a}{2} = 20 - 1.50 = 18.50$ in.

Trial:

$A_s = \dfrac{M_u}{\phi f_y (d - a/2)} = \dfrac{6400}{0.90 \times 60 \times 18.50} = 6.40$ in.^{2}

$\rho = \dfrac{A_s}{bd} = \dfrac{6.40}{47 \times 20} = 0.00681$

$a = \dfrac{\rho f_y d}{0.85 f_c'} = \dfrac{0.00681 \times 60 \times 20}{0.085 \times 3} = 3.20$ in.

Since a is greater than h_f, a T-beam analysis is indicated.

$$A_{sf} = \frac{0.85f'_c(b - b_w)h_f}{f_y} = \frac{0.85 \times 3 \times 36 \times 3}{60} = 4.58 \text{ in.}^2$$

$$M_1 = \phi A_{sf}f_y\left(d - \frac{h_f}{2}\right) = 0.90 \times 4.58 \times 60 \times 18.50 = 4570 \text{ in.-kips}$$

$$M_2 = M_u - M_1 = 6400 - 4570 = 1830 \text{ in.-kips}$$

Assume $a = 4.00$ in.

$$(A_s - A_{sf}) = \frac{M_2}{\phi f_y(d - a/2)} = \frac{1830}{0.90 \times 60 \times 18.00} = 1.88 \text{ in.}^2$$

Check:

$$a = \frac{(A_s - A_{sf})f_y}{0.85f'_c b_w} = \frac{1.88 \times 60}{0.85 \times 3 \times 11} = 4.02 \text{ in.}$$

This is satisfactorily close to the assumed value of 4 in.* Then

$$A_s = A_{sf} + (A_s - A_{sf}) = 4.58 + 1.88 = 6.46 \text{ in.}^2$$

Checking to ensure that tensile yielding will control the strength of the section,

$$\rho_w = 6.46/220 = 0.0294$$
$$\rho_f = 4.58/220 = 0.0208$$

$$\rho_b = 0.85 \times 0.85 \times \frac{3}{60} \times \frac{87}{147} = 0.0214$$

$$\rho_{w,\max} = 0.75(0.0214 + 0.0208) = 0.0316$$

indicating that the actual ρ_w is satisfactorily low. The close agreement should be noted between the approximate tensile-steel area of 6.40 in.² found by assuming the stress-block depth equal to the flange thickness and the more exact value of 6.46 in.² found by T-beam analysis. The approximate solution would be satisfactory in many cases.

PROBLEMS

1. A tensile-reinforced T beam is to be designed to carry a uniformly distributed load on a 20-ft simple span. The total design moment at midspan due to all loads is 6240 in.-kips. Concrete dimensions, as governed by web-shear and clearance considerations, are $b = 20$ in., $b_w = 10$ in., $h_f = 5$ in., and $d = 20$ in. If $f_y = 60$ ksi and $f'_c = 4$ ksi, what tensile reinforcement is required at midspan?

* Alternatively, from Graph 1 of the Appendix, with

$$\frac{M_u}{\phi bd^2} = \frac{1830 \times 1000}{11 \times 400 \times 0.9} = 462$$

the steel ratio $(\rho_w - \rho_f)$ is found to be 0.0085. Then

$$(A_s - A_{sf}) = (\rho_w - \rho_f)b_w d = 0.0086 \times 11 \times 20 = 1.88 \text{ in.}^2$$

2. A T beam has an effective flange width of 30 in., web width of 14 in., slab thickness of 4 in., and effective depth to the steel centroid of 20 in. If steel having a yield stress of 60 ksi is used with concrete of $f'_c = 2.5$ ksi, what is the moment capacity of the beam for $A_s = 3$ No. 11 bars?

3. A single precast T beam is to be used as a bridge over a small roadway. Concrete dimensions are $b = 48$ in., $b_w = 12$ in., $h_f = 5$ in., and $h = 24$ in. The effective depth to the steel centroid is 20 in. Concrete and steel strengths are 4 and 60 ksi, respectively. Using as nearly as possible the maximum tensile reinforcement permitted under the ACI Code (select actual bar size and number), what is the useful ultimate-moment capacity of the girder? If the beam is to be used on a 30-ft simple span, and if in addition to its own weight it must support railings, curbs, and suspended loads totaling 0.475 klf, what service-live-load limit should be posted?

3.5 DESIGN FOR SHEAR AND DIAGONAL TENSION

In addition to meeting flexural requirements, beams must be safe against premature failure due to diagonal tension in the concrete, resulting from combined shear and longitudinal flexural stress. This mechanism of failure was discussed in Art. 2.5. Beams may be designed with cross sections sufficiently large so that the concrete can resist all the diagonal tension. However, a more economical design will usually result if a smaller cross section is employed, with supplementary steel provided to reinforce the beam web. In addition, such reinforcement improves the ductility of the member by restraining the growth of inclined cracks, thus providing warning of impending failure.

It was pointed out in Art. 2.5 that the average shear stress

$$v = \frac{V}{bd} \tag{2.32}$$

can be used as a measure of the maximum intensity of shear stress at a section. For design based on conditions at failure, the shear force V is found using factored loads, as described in Art. 3.3, and may be written V_u. Furthermore, Eq. (2.32) applied only to rectangular beams. For T beams it has been found that most of the shear force is carried by the web. Calculations may therefore be based on the web area $b_w d$, which may be substituted for bd in Eq. (2.32) with the understanding that $b_w = b$ for rectangular sections. According to ACI Code procedures, the nominal shear stress v_u is thus calculated from

$$v_u = \frac{V_u}{\phi b_w d} \tag{3.23}$$

in which ϕ is a capacity reduction factor equal to 0.85 for shear. (The slight additional conservatism, as compared with the use of $\phi = 0.90$ for flexure, reflects both the sudden nature of diagonal tension failure and the imperfect understanding of that failure mode.)

Note that the use of ϕ in Eq. (3.23) has the effect of increasing the shear force for which a section must be designed. The stress v_u found in this way must be compared against the calculated shear strength of the member, with no additional capacity reduction factor applied. While this procedure tends to obscure the fact that ϕ is a capacity reduction factor, rather than the reciprocal of a load factor, it is evident that the net result is the same.

Recognizing the beneficial effect of vertical compression due to support reactions, the Code permits sections located less than a distance d from the face of a support to be designed for the same v_u as that computed at a distance d.

a. Beams with no web reinforcement Beams which are to be reinforced only for flexural tension must be designed so that shearing stresses throughout the beam are well below the value that would result in diagonal tension cracking of the web. On the basis of Art. 2.5b, that shear stress v_c can be estimated by the equation

$$v_c = v_{cr} = 1.9 \sqrt{f_c'} + 2500\rho_w \frac{V_u d}{M_u} \leq 3.5 \sqrt{f_c'} \qquad (3.24a)$$

where ρ_w = longitudinal tensile-steel ratio (A_s/bd for rectangular beams and $A_s/b_w d$ for T beams)

   d = effective depth to tensile steel

   V_u = external shear force at section under consideration

   M_u = external moment at section under consideration, not to be taken less than $V_u d$

With V_u, d, and M_u in self-consistent units and $\sqrt{f_c'}$ in psi units, v_c is given in psi units. As an alternative to Eq. (3.24a), the Code permits use of the simpler, more conservative, but less accurate, equation

$$v_c = 2 \sqrt{f_c'} \qquad (3.24b)$$

Equation (3.24b) is adequate for most design purposes.

If v_u, calculated by Eq. (3.23), is no larger than v_c, calculated by Eq. (3.24a), or alternatively by Eq. (3.24b), then theoretically no web reinforcement is required. Even in such a case, however, the Code requires provision of a minimum area of web reinforcement equal to

$$A_v = \frac{50 b_w s}{f_y} \qquad (3.25)$$

where s = longitudinal spacing of web reinforcement

   A_v = total cross-sectional area of web steel within distance s

   f_y = yield strength of steel

This provision holds unless v_u is less than one-half of v_c or unless the total beam depth does not exceed either 10 in., $2\frac{1}{2}$ times the flange thickness, or $\frac{1}{2}$ the width of the web.

Example 1: beam without web reinforcement   A rectangular beam is to be designed to carry a shear force of 30 kips. No web reinforcement is to be used, and f'_c is 4 ksi. What is the minimum cross section if controlled by shear?

      If no web reinforcement is to be used, it is necessary that the cross-sectional dimensions be selected so that the shear stress given by Eq. (3.23) be no larger than one-half the value permitted on the unreinforced web. From Eq. (3.24b),

$$v_c = 2\sqrt{4000} = 126 \text{ psi}$$

Thus,

$$b_w d = \frac{V_u}{\phi v_c \times 1/2} = \frac{30{,}000}{0.85 \times 126 \times 1/2} = 560 \text{ in.}^2$$

A beam with $b_w = 18$ in. and $d = 31$ in. is required. Alternatively, if the minimum amount of web reinforcement given by Eq. (3.25) is used, the concrete shear stress can assume its full value v_c, and it is easily confirmed that a beam with $b_w = 12$ in. and $d = 23.5$ in. will be sufficient.

b. Region in which web reinforcement is required   The portion of any span through which web reinforcement is theoretically necessary can be found by drawing the shear-stress diagram for the span and superimposing a plot of the shear strength of the concrete. When the unit shear stress v_u, given by Eq. (3.23), exceeds the capacity of the unreinforced web, as given by Eq. (3.24a) or (3.24b), then web reinforcement is required.

      In addition to the theoretical requirement, according to the Code, web reinforcement at least equal to that given by Eq. (3.25) must be provided elsewhere in the span, unless the calculated shear stress is less than one-half of v_c.

Example 2: limits of web reinforcement   A simply supported rectangular concrete beam 16 in. wide, having an effective depth of 22 in., carries a total design load of 7.9 kips per ft on a 20-ft clear span. It is reinforced with 9.86 in.² of tensile steel, which continues uninterrupted into the supports. If $f'_c = 3$ ksi, through what part of the beam is web reinforcement required?

      The maximum external shear force occurs at the end of the span, where $V_u = 7.9 \times \frac{20}{2} = 79.0$ kips. At the critical section for shear, a distance d from the support, $V_u = 79.0 - 7.9 \times 1.83 = 64.5$ kips. The shear force varies linearly to zero at midspan. For convenience, the applied shear is plotted in terms of nominal shear stress, obtained by dividing V_u by $\phi b_w d$, or $0.85 \times 16 \times 22$, with the results shown in **Fig. 3.7a**. Adopting Eq. (3.24b),

$$v_c = 2\sqrt{3000} = 109 \text{ psi}$$

This value is superimposed on the shear-stress diagram, and from geometry, the point at which web reinforcement theoretically is no longer required is

$$10 \times \frac{264 - 109}{264} = 5.86 \text{ ft}$$

from the support face. However, according to the Code, at least a minimum amount

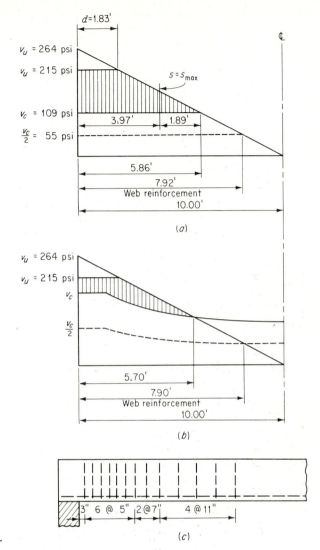

Fig. 3.7 Shear design example.

of web reinforcement is required wherever the shear stress exceeds one-half of v_c, or 55 psi in this case. As seen from Fig. 3.7a, this controls at a distance

$$10 \times \frac{264 - 55}{264} = 7.92 \text{ ft}$$

from the support face. Summarizing, at least the minimum web steel must be provided within a distance of 7.92 ft from the supports, and within 5.86 ft the web steel must provide for the shear force corresponding to the shaded area.

If the alternative, Eq. (3.24a), is used, the variation along the span of ρ_w, V_u, and M_u must be known so that V_c can be calculated. This is best done in tabular form as follows:

Distance from support, ft	M_u, ft-kips	V_u, kips	$v_c = 1.9 \sqrt{f_c'} + 2500 \rho_w \dfrac{V_u d}{M_u} \leq 3.5 \sqrt{f_c'}$
0	0	79.0	174
1	76	71.1	174
2	142	63.2	161
3	201	55.3	139
4	252	47.4	128
5	296	39.5	121
6	332	31.6	116
7	360	23.7	112
8	379	15.8	109
9	391	7.9	107
10	395	0	104

The applied ultimate shear stress v_u and the shear capacity v_c are plotted in Fig. 3.7b. From the graph it is found that stirrups are theoretically no longer required 5.70 ft from the support face. However, from the plot of $v_c/2$ it is found that at least the minimum web steel is to be provided within a distance of 7.90 ft.

Comparing Fig. 3.7a and b, it is evident that the lengths over which web reinforcement is needed are nearly the same for this example whether Eq. (3.24a) or (3.24b) is used. However, the smaller shaded area of Fig. 3.7b indicates that substantially less web-steel area would be needed within that required distance if the more accurate Eq. (3.24a) were adopted.

c. Types of web reinforcement

Reinforcement to resist diagonal tensile stresses in a beam may take several forms. A part of the longitudinal steel may be bent up where it is no longer needed to resist flexural tension, as shown in Fig. 3.8a. While the ACI Code requires only that the inclined part of any such bar make an angle of at least 30° with the longitudinal part, usually such bars are bent at a 45° angle. Only the center three-fourths of the inclined part of such a bar is to be considered effective as web reinforcement.

Alternatively, separate web steel may be used, as shown in Fig. 3.8b. These separate bars, called *stirrups*, are usually placed perpendicular to the axis of the beam, although they may sometimes be inclined. Where inclined stirrups are used, the Code requires that they make an angle of at least 45° with the longitudinal reinforcement and that they be securely anchored against slipping. Because of the relatively short length of stirrup embedded in the compression zone of a beam, bond requirements (see Art. 3.6) usually dictate the use of special anchorage, obtained by hooking the end of the stirrup. Various forms of stirrups are shown in Fig. 3.8c. The 180° and 90° hooks shown

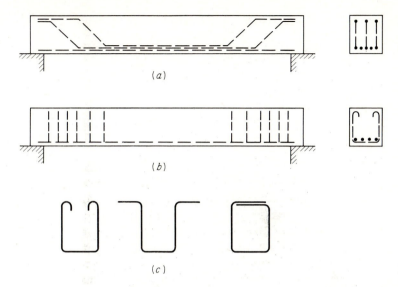

Fig. 3.8 Types of web reinforcement.

in the first two sketches are most common. The closed stirrup shown in the
third sketch is convenient when separate straight bars are used for negative
flexural reinforcement. These bars can be wired directly to the upper part of
the stirrups for temporary support while concrete is poured.

Even though the longitudinal bars at the bottom of a continuous beam
are often bent up to provide tensile reinforcement at the top of the beam over
the supports, the requirements for longitudinal reinforcement often conflict
with those for diagonal tension. Because the saving in steel is usually small,
many designers prefer to include separate stirrups to carry all the excess shear,
counting on the bent part of the longitudinal bars only to increase the overall
safety against diagonal tension failure.

In addition to bent bars or stirrups, welded wire mesh or continuous
spirals are sometimes used for web reinforcement.

To avoid excessive crack widths in beam webs subject to diagonal tension,
the ACI Code limits the design yield strength of web reinforcement to 60 ksi.

d. Design of web reinforcement   For beams in which the shear stress through
the critical region exceeds that which can be resisted by the unreinforced web,
additional reinforcement must be provided. For reasons given in Art. 2.5c, the
shear resistance of the concrete after cracking, V_c, can be taken equal to the
shear force which caused the first crack to form, V_{cr}, and web reinforcement
need be provided only for the excess shear above that value.

For vertical stirrups and for bent bars (the most common types of web reinforcement) the following equations were developed in Art. 2.5c:

For vertical stirrups: $v'_u = v_{cr} + \dfrac{A_v f_y}{bs}$ $\qquad\qquad$ (2.36a)

For inclined bars: $\qquad v'_u = v_{cr} + \dfrac{A_v f_y (\sin \alpha + \cos \alpha)}{bs}$ $\qquad$ (2.37a)

where v'_u = nominal ultimate shear stress capacity = V'_u/bd
$\qquad v_{cr}$ = nominal shear stress causing cracking = V_{cr}/bd
$\qquad\ s$ = spacing of web reinforcement in direction of beam axis
$\qquad A_v$ = total cross-sectional area of web steel in distance s
$\qquad f_y$ = yield stress of web steel
$\qquad\ d$ = effective beam depth
$\qquad\ \alpha$ = angle of inclination of web steel

Since $V_c = V_{cr}$, it follows that $v_c = v_{cr}$, and this substitution will be made in the above equations. In addition, the equations may be generalized to include T beams as well as rectangular beams by writing b_w for b. Finally, according to present ACI Code procedures, capacity reduction factors for shear strength are applied in an inverse way to increase design ultimate load. As a result v_u should be taken equal to v'_u with no further reduction of capacity. With these modifications Eqs. (2.36a) and (2.37a) become

For vertical stirrups: $v_u = v_c + \dfrac{A_v f_y}{b_w s}$ $\qquad\qquad$ (3.26)

For inclined bars: $\qquad v_u = v_c + \dfrac{A_v f_y (\sin \alpha + \cos \alpha)}{b_w s}$ $\qquad$ (3.27)

where v_u = nominal ultimate shear-stress capacity to be used for design
$\qquad v_c$ = nominal shear-stress capacity of the concrete, computed from either Eq. (3.24a) or (3.24b)
$\qquad b_w$ = width of web, equal to b for rectangular sections
and all other terms are as previously defined.

Usually, in designing, it is convenient to select a trial web-steel area A_v based on standard bar sizes, for which the required spacing s may be found. For this purpose Eqs. (3.26) and (3.27) may be rewritten

For vertical stirrups: $s = \dfrac{A_v f_y}{(v_u - v_c)b_w}$ $\qquad\qquad$ (3.26a)

For inclined bars: $\qquad s = \dfrac{A_v f_y (\sin \alpha + \cos \alpha)}{(v_u - v_c)b_w}$ $\qquad$ (3.27a)

It should be noted that when conventional U stirrups are used, the web area A_v provided by each stirrup is twice the cross-sectional area of the bar.

It is undesirable to space vertical stirrups closer than about 4 in.; the size of the stirrups should be chosen so as to avoid a closer spacing. When

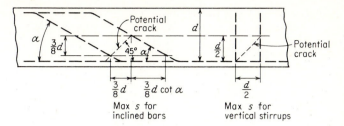

Fig. 3.9

vertical stirrups are required over a comparatively short distance, it is good practice to space them uniformly over the entire distance, the spacing being calculated for the point of greatest shear (minimum spacing). If the web reinforcement is required over a long distance, and if the shear varies materially throughout this distance, it is more economical to compute the spacings required at several sections and to place the stirrups accordingly, in groups of varying spacing.

Where web reinforcement is needed, the Code requires it to be spaced so that every 45° line, representing a potential diagonal crack and extending from the middepth $d/2$ of the member to the longitudinal tension bars, is crossed by at least one line of web reinforcement; in addition, the Code specifies a maximum spacing of 24 in. When the excess shear stress $(v_u - v_c)$ exceeds $4\sqrt{f_c'}$, these maximum spacings are halved. These limitations are shown in Fig. 3.9 for both vertical stirrups and inclined bars, for situations in which the excess shear does not exceed the stated limit.

In no case, according to the Code, is the excess shear stress $(v_u - v_c)$ to exceed $8\sqrt{f_c'}$ regardless of the amount of web steel used.

Example 3: design of web reinforcement   Using vertical U stirrups, with $f_y = 40$ ksi, design the web reinforcement for the beam of Example 2.

The solution will be based on the shear diagram in Fig. 3.7a. The stirrups must be designed to resist that part of the shear shown shaded. With No. 3 stirrups used for trial, the three maximum spacing criteria are first applied. For $(v_u - v_c) = 215 - 109 = 106$ psi, which is less than $4\sqrt{f_c'}$, the maximum spacing must exceed neither $d/2 = 11$ in. nor 24 in. Also, from Eq. (3.25),

$$s_{max} = \frac{A_v f_y}{50 b_w} = \frac{0.22 \times 40,000}{50 \times 16} = 11 \text{ in.}$$

The first and third criteria identically control in this case, and a maximum spacing of 11 in. is adopted. From the support to a distance d from the support, the excess shear stress $(v_u - v_c)$ is 106 psi. In this region the required spacing is

$$s = \frac{A_v f_y}{(v_u - v_c) b_w} = \frac{0.22 \times 40,000}{106 \times 16} = 5.2 \text{ in.}$$

This is neither so small that placement problems would result nor so large that maximum spacing criteria would control, and the choice of No. 3 stirrups is confirmed. Solving Eq. (3.26a) for the excess shear at which the maximum spacing can be used,

$$(v_u - v_c) = \frac{A_v f_y}{s b_w} = \frac{0.22 \times 40{,}000}{11 \times 16} = 50 \text{ psi}$$

With reference to Fig. 3.7a, this is attained at a distance x_1 from the point of zero excess shear, where $x_1 = 5.86 \times 50/155 = 1.89$ ft. This is 3.97 ft from the support face. With this information, a satisfactory spacing pattern can be selected. The first stirrup is usually placed at a distance $s/2$ from the support. The following spacing pattern is satisfactory:

```
1 space  at   3 in. =  3 in.
6 spaces at   5 in. = 30 in.
2 spaces at   7 in. = 14 in.
4 spaces at  11 in. = 44 in.
                      ───────
          Total  =   91 in. = 7 ft 7 in.
```

The resulting stirrup pattern is shown in Fig. 3.7c. As an alternative solution, it is possible to plot a curve showing required spacing as a function of distance from the support. Once the required spacing at some reference section, say at the support, is determined,

$$s_0 = \frac{0.22 \times 40{,}000}{155 \times 16} = 3.55 \text{ in.}$$

it is easy to obtain the required spacings elsewhere by a single setting of the slide rule. In Eq. (3.26a) only $(v_u - v_c)$ changes with distance from the support. For uniform load this quantity is a linear function of distance from the point of zero excess shear, 5.86 ft from the support face. Hence, at 1-ft intervals,

$$s_1 = 3.55 \times 5.86/4.86 = 4.27 \text{ in.}$$
$$s_2 = 3.55 \times 5.86/3.86 = 5.39 \text{ in.}$$
$$s_3 = 3.55 \times 5.86/2.86 = 7.27 \text{ in.}$$
$$s_4 = 3.55 \times 5.86/1.86 = 11.18 \text{ in.}$$
$$s_5 = 3.55 \times 5.86/0.86 = 24.20 \text{ in.}$$

This is plotted in Fig. 3.10, together with the maximum spacing of 11 in., and a practical spacing pattern is selected. The spacing at a distance d from the support face is selected as the minimum requirement, in accordance with the ACI Code. The pattern of No. 3 U-shaped stirrups selected (shown on the graph) is identical with the previous solution. In many cases the experienced designer would find it unnecessary actually to plot the spacing diagram of Fig. 3.10 and would select a spacing pattern directly after calculating the required spacing at intervals along the beam.

If the web steel were to be designed on the basis of the excess-shear diagram of Fig. 3.7b, the second approach illustrated above would necessarily be selected, and spacings would be calculated at intervals along the span. In this particular case, the maximum permitted spacing of 11 in. is less than that required by excess shear anywhere between the point of zero excess shear and a distance d from the support. Consequently, the following spacing could be used:

```
1 space  at   5 in. =  5 in.
8 spaces at  11 in. = 88 in.
                      ───────
          Total  =   93 in. = 7 ft 9-in.
```

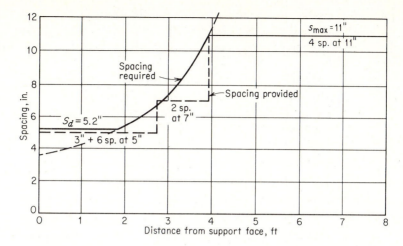

Fig. 3.10

Nine No. 3 stirrups would be used, rather than the 13 previously calculated, in each half of the span. In designs for which the stress governs the stirrup spacing for both methods rather than the maximum spacing requirements, the saving obtained by the second solution would be even more significant.

PROBLEMS

1. A beam is to be designed to carry a total factored shear of 67,600 lb, using a concrete with $f'_c = 3$ ksi. Proceeding on the basis that the beam cross section will be determined by diagonal tension requirements, select an appropriate width and effective depth (*a*) for a beam in which no web reinforcement is to be used and (*b*) for a beam in which stirrups are used to carry excess shear.

2. A beam of 11 in. width and effective depth of 16 in. carries a factored, uniformly distributed load of 6.0 kips per ft, including its own weight, in addition to a central concentrated, factored load of 14 kips. It spans 18 ft, and restraining end moments at full-factored load are 155 ft-kips at each support. It is reinforced with three No. 9 bars for both positive and negative bending. If f'_c is 4 ksi, through what part of the beam is web reinforcement required (*a*) if $v_c = 2\sqrt{f'_c}$ and (*b*) if $v_c = 1.9\sqrt{f'_c} + 2500\rho_w V_u d/M_u \leq 3.5\sqrt{f'_c}$?

3. In Prob. 2*a*, what effect would a clockwise moment of 200 ft-kips at the right support have on the placement of shear reinforcement previously determined?

4. Design the web reinforcement for the beam of Prob. 2*a* using (*a*) No. 2 stirrups with $f_y = 4$ ksi and (*b*) No. 3 stirrups with $f_y = 4$ ksi. Compare the total weight of web reinforcement for these two alternative designs. Use standard hooks (see Fig. 3.12) at the upper ends of the stirrups. What is your recommendation for the final design?

e. Shear design using lightweight concrete

Lightweight aggregate concrete (see Art. 1.11) is used increasingly in construction, particularly for precast and prestressed work. These special concretes, normally having a density of 100 to 110 pcf, can be proportioned to produce compressive strengths as high as those of normal-aggregate concretes. Their tensile strength, of particular importance in shear and diagonal tension calculations, has not been thoroughly investigated. It is apt to be lower than that of ordinary concrete of the same compressive strength.

It is advisable, when designing for lightweight concrete, to obtain an accurate estimate of the tensile strength of the material. The split-cylinder strength f_{ct} (see Art. 1.8) is not identical with the direct tensile strength, but it serves as a convenient and reliable measure.

For normal concrete the split-cylinder strength is often taken equal to $6.7 \sqrt{f_c'}$. Accordingly, the ACI Code specifies that $f_{ct}/6.7$ shall be substituted for $\sqrt{f_c'}$ in all equations for v_c, with the further restriction that $f_{ct}/6.7$ shall not exceed $\sqrt{f_c'}$. If the split-cylinder strength is not available, values of v_c calculated using $\sqrt{f_c'}$ shall be multiplied by 0.75 for "all-lightweight" concrete and by 0.85 for "sand-lightweight" concrete. All other shear provisions remain unchanged.

f. Shear design for members subject to axial forces Reinforced-concrete beams may be subjected to axial forces due to a variety of causes, including prestressing, restrained shrinkage, and temperature effects, as well as directly applied load. The design of prestressed-concrete members for shear will be treated separately in Chap. 11. Nonprestressed members may have their strength in shear significantly modified in the presence of axial tension or compression, as is evident from a review of Art. 2.5a.

A longitudinal compressive force N_u acting on a flexural member will increase the nominal shear stress $v_c = v_{cr}$ associated with formation of the first diagonal tension crack. This can be accounted for in design by suitable modification of Eqs. (3.24a) and (3.24b). According to ACI Code procedures,[1] M_u in Eq. (3.24a) is replaced by

$$M_m = M_u - N_u \frac{4h - d}{8}$$

in which h is the total depth of the member, with the additional stipulation that M_m is permitted to assume values less than $V_u d$. Thus, in the presence of axial compression,

$$v_c = 1.9 \sqrt{f_c'} + 2500\rho_w \frac{V_u d}{M_u - \tfrac{1}{8}N_u(4h - d)} \qquad (3.28a)$$

not to exceed

$$v_c \leq 3.5 \sqrt{f_c'} \sqrt{1 + \frac{0.002 N_u}{A_g}}$$

where A_g is the gross area of the concrete section, and N_u/A_g is expressed in

[1] Derived in Ref. 2.5 and supported by comparison with test data in Ref. 3.4.

psi units. As an alternative to Eq. (3.28a), which is difficult to apply in practice, the Code permits calculation of v_c by the simpler but less accurate expression

$$v_c = 2\left(1 + \frac{0.0005N_u}{A_g}\right)\sqrt{f_c'} \qquad\qquad (3.28b)$$

Equation (3.28b) is sufficiently accurate for most purposes.

In contrast to compression, longitudinal tension reduces the nominal shear stress associated with cracking. This influence may be recognized, according to the Code, by replacing Eqs. (3.24a) and (3.24b) by

$$v_c = 2\left(1 + \frac{0.002N_u}{A_g}\right)\sqrt{f_c'} \qquad\qquad (3.28c)$$

in which N_u is negative for tension. Alternatively, for members subject to longitudinal tension, the web steel may be designed to carry all the shear; that is, v_c is taken equal to zero.

g. Shear design for members of variable depth   If a beam has a varying depth, the inclination of the compressive- and tensile-stress resultants may significantly affect the shear for which the member should be designed. Figure 3.11a shows a cantilevered beam loaded at its free end and having a tapered profile. In such case the tensile force and the compressive-stress resultant are inclined, and introduce components transverse to the axis of the member. If the slope of the top surface of the beam is θ_1 and the slope of the bottom surface is θ_2, then the net shear force $\bar{V}_u$ for which the member should be designed is very nearly equal to

$$\bar{V}_u = V_u - T\tan\theta_1 - C\tan\theta_2$$

where V_u is the external shear force acting. C, which is equal to T, is given by M_u/z, where M_u is the applied moment and z is the internal lever arm of the

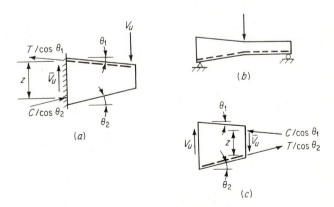

Fig. 3.11   Effect of variable beam depth on shear.

resisting couple. Thus, in a case such as that shown in Fig. 3.11a, for which the beam depth increases in the direction of increasing moment, the shear for which the member should be designed is, approximately,

$$\bar{V}_u = V_u - \frac{M_u}{z}(\tan\theta_1 + \tan\theta_2) \qquad (3.29a)$$

For a case such as that shown in Fig. 3.11b, in which the member depth decreases in the direction of increasing moment, the corresponding expression is

$$\bar{V}_u = V_u + \frac{M_u}{z}(\tan\theta_1 + \tan\theta_2) \qquad (3.29b)$$

as is easily confirmed by summation of forces shown in Fig. 3.11c.

These equations are only approximate, since the direction of the stress resultants is not exactly as assumed; however, they may be used for design provided the sum of the slope angles θ_1 and θ_2 does not exceed about $30°$, according to Ref. 3.5.

3.6 BOND, ANCHORAGE, AND DEVELOPMENT LENGTH

Flexural bond stress U exists along a reinforcing bar embedded in a loaded concrete beam, due to the change in bending moment along the axis. In Art. 2.6 it was shown that the magnitude of U per unit length of bar is given by

$$U = \frac{V}{z} \qquad (3.30)$$

in which V is the external shear force, and z is the internal lever arm of the resisting moment at the section considered. It was shown also that local bond stresses much higher than those predicted by Eq. (3.30) occur at cracks, due to local high-stress gradients in the bar at crack locations. Current design methods disregard such local high bond stresses, even though they may result in localized slip between steel and concrete adjacent to the cracks. Instead, attention is directed toward providing adequate length of embedment, past the location at which the bar is fully stressed, to develop the full strength of the bar.

It was shown in Art. 2.6 that, on the basis of experimental evidence, the development length l_{du} for bars of common sizes is, approximately,

$$l_{du} = \frac{0.028 A_b f_y}{\sqrt{f'_c}} \qquad (2.41a)$$

For design purposes this experimental development length should be increased somewhat to ensure safety of the member. In addition, Eq. (2.41a) tends to overestimate l_{du} for large-diameter No. 14 and No. 18 bars. Accordingly, the ACI Code recommends the basic development lengths given by Table 3.2.

Table 3.2   Development length l_d in tension

A. Basic development length l_d

    No. 11 bars or smaller $0.04 A_b f_y / \sqrt{f'_c}$ but $\geq 0.0004 d_b f_y$

    No. 14 bars $0.085 f_y / \sqrt{f'_c}$

    No. 18 bars $0.11 f_y / \sqrt{f'_c}$

    Deformed wire $0.03 d_b f_y / \sqrt{f'_c}$

B. Modification factors to be applied to l_d

Top bars (horizontal reinforcement so placed that more than 12 in. of concrete is cast in the member below the bar)	1.4
Bars with f_y greater than 60,000 psi	$2 - 60,000/f_y$
"All-lightweight" concrete	1.33
"Sand-lightweight" concrete	1.18
Reinforcement spaced laterally at least 6 in. on centers and at least 3 in. from the side face of the member	0.8
Reinforcement in excess of that required	A_s required$/A_s$ provided
Bars enclosed within a spiral which is not less than $\frac{1}{4}$ in. in diameter and not more than 4 in. in pitch	0.75

The basic lengths shown in part *A* of Table 3.2 are to be modified by the applicable factors of part *B*, and in no case is l_d to be less than 12 in.

For example, for No. 8 top bars having f_y of 60,000 psi, used with sand-lightweight concrete having f'_c of 4000 psi, the development length to be used in design is

$$l_d = (0.04 \times 0.79 \times 60,000/\sqrt{4000}) \times 1.4 \times 1.18 = 50 \text{ in.}$$

The basic development lengths of part *A* of Table 3.2 are easily tabulated for bars of given size and strength and for particular concrete strengths. These are given in Table 8 of the Appendix.

For bars in compression, l_d is computed as $0.02 f_y d_b / \sqrt{f'_c}$ but not less than $0.0003 f_y d_b$, or 8 in., according to the Code.

In the event that the desired stress in a bar cannot be developed by bond alone, it is necessary to provide mechanical anchorage at the end of the bar, usually by means of a 90° bend or a 180° hook. The dimensions and bend radii for hooks have been standardized by the ACI Code as follows (see Fig. 3.12):

1. A semicircular turn plus an extension of at least 4 bar diameters, but not less than $2\frac{1}{2}$ in. at the free end of the bar, or
2. A 90° turn plus an extension of at least 12 bar diameters at the free end of the bar, or
3. For stirrup and tie anchorage only, either a 90° or 135° turn plus an extension of at least 6 bar diameters, but not less than $2\frac{1}{2}$ in. at the free end of the bar.

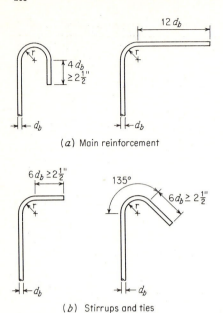

(*a*) Main reinforcement

(*b*) Stirrups and ties

Fig. 3.12 Standard hooks.

The minimum radius of bend, measured on the inside of the bar, for standard hooks other than stirrup or tie hooks, should not be less than the values shown in Table 3.3, except that, for sizes No. 6 to No. 11, in Grade 40 bars with 180° hooks only, the minimum diameter is 5 bar diameters. For stirrup and tie hooks and bends other than standard hooks, the inside diameter of bends should not be less than $1\frac{1}{2}$ in. for No. 3 bars, 2 in. for No. 4 bars, and $2\frac{1}{2}$ in. for No. 5 bars, according to the Code.

Standard hooks in tension may be considered to develop a tensile stress in the bar of $f_h = \xi \sqrt{f'_c}$, where ξ is given by Table 3.4. According to the Code, the values of ξ in Table 3.4 may be increased 30 per cent where enclosure is provided perpendicular to the plane of the hook. Hooks are not effective in adding to the compressive resistance of reinforcement.

Special attention should be given to proper anchorage of web reinforcement. It should be carried as close as possible to the compression face of the member in order better to contain potential diagonal tension cracks. The ACI

Table 3.3   Minimum diameters of bend for standard hooks

Bar size	Minimum diameter
Nos. 3 to 8	6 bar diameters
Nos. 9 to 11	8 bar diameters
Nos 14 and 18	10 bar diameters

Table 3.4   Coefficient ξ for standard hooks in tension

Bar size	$f_y = 60\ ksi$		$f_y = 40\ ksi,$ all bars
	Top bars	Other bars	
Nos. 3 to 5	540	540	360
No. 6	450	540	360
Nos. 7 to 9	360	540	360
No. 10	360	480	360
No. 11	360	420	360
No. 14	330	330	330
No. 18	220	220	220

Code includes special provisions for anchorage of web reinforcement. The ends of single-leg, simple-U, or multiple-U stirrups are to be anchored by one of the following means:

1. A standard hook plus an effective embedment of $0.5l_d$. The effective embedment is taken as the distance between the middepth of the member and the start of the hook.
2. Embedment above or below the middepth of the beam on the compression side for a full development length l_d but not less than 24 bar diameters.
3. Bending around the longitudinal reinforcement through at least 180°.

3.7 SELECTION OF BARS AND BAR SPACING

The standard reinforcing-bar sizes and types available to the designer were described in Art. 1.12b. Common bars range in size from No. 2 to No. 11, the bar number corresponding approximately to the number of $\frac{1}{8}$ in. of bar diameter. The larger sizes, No. 14 ($1\frac{3}{4}$-in. diameter) and No. 18 ($2\frac{1}{4}$-in. diameter), are used primarily for column reinforcement. Table 1 of the Appendix gives areas, perimeters, and weights of standard bars; Tables 2 and 3 give information for groups of bars.

It is often desirable to mix bar sizes in order to meet steel-area requirements more closely. In general, mixed bars should be of comparable sizes; for example, it would be bad practice to use No. 11 bars in conjunction with No. 3 bars, while mixing No. 10 bars with No. 8 bars would be perfectly acceptable. There is some practical advantage to minimizing the number of different bar sizes used for a given structure.

Normally, it is necessary to maintain a certain minimum distance between adjacent bars in order to ensure proper placement of concrete around them. Air pockets below the steel are to be avoided, and full surface contact between the bars and the concrete is desirable to optimize bond strength. The ACI Code specifies that the minimum clear distance between adjacent bars shall not be less than the nominal diameter of the bars, or 1 in. Where reinforcement is placed in two or more layers, the clear distance between layers must not be less than 1 in., and the bars in the upper layer should be placed directly above those in the bottom layer.

To provide the steel with adequate concrete protection against fire and corrosion, the designer must maintain a certain minimum thickness of concrete cover outside the outermost steel. The thickness required will vary, depending upon the type of member and conditions of exposure. According to the ACI Code, for cast-in-place concrete, concrete protection at surfaces not exposed directly to the ground or weather should not be less than $\frac{3}{4}$ in. for slabs and walls and $1\frac{1}{2}$ in. for beams and girders. If the concrete surface is to be exposed to the weather or in contact with the ground, a protective covering of at least 2 in. is required ($1\frac{1}{2}$ in. for No. 5 bars and smaller), except that if the concrete is to be poured in direct contact with the ground, without the use of forms, a covering of at least 3 in. must be furnished.

In general, the centers of bars in beams should be placed at least $2\frac{1}{2}$ in. from the bottom surface of the beam, in order to furnish at least $1\frac{1}{2}$ in. of clear insulation below the bars and the stirrups (see Fig. 3.13). In slabs, 1 in. to the center of the bar is ordinarily sufficient to give the required $\frac{3}{4}$-in. insulation. Although the distances shown in Fig. 3.13 are not the exact distances required in all cases to furnish the clear insulation specified by the Code, they will almost always satisfy the minimum requirements, and they are sufficiently exact for all design purposes. Total depths of beams should be taken in multiples of not less than $\frac{1}{2}$ in., and preferably 1 in.

Recognizing the closer tolerances that can be maintained under plant-control conditions, the Code permits some reduction in concrete protection for reinforcement in precast concrete.

In some cases bars are placed in direct contact with one another, i.e., "bundled," in order to save space. These bars may be assumed to act as a unit, with not more than four in any bundle, provided that stirrups or ties enclose the bundle. Individual bars in a bundle, cut off within the span of flexural members, should terminate at different points. The Code requires at least 40 bar diameters stagger. Where spacing limitations and minimum clear cover are based on bar size, a unit of bundled bars may be treated as a single bar of a

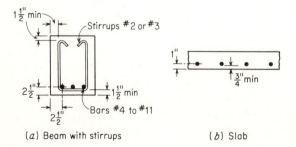

(a) Beam with stirrups   (b) Slab

Fig. 3.13 Requirements for concrete cover.

diameter derived from the total equivalent area. In calculating development length of bundled bars (see Art. 3.6) the Code requires a 20 per cent increase for a three-bar bundle and a 33 per cent increase for a four-bar bundle, in recognition of the reduced effective perimeter in such cases. Bars larger than No. 11 are not to be bundled.

3.8 BAR CUTOFF AND BEND POINTS

Article 3.4 has dealt with moments, stresses, concrete dimensions, and bar areas at the critical-moment sections of beams. The critical-moment sections are at the face of supports (negative bending) and near the middle of the span (positive bending). Occasionally, "haunched" members of variable depth or width are used so that the concrete flexural capacity will agree more closely with the variation in bending moment along a span or series of spans. Usually, however, prismatic members with constant concrete dimensions are used to simplify formwork and reduce cost.

The steel reinforcement, on the other hand, is easily varied in accordance with requirements for flexure, and it is common practice either to cut off bars where they are no longer required to resist stress, or in the case of continuous beams, to bend up the bottom steel (usually at 45°) so that it provides tensile reinforcement at the top of the beam over the supports.

The tensile force to be resisted by the reinforcement at any cross section is

$$T = A_s f_s = \frac{M}{z}$$

where M is the value of bending moment at that section, and z is the internal lever arm of the resisting moment. The lever arm z varies only within narrow limits and is never less than the value obtained at the maximum-moment section. Consequently, the tensile force can be taken with good accuracy directly proportional to the bending moment. Since it is desirable to design so that the steel everywhere in the beam is as nearly fully stressed as possible, it follows that the required steel area is very nearly proportional to the bending moment.

To illustrate, the moment diagram for a uniformly loaded simple-span beam shown in Fig. 3.14a can be used as a steel-requirement diagram. At the maximum-moment section, 100 per cent of the tensile steel is required (zero per cent can be discontinued or bent), while at the supports, zero per cent of the steel is theoretically required (100 per cent can be discontinued or bent). The percentage of bars which could be bent elsewhere along the span is obtainable directly from the moment diagram, drawn to scale. To facilitate the determination of bend points for simple spans, Graph 10 of the Appendix has been prepared. It represents a half-moment diagram for a uniformly loaded simple span.

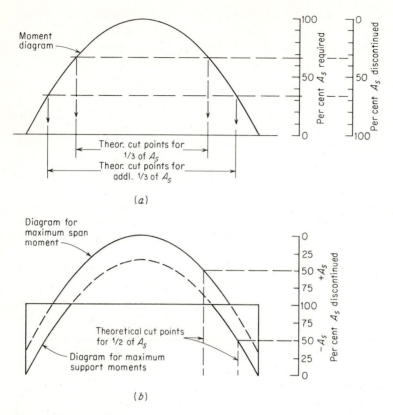

Fig. 3.14 Bar-cutoff points from moment diagrams.

To determine bend points for continuous beams, the moment diagrams resulting from loading for maximum span moment and maximum support moment are drawn. A moment envelope results, which defines the range of values of moment at any section. Bend points can be found from the appropriate moment curve as for simple spans. Figure 3.14b illustrates, for example, a continuous beam with moment envelope resulting from alternate loadings to produce maximum span and maximum support moments. The locations of the points at which 50 per cent of the bottom and top steel may theoretically be bent are shown.

According to the ACI Code, uniformly loaded, continuous reinforced-concrete beams of fairly regular span may be designed using moment coefficients (see Table 7.1). These coefficients, analogous to the numerical constant in the expression $\frac{1}{8}wL^2$ for simple-beam bending moment, give a conservative approximation of span and support moments for continuous beams. When such coefficients are used in design, bend points may conveniently be found from Graph 11 of the Appendix. Moment curves corresponding to the various span- and support-moment coefficients are given at the top and bottom of the chart, respectively.

Actually, in no case should the tensile steel be discontinued exactly at the theoretically described points. As described in Art. 2.5*b* and shown in Fig. 2.18*b*, when diagonal tension cracks form, an internal redistribution of forces occurs in a beam. Prior to cracking, the steel tensile force at any point is proportional to the moment at a vertical section passing through that point. However, after the crack has formed, the tensile force in the steel at the crack is governed by the moment at a section nearer midspan, which may be much larger. Furthermore, the actual moment diagram may differ from that used as a design basis, due to approximation of the real loads, approximations in the analysis, or the superimposed effect of settlement or lateral loads. Recognizing these facts, the ACI Code specifies that every bar should be continued at least a distance equal to the effective depth of the beam or 12 bar diameters (whichever is larger) beyond the point at which it is theoretically no longer required to resist stress.

In addition, it is necessary that the calculated stress in the steel at each section be developed by adequate embedded length or end anchorage, or a combination of the two. For the usual case, with no special end anchorage, this means that the full development length l_d must be provided beyond critical sections at which peak stress exists in the bars. These critical sections are located at points of maximum moment and at points where adjacent terminated reinforcement is no longer needed to resist bending.[1]

Further reflecting the possible change in peak-stress location, the ACI Code requires that at least one-third of the positive-moment steel (one-fourth in continuous spans) must be continued uninterrupted along the same face of the beam a distance at least 6 in. into the support. At least one-third of the total reinforcement provided for negative moment at the support must be extended beyond the extreme position of the point of inflection a distance not less than one-sixteenth the clear span or d or $12d_b$, whichever is greater.

Requirements for bar cutoff or bend-point locations are summarized in Fig. 3.15. If negative bars L are to be cut off, they must extend a full development length l_d beyond the face of the support. In addition, they must extend a distance d or $12d_b$ beyond the theoretical point of cutoff defined by the moment diagram. The remaining negative bars M (at least one-third of the total negative area) must extend at least l_d beyond the theoretical point of

---

[1] The ACI Code is somewhat ambiguous as to whether or not the extension length (d or $12d_b$) is to be added to the required development length (l_d). The Code Commentary presents the view that these requirements need not be superimposed, and Fig. 3.15 has been prepared on that basis. However, the argument just presented regarding possible shifts in moment curves or steel stress distribution curves leads to the conclusion that these requirements should be superimposed. In such case, each bar should be continued a distance l_d plus the greater of d or $12d_b$ beyond the peak stress location. Examples in this chapter and later chapters follow this more conservative interpretation.

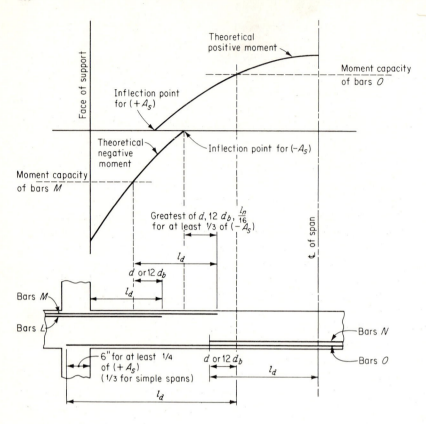

Fig. 3.15 Bar-cutoff requirements.

cutoff of bars L, and in addition must extend d, $12d_b$, or $l_n/16$ (whichever is greatest) past the point of inflection of the negative-moment diagram.

If the positive bars N are to be cut off, they must project l_d past the point of theoretical maximum moment, as well as d or $12d_b$ beyond the cutoff point from the positive-moment diagram. The remaining positive bars O must extend l_d past the theoretical point of cutoff of bars N and must extend at least 6 in. into the face of the support.

When bars are cut off in a tension zone, there is a tendency toward the formation of premature flexural and diagonal tension cracks in the vicinity of the cut end. This may result in a reduction of shear capacity and a loss in overall ductility of the beam. The ACI Code requires special precautions, specifying that no flexural bar shall be terminated in a tension zone unless *one* of the following conditions is satisfied:

1. The shear is not over two-thirds that normally permitted, including allowance for shear reinforcement if any.

2. Stirrups in excess of those normally required are provided over a distance from the point of cutoff equal to $\frac{3}{4}d$. These "binder" stirrups shall provide an area A_v such that $A_v f_y / b_w s$ is not less than 60 psi. In addition, the stirrup spacing shall not exceed $d/8\beta_b$, where β_b is the ratio of the area of bars cut to the total area of bars at the section.

3. The continuing bars provide twice the area required for flexure at that point, and the shear does not exceed three-quarters that permitted.

As an alternative to cutting off the steel, tension bars may be anchored by bending them across the web and making them continuous with the reinforcement on the opposite face. Although this leads to some complication in detailing the steel, many engineers prefer the arrangement because added insurance is provided against the spread of diagonal tension cracks (see Art. 3.5). Such a bent bar may be considered effective, in satisfying the requirements of extension, to the point at which it crosses the middepth $d/2$ of the member. Thus bars may be bent a distance $d/2$ ahead of the cutoff points shown in Fig. 3.15.

In some cases, particularly for relatively deep beams in which a large percentage of the total bottom steel is to be bent, it may be impossible to locate the bend-up point for bottom bars far enough from the support for the same bars to meet the requirements for top steel. The theoretical points of bend should be checked carefully for both bottom and top steel.

Since the determination of bend points may be a rather tedious matter, particularly for frames which have been analyzed by elastic methods rather than by moment coefficients, many designers specify that bars be bent at more or less arbitrarily defined points which experience has indicated are safe. For nearly equal spans, uniformly loaded, in which not more than about one-half the tensile steel is to be bent, the bend locations shown in Fig. 3.16 are satisfactory.

It was shown in Art. 2.6 that the use of a larger number of small bars to provide a given A_s, rather than a smaller number of large bars, was advantageous in reducing the intensity of bond stress U per unit length along the steel. While present design methods emphasize development length rather than calculation of bond-stress intensity, the use of small bars is encouraged by the

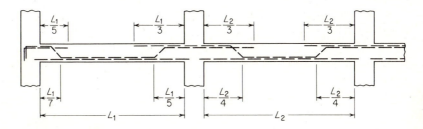

Fig. 3.16 Approximate bend points.

Code with the requirement that, at simple supports and at points of inflection, positive-moment-tension reinforcement shall be limited to a diameter such that l_d does not exceed

$$l_d \leq \frac{M_t}{V_u} + l_a \tag{3.31}$$

where M_t = computed flexural strength assuming all reinforcement at section to be stressed to f_y

V_u = maximum applied shear at section

l_a = embedded length of steel beyond center of support or past point of inflection (limited to greater of d or $12d_b$ at a point of inflection).

The value of M_t/V_u in Eq. (3.31) may be increased 30 per cent when the ends of the reinforcement are confined by a compressive reaction, such as at the end of a conventionally supported simple-span beam, in recognition of the beneficial effect of the reaction in preventing splitting of the beam.

The meaning of Eq. (3.31) may be clarified by Fig. 3.17, which shows shear and moment diagrams at ultimate load for a uniformly loaded continuous beam. From the inflection point, the distance a at which the bars in question must be fully stressed is defined by the intersection of the line for the calculated M_t with the actual moment diagram. Since the change in moment in the distance a is equal to the area under the shear diagram in that distance, a conservatively low estimate of a is M_t/V_u. Thus, if the bars are to be fully effective, they must be of sufficiently small diameter so that their development length l_d is contained within the distance a plus an arbitrarily allowed extension l_a equal to d or $12d_b$ beyond the point of inflection.

Although the intent of this Code provision (to encourage the use of smaller bars) cannot be faulted, the arbitrary definition of l_a (which disregards the actual extension of the steel into the compression zone in the negative bending region of the beam) cannot be justified in physical terms.

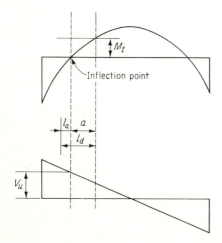

Fig. 3.17   Basis of bar-diameter limitation.

3.9 INTEGRATED DESIGN EXAMPLE

A floor system consists of single-span T beams 8 ft on centers, supported by 12-in. masonry walls spaced at 25 ft between inside faces. The general arrangement is shown in Fig. 3.18a. A 5-in. monolithic slab carries a uniformly distributed service live load of 165 psf. The T beams, in addition to the slab load and their own weight, must carry two 16,000-lb equipment loads applied 3 ft from the span center line as shown. A complete design is to be provided for the T beams, using concrete of 4000-psi strength and bars with 60,000-psi yield stress.

According to the Code, the span length is to be taken as the clear span plus the beam depth, but need not exceed the distance between centers of the supports. The latter provision controls in this case, and the effective span is 26 ft. Estimating the

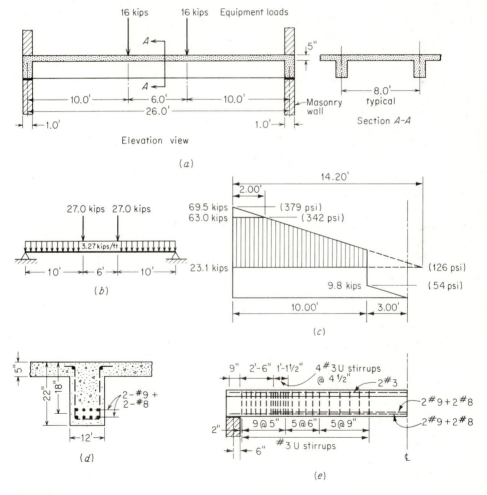

Fig. 3.18

beam-web dimensions to be 12×24 in., the calculated and factored dead loads are

Slab: $5/12 \times 150 \times 7 = 440$ plf

Beam: $\dfrac{12 \times 24}{144} \times 150 = 300$

$$D = 740 \text{ plf}$$
$$1.4D = 1040 \text{ plf}$$

The uniformly distributed live load is

$$L = 165 \times 8 = 1310 \text{ plf}$$
$$1.7L = 2230 \text{ plf}$$

Live-load overload factors are applied to the two concentrated loads to obtain $P_u = 16{,}000 \times 1.7 = 27{,}000$ lb. Factored loads are summarized in Fig. 3.18b.

In lieu of other controlling criteria, the beam-web dimensions will be selected on the basis of shear. The left and right reactions under factored load are $27.0 + 3.27 \times 13 = 69.5$ kips. With the effective beam depth estimated to be 20 in., the maximum shear which need be considered in design is $69.5 - 3.27(0.50 + 1.67) = 62.4$ kips. Although the ACI Code permits $(v_u - v_c)$ as high as $8\sqrt{f_c'}$, this would require very heavy web reinforcement. A lower limit of $4\sqrt{f_c'}$ will be adopted. With $v_c = 2\sqrt{f_c'}$, this results in a maximum $v_u = 6\sqrt{f_c'} = 380$ psi. Then $b_w d = V_u/\phi v_u = 62{,}400/0.85 \times 380 = 194$ in.^{2} Cross-sectional dimensions $b_w = 12$ in. and $d = 18$ in. are selected, providing a total beam depth of 22 in. The assumed dead load of the beam need not be revised.

According to the Code, the effective flange width b is the smallest of the three quantities

$$\frac{L}{4} = \frac{26 \times 12}{4} = 78 \text{ in.}$$
$16h_f + b_w = 80 + 12 = 92$ in.
℄ spacing $= 96$ in.

The first controls in this case. The maximum moment is at midspan, where

$$M_u = \tfrac{1}{8} \times 3.27 \times 26^2 + 27.0 \times 10 = 546 \text{ ft-kips}$$

Assuming for trial that the stress-block depth will equal the slab thickness,

$$A_s = \frac{M}{\phi f_y(d - a/2)} = \frac{546 \times 12}{0.90 \times 60 \times 15.5} = 7.82 \text{ in.}^2$$

Then

$$a = \frac{A_s f_y}{0.85 f_c' b} = \frac{7.82 \times 60}{0.85 \times 4 \times 78} = 1.78 \text{ in.}$$

The stress-block depth is seen to be less than the slab depth; hence, rectangular-beam equations are valid. An improved estimate of A_s is

$$A_s = \frac{546 \times 12}{0.90 \times 60 \times 17.11} = 7.10 \text{ in.}^2$$

A check by Eqs. (3.2) and (3.3) confirms that this is well below the maximum permitted steel ratio. Four No. 9 plus four No. 8 bars will be used, providing a total area of 7.14 in.^{2} They will be arranged in two rows, as shown in Fig. 3.18d, with No. 9 bars at the outer end of each row.

While the Code permits discontinuation of two-thirds of the longitudinal reinforcement for simple spans, in the present case it is convenient to discontinue only the upper layer of steel, consisting of one-half of the total area. The moment capacity of the member after half the bars have been discontinued is then found:

$$a = \frac{3.57 \times 60}{0.85 \times 4 \times 78} = 0.81 \text{ in.}$$

$$M_u = \phi A_s f_y \left(d - \frac{a}{2} \right) = 0.90 \times 3.57 \times 60 \times 17.60 \times \tfrac{1}{12} = 283 \text{ ft-kips}$$

For the present case, with a moment diagram resulting from combined distributed and concentrated loads, the point at which the applied moment is equal to this amount must be calculated. (In the case of uniformly loaded beams, Graphs 10 and 11 of the Appendix are helpful). If x is the distance from the support center line to the point at which the moment is 283 ft-kips, then

$$69.5x - \frac{3.27x^2}{2} = 283$$
$$x = 4.50 \text{ ft}$$

The bars must be continued at least d or $12d_b$, 1.50 ft or 1.13 ft, beyond this theoretical point of cutoff. In addition, the full development length l_d plus d or $12d_b$ will be provided past the maximum-moment section at which peak stress occurs in the bars to be cut. The length l_d is found from Table 8 of the Appendix to be 38 in., or 3.17 ft. Thus the bars will be continued at least $3.17 + 1.50 = 4.67$ ft past the midspan point, but in addition must continue to a point $4.50 - 1.50 = 3.00$ ft from the support center line. The second requirement controls, and the upper bars will be terminated, as shown in Fig. 3.18e.

The full length l_d plus extension, totaling 4.67 ft, will also be provided for the continuing steel past the theoretical cutoff point for the discontinued bars. The lower bars will be continued 9 in. into the support as shown, providing 4.75 ft.

Checking by Eq. (3.31) to ensure that the continuing steel is of sufficiently small diameter,

$$l_d \leq \frac{M_t}{V_u} + l_a$$
$$\leq 283 \times 12/69.5 + 3 = 52 \text{ in.}$$

The actual l_d of 38 in. meets this restriction.

Since the cut bars are located in the tension zone, special binding stirrups will be used to control cracking; these will be selected after the normal shear reinforcement has been determined.

The shear diagram resulting from application of factored loads is shown in Fig. 3.18c. The shear force V_u is converted to nominal unit shear stress v_u by dividing all ordinates by $\phi b_w d = 0.85 \times 12 \times 18 = 184$ in.^{2}, and the results shown in parentheses. The shear contribution of the concrete is

$$v_c = 2 \sqrt{f_c'} = 2 \sqrt{4000} = 126 \text{ psi}$$

Thus web reinforcement must be provided for that part of the shear diagram shown shaded.

No. 3 stirrups will be selected. The maximum spacings must not exceed $d/2 = 9$ in., 24 in., or $A_v f_y/50b_w = 0.22 \times 60,000/50 \times 12 = 22$ in. The first criterion controls here. For reference, from Eq. (3.26a), the hypothetical stirrup spacing at the support is

$$s_0 = \frac{0.22 \times 60,000}{253 \times 12} = 4.35 \text{ in.}$$

and at 2-ft intervals along the span

$s_2 = 5.06 \text{ in.}$ $s_8 = 9.96 \text{ in.}$
$s_4 = 6.05 \text{ in.}$ $s_{10} = 14.70 \text{ in.}$
$s_6 = 7.53 \text{ in.}$

The spacing need not be closer than that required 2.00 ft from the support center line. In addition, stirrups are not required past the point of application of concentrated load, since beyond that point the shear is less than half of v_c. The final spacing of vertical stirrups selected is

1 space at 2 in. =   2 in.
9 spaces at 5 in. =   45 in.
5 spaces at 6 in. =   30 in.
5 spaces at 9 in. =   45 in.
           Total = 122 in. = 10 ft 2 in.

For convenience during construction, two No. 3 longitudinal bars will be added to fix the top of the stirrups.

In addition to the shear reinforcement just specified, it is necessary to provide extra web reinforcement over a distance equal to $\frac{3}{4}d$, or 13.5 in., from the cut ends of the discontinued steel. The spacing of this extra web reinforcement must not exceed $d/8\beta_b = \dfrac{18}{8 \times \frac{1}{2}} = 4.5$ in. In addition, the area of added steel within the distance s must not be less than $60b_w s/f_y = 60 \times 12 \times 4.5/60,000 = 0.054 \text{ in.}^2$ For convenience, No. 3 stirrups will be used for this purpose also, providing an area of 0.22 in.^{2} in the distance s. The placement of the four extra stirrups is shown in Fig. 3.18d.

3.10 BAR SPLICES IN FLEXURAL MEMBERS

In general, reinforcing bars are stocked by fabricators in lengths of 60 ft. For this reason, and because it is often more convenient to work with shorter bar lengths, it is frequently necessary to splice bars in the field. Splices in reinforcement at points of maximum stress should be avoided whenever possible, and splices where used should be staggered.

Splices for No. 11 bars and smaller are most often made simply by lapping the bars a sufficient distance to transfer stress by bond from one bar to the other. The lapped bars are usually placed in contact, and lightly wired so that they stay in position as the concrete is poured.

Alternatively, splicing may be accomplished by welding or by sleeves or mechanical devices which provide full positive connection between the bars. According to the Code, welded splices must be butted and welded, and other full positive connections made, so that the connection will develop in tension at least 125 per cent of the specified yield strength of the bar. Connections not

Table 3.5 Lap splices for tension bars

Class	Description		Minimum lap length
	Splices in regions where f_s equals or exceeds $0.50f_y$	*Splices in regions where f_s is less than $0.50f_y$*	
A		No more than three-fourths of bars spliced within required lap length	$1.0l_d$
B	No more than one-half of bars spliced within required lap length	More than three-fourths of bars spliced within required lap length	$1.3l_d$
C	More than one-half of bars spliced within required lap length		$1.7l_d$
D	Splices in tension tie members; bars must be enclosed within spirals not less than $\frac{1}{4}$ in. in diameter and not more than 4 in. in pitch; ends of bars larger than No. 4 must be hooked 180°.		$2.0l_d$

meeting this requirement may be used at points of less than maximum tensile stress.

The required length of lap for tension splices, established by test, may be stated in terms of the development length l_d for the reinforcement (see Art. 3.6). Different classifications of lap splices in tension have been established, depending upon the type of member, the proportion of total steel to be spliced, and the location of the splice relative to the point of maximum tensile stress. Code requirements for each class are summarized in Table 3.5. Lap splices of bundled bars are based on the same standards as summarized in Table 3.4, but must be increased in length by 20 per cent for three-bar bundles and by 33 per cent for four-bar bundles, because of the reduced effective perimeter.

Compression bars may be spliced by lapping, by direct end bearing, or by welding or mechanical devices which provide full positive connection. According to the Code, the minimum length of lap for compression splices is l_d, but must not be less than $0.0005f_yd_b$ for f_y of 60,000 psi or less or $(0.0009f_y - 24)d_b$ for f_y greater than 60,000 psi. If lateral ties are used having an area of at least $0.0015hs$, these lengths may be reduced by 17 per cent. If spiral reinforcement confines the splice, the lengths required may be reduced by 25 per cent, according to the Code. In no case is a compression splice to be less than 12 in. in length.

Direct end bearing of the bars has been found by test and experience to be an effective means for transmitting compression. In such a case, the bars must be held in proper alignment by a suitable device, and the bar ends must terminate in flat surfaces within specified tolerances (within $1\frac{1}{2}°$ of a right angle according to the ACI Code). Ties, closed stirrups, or spirals must be used.

If compression bars are spliced by welding or other positive means, the connection must be such as to develop 125 per cent of the specified yield strength of the bars, as for tension splices.

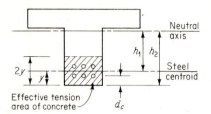

Fig. 3.19

3.11 CONTROL OF CRACKING

With the increased use of steel with yield stress above 40 ksi, the question of cracking of the tensile portions of reinforced-concrete beams is receiving more attention than it has in the past. The primary concern is not the number of cracks which form, but rather the width of those cracks. It is desirable to minimize the crack width (1) from the point of view of appearance and (2) because of the detrimental corrosion of the reinforcing steel that may occur if it is exposed by excessively wide cracks.

Present methods of crack control are based largely on experimental studies. These have demonstrated that:

1. Crack widths are minimized through use of deformed bars.
2. The maximum width of crack due to load is approximately proportional to stress in the reinforcement.
3. Flexural crack widths are minimized if the reinforcement is well distributed over the concrete tension zone.
4. Crack width at the surface of the concrete is proportional to the amount of concrete cover provided for the bars.

It follows from 3 and 4 that a larger number of smaller bars, well distributed in the tension zone of the cross section, is more effective in minimizing crack width than a smaller number of larger bars having the same total area.

Based on their research at Cornell University (Ref. 3.6), Gergely and Lutz proposed the following equation for predicting the maximum width of crack at the tension face of a beam:

$$w = 0.076 R f_s \sqrt[3]{d_c A} \tag{3.32}$$

in which w is the maximum width of crack, in thousandth-inches, and f_s is the steel stress at the load for which crack width is to be determined, measured in ksi. The geometric parameters are shown in Fig. 3.19 and are as follows:

d_c = thickness of concrete cover measured from tension face to center of bar closest to that face, in.

R = ratio of distances from tension face and from steel centroid to neutral axis, equal to h_2/h_1

A = concrete area surrounding one bar, equal to total effective tension area of concrete surrounding reinforcement and having same centroid, divided by number of bars, in.^{2}

In view of the random nature of the cracking process and the wide scatter of maximum-crack-width measurements, even under laboratory conditions, excessive precision in calculating crack width is not justified. Accordingly, Eq. (3.32) may be simplified by adopting a representative value of R equal to 1.2. Then a parameter z may be defined:

$$z = f_s \sqrt[3]{d_c A} \qquad\qquad (3.33)$$

in which

$$z = \frac{w}{0.076 \times 1.2} = \frac{w}{0.091}$$

Control of maximum crack width may be thus obtained by setting an upper limit on the parameter z. The ACI Code specifies that z shall not exceed 175 for interior exposure and 145 for exterior exposure. These limits correspond to maximum crack widths of 0.016 and 0.013 in. respectively. In addition, the Code specifies that only deformed reinforcement shall be used and that tension reinforcement shall be well distributed in the zone of maximum concrete tension. It further states that designs shall not be based on a yield strength f_y in excess of 80,000 psi.

When concrete flanges are in tension, as in the negative-moment region of continuous T beams, concentration of the reinforcement over the web may result in excessive crack width in the slab adjacent to the web, even though cracks directly over the web are fine and well distributed. To prevent this, the tensile reinforcement should be distributed over the width of the flange, rather than concentrated. However, because of shear lag, the outer bars in such a distribution would be considerably less highly stressed than those directly over the web, producing an uneconomical design. As a reasonable compromise, the ACI Code requires that the tension reinforcement in such cases should be distributed over the effective flange width or a width equal to one-tenth the span, whichever is smaller. If the effective flange width exceeds one-tenth the span, some longitudinal reinforcement must be provided in the outer portions of the flange. The amount of such additional reinforcement is left to the discretion of the designer; it should at least be the equivalent of temperature reinforcement for the slab (see Art. 4.3), and is often taken as twice that amount.

Example: check for satisfactory crack widths   The T beam of Example 1, Art. 3.4c, carries a service-load moment of 5850 in.-kips. Estimate the maximum width of crack to be expected at the bottom surface of the member at full service load and determine if reinforcing details, with respect to cracking, are satisfactory for exterior exposure according to the ACI Code.

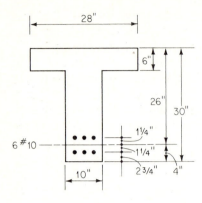

Fig. 3.20

The geometry of the T beam of Example 1 is shown in Fig. 3.20. The total tensile-steel area provided by the six No. 10 bars is 7.59 in.² The steel stress at service load can be estimated closely by taking the internal lever arm equal to the distance $(d - t/2)$:

$$f_s = \frac{M_s}{A_s(d - t/2)} = \frac{5850}{7.59 \times 23} = 33.6 \text{ ksi}$$

(Alternatively, the ACI Code permits using $f_s = 0.60f_y$, giving 36.0 ksi.) The distance from the steel centroid to the tensile face of the beam is 4 in.; hence, the total effective concrete area for purposes of cracking calculations is $4 \times 2 \times 10 = 80$ in.² and

$$A = \tfrac{80}{6} = 13.30 \text{ in.}^2$$

Then, by Eq. (3.32), with $d_c = 2\tfrac{3}{4}$ in.,

$$w = 0.076 \times 1.2 \times 33.6 \sqrt[3]{2.75 \times 13.30}$$
$$= 10 \text{ thousandth-in.} = 0.010 \text{ in.}$$

Alternatively, by Eq. (3.33),

$$z = 33.6 \sqrt[3]{2.75 \times 13.30} = 112$$

well below the limit of 145 imposed by the Code for exterior construction. If the results had been unfavorable, a redesign using a larger number of smaller-diameter bars, thus reducing the value of A, would have been indicated.

3.12 CONTROL OF DEFLECTION

Increased use of refined design methods, together with use of higher-strength materials, has resulted in members that are more slender than those commonly built in the past. The deflection under load of such members may in some cases exceed desirable limits. While in the past deflection control was obtained indirectly by limiting stresses for steel and concrete under service load, such methods are no longer adequate, and in many cases an explicit check of deflection should be made.

The designer may wish to calculate deflection of a member under one or several loading conditions. For example, it may be important to know the maximum instantaneous deflection under dead load plus full service live load. Or it may be that only the live-load deflection will be of interest. On the other hand, it may be of greatest significance to know the long-term deflection of a member under sustained load. Usually, the specified service live load will act only infrequently on a member, and the sustained-load deflection should be calculated for dead load plus some fraction of the live load. The fractional part of the live load to be considered depends upon the type of occupancy. For example, in calculating long-term deflections for a beam carrying the floor of a residential apartment, the designer might reasonably consider about 20 per cent of the live load to be sustained. Longtime deflection for a storage warehouse, on the other hand, should probably be based on 100 per cent of design live load.

Methods for predicting instantaneous and longtime deflections were discussed in Art. 2.8. Instantaneous deflections can be calculated by the usual equations of mechanics and strength of materials, the only difficulty being in the choice of a suitable value for the moment of inertia for the member, which will normally be partially cracked at service load.

The cracking moment for a beam of given cross section can be calculated by Eq. (2.60). The moment of inertia of the gross concrete section neglecting reinforcement, I_g, may be used in place of I_{ut} without serious error, in which case Eq. (2.60) becomes

$$M_{cr} = \frac{f_r I_g}{y_t} \tag{3.34}$$

where y_t is the distance from neutral axis to tension face, and f_r is the modulus of rupture. The modulus of rupture f_r for normal concrete is taken as $7.5 \sqrt{f'_c}$. For lightweight-aggregate concretes the modulus of rupture may be calculated using $f_{ct}/6.7$ for $\sqrt{f'_c}$, where the split cylinder strength f_{ct} is determined by test. The value of $f_{ct}/6.7$ should not be taken larger than $\sqrt{f'_c}$, however. In lieu of test information on tensile strength, f_r may be calculated by the usual equation multiplied by 0.75 for "all-lightweight" concrete and 0.85 for "sand-light-weight" concrete.

The effective moment of inertia for a partially cracked beam may be approximated satisfactorily by Eq. (2.62). Substituting I_g for I_{ut} as before and adopting the notation I_{cr} rather than I_{ct} for the moment of inertia of the cracked transformed section, Eq. (2.62) becomes

$$I_e = \left(\frac{M_{cr}}{M_{max}}\right)^3 I_g + \left[1 - \left(\frac{M_{cr}}{M_{max}}\right)^3\right] I_{cr} \leq I_g \tag{3.35}$$

For continuous spans, it is satisfactory to use the average of values obtained from Eq. (3.35) for the positive- and negative-moment regions.

The effect of longtime loads can be estimated using Eq. (2.63), which gives the multiplier λ, to be applied to instantaneous deflection to estimate the value of *additional* longtime deflection:

$$\lambda = 2 - 1.2 \left(\frac{A'_s}{A_s} \right) \geq 0.6 \qquad\qquad (2.63)$$

The presence of compressive reinforcement is seen to be significant in reducing longtime deflection, and it may sometimes be included for this reason alone. For continuous beams, in which different values of λ will ordinarily be obtained at positive- and negative-bending regions, an average value may be used.

To ensure satisfactory deflection characteristics, the ACI Code imposes certain limits on deflections computed according to the above procedures. These limits are given in Table 3.6. The last two limits given may be exceeded under special conditions.

In lieu of explicit calculation of deflections, for members of lesser importance it is satisfactory to control deflection indirectly by limiting the span-depth ratio. According to the Code, beams and one-way slabs must not have thicknesses less than those specified by Table 3.7 unless deflections are calculated. For members using lightweight concrete, having unit weight in the range of from 90 to 120 pcf, the values of Table 3.7 should be multiplied by $(1.65 - 0.005w) \geq 1.09$, where w is the unit weight in lb per ft³. For yield strengths other than 60 ksi, the values of Table 3.7 should be multiplied by $(0.4 + 0.01f_y)$, where f_y is expressed in ksi.

Table 3.6   Maximum allowable computed deflections

Type of member	*Deflection to be considered*	*Deflection limitation*
Flat roofs not supporting or attached to nonstructural elements likely to be damaged by large deflections	Immediate deflection due to the live load L	$\dfrac{l}{180}$
Floors not supporting or attached to nonstructural elements likely to be damaged by large deflections	Immediate deflection due to the live load L	$\dfrac{l}{360}$
Roof or floor construction supporting or attached to nonstructural elements likely to be damaged by large deflections	That part of the total deflection which occurs after attachment of the nonstructural elements, the sum of the longtime deflection due to all sustained loads and the immediate deflection due to any additional live load	$\dfrac{l}{480}$
Roof or floor construction supporting or attached to nonstructural elements not likely to be damaged by large deflections		$\dfrac{l}{240}$

Table 3.7 Minimum thickness of beams or one-way slabs unless deflections are computed

| Member | Minimum thickness, h | | | |
| | Simply supported | One end continuous | Both ends continuous | Cantilever |
	Members not supporting or attached to partitions or other construction likely to be damaged by large deflections			
Solid one-way slabs	$l/20$	$l/24$	$l/28$	$l/10$
Beams or ribbed one-way slabs	$l/16$	$l/18.5$	$l/21$	$l/8$

Example: deflection calculation   The beam shown in Fig. 3.21a is a part of the floor system of an apartment house and is designed to carry calculated dead load D of 1.1 klf and a service live load L of 2.2 klf. Of the total live load, 20 per cent is sustained in nature,

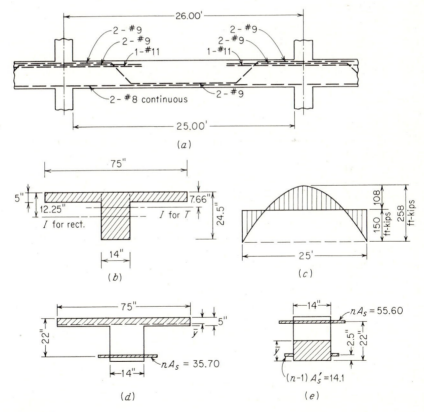

Fig. 3.21

while 80 per cent will be applied only intermittently over the life of the structure. Under full dead and live load (D plus L), the moment diagram is as shown in Fig. 3.21c. The beam will support nonstructural partitions which would be damaged if large deflections were to occur. They will be installed shortly after construction shoring is removed and dead loads take effect. It is required to calculate that part of the total deflection that would adversely affect the partitions, i.e., the sum of longtime deflection due to dead and partial life load plus the immediate deflection due to the non-sustained part of the live load. Materials having strength $f'_c = 2500$ psi and $f_y = 40$ ksi will be used.

For the specified materials $E_c = 57,000 \sqrt{2500} = 2.85 \times 10^6$ psi, and with $E_s = 29 \times 10^6$ psi, the modular ratio $n = 10$. The modulus of rupture $f_r = 7.5 \sqrt{2500} = 375$ psi. The effective moment of inertia will be calculated for the moment diagram shown in Fig. 3.21c, corresponding to full service load, on the basis that the extent of cracking will be governed by full service load, even though that load is intermittent. In the positive-moment region, the centroidal axis of the uncracked T section of Fig. 3.21b is found, by taking moments about the top surface, to be at 7.66 in. depth, and $I_g = 33,160$ in.[4] By similar means, the centroidal axis of the cracked transformed T section shown in Fig. 3.21d is located 4.14 in. below the top of the slab and $I_{cr} = 13,180$ in.[4] The cracking moment is then found by means of Eq. (3.34).

$$M_{cr} = 375 \times 33,160/16.84 \times 12,000$$
$$= 62 \text{ ft-kips}$$

With $M_{cr}/M_{max} = \frac{62}{108} = 0.574$, the effective moment of inertia in the positive bending region is found from Eq. (3.35) to be

$$I_e = 0.574^3 \times 33,160 + (1 - 0.574^3) \times 13,180$$
$$= 16,970 \text{ in.}[4]$$

In the negative bending region, the gross moment of inertia will be based on the rectangular section shown dotted in Fig. 3.21b. For this area, the centroid is 12.25 in. from the top surface and $I_g = 17,200$ in.[4] For the cracked transformed section shown in Fig. 3.21e the centroidal axis is found, taking moments about the bottom surface, to be 9.33 in. from that level, and $I_{cr} = 13,368$ in.[4] Then

$$M_{cr} = 375 \times 17,200/12.25 \times 12,000 = 44 \text{ ft-kips}$$

giving $M_{cr}/M_{max} = \frac{44}{150} = 0.293$. Thus, for the negative-moment regions,

$$I_e = 0.293^3 \times 17,200 + (1 - 0.293^3) \times 13,368$$
$$= 13,530 \text{ in.}[4]$$

The average value of I_e to be used in calculation of deflection is

$$I_{e,avg} = \tfrac{1}{2}(16,970 + 13,530) = 15,250 \text{ in.}[4]$$

It is next necessary to calculate the sustained-load deflection multipliers given by Eq. (2.64). For the positive bending zone, with no compression reinforcement, $\lambda_{pos} = 2.00$, while for the negative zone, with $A'_s/A_s = 1.57/5.56 = 0.283$, the multiplier is $\lambda_{neg} = 2.00 - 1.2 \times 0.283 = 1.66$, giving an average value of

$$\lambda_{avg} = \tfrac{1}{2}(2.00 + 1.66) = 1.83$$

For convenient reference, the deflection of the member under full dead plus live load of 3.3 klf, corresponding to the moment diagram of Fig. 3.21c, will be found.

Making use of the moment-area principles,

$$\delta_{D+L} = \frac{1}{EI} [(\tfrac{2}{3} \times 258 \times 12.5 \times \tfrac{5}{8} \times 12.5) - (150 \times 12.5 \times 6.25)]$$

$$= \frac{5100}{EI}$$

$$= \frac{5100 \times 1728}{2850 \times 15{,}250} = 0.203 \text{ in.}$$

Using this figure as a basis, the time-dependent portion of dead-load deflection (the only part of the total which would affect the partitions) is

$$\delta_D = 0.203 \times \frac{1.1}{3.3} \times 1.83 = 0.124 \text{ in.}$$

while the sum of the immediate and time-dependent deflection due to the sustained portion of the live load is

$$\delta_{0.20L} = 0.203 \times \frac{2.2}{3.3} \times 0.20 \times 2.83 = 0.077 \text{ in.}$$

and the instantaneous deflection due to application of the short-term portion of the live load is

$$\delta_{0.80L} = 0.203 \times \frac{2.2}{3.3} \times 0.80 = 0.108 \text{ in.}$$

Thus the total deflection which would adversely affect the partitions, from the time they are installed until all longtime and subsequent instantaneous deflections have occurred, is

$$\delta = 0.124 + 0.077 + 0.108 = 0.309 \text{ in.}$$

For comparison, the limitation imposed by the ACI Code in such circumstances is $l/480 = 26 \times \frac{12}{480} = 0.650$ in., indicating that the stiffness of the proposed member is sufficient.

It may be noted that relatively little error would have been introduced in the above solution if the cracked-section moment of inertia had been used for both positive and negative sections, rather than I_e. Significant saving in computational effort would have resulted. If M_{cr}/M_{max} is less than $\tfrac{1}{3}$, use of I_{cr} would almost always be acceptable. It should be noted further that computation of moment of inertia for both uncracked and cracked sections is greatly facilitated by design aids such as are included in Ref. 3.2.

3.13 DESIGN FOR TORSION

Until recently, torsion generally was not taken into account in designing concrete structures. Safety factors incorporated in ordinary design procedures were sufficient to accommodate the effects of torsion in all but special cases, such as curved beams or spiral stairways. More refined design methods now permit reduced overall safety factors, with the result that torsion must, in many cases, be accounted for explicitly. The 1971 edition of the ACI Code includes, for the first time, specific recommendations for torsional design.

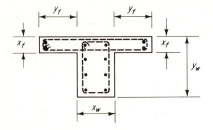

Fig. 3.22 Basis of torsional section properties and typical reinforcement.

The basic principles upon which Code provisions are based have been discussed in Art. 2.7. For rectangular members, the maximum tensile or shear stress resulting from torsion is given by Eq. (2.43). For flanged sections, the torsional strength may be taken conservatively as the sum of the torsional strengths of the web and the projecting flanges. Accordingly, the term x^2y in Eq. (2.43) is replaced by Σx^2y, where x and y are, respectively, the smaller and larger side dimensions of each of the component rectangles of the section, partitioned as shown in Fig. 3.22. For torsion, tests indicate that the effective width of the projecting flanges should not be larger than 3 times the thickness.

If the ultimate torque produced by factored loads is T_u, the nominal design torsional stress v_{tu} is

$$v_{tu} = \frac{3T_u}{\phi \Sigma x^2 y} \tag{3.36}$$

in which the capacity reduction factor ϕ is taken equal to 0.85 for torsion, as for shear. According to the Code, torsional effects may be neglected if v_{tu} calculated according to Eq. (3.36) is no larger than $1.5\sqrt{f'_c}$; otherwise specific consideration of torsion must be included.

Ordinarily, reinforced-concrete members subject to torsion are subject to shear and bending stress as well. It was found experimentally, as discussed in Art. 2.7c, that in such cases a circular interaction diagram conservatively relates shear capacity and torsion capacity. By analogy, the torsional shear stress and flexural shear stress contributed by the partially cracked concrete in members with stirrups were related by a similar circular interaction curve. This produced Eqs. (2.56) and (2.58). With simplifying notational changes appropriate for design, these equations may be written

$$v_{tc} = \frac{2.4\sqrt{f'_c}}{\sqrt{1 + (1.2v_u/v_{tu})^2}} \tag{3.37}$$

$$v_c = \frac{2.0\sqrt{f'_c}}{\sqrt{1 + (v_{tu}/1.2v_u)^2}} \tag{3.38}$$

in which v_{tc} and v_c are, respectively, the nominal torsion stress carried by the concrete and the nominal shear stress carried by the concrete, and v_{tu} and v_u

are, respectively, the nominal total torsional stress and the nominal total shear stress for which the member must be designed. It is useful to note, for any particular ratio of v_{tu}/v_u, that

$$\frac{v_c}{v_u} = \frac{v_{tc}}{v_{tu}}$$

If the torque T_u produces higher torsion stress v_{tu} than can be carried by the concrete according to Eq. (3.37), then reinforcement must be provided to carry the excess torsion stress $(v_{tu} - v_{tc})$, just as stirrups are provided to carry the excess shear stress $(v_u - v_c)$. Even if torsion reinforcement is provided, in order to ensure ductile rather than brittle behavior if failure should occur, the nominal torsion stress is limited to

$$v_{tu} \leq \frac{12\sqrt{f_c'}}{\sqrt{1 + (1.2v_u/v_{tu})^2}} \qquad (3.39)$$

according to the ACI Code.

Torsion reinforcement comprises two types: transverse stirrups and additional longitudinal reinforcement. Both must be provided, for reasons discussed in Art. 2.7b.

The transverse stirrups used for torsional reinforcement *must be of closed form*, as shown in Fig. 3.22, since principal tensile stress results on each of the four faces of a beam in torsion. U-shaped stirrups commonly used for transverse shear reinforcement are not suitable for torsional reinforcement. Good anchorage is provided by hooking the stirrup bar ends around the longitudinal reinforcement as shown. Alternatively, pairs of U stirrups may be placed so as to form a closed tie with vertical legs lapped at least $2l_d$ (see Table 3.2). If flanges are included in the computation of torsional strength of T or L beams, supplementary slab reinforcement should be provided, as shown in Fig. 3.22.

The required spacing of closed stirrups is given by Eq. (2.57). For T or L sections, x^2y in that equation is replaced by Σx^2y. With simplifying notational changes as made earlier in this article, Eq. (2.57) can be written

$$A_t = \frac{(v_{tu} - v_{tc})s\Sigma x^2 y}{3\alpha_t x_1 y_1 f_y} \qquad (3.40)$$

where A_t = area of *one leg* of closed stirrup resisting torsion within distance s
  x_1, y_1 = shorter and longer center-to-center stirrup dimensions, respectively
    α_t = a coefficient given by Eq. (2.47):

$$\alpha_t = 0.66 + 0.33 \left(\frac{y_1}{x_1}\right) \leq 1.50 \qquad (2.47)$$

In order properly to control spiral cracking, the maximum spacing of torsional stirrups should not exceed $(x_1 + y_1)/4$ or 12 in., whichever is smaller. In addition, for members requiring both shear and torsion reinforcement, the minimum area of closed stirrups must be such that $A_v + 2A_t \geq 50b_w s/f_y$, according to the ACI Code.

The longitudinal-bar area required by torsion is given by Eq. (2.51),

$$A_l = 2A_t \frac{x_1 + y_1}{s} \qquad (2.51)$$

or by

$$A_l = \left(\frac{400xs}{f_y} \frac{v_{tu}}{v_{tu} + v_u} - 2A_t\right)\frac{x_1 + y_1}{s}$$

whichever is greater. The value of A_l computed by the latter equation need not exceed that obtained by substituting $50b_w s/f_y$ for $2A_t$, according to the Code. The spacing of the longitudinal bars should not exceed 12 in. The bars should be well distributed around the perimeter of the cross section to control cracking. According to the Code, the bars must not be less than No. 3 in size, and at least one bar must be placed in each corner of the stirrups. Careful attention must be paid to the anchorage of longitudinal torsional reinforcement so that it is able to develop its yield strength at the face of the supporting columns, where torsional moments are often maximum.

Reinforcement required for torsion may be combined with that required for other forces, provided that the area furnished is the sum of the individually required areas and that the most restrictive requirements for spacing and placement are met. According to the Code, torsional reinforcement must be provided at least a distance $d + b$ beyond the point theoretically required. Sections located less than a distance d from the face of a support may be designed for the same torsional stress as that computed at a distance d, recognizing the beneficial effects of support compression.

The effect of axial tension on torsional resistance has not been studied experimentally. Conservatively, the designer may provide torsional reinforcement to carry the total torque, disregarding the contribution of the concrete. On the basis that the effect of axial force on torsional strength should be similar to its effect on shear strength, the ACI Code permits that Eqs. (3.37) and (3.38) be used, but that the resulting values of v_{tc} and v_c be reduced by the factor $(1 + 0.002N_u/A_g)$, where N_u is the axial force, negative for tension.

Example: design for torsion with shear The 28-ft span beam shown in Fig. 3.23a and b carries a monolithic slab cantilevering 6 ft past the beam center line as shown in the section. The resulting L beam supports a live load of 50 psf uniformly distributed over its upper surface. The effective depth to the flexural-steel centroid is 21.5 in., and the distance from the beam surfaces to the centroid of stirrup steel is $1\frac{3}{4}$ in. Material strengths are $f_c' = 4000$ psi and $f_y = 60,000$ psi. Design the torsional and shear reinforcement for the beam.

Applying ACI load factors, the slab load is

$1.4D = 75 \times 5.5 \times 1.4 = 580$ plf
$1.7L = 50 \times 1.7 \times 5.5 = 470$ plf
Total $= 1050$ plf at 3.25-ft eccentricity

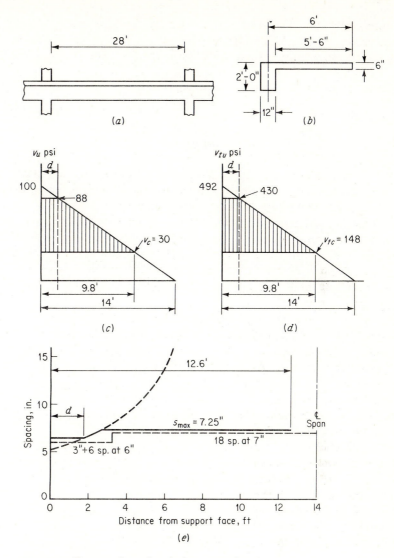

Fig. 3.23 Shear and torsion design example.

while the beam carries directly

$$1.4D = 300 \times 1.4 = 420 \text{ plf}$$
$$1.7L = 50 \times 1.7 = \underline{85} \text{ plf}$$
$$\text{Total} = \overline{505} \text{ plf}$$

Thus, the uniformly distributed load on the beam is 1555 plf, acting together with a uniformly distributed torque of $1050 \times 3.25 = 3410$ ft-lb/ft. At the face of the column the design shear force is $V_u = 1.555 \times \frac{28}{2} = 21.8$ kips, and the nominal shear stress is found by Eq. (3.23):

$$v_u = \frac{21,800}{0.85 \times 12 \times 21.5} = 100 \text{ psi}$$

At the same location, the design torsional moment of $T_u = 3.410 \times \frac{28}{2} = 47.7$ ft-kips.
With $\Sigma x^2 y = 12^2 \times 24 + 6^2 \times 18 = 4098$ in.³., Eq. (3.36) gives the nominal torsional stress,

$$v_{tu} = \frac{3 \times 47,700 \times 12}{0.85 \times 4098} = 492 \text{ psi}$$

The variation of v_u and v_{tu} with distance from the face of the supporting column is
given by Fig. 3.23c and d, respectively. The values of v_u and v_{tu} at the critical design
section, a distance d from the column face, are

$$v_u = 100 \times 12.21/14 = 88 \text{ psi}$$
$$v_{tu} = 492 \times 12.21/14 = 430 \text{ psi}$$

According to the Code, torsion may be neglected if $v_{tu} \leq 1.5 \sqrt{4000} = 95$ psi. Torsion must clearly be considered in the present case. With $v_{tu}/v_u = \frac{430}{88} = 4.90$, the
maximum torsion stress permitted under any circumstances is given by Eq. (3.39):

$$v_{tu} \leq \frac{12 \sqrt{4000}}{\sqrt{1 + (1.2/4.90)^2}} = 740 \text{ psi}$$

The present upper limit of 430 psi is well below this permitted value, and so the beam
section is satisfactory if properly reinforced.

From Eq. (3.38) the nominal shear stress carried by the concrete is

$$v_c = \frac{2.0 \sqrt{4000}}{\sqrt{1 + (4.90/1.2)^2}} = 30 \text{ psi}$$

while from Eq. (3.37) the nominal torsion stress carried by the concrete is

$$v_{tc} = \frac{2.4 \sqrt{4000}}{\sqrt{1 + (1.2/4.90)^2}} = 148 \text{ psi}$$

Note that, for this very common case of uniformly distributed transverse load coupled
with uniformly distributed applied torque, both T_u and V_u decrease linearly to zero
at midspan. It follows that v_{tu}/v_u remains constant and, in turn, that v_c and v_{tc} are constant, as is shown by Fig. 3.23c and d. Furthermore, as was observed earlier, $v_c/v_u =
v_{tc}/v_{tu}$ if v_{tu}/v_u is constant; as a result, the point at which torsion reinforcement is no
longer required is also the point at which shear reinforcement is no longer required, in
the present case 9.8 ft from the support face.

With $1\frac{3}{4}$ in. cover to the center of the stirrup bar from all faces, $x_1 = 12.0 -
3.5 = 8.5$ in. and $y_1 = 24.0 - 3.5 = 20.5$ in. The coefficient $\alpha_t = 0.66 + 0.33 \times
20.5/8.5 = 1.46$, not to be taken greater than 1.50. Thus, from Eq. (3.40), the web
reinforcement required for torsion, computed at the column face for reference only, is

$$A_t = \frac{(492 - 148) \times 4098}{3 \times 1.46 \times 8.5 \times 20.5 \times 60,000} \times s = 0.0309s$$

for one leg of a closed vertical stirrup, at a spacing s, or $0.0618s$ for two legs. From Eq. $(3.26a)$ the web reinforcement required for transverse shear, again computed at the column face, is

$$A_v = \frac{(100 - 30) \times 12}{60,000} \times s = 0.0140s$$

to be provided in the two vertical legs. Thus the total area to be provided by the two vertical legs, for combined shear and torsion reinforcement at spacing s, at the face of the support is

$$2A_t + A_v = (0.0618 + 0.0140)s = 0.0758s$$

No. 4 closed stirrups will provide a total area in the two vertical legs of 0.40 in.^{2}; hence, the reference spacing at the column face would be $s_0 = 0.40/0.0758 = 5.28$ in. Since all coefficients of s above are proportional to distance from the point of zero excess shear stress or zero excess torsion stress, 9.8 ft from the column face, the required spacings at d and at 2-ft intervals along the span can be found by proportion:

$$
\begin{aligned}
s_0 &= 5.28 \text{ in.} \\
s_d &= 5.28 \times 9.8/8.0 = 6.47 \text{ in.} \\
s_2 &= 5.28 \times 9.8/7.8 = 6.62 \text{ in.} \\
s_4 &= 5.28 \times 9.8/5.8 = 8.91 \text{ in.} \\
s_6 &= 5.28 \times 9.8/3.8 = 13.6 \text{ in.} \\
s_8 &= 5.28 \times 9.8/1.8 = 28.7 \text{ in.}
\end{aligned}
$$

These values of s are plotted in Fig. $3.23e$. Code provisions for maximum spacing should now be checked. For torsion reinforcement the maximum spacing is the smaller of

$$\frac{x_1 + y_1}{4} = \frac{8.5 + 20.5}{4} = 7.25 \text{ in.}$$

or 12 in., while by the shear provisions the maximum spacing is $d/2 = 10.75$ in. The most restrictive provision is the first, and the maximum spacing of 7.25 in. is plotted in Fig. $3.23e$. In addition, it is noted that the stirrups need not be spaced more closely than the requirement at a distance d from the column face. The resulting spacing requirements are shown by the solid line in the figure. These requirements are met in a practical way by No. 4 closed stirrups, the first 3 in. from the column face, followed by 6 at 6-in. spacing and 18 at 7-in. spacing. While stirrups could be discontinued $(d + b)$, or 2.8 ft, past the point of zero excess shear or torsion, the resulting 12.6-ft distance is sufficiently close to the half span that reinforcement will be carried throughout. The minimum web steel provided, 0.40 in.^{2}, satisfies the Code minimum of $50b_w s/f_y = 50 \times 12 \times 7/60,000 = 0.07$ in.^{2}

The longitudinal steel required for torsion at a distance d from the column face is next computed. At that location

$$A_t = 0.0309 \times 5.28 \times 8.01/9.8 = 0.134 \text{ in.}^2$$

and from Eq. (2.51)

$$A_l = 2 \times 0.134 \times (8.5 + 20.5)/6.47 = 1.20 \text{ in.}^2 \text{ total not to be less than}$$

$$A_l = \left(\frac{400 \times 12 \times 6.47}{60,000} \times \frac{430}{430 + 88} - 2 \times 0.134 \right) \times \frac{29}{6.47} = 0.74 \text{ in.}^2$$

According to the Code, the spacing must not exceed 12 in. Reinforcement will be placed at the top, middepth, and bottom of the member, each level to provide not less than $1.20/3 = 0.40$ in.^{2} Two No. 4 bars will be used at the middepth, while reinforcement to be placed for flexure will be increased by 0.40 in.^{2} at the top and bottom of the member.

While A_l reduces in direct proportion to A_t, and hence becomes zero at 9.8 ft from the column, for simplicity of construction the bars will be carried throughout the length of the member. Adequate embedment will be provided past the face of the column to fully develop f_y in the bars at that location.

3.14 DESIGN BASED ON ALLOWABLE STRESSES

The design of reinforced-concrete members can also be based on conditions at service load with load factors of unity applied (see Arts. 3.1 and 3.2). At this stage, stresses in the concrete and the steel must be kept within certain allowable limits, sufficiently low so that an approximately linear relation exists between stress and strain in the concrete as well as in the steel. According to the ACI Code, members designed on the basis of service loads are to be proportioned so that the compressive stress in the extreme fiber of the concrete does not exceed $0.45f'_c$. The allowable tensile stress in the reinforcement must not exceed 20 ksi for Grade 40 steel or Grade 50 steel and must not exceed 24 ksi for Grade 60 steel or steels with yield strength greater than 60 ksi.

a. Singly reinforced rectangular beams The following relations have already been derived in Art. 2.4d with reference to Fig. 2.8b:

$$M = A_s f_s jd \tag{2.15}$$

$$M = \frac{f_c}{2} kjbd^2 \tag{2.17}$$

$$j = 1 - \frac{k}{3} \tag{2.19}$$

If the dimensional properties of the cross section are given, the value of k, which locates the neutral axis, can be found by means of Eq. (2.13). Substituting $\rho = A_s/bd$ in that equation, one obtains

$$b \frac{(kd)^2}{2} - \rho nbd(d - kd) = 0$$

Dividing through by bd^2,

$$\frac{k^2}{2} - \rho n(1 - k) = 0$$

from which

$$k = \sqrt{2\rho n + (\rho n)^2} - \rho n \tag{3.41}$$

This expression for k is useful in reviewing the moment capacity of a given beam cross section for which the modular ratio and the steel ratio are known, or for calculating unit stresses in the steel and concrete when a known moment is applied.

If one is *designing* a cross section to resist a given moment, it is often convenient to express k in terms of the desired stresses in the concrete and steel. From the geometry of the strain diagram in Fig. 2.5e one obtains

$$\frac{\epsilon_c}{\epsilon_s} = \frac{kd}{d - kd}$$

Substituting $\epsilon_c = f_c/E_c$ and $\epsilon_s = f_s/E_s$,

$$\frac{f_c}{E_c}\frac{E_s}{f_s} = \frac{k}{1 - k}$$

Introducing the modular ratio $n = E_s/E_c$ and the stress ratio $r = f_s/f_c$,

$$\frac{n}{r} = \frac{k}{1 - k}$$

and solving for k,

$$k = \frac{n}{n + r} \tag{3.42}$$

In this case, k is expressed in terms of the modular ratio and the ratio of simultaneous stresses (i.e., the stresses in the concrete and steel at any particular stage of loading). In the design, these stresses would normally be taken as the maximum allowable stresses in the two materials, and the value of k could then be calculated directly from (3.42).

It may be of interest in certain situations to know the value of ρ which should be used in designing a beam such that the steel and the concrete will reach their maximum allowable stresses for the same applied moment. This is known as *balanced-stress design*. Equating Eqs. (2.14a) and (2.14b) and substituting $\rho = A_s/bd$,

$$\frac{f_c}{2} bkd = \rho f_s bd$$

Dividing both sides by bd and solving for ρ, which is then designated as ρ_e,

$$\rho_e = \frac{k f_c}{2 f_s} = \frac{k}{2r}$$

With the desired simultaneous stress values known, Eq. (3.42) can be used to eliminate k, giving

$$\rho_e = \frac{n}{2r(n + r)} \tag{3.43}$$

The ratio ρ_e is called the *balanced-stress steel ratio*. If $\rho < \rho_e$, the steel will reach its allowable stresses at a lower load than the concrete, and the allowable moment will be determined by Eq. (2.15). If $\rho > \rho_e$, the concrete will reach its allowable stress first, and Eq. (2.17) will control. Values of k, j, ρ_e and the flexural coefficient $K = f_c kj/2$ to be used for design are given in Table 5 of the Appendix. Values of k and j to be used for the review of known cross sections are given in Table 6.

The relation between beam behavior implied by linear elastic analysis and actual performance as given closely by the strength equations of Art. 3.4 is illustrated by Fig. 3.24, which plots resisting moment vs. steel ratio for a singly reinforced beam for $f_y = 40{,}000$ psi and $f'_c = 3000$ psi. Values of M_u predicted by strength equations have been divided by a safety factor of 2 to permit direct comparison with elastic analysis.

It is evident from Fig. 3.24 that strength theory does not differ significantly from elastic theory for values of ρ smaller than ρ_e. At that value elastic theory assumes the resisting moment limit of steel and concrete to be equal. The bending moment corresponding to ρ_e can, by that theory, be increased by the addition of tension steel, but this is indicated to be uneconomical, since the steel would be stressed below its allowable value. Any amount of steel in excess of ρ_e, by elastic theory, provides half as much or less additional strength as does the same amount of steel added to an underreinforced beam with ρ less than ρ_e.

This implication of elastic analysis is false. The actual behavior is depicted by the upper curve of Fig. 3.24. It shows that additional steel increases

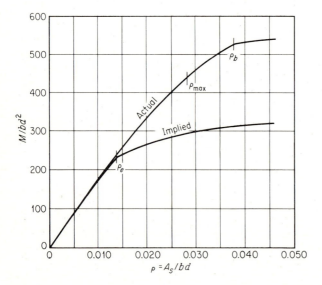

Fig. 3.24

the strength of the beam at a practically constant rate until the much larger critical steel ratio ρ_b is reached. (The maximum steel ratio to be used, according to the ACI Code, is $\rho_{\max}$, which is also shown in the figure.)

b. Doubly reinforced rectangular beams If both steel and concrete are completely elastic, the stress in the compression steel is n times the concrete stress at the same level, since the unit strains in the steel and the adjacent concrete are equal. This allows the compression steel stress to be computed. However, the stresses and strains in the concrete are proportional only at relatively low strains; at higher strains, the stresses no longer increase proportionately. Since the strains in the compression steel and the adjacent concrete remain equal, this means that at higher strain levels the unit stress in the steel, being proportional to the strain, will be larger than it would be if the concrete behaved elastically. This increase in the steel stress over that computed by assuming elastic behavior is accentuated by the fact that concrete, to a certain extent, compresses under constant load or stress (flow or creep). In contrast, steel, if stressed below the yield point, maintains its length essentially unchanged under constant stress. As a result, in a beam reinforced for compression, in the course of time the concrete, by minute flow, transfers part of its compression stress to the steel; the actual stress in the steel becomes higher than that computed on the basis of elastic behavior.

To approximate the effects of the nonlinear concrete stress-strain curve and of plastic flow of the concrete, the ACI Code specifies that an effective modular ratio of $2E_s/E_c$ be used to transform the compression reinforcement for stress computations. The allowable compressive stress in such reinforcement is not to exceed the allowable stress in tension. This provision is an attempt to compensate in part for the underestimate of the flexural compression strength inherent in working-stress design, as discussed previously.

In Fig. 3.25a, a rectangular-beam cross section is shown with compression steel A'_s placed a distance d' from the compression face and with tensile steel A_s at an effective depth d. The stress in the tensile steel is f_s, while that in the compressive steel is f'_s. It is a design convenience to divide the total resisting

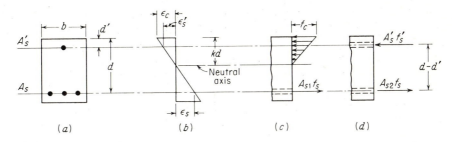

Fig. 3.25 Doubly reinforced rectangular beam.

moment into two parts. If M is the total resisting moment, and

M_1 = moment that can be developed by given cross section of concrete without compression reinforcement, working with a partial steel area A_{s1} to balance concrete compression

M_2 = moment in excess of M_1, developed by compression reinforcement, acting with additional tensile-steel area A_{s2}

then

$$M = M_1 + M_2 \qquad \text{and} \qquad A_s = A_{s1} + A_{s2}$$

The moment M_1 depends on the concrete dimensions and allowable stresses and is

$$M_1 = \frac{f_c}{2} kjbd^2 \tag{3.44}$$

where k and j are computed from Eqs. (3.42) and (2.19). The area of tensile steel required for balanced-stress design of a singly reinforced beam having a resisting moment M_1 is

$$A_{s1} = \frac{M_1}{f_s jd} \tag{3.45}$$

A moment M_2 is provided by compressive steel and the remaining tensile steel $A_{s2} = A_s - A_{s1}$, acting with an internal lever arm $(d - d')$:

$$M_2 = A_{s2}f_s(d - d') = A'_s f'_s(d - d') \tag{3.46}$$

The additional moment M_2 may be governed by either the tensile or compressive steel, depending on the stresses in each and their relative areas. From the geometry of Fig. 3.25b,

$$\frac{\epsilon_s}{\epsilon'_s} = \frac{d - kd}{kd - d'} = \frac{f_s}{f'_s}$$

Then

$$f'_s = f_s \frac{kd - d'}{d - kd}$$

$$= f_s \frac{k - d'/d}{1 - k}$$

However, the effect of ACI Code provisions is that f'_s be taken equal to twice this value, but not greater than f_s. Thus,

$$f'_s = 2f_s \frac{k - d'/d}{1 - k} \leq f_s \tag{3.47}$$

The moment contribution M_2 can then easily be found from Eq. (3.46).

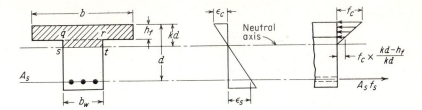

Fig. 3.26   T beam.

The above analysis is approximate in that it does not account for the shift in the neutral axis as the compression steel A_s' and the tensile steel A_{s2} are added. The error introduced is so small that it is safely neglected.

c. T beams   The general discussion of Art. 3.4c, with respect to effective width of flange and position of the neutral axis, applies to T beams designed by service load methods as well as those designed by strength methods. When designing by service load methods, however, it is necessary to account for the variable concrete stress in the compression concrete zone. Figure 3.26 shows a T-beam cross section. The compressive force in the web, represented by the area $qrst$, is usually small in comparison with that in the flange and is neglected in the derivation of equations for ordinary design.

On the basis of the geometry of the strain diagram and assumed perfect elasticity of both materials, Eq. (3.42), obtained for rectangular beams, applies here:

$$k = \frac{n}{n + r} \qquad\qquad (3.42)$$

This permits the value of k to be calculated easily, provided both n and r are known. Unfortunately, for T beams, the actual $r = f_s/f_c$ is usually not known, even though the maximum allowable stresses are. In a T beam which is a part of a floor system, the compressive area provided by the slab is so large that the actual f_c will be some unknown fraction of its allowable value, and Eq. (3.42) cannot be used directly. In Fig. 3.26 the total tension force is $A_s f_s$, while the total compression force in the flange is

$$\frac{f_c + f_c(kd - h_f)/kd}{2}\, bh_f = f_c\, \frac{2kd - h_f}{2kd}\, bh_f$$

For horizontal equilibrium, the tension force must equal the compression force, and

$$(a) \quad A_s f_s = \rho b d f_s = f_c\, \frac{2kd - h_f}{2kd}\, bh_f$$

From Eq. (3.42),

$$k = \frac{n}{n+r} = \frac{n}{n+f_s/f_c}$$

from which

(b) $f_c = f_s \dfrac{k}{n(1-k)}$

Substituting this expression for f_c into Eq. (a) to eliminate unit stresses,

$$k = \frac{n\rho + \frac{1}{2}(h_f/d)^2}{n\rho + h_f/d} \tag{3.48}$$

The distance to the center of compression (center of gravity of the trapezoid) from the upper face of the beam is

$$z = \frac{3kd - 2h_f}{2kd - h_f} \times \frac{h_f}{3} \tag{3.49}$$

and the lever arm of the couple formed by the tensile and compressive forces is

$$jd = d - z \tag{3.50}$$

From the above analysis,

$$j = \frac{6 - 6h_f/d + 2(h_f/d)^2 + (h_f/d)^3(1/2\rho n)}{6 - 3h_f/d} \tag{3.51}$$

The evaluation of k and j from Eqs. (3.48) and (3.51), respectively, is greatly facilitated by Graph 12 of the Appendix. The resisting moments of the steel and concrete are equal to the product of the lever arm jd of the internal stress couple and the total tension and compression, respectively; hence,

$$M = A_s f_s jd \tag{3.52a}$$

or

$$M = f_c \left(1 - \frac{h_f}{2kd}\right) bh_f jd \tag{3.52b}$$

Approximate equations for resisting moments can be developed as follows: Since the center of gravity of the compression-stress trapezoid is above the middle of the slab, the lever arm jd of the resisting couple is never less than $d - h_f/2$. The average unit compressive stress

$$f_c \left(1 - \frac{h_f}{2kd}\right)$$

is never as small as $f_c/2$, except when the neutral axis is at or above the bottom of the slab, in which case rectangular-beam equations apply. Equations

(3.52) can be approximated by substituting these limiting values for jd and $f_c(1 - h_f/2kd)$. Then

$$M = A_s f_s \left(d - \frac{h_f}{2} \right)$$ (3.53a)

or

$$M = \frac{f_c}{2} b h_f \left(d - \frac{h_f}{2} \right)$$ (3.53b)

The use of Eqs. (3.53) in *design* is justified for all practical purposes. They must not be used in *review* problems.

d. Shear and torsion   In designing by the service-load method, the ACI Code requires that the allowable concrete stresses and the limiting maximum stresses for shear and torsion (see Arts. 3.5 and 3.13) be taken equal to 55 per cent of the value used for strength design for beams, joists, walls, and one-way slabs, and 50 per cent of the values used for strength design for two-way slabs and footings. However, the yield stress f_y is to be used in computing minimum area of reinforcement by Eq. (3.25).

e. Development of reinforcement   Requirements for development of reinforce-ment, by service-load design, are the same as those for strength design given in Art. 3.6, except that computed shears must be multiplied by 2 prior to sub-stitution for V_u in Eq. (3.31). In computing M_t in the same equation, the quantity $(d - a/2)$ is to be taken equal to $0.85d$, according to the Code.

REFERENCES

3.1. *Univ. Illinois Eng. Exp. Sta. Bull.* 399, 1951.
3.2. Ultimate Strength Design Handbook, *ACI Publ.* SP-17, 1967.
3.3. C. W. Dolan: Ultimate Capacity of Reinforced Concrete Sections Using a Continuous Stress-Strain Function, M.S. thesis, Cornell University, Ithaca, N.Y., June, 1967.
3.4. J. G. MacGregor and J. M. Hanson: Proposed Changes in Shear Provisions for Rein-forced and Prestressed Concrete Beams, *ACI J.*, vol. 66, no. 4, pp. 276–288, April, 1969.
3.5. Recommended Practice and Standard Specifications for Concrete and Reinforced Concrete, *Proc. ASCE*, vol. 66, no. 6, June, 1940.
3.6. P. Gergely and L. A Lutz: Maximum Crack Width in Reinforced Concrete Flexural Members, in "Causes, Mechanisms and Control of Cracking in Concrete," ACI Publ. SP-20, pp. 1–17, 1968.
3.7. Deflections of Reinforced Concrete Flexural Members, Report of ACI Committee 435, *ACI J.*, vol. 63, no. 6, pp. 637–674, June, 1966.

4
Slabs

4.1 TYPES OF SLABS

In reinforced-concrete construction, slabs are used to provide flat, useful surfaces. A reinforced-concrete slab is a broad, flat plate, usually horizontal, with top and bottom surface parallel or nearly so. It may be supported by reinforced-concrete beams (and is usually poured monolithically with such beams), by masonry or reinforced-concrete walls, by structural-steel members, directly by columns, or continuously by the ground.

Slabs may be supported on two opposite sides only, as in Fig. 4.1a, in which case the structural action of the slab is essentially *one-way*, the loads being carried by the slab in the direction perpendicular to the supporting beams. On the other hand, there may be beams on all four sides, as in Fig. 4.1b, so that *two-way-slab* action is obtained. Intermediate beams, as shown in Fig. 4.1c, may be provided. If the ratio of length to width of one slab panel is larger than about 2, most of the load is carried in the short direction to the supporting beams, and one-way action is obtained in effect, even though supports are provided on all sides.

Concrete slabs may in some cases be carried directly by columns, as in Fig. 4.1d, without the use of beams or girders. Such slabs are described as

197

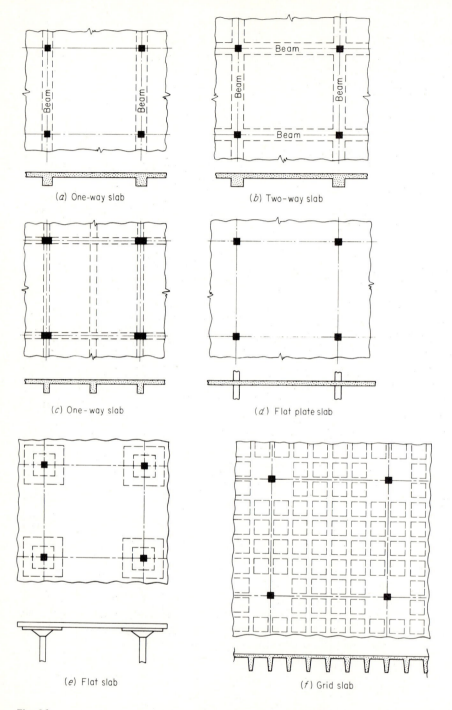

(a) One-way slab

(b) Two-way slab

(c) One-way slab

(d) Flat plate slab

(e) Flat slab

(f) Grid slab

Fig. 4.1

flat plates and are commonly used where spans are not large and loads not particularly heavy. *Flat-slab* construction, shown in Fig. 4.1e, is also beamless, but incorporates a thickened slab region in the vicinity of the column and often employs flared column tops. Both are devices to reduce stresses due to shear and negative bending around the columns. They are referred to as *drop panels* and *column capitals*, respectively. Closely related to the flat-plate slab is the *grid slab*, shown in Fig. 4.1f. To reduce the dead load of solid-slab construction, voids are formed in a rectilinear pattern through use of metal, wood, or cardboard form inserts. A two-way ribbed construction results. Often inserts are omitted near the columns, forming a solid slab to better resist moments and shears in these areas.

In addition to the column-supported types of construction shown in Fig. 4.1, many slabs are supported continuously on the ground, as for highways, airport runways, and basement slabs in buildings. In such cases, a well-compacted layer of crushed stone or gravel is usually provided to ensure uniform support and to allow for proper subgrade drainage.

Reinforced-concrete slabs are usually designed for a uniform load covering an entire panel area. Concentrated loads are supported by a width of slab greater than the contact width. Methods of computing the probable distribution of concentrated loads as applied by the wheels of trucks are given in Chap. 12. Very heavy, fixed concentrated loads often require supporting beams.

Reinforcing steel for slabs is placed primarily parallel to the surfaces. Straight-bar reinforcement may be used, although in continuous slabs bottom bars are often bent up to provide for negative moment over the supports. Welded wire mesh is commonly used for reinforcement of slabs on ground. Bar or rod mats are available for the heavier reinforcement of highway slabs and airport runways. Slabs may be prestressed, using high-tensile-strength wires or strands.

ONE - WAY SLABS

4.2 FLEXURAL ANALYSIS

A one-way slab is essentially a rectangular beam of comparatively large ratio of width to depth. There are, however, certain factors entering into the design of such slabs which were not considered in the design of rectangular beams. A unit strip of slab cut out at right angles to the supporting beams (shaded area in Fig. 4.2) may be considered as a rectangular beam of unit width, with a depth equal to the thickness of the slab and a length equal to the distance between supports. This strip could be analyzed by the methods which were used in problems dealing with rectangular beams, the bending moment being computed for a unit width, for example. The load per unit area on the slab would then be the load per unit length on the imaginary beam. Since all the load on the slab must be transmitted to the two supporting beams, it follows that all the reinforcing steel should be placed at right angles to these beams,

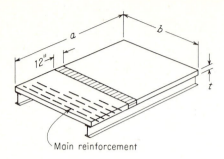

Main reinforcement   **Fig. 4.2**

with the exception of any bars that may be placed in the other direction to
carry shrinkage and temperature stresses. A one-way slab thus consists of a
series of rectangular beams side by side. This simplified analysis, which assumes
Poisson's ratio to be zero, is slightly conservative. Actually, longitudinal
flexural compression will result in lateral expansion unless the compressed
material is restrained. In a one-way slab, this lateral expansion is resisted by
the adjacent beam strips, which tend to expand also. The net result is a slight
strengthening and stiffening in the span direction, but this effect is small and is
almost always disregarded.

   The ratio of steel in a slab may be determined by dividing the sectional
area of one bar by the area of concrete between two successive bars, the latter
area being the product of the depth to the center of the bars and the distance
between them, center to center. The ratio of steel may also be determined by
dividing the average area of steel per foot of width by the effective area of
concrete in a 1-ft strip. The average area of steel per foot of width is equal to
the area of one bar times the average number of bars in a 1-ft strip (12 divided
by the spacing in inches), and the effective area of concrete in a 1-ft (or 12-in.)
strip is equal to 12 times the effective depth d.

   To illustrate the latter method of obtaining the steel ratio ρ, assume a 5-in.
slab with an effective depth of 4 in., with No. 4 bars spaced $4\frac{1}{2}$ in. center to center. The average number of bars in a 12-in. strip of slab is $12/4.5 = 2.7$ bars,
and the average steel area in a 12-in. strip is $2.7 \times 0.20 = 0.54$ in.^{2} Hence
$\rho = 0.54/(12 \times 4) = 0.0112$. By the other method,

$$\rho = \frac{0.20}{(4.5 \times 4)} = 0.0112$$

   The spacing of bars, which is necessary to furnish a given area of steel
per foot of width, is obtained by dividing the number of bars required to
furnish this area into 12. For example, to furnish an average area of 0.46 in.^{2}
per ft, with No. 4 bars, requires $0.46/0.20 = 2.3$ bars per foot; the bars must
be spaced not more than $12/2.3 = 5.2$ in. center to center.[1]

   Design moments and shears in one-way slabs may be found either by

---

[1] The determination of slab-steel areas for various combinations of bars and spacing is
greatly facilitated by Table 4 of the Appendix.

Table 4.1   Minimum slab thickness h

Simply supported	$l/20$
One end continuous	$l/24$
Both ends continuous	$l/28$
Cantilever	$l/10$

elastic analysis or through use of moment coefficients as for beams (see Chap. 7). If the slab rests freely on its supports, the span length may be taken equal to the clear span plus the depth of the slab, but need not exceed the distance between centers of supports. In general, center-to-center distances should be used in continuous slab analysis and some reduction allowed in negative moment to account for support width. For slabs with clear spans not more than 10 ft that are built integrally with their supports, the ACI Code permits analysis as a continuous slab on knife-edge supports with spans equal to the clear spans and the width of beams otherwise neglected. If moment coefficients are employed, moments should be based on clear spans.

Concrete compression will seldom control the flexural design of one-way slabs. Flexural design will ordinarily consist in selecting a slab depth which will permit the use of an economically low steel ratio and which will not allow unsightly or damaging deflections. The ACI Code specifies the minimum thicknesses in Table 4.1 for nonprestressed slabs of normal-weight concrete ($w = 145$ pcf) using 60-ksi grade steel, unless calculation of deflections proves that lesser thicknesses may be used without adverse effect. For concrete of density w from 90 to 120 pcf, the tabulated values should be multiplied by the factor $(1.65 - 0.005w)$, but no less than 1.09. For reinforcement having yield stress f_y other than 60,000 psi, the tabulated values should be multiplied by $(0.4 + f_y/100,000)$.

Although diagonal tension and bond stresses will seldom influence the design of one-way slabs, these stresses should be checked, particularly for slabs of unusual proportions or loads.

The total slab thickness is usually rounded to the next higher $\frac{1}{4}$ in. for slabs up to 6 in. thickness, and to the next higher $\frac{1}{2}$ in. for thicker slabs. The concrete protection below the reinforcement should follow the recommendations of the ACI Code unless conditions warrant some change. In the average slab, a depth of 1 in. below the center of the steel may be used. The lateral spacing of the bars, except those used only to prevent shrinkage and temperature cracks, should not exceed 3 times the thickness of the slab nor 18 in., according to the Code. The spacing should not be less than 1 bar diameter or 1 in. under any condition.

4.3 TEMPERATURE AND SHRINKAGE REINFORCEMENT

Concrete shrinks as the cement paste hardens, as was pointed out in Art. 1.10. It is advisable to minimize such shrinkage by using concretes with the smallest possible amounts of water and cement compatible with other requirements, such as strength and workability, and by thorough moist-curing of sufficient duration. However, no matter what precautions are taken, a certain amount of shrinkage is usually unavoidable. If a slab of moderate dimensions rests freely on its supports, it can contract to accommodate the shortening of its length produced by shrinkage. Usually, however, slabs and other members are joined rigidly to other parts of the structure and cannot contract freely. This results in tension stresses known as *shrinkage stresses*. A decrease in temperature relative to that at which the slab was poured, particularly in outdoor structures such as bridges, may have an effect similar to shrinkage. That is, the slab tends to contract and, if restrained from doing so, becomes subject to tensile stresses.

Since concrete is weak in tension, these temperature and shrinkage stresses are likely to result in cracking. Cracks of this nature are not detrimental, provided their size is limited to what are known as *hairline cracks*. This can be achieved by placing reinforcement in the slab to counteract contraction and distribute the cracks uniformly. As the concrete tends to shrink, such reinforcement resists the contraction, and consequently becomes subject to compression. The total shrinkage in a slab so reinforced is less than that in one without reinforcement; in addition, whatever cracks do occur will be of smaller width and more evenly distributed by virtue of the reinforcement.

In one-way slabs the reinforcement provided for resisting the bending moments has the desirable effect of reducing shrinkage and distributing cracks. However, as contraction takes place equally in all directions, it is necessary to provide special reinforcement for shrinkage and temperature contraction in the direction perpendicular to the main reinforcement. This added steel is known as *temperature* or *shrinkage reinforcement*.

Reinforcement for shrinkage and temperature stresses normal to the principal reinforcement should be provided in a structural slab in which the principal reinforcement extends in one direction only. The ACI Code specifies the following minimum ratios of reinforcement area to gross concrete area, but in no case shall such reinforcing bars be placed farther apart than 5 times the slab thickness or more than 18 in.:

Slabs where Grade 40 or 50 deformed bars are used	0.0020
Slabs where Grade 60 deformed bars or welded wire fabric, deformed or plain, are used	0.0018
Slabs where reinforcement with yield strength exceeding 60,000 psi measured at yield strain of 0.35 per cent is used	$\dfrac{0.0018 \times 60{,}000}{f_y}$

In no case is the steel ratio to be less than 0.0014.

4.4 Example: one-way-slab design A reinforced-concrete slab is built integrally with its supports and consists of two equal spans, each with a clear span of 15 ft. The service live load is 100 psf, and 4000-psi concrete is specified for use with steel of yield stress equal to 60,000 psi. Design the slab, following the provisions of the ACI Code.

The thickness of the slab is first estimated, based on the minimum thickness of Table 4.1: $l/28 = 15 \times \frac{12}{28} = 6.43$ in. A trial thickness of 6.50 in. will be used, for which the weight is $150 \times 6.50/12 = 81$ psf. The specified live load and computed dead load are multiplied by the ACI overload factors to obtain design loads:

Dead load: $81 \times 1.4 = 113$ psf
Live load: $100 \times 1.7 = 170$ psf
Total $= \overline{283}$ psf

For this case, design moments at critical sections may be found using the ACI moment coefficients (see Table 7.1):

At interior support: $-M = \frac{1}{9} \times 0.283 \times 15^2 = 7.06$ ft-kips
At midspan: $+M = \frac{1}{14} \times 0.283 \times 15^2 = 4.53$ ft-kips
At exterior support: $-M = \frac{1}{24} \times 0.283 \times 15^2 = 2.65$ ft-kips

The maximum steel ratio permitted by the Code is [Eq. (3.3)]

$$0.75\rho_b = 0.75 \times 0.85\beta_1 \frac{f'_c}{f_y} \frac{87,000}{87,000 + f_y}$$
$$= 0.75 \times 0.85^2 \times \tfrac{4}{40} \times \tfrac{87}{147} = 0.021$$

The minimum required effective depth, as controlled by the negative moment at the interior support, is found from Eq. (3.6b) to be

$$d^2 = \frac{M_u}{\phi\rho f_y b(1 - 0.59\rho f_y/f'_c)}$$
$$= \frac{7.06 \times 12}{0.90 \times 0.021 \times 60 \times 12(1 - 0.59 \times 0.21 \times \tfrac{6.0}{4})}$$
$$= 7.67 \text{ in.}^2$$
$$d = 2.77 \text{ in.*}$$

This is less than the effective depth of $6.50 - 1.00 = 5.50$ in. resulting from application of Code restrictions, and the latter figure will be adopted. At the interior support, assuming the stress-block depth $a = 1.00$ in., the area of steel required per foot of width in the top of the slab is [Eq. (3.6a)]

$$A_s = \frac{M_u}{\phi f_y(d - a/2)} = \frac{7.06 \times 12}{0.90 \times 60 \times 5.00} = 0.31 \text{ in.}^2$$

Checking the assumed depth a by Eq. (3.4),

$$a = \frac{A_s f_y}{0.85 f'_c b} = \frac{0.31 \times 60}{0.85 \times 4 \times 12} = 0.46 \text{ in.}$$

A second trial will be made with $a = 0.46$ in. Then

$$A_s = \frac{7.06 \times 12}{0.90 \times 60 \times 5.27} = 0.30 \text{ in.}$$

for which $a = 0.46 \times 0.30/0.31 = 0.45$ in. No further revision is necessary. At other critical-moment sections, it will be satisfactory to use the same lever arm to deter-

---

* This depth is more easily found using Graph 1 of the Appendix. For $\rho = \rho_{\max}$, $M_u/\phi bd^2 = 1050$, from which $d = 2.75$ in.

mine steel areas, and

At midspan: $A_s = \dfrac{4.53 \times 12}{0.90 \times 60 \times 5.27} = 0.19$ in.^{2}

At exterior support: $A_s = \dfrac{2.65 \times 12}{0.90 \times 60 \times 5.27} = 0.11$ in.^{2}

The minimum reinforcement is that required for control of shrinkage and temperature cracking. This is

$A_s = 0.0018 \times 12 \times 6.50 = 0.14$ in.^{2}

per 12-in. strip. This requires a small increase in the amount of steel used at the exterior support.

The maximum shear force, at a distance d from the face of the interior support, is

$V_u = 1.15 \times \dfrac{283 \times 15}{2} - 283 \times \dfrac{5.50}{12} = 2310$ lb

and by Eq. (3.23),

$v_u = \dfrac{2310}{0.85 \times 12 \times 5.50} = 42$ psi

This is well below the permissible value of shear on the unreinforced concrete, which is $v_c = 2\sqrt{f_c'} = 2 \times \sqrt{4000} = 126$ psi.

The required tensile-steel areas may be provided in a variety of ways, but whatever the selection, due consideration must be given to actual placing of the steel during construction. The arrangement should be such that the steel may be placed rapidly with the minimum of labor costs even though excess steel is necessary to achieve this end.

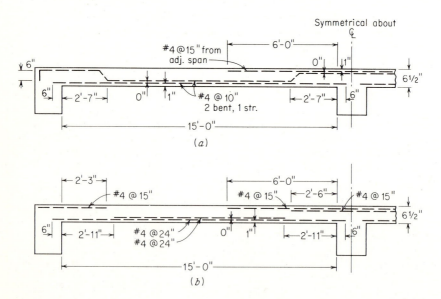

Fig. 4.3

Two possible arrangements are shown in Fig. 4.3. In Fig. 4.3a bent bars are used, while in Fig. 4.3b all bars are straight. The arrangement in (a) requires slightly more steel and the additional cost of bending, but on the other hand, requires fewer supporting chairs and permits easier placement, with less likelihood of bar omissions.

In the arrangement of Fig. 4.3a, No. 4 bars at 10 in. furnish 0.24 in.² of steel at midspan, slightly more than required. If two-thirds of these bars are bent upward for negative reinforcement over the interior support, the average spacing of such bent bars at the interior support will be $(10 + 20)/2 = 15$ in. Since an identical pattern of bars is bent upward from the other side of the support, the effective spacing of the No. 4 bars over the interior support is $7\frac{1}{2}$ in. This pattern closely satisfies the required steel area of 0.30 in.² per ft width of slab over the support. The bars bent at the interior support will also be bent upward for negative reinforcement at the exterior support, providing reinforcement equivalent to No. 4 bars at 15 in., or 0.16 in.² of steel.

Note that it is not necessary to achieve uniform spacing of reinforcement in slabs, and that the steel provided can be calculated safely on the basis of average spacing as in the example. Care should be taken to satisfy requirements for both minimum and maximum spacing of principal reinforcement, however,

The locations of bend and cutoff points shown in Fig. 4.3a were obtained using Graph 11 of the Appendix, as explained in Art. 3.8 and Table 8 of the Appendix (see also Fig. 3.15).

The arrangement of Fig. 4.3b uses only straight bars. Although satisfactory according to the ACI Code (since the shear stress does not exceed two-thirds of that permitted), cutting off the shorter positive and negative bars as shown leads to an undesirable condition at the ends of those bars, where there will be concentrations of stress in the concrete. The design would be improved if the negative bars were cut off at 3 ft from the face of supports, rather than 2 ft 3 in. and 2 ft 6 in. as shown, and if the positive steel were cut off at 2 ft 2 in., rather than at 2 ft 11 in. This would result in an overlap of approximately 2d of the cut positive and negative bars.

The required area of steel to be placed normal to the main reinforcement for purposes of temperature and shrinkage crack control is 0.14 in.² This will be provided by No. 4 bars at 16 in. spacing, placed directly on top of the main reinforcement in the positive-moment region and below the main steel in the negative-moment zone.

PROBLEM

A small bridge consisting of a concrete slab supported by steel stringers is to carry a uniformly distributed service load of 300 psf in addition to its own weight. The four stringers of the bridge, spanning in the long direction, are spaced 8 ft on centers. The concrete slab spans in the transverse direction and is continuous over the two interior stringers. Find the required thickness of the slab, and design and detail the bar reinforcement, using $f_y = 50,000$ psi and $f'_c = 3,000$ psi. Bent bars will be used in preference to all-straight-bar reinforcement. The ACI moment coefficients do not apply. Use overload factors of 1.4 and 1.7 applied to dead and live loads, respectively. Use a maximum steel ratio of $0.50\rho_b$.

TWO - WAY SLABS

4.5 BEHAVIOR OF TWO-WAY SLABS

The slabs discussed in Arts. 4.2 to 4.4 deform under load into a cylindrical surface. The main structural action is one-way in such cases, in the direction normal to supports on two opposite edges of a rectangular panel. In many cases, however, rectangular slabs are of such proportions and are supported in such a way that two-way action results. When loaded, such slabs bend into a dished surface rather than a cylindrical one. This means that at any point the slab is curved in both principal directions, and since bending moments are proportional to curvatures, moments also exist in both directions. To resist these moments, the slab must be reinforced in both directions, by mutually perpendicular layers of bars perpendicular, respectively, to two pairs of edges. The slab must be designed to take a proportionate share of the load in each direction.

   Types of reinforced-concrete construction which are characterized by two-way action include slabs supported by walls or beams on all sides (Fig. 4.1*b*), flat plates (Fig. 4.1*d*), flat slabs (Fig. 4.1*e*), and grid slabs (Fig. 4.1*f*).

a. Two-way slabs on unyielding walls   The simplest case is that of a rectangular concrete slab supported on all four sides by simple unyielding supports, as for example by masonry walls. Such a slab is shown in Fig. 4.4*a*. To visualize the flexural performance of such a slab, it is convenient to think of it as consisting of two sets of parallel strips, in each of the two directions, intersecting each other. Evidently, part of the load is carried by one set and transmitted to one pair of edge supports, and the remainder by the other.

   Figure 4.4*a* shows the two center strips of a rectangular plate with short span A and long span B. If the uniform load is w per ft² of slab, each of the two strips acts approximately like a simple beam uniformly loaded by its share of w. Because these imaginary strips actually are part of the same monolithic slab, their deflections at the intersection point must be the same. Equating the center deflections of the short and long strips,

$$(a) \qquad \frac{5w_s A^4}{384EI} = \frac{5w_l B^4}{384EI}$$

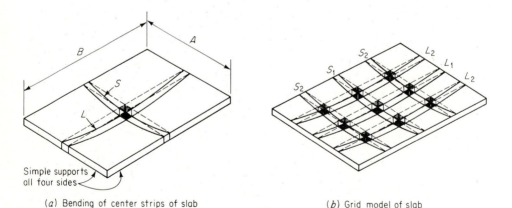

(*a*) Bending of center strips of slab   (*b*) Grid model of slab

Fig. 4.4   Two-way slab on simple edge supports.

where w_s is the share of the load w carried in the short direction, and w_l is the share of the load w carried in the long direction. Consequently,

(b) $$\frac{w_s}{w_l} = \frac{B^4}{A^4}$$

One sees that the larger share of the load is carried in the short direction, the ratio of the two portions of the total load being proportional to the fourth power of the ratio of the spans.

This result is approximate because the actual behavior of a slab is more complex than that of the two intersecting strips. An understanding of the behavior of the slab itself can be gained from Fig. 4.4b, which shows a slab model consisting of two sets of three strips each. It is seen that the two central strips S_1 and L_1 bend in a manner similar to that of Fig. 4.4a. The outer strips S_2 and L_2, however, are not only bent, but also twisted. Consider, for instance, one of the intersections of S_2 with L_2. It is seen that at the intersection the exterior edge of strip L_2 is at a higher elevation than the interior edge, while at the nearby end of strip L_2, both edges are at the same elevation; the strip is twisted. This twisting results in torsional stresses and torsional moments which are seen to be most pronounced near the corners. Consequently, the total load on the slab is carried not only by the bending moments in two directions, but also by the twisting moments. For this reason bending moments in elastic slabs are smaller than would be computed for sets of unconnected strips loaded by w_s and w_l. For instance, for a simply supported square slab, $w_s = w_l = w/2$. If only bending were present, the maximum moment in each strip would be

(c) $$\frac{(w/2)A^2}{8} = 0.0625wA^2$$

The exact theory of bending of elastic plates shows that, actually, the maximum moment in such a square slab is only $0.048wA^2$, so that in this case the twisting moments relieve the bending moments by about 25 per cent.

The largest moment in the slab occurs where the curvature is sharpest. Figure 4.4b shows this to be the case at midspan of the short strip S_1. It is evident that the curvature, hence the moment, in the short strip S_2 is less than at the corresponding location of strip S_1. Consequently, a variation of short-span moment occurs in the long direction of the span. This variation is shown qualitatively in Fig. 4.5. The short-span-moment diagram in Fig. 4.5a is valid only along the center strip at 1-1. Elsewhere the maximum-moment value is less, as shown in Fig. 4.5b; all other moment ordinates are reduced proportionately. Similarly, the long-span-moment diagram in Fig. 4.5c applies only at the longitudinal center line of the slab; elsewhere ordinates are reduced according to the variation shown in Fig. 4.5d. These variations in maximum moment across the width and length of a rectangular slab are accounted for in an

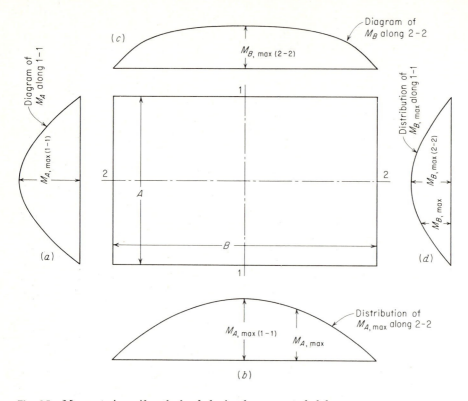

Fig. 4.5 Moments in uniformly loaded, simply supported slab.

approximate way in most practical design methods by designing for a reduced moment in the outer quarters of the slab span in each direction.

b. Two-way slabs supported by columns When two-way-slab construction is supported by columns, rather than by unyielding continuous wall supports on the four sides, several new considerations are introduced. Figure 4.6a shows a portion of a floor system in which a rectangular-slab panel is supported by beams on four sides. The beams, in turn, are supported by columns at the intersections of their center lines. If a surface load w is applied, that load is shared between imaginary slab strips S in the short direction and L in the long direction, as before. Note that the portion of the load that is carried by the long strips L is delivered to the beams B spanning in the short direction of the panel. This portion carried by the beams B, plus that carried directly in the short direction by the slab strips S, sums up to 100 per cent of the load applied to the panel. Similarly, the short-direction slab strips S deliver a part of the load to long-direction girders G. That load, plus load carried directly in the long direction by the slab, includes 100 per cent of the applied load. It is clearly a requirement of statics that, for column-supported construction,

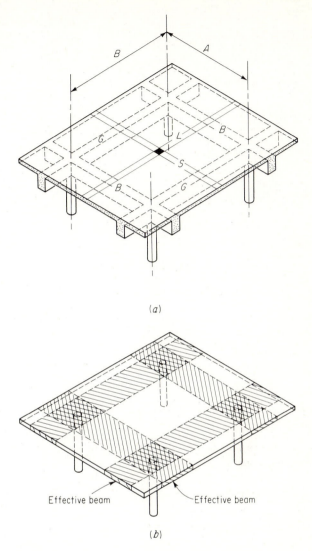

(a)

(b)

Effective beam   Effective beam

Fig. 4.6 Column-supported two-way slabs. (*a*) Two-way slab with beams and columns; (*b*) two-way slab without beams.

100 *per cent of the applied load must be carried in each direction*, jointly by the slab and its supporting beams.

A similar situation is obtained in the flat-plate floor shown in Fig. 4.6*b*. In this case beams are omitted. However, broad strips of the slab centered on the column lines in each direction serve the same function as the beams of Fig. 4.6*a*; for this case, also, the full load must be carried in each direction. The

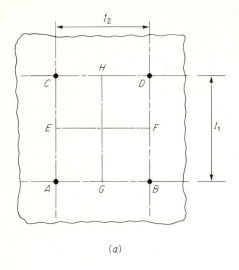

(a)

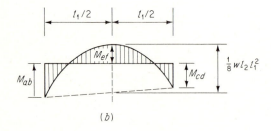

(b)

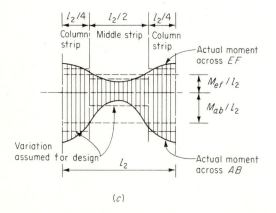

(c)

Fig. 4.7 Deflections and moment variation in column-supported two-way slabs. (a) Moment sections; (b) moment variation along span; (c) moment variation across width of critical sections.

presence of dropped panels or column capitals in the double-hatched zone near the columns does not modify this requirement of statics.

Figure 4.7a shows a flat-plate floor support by columns at A, B, C, and D. Figure 4.7b shows the moment diagram for the direction of span l_1. In this direction, the slab may be considered as a broad, flat beam of width l_2. Accordingly, the load per foot of span is wl_2. In any span of a continuous beam, the sum of the midspan positive moment and the average of the negative moments at adjacent supports is equal to the midspan positive moment of a corresponding simply supported beam. In terms of the slab, this requirement of statics may be written

$$(a) \quad \tfrac{1}{2}(M_{ab} + M_{cd}) + M_{ef} = \tfrac{1}{8}wl_2l_1{}^2$$

A similar requirement exists in the perpendicular direction, leading to the relation

$$(b) \quad \tfrac{1}{2}(M_{ac} + M_{bd}) + M_{gh} = \tfrac{1}{8}wl_1l_2{}^2$$

These results disclose nothing about the relative magnitudes of the support moments and span moments. The proportion of the total static moment which exists at each critical section can be found from an elastic analysis which considers the relative span lengths in adjacent panels; the loading pattern; and the relative stiffness of the supporting beams, if any, and that of the columns. Alternatively, empirical methods which have been found to be reliable under restricted conditions may be adopted.

The moments across the width of critical sections such as AB or EF are not constant, but vary as shown qualitatively in Fig. 4.7c. The exact variation depends on the presence or absence of beams on the column lines, the existence of dropped panels and column capitals, as well as on the intensity of the load. For design purposes it is convenient to divide each panel as shown in Fig. 4.7c into column strips, having a width of one-fourth the panel width, on each side of the column center lines, and middle strips in the one-half panel width between two column strips. Moments may be considered constant within the bounds of a middle strip or column strip, as shown, unless beams are present on the column lines. In the latter case, while the beam must have the same curvature as the adjacent slab strip, the beam moment will be larger in proportion to its greater stiffness, producing a discontinuity in the moment-variation curve at the lateral face of the beam. Since the total moment must be the same as before, according to statics, the slab moments must be correspondingly less.

4.6 MOMENT DETERMINATION IN TWO-WAY SLABS BY THE ACI CODE

Chapter 13 of the ACI Code deals in a unified way with two-way systems. Its provisions apply to slabs supported by walls or beams and to flat slabs and flat plates, as well as to two-way ribbed or grid slabs. While permitting design " . . . by any procedure satisfying the conditions of equilibrium and geometrical compatibility . . . ," specific reference is made to two alternative approaches: a semiempirical *direct design method* and an approximate elastic analysis known as the *equivalent-frame method*.

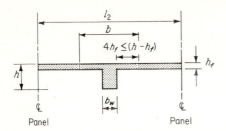

Fig. 4.8 Cross section of slab and effective beam.

In either case a typical panel is divided, for purposes of design, into *column strips* and *middle strips*. A column strip is defined as a strip of slab having a width on each side of the column center line equal to one-fourth the smaller of the panel dimensions l_1 and l_2. Such a strip includes column-line beams, if present. A middle strip is a design strip bounded by two column strips. In the case of monolithic construction, beams are defined to include that part of the slab on each side of the beam extending a distance equal to the projection of the beam above or below the slab (whichever is greater) but not greater than 4 times the slab thickness (see Fig. 4.8).

a. Direct design method Moments in two-way slabs may be found using a semiempirical direct design method subject to the following restrictions:

1. There shall be a minimum of three continuous spans in each direction.
2. The panels shall be rectangular, with the ratio of the longer to the shorter spans within a panel not greater than 2.
3. The successive span lengths in each direction shall not differ by more than one-third the longer span.
4. Columns may be offset a maximum of 10 per cent of the span, in the direction of the offset, from either axis between center lines of successive columns.
5. The live load shall not exceed 3 times the dead load.
6. If beams are used on the column lines, the relative stiffness of the beams in the two perpendicular directions, given by the ratio $\alpha_1 l_2{}^2/\alpha_2 l_1{}^2$, must be between 0.2 and 5.0 (see below for definitions).

For purposes of calculating the total static moment M_0 in a panel, l_n, the clear span in the direction of moments is used. The clear span is defined to extend from face to face of the columns, capitals, brackets, or walls, but is not to be less than 0.65 of the center-to-center span distance l_1. The total static design moment in a span, for a strip bounded laterally by the center line of the panel on each side of the center line of supports, is

$$M_0 = \frac{w l_2 l_n{}^2}{8}$$

in which l_2 is the length of span transverse to l_1 measured center to center of supports.

For interior spans, this total static moment is divided between the critical positive and negative bending sections (the latter taken at the face of rectangular supports):

Negative design moment: $M_{neg} = 0.65M_0$ (4.1a)
Positive design moment: $M_{pos} = 0.35M_0$ (4.1b)

In the case of end spans, the relative stiffness of the columns and the slab is of greater importance. This relative stiffness is incorporated explicitly in apportioning the static moment in the following way. Referring to Fig. 4.9, it is clear that the rotational restraint provided at the exterior end of the slab spanning in the direction l_1 is influenced not only by the flexural stiffness of the exterior column, but by the torsional stiffness of the edge beam AC. With distributed torque m_t applied by the slab and resisting torque M_t provided by the column, the edge-beam sections at A and C will rotate to a greater degree than the section at B, owing to torsional deformation of the edge beam. To allow for this effect, the actual column and edge beam are replaced by an equivalent column, so defined that the total flexibility (inverse of stiffness) of the equivalent column is the sum of the flexibilities of the actual column and edge beam. Thus

$$\frac{1}{K_{ec}} = \frac{1}{\Sigma K_c} + \frac{1}{K_t} \qquad\qquad (4.2)$$

where K_{ec} = flexural stiffness of equivalent column
      K_c = flexural stiffness of actual column
      K_t = torsional stiffness of edge beam
all expressed in terms of moment per unit rotation.

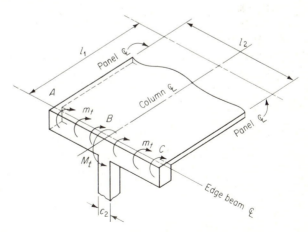

Fig. 4.9   Torsion of edge beam.

The flexural stiffness of members can be calculated by the usual equations of mechanics, and for members having a uniform cross section is equal to $4EI/l$. The moment of inertia may be based on the concrete section, neglecting reinforcement, and (in applying the direct method of analysis) variations due to column capitals and drop panels may be neglected.

According to the Code, the effective cross section of the torsional member consists of the largest of:

1. A portion of the slab having a width equal to that of the column or capital in the direction in which moments are taken
2. The portion of the slab specified in item 1 plus that part of any transverse beam above and below the slab
3. The transverse beam, including that part of the slab on each side of the beam extending a distance equal to the projection of the beam above or below the slab (whichever is greater), but not greater than 4 times the slab thickness (see Fig. 4.8)

The torsional stiffness of the edge beam may be calculated by the expression

$$K_t = \sum \frac{9E_{cs}C}{l_2(1 - c_2/l_2)^3} \qquad (4.3)$$

where E_{cs} = modulus of elasticity of slab concrete
c_2 = size of rectangular column, capital, wall, or bracket in direction l_2
C = a cross-sectional constant

The summation applies to the typical case in which there are edge beams on both sides of the column. The constant C pertains to the torsional rigidity of the effective beam cross section. It is found by dividing the section into its component rectangles, each having smaller dimension x and larger dimension y, and summing the contribution of all the parts by means of the equation

$$C = \sum \left(1 - 0.63 \frac{x}{y}\right) \frac{x^3 y}{3} \qquad (4.4)$$

The subdivision may be done in such a way as to maximize C.

If a panel contains a beam parallel to the direction in which moments are being determined, the value of K_t obtained from Eq. (4.3) leads to values of K_{ec} which are too low. Accordingly, it is recommended that in such cases the value of K_t found by Eq. (4.3) be multiplied by the ratio of the moment of inertia of the slab with such beam to the moment of inertia of the slab without it.

Having obtained the flexural stiffness of the equivalent column, K_{ec}, the total static moment M_0 in an end span is divided between the critical positive and negative bending sections according to the following relationships:

Interior negative design moment: $\quad M_{neg} = \left(0.75 - \dfrac{0.10}{1 + 1/\alpha_{ec}}\right) M_0 \qquad (4.5a)$

Positive design moment: $\qquad\qquad M_{pos} = \left(0.63 - \dfrac{0.28}{1 + 1/\alpha_{ec}}\right) M_0 \qquad (4.5b)$

Exterior negative design moment: $\quad M_{neg} = \left(\dfrac{0.65}{1 + 1/\alpha_{ec}}\right) M_0 \qquad\qquad (4.5c)$

in which

$$\alpha_{ec} = \frac{K_{ec}}{\Sigma(K_s + K_b)} \qquad\qquad (4.6)$$

In Eq. (4.6), K_s is the flexural stiffness of the slab of width l_2, depth h, and span l_1, and K_b is the flexural stiffness of the column-line beam spanning in the direction l_1 and having effective dimensions shown in Fig. 4.8. For T beams of usual proportions, the moment of inertia may be taken equal to twice that of the rectangular web.

To summarize, to compute the distribution of the static moment M_0 in an end panel, the following steps are followed:

1. Compute K_s, the flexural stiffness of the slab for bending in the direction l_1, neglecting the presence of any beam on the column line.
2. Compute K_b, the flexural stiffness of the column-line beam, for bending in the direction l_1.
3. Compute K_c, the flexural stiffness of the actual column.
4. Compute C, the torsional constant for the edge beam, including the adjacent contributing slab, from Eq. (4.4).
5. Compute K_t, the torsional stiffness of the edge beam, from Eq. (4.3).
6. Compute K_{ec}, the flexural stiffness of the equivalent column, by Eq. (4.2).
7. Compute α_{ec}, the flexural stiffness ratio, by Eq. (4.6).
8. Apply Eqs. (4.5a), (4.5b), and (4.5c) to distribute M_0 to the negative and positive bending sections.

This rather complicated procedure is necessary, according to the research reported in Ref. 4.1, in order properly to reflect the influence of torsional deformation of the edge beam. The evaluation of Eqs. (4.5) is facilitated through use of Graph 13 of the Appendix.

Having distributed the moment M_0 to the positive- and negative-moment sections as just described, it still remains to distribute these design moments across the width of the critical sections. For design purposes, as discussed in Art. 4.5b, it is convenient to consider the moments constant within the bounds of a middle strip or column strip, unless there is a beam present on the column line. In the latter case, because of its greater stiffness, the beam will tend to take a larger share of the column-strip moment than will the adjacent slab.

The distribution of total negative or positive moment between slab middle strips, slab column strips, and beams depends upon the ratio l_2/l_1, the relative stiffness of the beam and the slab, and the degree of torsional restraint provided by the edge beam.

A convenient parameter defining the relative stiffness of the beam and slab spanning in the direction l_1 is

$$\alpha = \frac{E_{cb}I_b}{E_{cs}I_s} \tag{4.7}$$

in which E_{cb} and E_{cs} are the moduli of elasticity of the beam and slab concrete (usually the same), and I_b and I_s are the moments of inertia of the effective beam and the slab. Subscripted parameters α_1 and α_2 are used to identify α computed for the directions of l_1 and l_2, respectively.

The relative restraint provided by the torsional resistance of the edge beam is reflected by the parameter β_t, defined by

$$\beta_t = \frac{E_{cb}C}{2E_{cs}I_s} \tag{4.8}$$

where I_s, as before, is calculated for the slab spanning in direction l_1, and having width l_2.

With these parameters defined, the ACI Code direct method distributes the design negative and positive moments between column strips and middle strips, assigning to the column strips the percentages of design moments shown in Table 4.2. Linear interpolations are to be made between the values shown.

Implementation of these provisions is facilitated by the interpolation charts of Graph 14 of the Appendix. Interior negative- and positive-moment

Table 4.2   Column-strip moment, per cent of total design moment

		l_2/l_1		
		0.5	1.0	2.0
Interior negative design moment:				
$\alpha_1 l_2/l_1 = 0$		75	75	75
$\alpha_1 l_2/l_1 \geq 1.0$		90	75	45
Exterior negative design moment:				
$\alpha_1 l_2/l_1 = 0$	$\beta_t = 0$	100	100	100
	$\beta_t \geq 2.5$	75	75	75
$\alpha_1 l_2/l_1 \geq 1.0$	$\beta_t = 0$	100	100	100
	$\beta_t \geq 2.5$	90	75	45
Positive design moment:				
$\alpha_1 l_2/l_1 = 0$		60	60	60
$\alpha_1 l_2/l_1 \geq 1.0$		90	75	45

percentages may be read directly from the chart for known values of l_2/l_1 and $\alpha_1 l_2/l_1$. For exterior negative moment, the parameter β_t requires an additional interpolation, facilitated by the auxiliary diagram on the right side of the chart. To illustrate its use for $l_2/l_1 = 1.55$ and $\alpha_1 l_2/l_1 = 0.6$, the dotted line indicates moment percentages of 100 for $\beta_t = 0$ and 65 for $\beta_t = 2.5$. Projecting to the right as indicated by the arrow to find the appropriate vertical scale of 2.5 divisions for an intermediate value of β_t, say 1.0, then upward and finally to the left, the corresponding percentage of 86 is read on the main chart.

The column-line beam spanning in the direction l_1 is to be proportioned to resist 85 per cent of the column-strip moment if $\alpha_1 l_2/l_1$ is equal to or greater than 1.0. For values between 1.0 and zero the proportion to be resisted by the beam may be obtained by linear interpolation. Concentrated loads applied directly to such a beam should be accounted for separately.

The portion of the design moment not resisted by the column strip is proportionately assigned to the adjacent half middle strips. The middle strip adjacent to an edge supported by a wall should be proportioned to resist twice the moment assigned to its interior half.

Special attention must be given to providing the proper resistance to shear, as well as to moment, when designing by the direct method. According to the Code, beams with $\alpha_1 l_2/l_1$ equal to or greater than 1.0 must be proportioned to resist the shear caused by loads on a tributary area defined as shown in Fig. 4.10. For values of $\alpha_1 l_2/l_1$ between 1.0 and zero, the proportion of load carried by beam shear is to be found by linear interpolation. The remaining fraction of the load on the shaded area is assumed to be transmitted directly by the slab to the columns at the four corners of the panel, and the shear stress in the slab computed accordingly (see Art. 4.8).

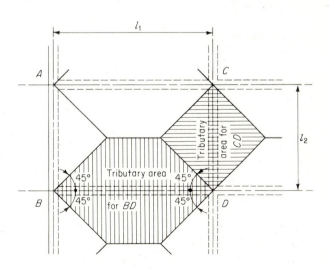

Fig. 4.10 Tributary areas for shear calculation.

Columns in two-way construction must be designed to resist the moments found from analysis of the slab-beam system. The column supporting an edge beam must provide a resisting moment equal to the moment applied from the edge of the slab (see Fig. 4.9). At interior locations, slab negative moments are found assuming that dead and full live loads act. For the column design, a more severe loading results from partial removal of live load. Accordingly, the Code requires that interior columns resist a moment

$$M = \frac{0.08[(w_d + 0.5w_l)l_2l_n{}^2 - w_d'l_2'(l_n')^2]}{1 + 1/\alpha_{ec}} \tag{4.9}$$

In Eq. (4.9), the primed quantities refer to the shorter of the two adjacent spans (assumed to carry dead load only), and the unprimed quantities refer to the longer span (assumed to carry dead load and half live load). In all cases, the moment is distributed to the upper and lower columns in proportion to their relative flexural stiffness.

When β_a, the ratio of dead to live load, is less than 2, *one* of the following two conditions must be satisfied:

1. The sum of the flexural stiffnesses of the columns above and below the slab must be such that $\alpha_c = \Sigma K_c/\Sigma(K_s + K_c)$ is not less than the minimum value given by Table 4.3.

Table 4.3 $\alpha_{min} = $ **minimum values of** $\alpha_c = \dfrac{\Sigma K_c}{\Sigma(K_s + K_b)}$

β_α	Aspect ratio l_2/l_1	Relative beam stiffness, $\alpha = \dfrac{E_{cb}I_b}{E_{cs}I_s}$				
		0	0.5	1.0	2.0	4.0
2.0	0.5–2.0	0	0	0	0	0
1.0	0.5	0.6	0	0	0	0
	0.8	0.7	0	0	0	0
	1.0	0.7	0.1	0	0	0
	1.25	0.8	0.4	0	0	0
	2.0	1.2	0.5	0.2	0	0
0.5	0.5	1.3	0.3	0	0	0
	0.8	1.5	0.5	0.2	0	0
	1.0	1.6	0.6	0.2	0	0
	1.25	1.9	1.0	0.5	0	0
	2.0	4.9	1.6	0.8	0.3	0
0.33	0.5	1.8	0.5	0.1	0	0
	0.8	2.0	0.9	0.3	0	0
	1.0	2.3	0.9	0.4	0	0
	1.25	2.8	1.5	0.8	0.2	0
	2.0	13.0	2.6	1.2	0.5	0.3

2. The positive design moment in the panels supported by these columns must be multiplied by the coefficient δ_s:

$$\delta_s = 1 + \frac{2 - \beta_a}{4 + \beta_a}\left(1 - \frac{\alpha_c}{\alpha_{\min}}\right) \qquad (4.10)$$

An example of the application of the ACI direct design method will be found in Art. 4.11.

b. Equivalent-frame method   The direct method of analysis of two-way slabs described in Art. 4.6a is useful provided each of the six restrictions on geometry and load is satisfied by the proposed structure. In other cases, a more general method is needed. One such method was proposed by Peabody in 1948 (Ref. 4.2), and was incorporated in subsequent editions of the ACI Code as *design by elastic analysis*. It appears, in refined form, in the 1971 ACI Code as the *equivalent-frame method*. Complete documentation is found in Ref. 4.3.

By the equivalent-frame method the structure is divided, for analysis, into continuous frames centered on the column lines and extending both longitudinally and transversely, as shown by the shaded strips in Fig. 4.11. Each frame is composed of a row of columns and a broad continuous beam. The beam includes the portion of the slab bounded by panel center lines on either side of the columns, together with column-line beams or drop panels if used. For vertical loading, each floor with its columns may be analyzed separately, the columns assumed fixed at the floors above and below. In calculating bend-

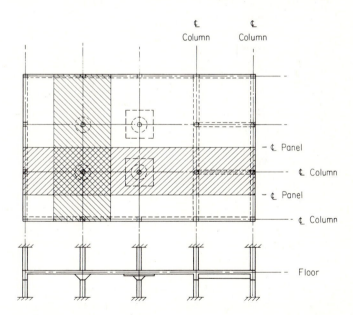

Fig. 4.11 Building idealization for equivalent-frame analysis.

ing moment at a support, it is convenient and sufficiently accurate to assume that the continuous frame is completely fixed at the support two panels removed from the one of interest.

Moments of inertia used for analysis may be based on the concrete cross section, neglecting reinforcement, but variations in cross section along the member axis should be accounted for.

For the beam strips, the first change from the midspan moment of inertia normally occurs at the edge of drop panels, if they exist. The next occurs at the edge of the column or column capital. While the stiffness of the slab strip could be considered infinite within the bounds of the column or capital, at locations close to the panel center lines (at each edge of the slab strip) the stiffness is much less. According to the Code, from the center of the column to the face of the column or capital, the moment of inertia of the slab is taken equal to that at the face of the column or capital, divided by the quantity $(1 - c_2/l_2)^2$, where c_2 and l_2 are measured transverse to the direction in which moments are being determined.

Accounting for these changes in moment of inertia produces a member, for analysis, in which the moment of inertia varies in a stepwise manner. Stiffness factors, carryover factors, and uniform-load fixed-end moments are given in Table 12a of the Appendix, for a slab without drop panels, and in Table 12b, for a slab with drop panels of depth equal to 1.25 times the slab depth and of length equal to $\frac{1}{3}$ the span length.

The concept of the equivalent column, introduced in the direct design method to account for the torsional deformation of the edge beams, is used for *all* columns for the equivalent-frame analysis. The stiffness K_{ec} of the equivalent column is found from Eq. (4.2) as before. For determining K_c for the present case, the moment of inertia of the actual column is taken to be infinite from the top of the slab to the bottom of the slab, beam, or capital. Stiffness and carryover factors for such a case are given in Table 12c. The stiffness of the torsional member, K_t, is given by Eq. (4.3), where the torsional constant C is found using Eq. (4.4).

With the effective stiffness of the slab strip and the columns found in this way, the analysis of the equivalent frame can proceed by any convenient means (see Art. 7.2).

In keeping with the requirements of statics (see Art. 4.5b), equivalent beam strips in each direction must each carry 100 per cent of the applied load. If the live load does not exceed three-quarters of the dead load, maximum moment may be assumed to obtain at all critical sections when the full live load (plus dead load) is on the entire slab. Otherwise pattern loadings must be used to maximize positive and negative moments (see Art. 7.3). Maximum positive moment is calculated with three-quarters live load on the panel and on alternate panels, while maximum negative moment at a support is calculated with three-quarters live load on the adjacent panels only. Use of three-quarters live load rather than the full value recognizes that maximum positive and negative moments cannot occur simultaneously (since they are found from different loadings) and that redistribution of moments to less highly stressed sections will take place before failure of the structure occurs.

Negative moments obtained from the analysis apply at the center lines of supports. Since the support provided is not a knife-edge, but is a rather broad band of slab spanning in the transverse direction, some reduction in the negative design moment is proper (see also Art. 7.5a). At interior supports, the critical section for negative bending, in both column and middle strips, may be taken at the face of the supporting column or capital, but in no case at a distance greater than $0.175l_1$ from the center of the column. To avoid excessive reduction of negative moment at, and perpendicular to, an edge (where the point of inflection is close), the critical section should be taken a distance from the column face not greater than one-half the projection of the column capital.

With positive and negative design moments obtained as just described, it still remains to distribute these moments across the widths of the critical sections. For design purposes, the total strip width is divided into column strip and adjacent middle strips, defined previously, and moments are assumed constant within the bounds of each. The distribution of moments to column and middle strips is done using the same percentages given in connection with the direct design method. These are summarized in Table 4.2 and by the interpolation charts of Graph 14 of the Appendix.

The distribution of moments and shears to column-line beams, if present, is in accordance with the procedures of the direct design method also. Restriction 6 of Art. 4.6a, pertaining to the relative stiffness of column-line beams in the two directions, applies here also if these distribution ratios are used.

4.7 FLEXURAL REINFORCEMENT

Consistent with the assumptions made in analysis, flexural reinforcement in two-way-slab systems is almost always placed in an orthogonal grid, with bars parallel to the sides of the panels. Bar diameters and spacings may be found as described in Art. 4.2. Straight bars may be used throughout, although in many cases positive-moment steel is bent up where no longer needed, in order to provide for part or all of the negative requirement. To provide for local concentrated loads, as well as to ensure that tensile cracks are narrow and well distributed, a maximum bar spacing at critical sections of 2 times the total slab thickness is specified by the Code. At least the minimum steel required for temperature and shrinkage crack control (see Art. 4.3) must be provided. For protection of the steel against damage from fire or corrosion, at least $\frac{3}{4}$-in. concrete cover must be maintained.

Because of the stacking that results when bars are placed in perpendicular layers, the inner steel will have an effective depth 1 bar diameter less than the outer steel. For rectangular panels, the short-direction bars are commonly placed so as to have the maximum d. For panels which are square or nearly so, the designer may calculate the steel area based on the average effective depth, obtaining the same bar size and spacing in each direction. This is slightly conservative for the outer steel, and slightly unconservative for the inner steel. Redistribution of loads and moments before failure would provide for the resulting difference in capacities.

If bent bars are used in two-way slabs, careful attention must be given

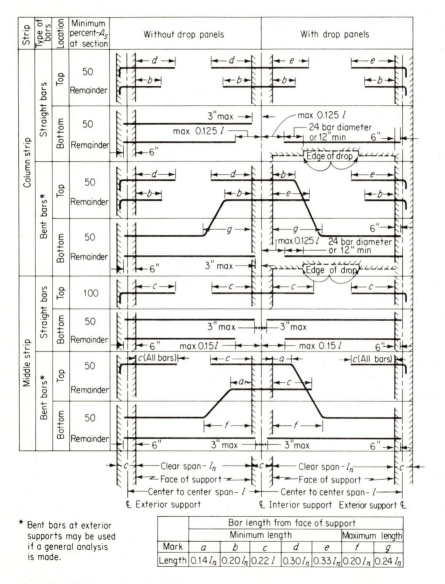

			Bar length from face of support					
			Minimum length				Maximum length	
Mark			a	b	c	d	e	f	g
Length			$0.14 l_n$	$0.20 l_n$	$0.22 l$	$0.30 l_n$	$0.33 l_n$	$0.20 l_n$	$0.24 l_n$

* Bent bars at exterior supports may be used if a general analysis is made.

Fig. 4.12 Minimum length of slab reinforcement, slabs without beams.

to the order of placement and detailing of bend dimensions, so that proper location of steel results.

Bar bend points or cutoff points may be found from moment envelopes, with due regard for development length (see Art. 3.8). When the direct method of analysis is used, moment envelopes and inflection points are not found explicitly. In such case (and in many cases where frame analysis is used as well), approximate bend and cutoff points from Fig. 4.12 may be used, according to the Code. Note that, in exterior spans, *all* positive reinforcement perpendicular to the discontinuous edge, as well as all exterior negative reinforcement, must extend to the edge of the slab and be suitably anchored.

When a slab supported on four sides is uniformly loaded, there is a tendency for the corners to raise. In practical cases they cannot do so, because the slab is cast monolithically with the supporting beams. As a result, corner slab moments exist which cause tension at the top of the slab in the direction of the panel diagonal. Also, in spite of the assumptions of the analysis, a slab tends to span diagonally at its corners between the supporting beams. This causes bending moment, producing tension at the bottom of the slab in the direction perpendicular to the panel diagonal.

Because of these conditions, special reinforcement is required at the *exterior* corners of two-way systems. According to the ACI Code, such reinforcement should be placed parallel to the panel diagonal at the top of the slab, and perpendicular to the panel diagonal at the bottom, and should be provided for a distance in each direction from the corner equal to one-fifth the longer span. In either the top or bottom of the slab, the reinforcement may be placed in a single band in the direction of the moment or in two bands parallel to the sides of the slab. The reinforcement in both the top and the bottom should be sufficient to resist a moment equal to the maximum positive moment per foot of width in the slab.

4.8 SHEAR DESIGN

If two-way slabs are supported by beams or walls, shear stress is normally low and will seldom be a controlling factor in design. The shear force V_u may be calculated using the tributary areas shown in Fig. 4.10. It is convenient to base calculations on slab strips of unit width perpendicular to the supported edge. The critical sections for shear may be taken a distance d from the face of the beam or wall, and the nominal shear stress at that section calculated, as for beams, using Eq. (3.23). The nominal shear stress should not exceed the value given by Eq. (3.24a) or (3.24b). Shear reinforcement is not normally used for edge-supported slabs.

In contrast, when two-way slabs are supported directly by columns as in flat slabs and flat plates, or when they carry concentrated loads as in footings, shear near the columns is of critical importance. Tests of flat-plate structures indicate that, in most practical cases, the capacity is governed by shear (Ref. 4.4).

a. Shear in column-supported slabs without special reinforcement   There are two kinds of shearing stress that may be critical in the design of flat slabs, flat plates, or footings. The first is the familiar beam-type shear leading to diagonal tension failure. Applicable particularly to long narrow slabs or footings, this analysis considers the slab to act as a wide beam, spanning between supports provided by the perpendicular column strips. A potential diagonal crack extends in a plane across the entire width of the slab. The critical section is taken a distance d from the face of the column or capital. As for beams, the nominal shear stress $V_u/\phi b_w d$ should not exceed the values given by Eq. (3.24a) or (3.24b).

Alternatively, failure may occur by *punching shear*, with the potential diagonal crack following the surface of a truncated cone or pyramid around the column, capital, or drop panel, as shown in Fig. 4.13a. The failure surface extends from the bottom of the slab, at the support, diagonally upward to the top surface. The angle of inclination with the horizontal, θ (see Fig. 4.13b), depends upon the nature and amount of reinforcement in the slab. It may range between about 20° and 45°. The critical section for shear is taken perpendicular to the plane of the slab and a distance $d/2$ from the periphery of the support, as shown. The nominal shear stress is

$$v_u = \frac{V_u}{\phi b_0 d} \tag{4.11}$$

where b_0 is the perimeter along the critical section, and $\phi = 0.85$ as usual for shear.

At such a section, in addition to the shearing stresses and horizontal compressive stresses due to negative bending moment, vertical or somewhat inclined compressive stress is present, owing to the reaction of the column. The simultaneous presence of vertical and horizontal compression increases the shear resistance of the concrete. Tests have indicated that, when punching-shear failure occurs, the shear stress computed on the perimeter of the critical

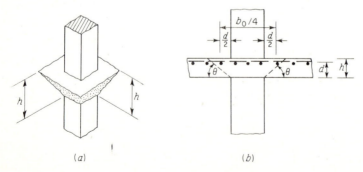

(a)   (b)

Fig. 4.13   Failure surface defined by punching shear.

section is larger than in beams or one-way slabs, and shear resistance may be taken equal to $4 \sqrt{f'_c}$.

b. Types of shear reinforcement for slabs Special shear reinforcement is usually desirable at the supports for flat plates, and sometimes for flat slabs as well. It may take several forms. A few common types are shown in Fig. 4.14.

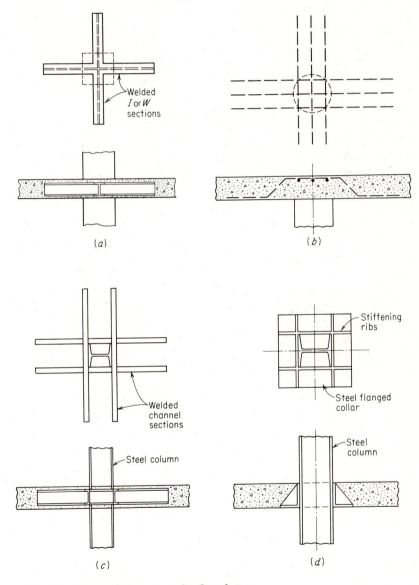

Fig. 4.14 Shear reinforcement for flat plates.

The *shearheads* shown in (a) and (c) consist of standard structural steel shapes embedded in the slab and projecting beyond the column. They serve to increase the effective perimeter b_0 of the critical section for shear. In addition, they may contribute to the negative bending resistance of the slab. The reinforcement shown in (a) is particularly suited for use with concrete columns. It consists of short lengths of I or wide-flange beams, cut and welded at the crossing point so that the arms are continuous through the column. Normal negative slab reinforcement passes over the top of the structural steel, while bottom bars are stopped short of the shearhead. Column bars pass vertically at the corners of the column without interference. The effectiveness of this type of shearhead has been well documented by testing at the laboratories of the Portland Cement Association (Ref. 4.4). The channel frame of (c) is very similar in its action, but is adapted for use with steel columns. The bent-bar arrangement of (b) is suited for use with concrete columns. The bars are usually bent at 45° across the potential diagonal tension crack, and extend along the bottom of the slab a distance sufficient to develop their strength by bond. The flanged collar of (d) is designed mainly for use with lift-slab construction (see Fig. 8.12). It consists of a flat bottom plate with vertical stiffening ribs. It may incorporate sockets for lifting rods, and usually is used in conjunction with shear pads welded directly to the column surfaces below the collar to transfer the vertical reaction.

c. Design of bar reinforcement   If shear reinforcement in the form of bars is used (Fig. 4.14b), the ultimate shear stress v_u, calculated at the critical section $d/2$ from the support face, may be increased to $6\sqrt{f_c'}$ according to the ACI Code. In this case, the shear resistance of the concrete, v_c, is taken equal to $2\sqrt{f_c'}$, and reinforcement must provide for the excess shear stress above that value. The total bar area A_v crossing the critical section at slope angle α is easily obtained by equating the vertical component of the steel force to the excess shear force to be accommodated:

(a) $A_v f_y \sin\alpha = V_u - V_c$

from which

(b) $A_v = \dfrac{V_u - V_c}{f_y \sin\alpha}$

This requirement may be expressed in terms of the nominal unit shear stresses on the section of area $b_0 d$:

$$A_v = \frac{(v_u - v_c)b_0 d}{f_y \sin\alpha} \tag{4.12}$$

Successive sections at increasing distances from the support must be investigated, and reinforcement provided wherever v_u exceeds $4\sqrt{f_c'}$. Only the center three-quarters of the inclined portion of the bent bars can be considered effective in resisting shear, and full development length must be provided past the location of peak stress in the steel.

d. Design of shearhead reinforcement If embedded structural-steel shapes are used, as shown in Fig. 4.14a and c, the limiting value of v_u may be increased to $7\sqrt{f_c'}$. Such a shearhead, provided it is sufficiently stiff and strong, has the effect of moving the critical section out away from the column, as shown in Fig. 4.15. According to the Code, this critical section crosses each arm of the shearhead at a distance equal to three-quarters of the projection beyond the face of the support, and is defined so that the perimeter is a minimum. It need not approach closer than $d/2$ to the face of the support.

Moving the critical section out in this way provides the double benefit of increasing the effective perimeter b_0 and decreasing the total shear force V_u for which the slab must be designed. The nominal shear stress at the new critical section must not exceed $4\sqrt{f_c'}$, according to the Code.

Tests reported in Ref. 4.4 indicate that, throughout most of the length of a shearhead arm, the shear is constant, and further, that the part of the total shear carried by the shearhead arm is proportional to α_v, its relative flexural stiffness, compared with that of the surrounding concrete section:

$$\alpha_v = \frac{E_s I_s}{E_c I_c}$$

The concrete section is taken with an effective width of $(c_2 + d)$, where c_2 is the width of the support measured perpendicular to the arm direction. Properties are calculated for the cracked, transformed section, including the shearhead. The observation that shear is essentially constant, at least up to the

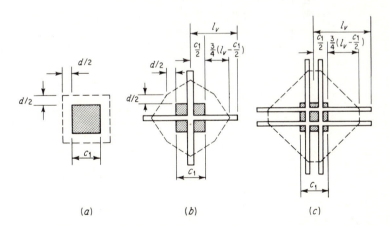

Fig. 4.15 Critical section for shear for flat plates. (a) No shearhead; (b) small shearhead; (c) large shearhead. (*After Ref. 4.4.*)

diagonal-cracking load, implies that the reaction is concentrated largely at the end of the arm. Thus, if the total shear at the support is V, the constant shear force in each arm is equal to $\alpha_v V/4$.

If load is increased past that which causes diagonal cracking around the column, tests indicate that the increased shear above the cracking shear V_c is carried mostly by the steel shearhead, and that the shear force in the projecting arm within a distance from the column face equal to h_v, the depth of the arm, assumes a nearly constant value greater than $\alpha_v V_c/4$. This increased value is very nearly equal to the total shear per arm, $V_u/4$, minus the shear carried by the partially cracked concrete. The latter term is equal to $(V_c/4)(1 - \alpha_v)$; hence, the idealized shear diagram of Fig. 4.16b is obtained.

The moment diagram of Fig. 4.16c is obtained by integration of the shear diagram. If V_c is equal to $\frac{1}{2}V_u$, as tests indicate for shearheads of common proportions, it is easily confirmed that the plastic moment at the face of the support, for which the shearhead arm must be proportioned, is

$$M_p = \frac{V_u}{8\phi}\left[h_v + \alpha_v\left(l_v - \frac{c_1}{2}\right)\right] \tag{4.13}$$

in which the capacity reduction factor ϕ is taken equal to 0.90 as usual for bending.

According to the Code, the value of α_v must be at least equal to 0.15; more flexible shearheads have proved ineffective. The compression flange must not be more than $0.3d$ from the bottom surface of the slab, and the steel shapes used must not be deeper than 70 times the web thickness.

For flexural design of the slab, moments found at the support center line by the equivalent-frame method are reduced to moments at the support face,

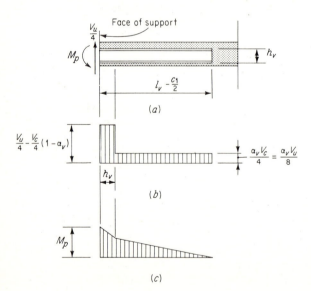

Fig. 4.16 Stress resultants in shearhead arm. (a) Shearhead arm; (b) shear; (c) moment.

assumed to be the critical section for moment. By the direct method, support-face moments are calculated directly through use of clear-span distance. If shearheads are used, they have the effect of reducing the design moment in the column strips still further by increasing the effective support width. This reduction is proportional to the share of the load carried by the shearhead, and to its size, and may be estimated conservatively (see Fig. 4.16b and c) by the expression

$$M_v = \frac{\phi \alpha_v V_u}{8} \left(l_v - \frac{c_1}{2} \right) \tag{4.14}$$

where $\phi = 0.90$. According to the Code, the reduction may not be greater than 30 per cent of the total design moment for the slab column strip, nor greater than the change in column-strip moment over the distance l_v, nor greater than M_p given by Eq. (4.13).

Limited test information pertaining to shearheads at a slab edge indicates that behavior may be substantially different due to torsional and other effects. If shearheads are to be used at an edge or corner column, special attention must be given to anchorage of the embedded steel within the column. The use of edge beams or a cantilevered slab edge may be preferred.

Example: design of shearhead reinforcement   A flat-plate floor slab $7\frac{1}{2}$ in. thick is supported by 10-in. square columns and is reinforced for negative bending with No. 5 bars 5 in. on centers in each direction, with an average effective depth d of 6 in. The concrete strength f_c' is 3,000 psi. The slab must transfer an ultimate shear V_u of 113,000 lb to the column. What special slab reinforcement is required, if any, at the column to transfer the required ultimate shear?

The nominal shear stress at the critical section $d/2$ from the face of the column is found from Eq. (4.11) to be

$$v_u = \frac{113,000}{0.85 \times 4(10 + 6)6} = 346 \text{ psi}$$

This is greater than $4\sqrt{3000} = 220$ psi, indicating that shear reinforcement is necessary. A shearhead similar to Fig. 4.14a will be used, fabricated from I-beam sections with $f_y = 36$ ksi. Maintaining $\frac{3}{4}$-in. clearance below such steel, bar clearance at the top of the slab permits use of an I beam of $4\frac{5}{8}$ in. depth; a nominal 4-in. section will be used. With such reinforcement, the upper limit of nominal shear stress on the critical section is $7\sqrt{3000} = 384$ psi, well above the calculated stress. The required perimeter b_0 can be found setting $v_u = 220$ psi in Eq. (4.11).

$$b_0 = \frac{113,000}{220 \times 0.85 \times 6} = 101 \text{ in.}$$

(Note that the actual shear force to be transferred at the critical section is slightly less than 113 kips, because a part of the floor load is within the effective perimeter b_0; however, the difference is small except for very large shearheads.) The required projecting length l_v of the shearhead arm is found from geometry, expressing b_0 in terms of l_v.

$$b_0 = 4\sqrt{2}\left[\frac{c_1}{2} + \frac{3}{4}\left(l_v - \frac{c_1}{2}\right)\right] = 101 \text{ in.}$$

from which $l_v = 22.2$ in. To determine the required plastic section modulus for the shear arm it is necessary to assume a trial value of the relative stiffness α_v. Selecting 0.25 for trial, the required moment capacity is found from Eq. (4.13).

$$M_p = \frac{113,000}{8 \times 0.90}[4 + 0.25(22.2 - 5)] = 130,000 \text{ in.-lb}$$

A standard I beam S4 × 7.7, with yield stress of 36 ksi, provides 126,000 in.-lb resistance, and will tentatively be adopted. The $E_s I_s$ value provided by the beam is 174×10^6 in.²-lb. The effective cross section of the slab strip is shown in Fig. 4.17. Taking moments of the composite cracked section about the bottom surface to locate the neutral axis,

$$y = \frac{8.90 \times 6 + 19.9 \times 2.75 + 8y^2}{8.90 + 19.9 + 16y}$$

from which $y = 2.29$ in. The moment of inertia of the composite section is

$$I_c = \tfrac{1}{3} \times 16 \times 2.29^3 + 8.90 \times 3.71^2 + 6 \times 9 + 19.9 \times 0.46^2$$
$$= 244 \text{ in.}^4$$

The flexural stiffness of the effective composite slab strip is

$$E_c I_c = 3.1 \times 10^6 \times 244 = 756 \times 10^6 \text{ in.}^2\text{-lb}$$

and from Eq. (4.12),

$$\alpha_v = \tfrac{174}{756} = 0.23$$

This is greater than the specified minimum of 0.15 and close to the 0.25 value assumed earlier. The revised value of M_p is

$$M_p = \frac{113,000}{8 \times 0.90}[4 + 0.23(22.2 - 5)] = 122,000 \text{ in.-lb}$$

The 4-in. I beam is adequate. The calculated length l_v of 22.2 in. will be increased to

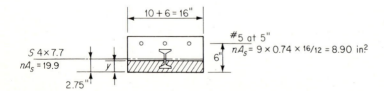

Fig. 4.17 Effective section of slab.

24 in. for practical reasons. The reduction in column-strip moment in the slab may be based on this actual length. From Eq. (4.14),

$$M_v = \frac{0.90 \times 0.23 \times 113,000}{8} (24 - 5) = 55,600 \text{ in.-lb}$$

This value is less than M_p, as required by specification, and must also be less than 30 per cent of the design negative moment in the column strip, and less than the change in the column-strip moment in the distance l_v.

4.9 OPENINGS IN SLABS

Almost invariably, slab systems must include openings. These may be of substantial size, as required by stairways and elevator shafts, or they may be of smaller dimensions, such as needed to accommodate heating, plumbing, and ventilating risers; floor and roof drains; and access hatches. Large openings should be framed by beams to restore, as nearly as possible, the continuity of the slab. The beams should be designed to carry a portion of the floor load, in addition to loads directly applied by partition walls, elevator support beams, stair slabs, etc.

Relatively small openings are usually not detrimental in beam-supported slabs. As a general rule, the equivalent of the interrupted reinforcement should be added at the sides of the opening. Additional diagonal bars are often added at the corners to control the cracking that will almost inevitably occur there.

The detrimental effect of openings in slabs supported directly by columns (flat slabs and flat plates) depends upon the location of the opening with respect to the columns. From a structural viewpoint, they are best located well away from the columns, preferably in the area common to the slab middle strips. Unfortunately, architectural considerations usually cause them to be located close to the columns. In such a case, the reduction in effective shear perimeter is of major concern, since such floors are usually shear-critical.

If the opening is close to the column (within 10 slab thicknesses or within the column strips, according to the Code), then that part of b_0 included within radial lines projecting from the opening to the centroid of the column should be considered ineffective. If shearheads are used under similar circumstances, the reduction in width of the critical section is found in a similar way, except that only one-half the perimeter included within the radial lines need be deducted.

With regard to flexural requirements, the total amount of steel required by calculation must be provided, regardless of openings. Any steel interrupted by holes should be matched with an equivalent amount of supplementary reinforcement on either side, properly lapped to transfer stress by bond. Concrete compression area to provide the required strength must be maintained; usually, this would be restrictive only near the columns. According to the Code, openings of any size may be placed in the area within the middle half of the span in each direction. In the area common to two column strips, not more than one-eighth the width of the strip in either span should be interrupted by openings. In the area common to one column strip and one middle strip, not more than one-quarter of the reinforcement in either strip should be interrupted by the opening.

4.10 DEPTH LIMITATIONS

Great progress has been made in recent years in developing methods which are both simple and accurate for predicting deflection of beams (see Art. 3.12). Efforts are now being made to develop corresponding methods for two-way systems. As yet, no means are available which are sufficiently accurate and at the same time simple enough for practical use.

In the absence of such methods, it continues to be necessary to rely, for deflection control, upon more or less arbitrary limitations on minimum slab thickness, developed from study of the observed deflections in actual structures. As a result of efforts to improve the accuracy of the limiting equations, they have become increasingly complex. It is to be hoped that they may soon be abandoned in favor of direct calculation of deflection for comparison with limiting values.

The ACI Code specifies that the slab thickness shall not be less than

$$h = \frac{l_n(800 + 0.005f_y)}{36,000 + 5000\beta[\alpha_m - 0.5(1 - \beta_s)(1 + 1/\beta)]} \tag{4.15}$$

nor less than

$$h = \frac{l_n(800 + 0.005f_y)}{36,000 + 5000\beta(1 + \beta_s)} \tag{4.16}$$

but in any case need not be more than

$$h = \frac{l_n(800 + 0.005f_y)}{36,000} \tag{4.17}$$

where l_n = clear span in long direction, in.
α_m = average value of α for all beams on edges of a panel [see Eq. (4.7)]
β_s = ratio of length of continuous edges to total perimeter of a slab panel
β = ratio of clear span in long direction to clear span in short direction
However, the thickness shall not be less than the following values:

For slabs without beams or drop panels:	5 in.
For slabs without beams but with drop panels:	4 in.
For slabs having beams on all four edges with a value of α_m at least equal to 2.0:	$3\frac{1}{2}$ in.

For slabs without beams but with drop panels having a length in each direction equal to at least one-third the clear-span length in that direction and a projection below the slab of at least $h/4$, the thickness required by Eq. (4.15) or (4.16) may be reduced by 10 per cent.

Unless an edge beam having a stiffness such that α is at least 0.80 is provided at all discontinuous edges, the minimum thickness must be increased by 10 per cent in the panel having the discontinuous edge.

4.11 Example: design of two-way slab A two-way reinforced-concrete building floor system is composed of slab panels measuring 20×25 ft in plan, supported by column-line beams cast monolithically with the slab, as shown in Fig. 4.18. Using concrete with $f'_c = 4000$ psi and steel having $f_y = 60,000$ psi, design a typical exterior slab panel to carry a service live load of 125 psf in addition to the self-weight of the floor.

The floor system satisfies all limitations stated in Art. 4.6a, and the ACI direct design method will be used. For illustrative purposes, only a typical exterior panel, as shown in Fig. 4.18, will be designed. The depth limitations of Art. 4.10 will be used as a guide to the desirable slab thickness. To use Eqs. (4.15) to (4.17), a trial value of $h = 7$ in. will be introduced, and beam dimensions 14×20 in. will be assumed, as shown in Fig. 4.18. The effective flange projection beyond the face of the beam webs is the lesser of $4h_f$ or $(h - h_f)$, and in the present case is 13 in. The moment of inertia of the T beams will be estimated as multiples of that of the rectangular portion as follows:

For the edge beams: $I = \frac{1}{12} \times 14 \times 20^3 \times 1.5 = 14,000$ in.⁴

For the interior beams: $I = \frac{1}{12} \times 14 \times 20^3 \times 2 = 18,600$ in.⁴

For the slab strips:

For the 13.1 ft width: $I = \frac{1}{12} \times 13.1 \times 12 \times 7^3 = 4500$ in.⁴

For the 20 ft width: $I = \frac{1}{12} \times 20 \times 12 \times 7^3 = 6900$ in.⁴

For the 25 ft width: $I = \frac{1}{12} \times 25 \times 12 \times 7^3 = 8600$ in.⁴

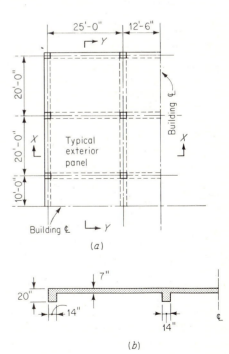

Fig. 4.18 Two-way-slab floor. (*a*) Partial floor plan; (*b*) section *X-X* (section *Y-Y* similar).

Thus, for the edge beam $\alpha = 14{,}000/4500 = 3.1$, for the two 25-ft-long beams $\alpha = 18{,}600/6900 = 2.7$, and for the 20-ft-long beam $\alpha = 18{,}600/8600 = 2.2$, producing an average value $\alpha_m = 2.7$. The ratio of continuous edges to total perimeter is $\beta_s = 70/90 = 0.78$, while the ratio of long to short clear spans is $\beta = 23.8/18.8 = 1.27$. Then the minimum thickness given by Eq. (4.15) is

$$h = \frac{286(800 + 0.005 \times 60{,}000)}{36{,}000 + 5000 \times 1.27(2.7 - 0.5 \times 0.22 \times 1.79)} = 6.08 \text{ in.}$$

However, the minimum thickness is not to be less than that given by Eq. (4.16):

$$h = \frac{286(800 + 0.005 \times 60{,}000)}{36{,}000 + 5000 \times 1.27 \times 1.78} = 6.65 \text{ in.}$$

and the limit need not be greater than the value from Eq. (4.17):

$$h = \frac{286(800 + 0.005 \times 60{,}000)}{36{,}000} = 8.75 \text{ in.}$$

The $3\frac{1}{2}$-in. limitation of Art. 4.10 clearly does not control in this case, and the 7-in. depth tentatively adopted will provide the basis of further calculation.

For a 7-in. slab, the dead load is $\frac{7}{12} \times 150 = 88$ psf. Applying the usual load factors to obtain design load,

$$w = 1.4 \times 88 + 1.7 \times 125 = 335 \text{ psf}$$

For the short-span direction, for the slab-beam strip centered on the interior column line, the total static design moment is

$$M_0 = \tfrac{1}{8} \times 0.335 \times 25 \times 18.8^2 = 371 \text{ ft-kips}$$

This is distributed as follows:

Negative design moment $= 371 \times 0.65 = 241$ ft-kips
Positive design moment $= 371 \times 0.35 = 130$ ft-kips

The column strip has a width of $2 \times \frac{20}{4} = 10$ ft. With $l_2/l_1 = \frac{25}{20} = 1.25$, and $\alpha_1 l_2/l_1 = 2.2 \times \frac{25}{20} = 2.75$, Graph 13 of the Appendix indicates that 68 per cent of the negative moment, or 163 ft-kips, is taken by the column strip, of which 85 per cent, or 139 ft-kips, is taken by the beam and 24 ft-kips by the slab. The remaining 78 ft-kips is allotted to the slab middle strip. Graph 13 also indicates that 68 per cent of the positive moment, or 88 ft-kips, is taken by the column strip, of which 85 per cent, or 75 ft-kips, is assigned to the beam and 13 ft-kips to the slab. The remaining 42 ft-kips is taken by the slab middle strip.

A similar analysis is performed for the short-span direction, for the slab-beam strip at the edge of the building, based on a total static design moment of

$$M_0 = \tfrac{1}{8} \times 0.335 \times 13.1 \times 18.8^2 = 194 \text{ ft-kips}$$

of which 65 per cent is assigned to the negative and 35 per cent to the positive bending sections as before. In this case, $\alpha_1 l_2/l_1 = 3.1 \times \frac{25}{20} = 3.9$. The distribution factor for column-strip moment, from Graph 13, is 68 per cent for positive and negative moments as before, and again 85 per cent of the column-strip moments are assigned to the beams.

In summary, the short-direction moments, in foot-kips, are as follows:

	Beam moment	Column-strip slab moment	Middle-strip slab moment
Interior slab-beam strip—20-ft span:			
Negative	139	24	78
Positive	75	13	42
Exterior slab-beam strip—20-ft span:			
Negative	73	13	40
Positive	39	7	22

The total static design moment in the long direction of the exterior panel is

$M_0 = \frac{1}{8} \times 0.335 \times 20 \times 23.8^2 = 475$ ft-kips

Before distributing this to negative- and positive-moment sections, it is necessary to find α_{ec}, the relative flexural stiffness of the equivalent exterior column, as described in Art. 4.6a. The eight-step procedure detailed in that section must be followed. (Since E_c is constant in the present example, it will be canceled from stiffness calculations.) For the slab, the flexural stiffness is

$K_s = \dfrac{4I}{l} = 4 \times \dfrac{6900}{300} = 92$

while for the column-line beam,

$K_b = 4 \times 18,600/300 = 248$

For the columns, assumed to be 14 in. $\times$ 14 in. $\times$ 12 ft. high, the moment of inertia $I = 3200$ in.^{4}, and

$K_c = 4 \times \dfrac{3200}{144} = 89$

The torsional constant for the edge beam is found from Eq. (4.4) for a 14- $\times$ 20-in. rectangular shape with a 7- $\times$ 13-in. projecting flange:

$C = \left(1 - 0.63 \times \dfrac{14}{20}\right) \dfrac{14^3 \times 20}{3} + \left(1 - 0.63 \times \dfrac{7}{13}\right) \dfrac{7^3 \times 13}{3} = 11,205$

and the torsional stiffness of the edge beam is obtained from Eq. (4.3):

$K_t = \dfrac{2 \times 9 \times 11,205}{240(1 - \frac{14}{240})^3} = 1010$

The ratio of moment of inertia of the slab with the column-line beam to that of the slab alone is 3.8 in the present case. In accordance with the Code, K_t is increased to

$K_t' = 3.8 \times 1010 = 3830$

From Eq. (4.2), the flexibility of the equivalent column is found from the sum of the flexibilities of the actual columns and the edge beams:

$\dfrac{1}{K_{ec}} = \dfrac{1}{2 \times 89} + \dfrac{1}{3830} = 0.0059$

from which $K_{ec} = 169.$* The flexural stiffness ratio α_{ec} is then found from Eq. (4.6):

$$\alpha_{ec} = \frac{169}{92 + 248} = 0.50$$

The total static moment may now be distributed according to Eq. (4.5) with the aid of Graph 14 of the Appendix. The moment ratios to be applied to obtain exterior negative, positive, and interior negative moments are, respectively, 0.20, 0.54, and 0.72. With $l_2/l_1 = 0.80$, $\alpha_1 l_2/l_1 = 2.7 \times \frac{20}{25} = 2.2$, and from Eq. (4.8), $\beta_t = 11{,}205/(2 \times 6900) = 0.82$, Graph 14 indicates that the column strip will take 93 per cent of the exterior negative moment, 81 per cent of the positive moment, and 81 per cent of the interior negative moment. As before, the column-line beam will account for 85 per cent of the column-strip moment. The results of applying these moment ratios are as follows:

	Beam moment	Column-strip slab moment	Middle-strip slab moment
Exterior Negative—25-ft span	75	13	7
Positive—25-ft span	177	31	48
Interior negative—25-ft span	236	41	65

It is convenient to tabulate the design of the slab reinforcement as in Table 4.4. In the 25-ft direction, the two half-column strips may be combined for purposes of calculation into one strip of 106 in. width. In the 20-ft direction the exterior half-column strip and the interior half-column strip will normally differ and are treated separately. Design moments from the previous distributions are summarized in column 3 of the table.

The short-direction positive steel will be placed first, followed by the long-direction positive bars. Allowing $\frac{3}{4}$ in. clear distance below the steel, and anticipating use of No. 4 bars, the effective depth in the short direction will be 6 in., while that in the long direction will be 5.5 in. A similar situation obtains for the top steel.

Following the calculation of design moment per foot strip of slab (column 6), the minimum effective slab depth required for flexure is found. For the material strengths to be used, the maximum permitted steel ratio is $0.75\rho_b = 0.0214$. For this ratio,

$$d^2 = \frac{M_u}{\phi \rho f_y b (1 - 0.59 \rho f_y / f'_c)}$$

$$= \frac{M_u}{0.90 \times 0.0214 \times 60{,}000 \times 12(1 - 0.59 \times 0.0214 \times \frac{60}{4})}$$

$$= \frac{M_u}{11{,}300}$$

* It is worth noting that, had the actual column stiffness $\Sigma K = 178$ been adopted here rather than $K_{ec} = 169$, very little difference in results would have obtained, although calculations would have been much simplified.

Table 4.4   Design of slab reinforcement

(1)	(2) Location	(3) M_u, ft-kips	(4) b, in.	(5) d, in.	(6) $M_u \times 12/b$, ft-kips per ft	(7) ρ	(8) A_s, in.²	(9) Number of No. 4 bars
25-ft span:								
Two half-column strips	Exterior negative	13	106	5.5	1.47	0.0023*	1.34	8
	Positive	31	106	5.5	3.51	0.0023	1.34	8
	Interior negative	41	106	5.5	4.64	0.0029	1.69	10
Middle strip	Exterior negative	7	120	5.5	0.70	0.0023*	1.52	9†
	Positive	48	120	5.5	4.80	0.0031	2.05	11
	Interior negative	65	120	5.5	6.50	0.0042	2.78	15
20-ft span:								
Exterior half-column strip	Negative	13	53	6	2.94	0.0021*	0.67	4
	Positive	7	53	6	1.58	0.0021*	0.67	4
Middle strip	Negative	78	180	6	5.20	0.0028	3.03	16
	Positive	42	180	6	2.80	0.0021*	2.27	13†
Interior half-column strip	Negative	12	53	6	2.71	0.0021*	0.67	4
	Positive	6.5	53	6	1.47	0.0021*	0.67	4

* Steel ratio controlled by shrinkage and temperature requirements.
† Number of bars controlled by maximum spacing requirement.

Hence, $d = \sqrt{M_u/11,300}$. Thus, the following minimum effective depths are needed:

In 25-ft direction: $d = \sqrt{6.50 \times 12,000/11,300} = 2.63$ in.
In 20-ft direction: $d = \sqrt{5.20 \times 12,000/11,300} = 2.35$ in.

both well below the depth dictated by deflection requirements. An underreinforced slab results. The required steel ratios (column 7) are conveniently found from Table 7 of the Appendix. Note that a minimum steel area equal to 0.0018 times the gross concrete area must be provided for control of temperature and shrinkage cracking. For a 12-in. slab strip the corresponding area is $0.0018 \times 7 \times 12 = 0.151$ in.² Expressed in terms of minimum steel ratio for actual effective depths, this gives

In 25-ft direction: $\rho_{min} = 0.151/5.5 \times 12 = 0.0023$
In 20-ft direction: $\rho_{min} = 0.151/6 \times 12 = 0.0021$

This requirement controls at the locations indicated in Table 4.4.
        The total steel area in each band is easily found from the steel ratio, and is given in column 8. Finally, with the aid of Table 2 of the Appendix, the required number of bars is obtained. Note that in two locations, the number of bars used is dictated by the maximum spacing requirement of $2 \times 7 = 14$ in.

The shear stress in the slab is checked on the basis of the tributary areas shown in Fig. 4.10. At a distance d from the face of the long beam,

$$V_u = 0.335[10 - 14/(2 \times 12) - \tfrac{6}{12}] = 2.99 \text{ kips}$$

producing a nominal unit shear stress of

$$v_u = 2990/0.85 \times 12 \times 6 = 49 \text{ psi}$$

This is well below the limiting value of $2\sqrt{4000} = 126$ psi.

Each beam must be designed for its share of the total static moment, as found in the above calculations, as well as the moment due to its own weight; this moment may be distributed to positive and negative bending sections, using the same ratios used for the static moments due to slab loads. Beam-shear design should be based on the loads from the tributary areas shown in Fig. 4.10. Since no new concepts would be introduced, the design of the beams will not be presented here, and the reader is referred to Chap. 3 for detailed procedures.

Since $0.85 \times 93 = 79$ per cent of the exterior negative moment in the long direction is carried directly to the column by the column-line beam in this example, torsional stresses in the spandrel beam are very low and may be disregarded. In other circumstances, the spandrel beams would be designed for torsion following the methods of Chap. 3.

PROBLEMS

1. Redesign the building floor of the preceding example, using the same loads and material strengths but omitting all beams to produce a flat-plate floor system. Shear reinforcement similar to Fig. 4.14a may be included if necessary.

2. A concrete slab roof is to be designed to cover a transformer vault. The outside dimensions of the vault are 17×20 ft, and walls are 8-in. brick. A service live load of 80 psf, uniformly distributed over the roof surface, will be assumed. Design the roof as a two-way slab, with $f'_c = 4000$ psi and $f_y = 50,000$ psi.

SLABS ON GROUND

4.12 USAGE

Concrete slabs are often poured directly on the ground; they receive more or less uniform support from the soil. Roadway and sidewalk slabs, airport runways, basement floors, and warehouse floors are common examples of this type of construction. Ordinarily, it is desirable to provide a base course of well-compacted crushed stone or gravel. The prepared subgrade, approximately 6 to 12 in. thick, serves (1) to provide more uniform support than if the slab were carried directly on the natural soil, and (2) to improve the drainage of water from beneath the slab. The latter is particularly important in outdoor locations subject to freezing temperatures.

Failures of concrete slabs on ground are not infrequent. Unequal settlement or overloading may cause cracking, as well as restrained shrinkage as volume changes occur. Passage of wheel loads over cracks or improperly made joints may lead to progressive failure by disintegration of the concrete. Failures are not spectacular and do not involve collapse in the usual sense. They may even pass unnoticed for a considerable period of time. Nevertheless, the function of the structure is often impaired, and repairs are both embarrassing and costly.

If the slab is loaded uniformly over its entire area and is supported by an absolutely uniform subgrade, stresses will be due solely to restrained volumetric changes. However, foundation materials are not uniform in their properties. In addition, most slabs are subjected to nonuniform loading. In warehouses, for example, the necessity for maintaining clear aisles for access to stored materials often results in a checkerboard-type load pattern. Wheel loads from trucks and other moving equipment may control the design.

Methods of analysis for such cases are similar to those developed for beams on elastic foundations. Usually, the slab is assumed to be homogeneous, isotropic, and elastic; the reaction of the subgrade is assumed to be only vertical and proportional to the deflection. The stiffness of the soil is expressed in terms of the modulus of subgrade reaction k, usually in units of lb per in.^{2} per in., or simply, lb per in.^{3} The numerical value of k varies widely for different soil types and degrees of consolidation and is generally based on experimental observations; typical values are given in Fig. 4.19. Shown for comparison are values of California bearing ratio, also widely used as a measure of soil consolidation under load.

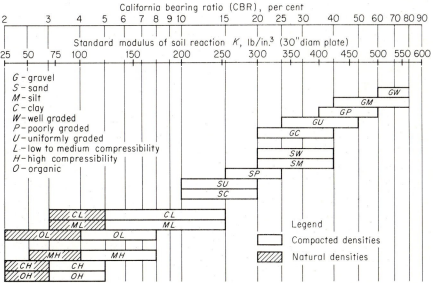

Note: Comparison of soil type to K, particularly in the L and H groups, should generally be made in the lower range of the soil type.

Fig. 4.19 Interrelationships of soil classifications and strength criteria.

4.13 ANALYSIS FOR CONCENTRATED LOADS

The analysis of slabs supporting concentrated loads is based largely on the work of Westergaard (Refs. 4.5 and 4.6). Three separate cases, differentiated on the basis of the location of the load with respect to the edge of the slab, are considered.

Case 1: wheel load close to the corner of a large slab With a load applied at the corner of a slab, the critical stress in the concrete is tension at the top surface of the slab. An approximate solution, due to A. T. Goldbeck, assumes a point load acting at the corner of the slab (see Fig. 4.20). At small distances from the corner, the upward reaction of the soil has little effect, and the slab is considered to act as a cantilever. At a distance x from the corner, the bending moment is Px; it is assumed to be uniformly distributed across the width of the section of slab at right angles to the bisector of the corner angle. For a 90° corner, the width of this section is $2x$, and the bending moment per unit width of slab is

$$\frac{Px}{2x} = \frac{P}{2}$$

If h is the thickness of the slab, the tensile stress at the top surface is

$$f_t = \frac{M}{S} = \frac{P/2}{h^2/6} = \frac{3P}{h^2} \qquad (4.18)$$

Equation (4.18) will give reasonably close results only in the immediate vicinity of the slab corner, and only if the load is applied over a small contact area.

In an analysis which considers the reaction of the subgrade, and which considers the load to be applied over a contact area of radius a (see Fig. 4.20b), Westergaard derives the expression for critical tension at the top of the slab,

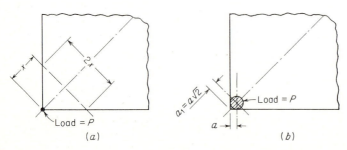

Fig. 4.20

occurring at a distance $2 \sqrt{a_1 L}$ from the corner of the slab:

$$f_t = \frac{3P}{h^2}\left[1 - \left(\frac{a\sqrt{2}}{L}\right)^{0.6}\right] \tag{4.19a}$$

in which L is the radius of relative stiffness, equal to

$$L = \sqrt[4]{\frac{Eh^3}{12(1-\mu^2)k}} \tag{4.19b}$$

where E = elastic modulus of concrete, psi
  μ = Poisson's ratio for concrete (approximately 0.15)
  k = modulus of subgrade reaction, lb/in.^{3}
  The value of L reflects the relative stiffness of the slab and the subgrade. It will be large for a stiff slab on a soft base and small for a flexible slab on a stiff base.

Case 2: wheel load a considerable distance from the edges of a slab When the load is applied some distance from the edges of the slab, the critical stress in the concrete will be tension at the bottom surface. This tension is greatest directly under the center of the loaded area and is given by the expression

$$f_b = 0.316\,\frac{P}{h^2}\,[\log h^3 - 4\log\,(\sqrt{1.6a^2 + h^2} - 0.675h) - \log k + 6.48]$$

$$\tag{4.20}$$

Case 3: wheel load at an edge of a slab, but removed a considerable distance from a corner When the load is applied at a point along an edge of the slab, the critical tensile stress is at the bottom of the concrete, directly under the load, and is equal to

$$f_b = 0.572\,\frac{P}{h^2}\,[\log h^3 - 4\log\,(\sqrt{1.6a^2 + h^2} - 0.675h) - \log k + 5.77]$$

$$\tag{4.21}$$

In the event that the tensile stress in the slab, as given by Eqs. (4.18), (4.20), and (4.21), exceeds the allowable tensile stress on the concrete, it is necessary either to increase the thickness of the slab or to provide reinforcement. Such reinforcement is usually designed to provide for all the tension indicated by the analysis of the assumed homogeneous, elastic slab. Its centroid should be no closer to the neutral axis than that of the tension concrete which it replaces.

Example: design of an unreinforced-concrete slab on grade A warehouse floor, consisting of an unreinforced slab on ground, is to be designed to support a maximum wheel load of 6000 lb from fork-lift trucks, assumed to apply the load over a contact area of 4-in. radius. Concrete of 4000-psi compressive strength is to be used, with an allowable tensile stress of 200 psi. The slab will be poured in 30-ft-wide strips with construction joints at each edge, and contraction joints will be located every 30 ft in the long direction of the strip. It will be conservatively assumed that no shear transfer is possible across the joints. The wheel load can be applied anywhere within each 900-ft² panel. The soil is a clay-gravel with a subgrade modulus of 300 lb per in.[3]

It will be assumed initially that the critical load location is at the corner of the slab. With

$$E_c = 33w^{1.5} \sqrt{f_c'} = 33 \times 145^{1.5} \sqrt{4000} = 3,640,000 \text{ psi}$$

and assuming the slab thickness h to be 8 in., the radius of relative stiffness is

$$L = \sqrt[4]{\frac{Eh^3}{12(1 - \mu^2)k}} = \sqrt[4]{\frac{3,640,000 \times 512}{12(1 - 0.15^2)300}} = 27 \text{ in.}$$

From Eq. (4.18),

$$h^2 = \frac{3P}{f_t} \left[1 - \left(\frac{a \sqrt{2}}{L} \right)^{0.6} \right]$$

$$= \frac{3 \times 6000}{200} \left[1 - \left(\frac{4 \sqrt{2}}{27} \right)^{0.6} \right] = 54.7 \text{ in.}^2$$

$$h = 7.4 \text{ in.}$$

Tentatively, a total depth of $7\frac{1}{2}$ in. is selected. If the load is applied centrally within a panel, the tensile stress at the bottom of a $7\frac{1}{2}$-in. slab is, from Eq. (4.20),

$$f_b = \frac{0.316 \times 6000}{56.2} (\log 421 - 4 \log 3.96 - \log 300 + 6.48)$$

$$= 144 \text{ psi} < 200 \text{ psi}$$

If the load is applied along the edge of a slab panel, the critical tension at the bottom of the concrete is, from Eq. (4.21),

$$f_b = \frac{0.572 \times 6000}{56.2} (\log 421 - 4 \log 3.96 - \log 300 + 5.77)$$

$$= 215 \text{ psi} > 200$$

Since this exceeds the allowable tension by a small amount, a total slab depth of 8 in. will be used.

Where the slab design must be based on pattern loading such as caused by tandem-axle trucks or many-wheeled aircraft landing gear, the determination of slab moments is greatly facilitated by influence charts developed by Pickett and Ray (Ref. 4.7). Based on the Westergaard equations, these permit rapid and accurate determination of bending moment resulting from loads at internal and edge locations. Simplified design charts for particular loading cases were prepared by Fordyce and Packard, and have been published by the Portland Cement Association (Ref. 4.8).

4.14 LOADS DISTRIBUTED OVER PARTIAL AREAS

In addition to concentrated loads, it may be that uniform loads distributed over partial areas of slabs will produce the critical design condition. In warehouses, heavy loads are often stacked about columns, leaving clear aisles mid-

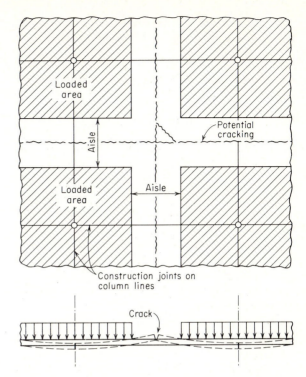

Fig. 4.21

way between the column lines for access, as shown in Fig. 4.21. For moderate column spacing, construction joints are located on column lines only. With the loading pattern shown, cracking is likely to occur along the center line of the aisles. In addition, as wheel loads pass over the cracked slab, secondary corner cracks may occur, as shown in one panel.

In an analysis based on the loading and joint arrangement of Fig. 4.21, Rice (Ref. 4.9) derived an expression for the critical negative moment in the slab, which occurs at the center line of the aisle:

$$M_c = \frac{w}{2\lambda^2} e^{-\lambda a} \sin \lambda a \qquad (4.22)$$

where M = slab moment, ft-lb per ft
$\quad \lambda = \sqrt[4]{k/4EI}$, in.$^{-1}$
$\quad a$ = half-aisle width, in.
$\quad k$ = modulus of subgrade reaction, lb per in.^{3}
$\quad w$ = uniform load, psi
$\quad e$ = base of natural logarithms

Recognizing that the width of the aisle cannot always be predicted exactly, Rice suggests that a "critical aisle width" be used. This width is such as to maximize Eq. (4.22) for bending moment. Tables are available (Ref. 4.9) which give critical aisle width and corresponding bending moment for various slab thicknesses and various values of subgrade modulus.

4.15 CONTINUOUSLY REINFORCED PAVEMENTS

Recent work has indicated the practicality of continuously reinforced concrete pavement, in which transverse expansion and contraction joints are eliminated. Many fine transverse cracks form, due to shrinkage and temperature effects, but they are well distributed and their width is controlled. In the case of moving traffic, they are usually not noticeable by sight, sound, or feel. Experience with pavements in service indicates that such cracks do not permit the passage of damaging amounts of water to the steel, and are capable of transmitting vertical shear from one slab segment to the next by aggregate interlock and dowel action of the reinforcement. In general, somewhat thinner slabs may be used, as compared with conventional jointed pavement, because of the elimination of the end-loaded slab condition (see Ref. 4.8) which otherwise would control the design.

Thickness of such slabs can be based on the work by Westergaard et al. described in Art. 4.14. Experience indicates that the minimum longitudinal steel ratio should be 0.006 times the gross concrete area, and that a spacing of deformed bars between 4 and 9 in. should be used. Alternatively, welded wire mesh may be employed. In either case, reinforcement should have a yield stress not less than 60 ksi, and should be placed from middepth of the slab to within $2\frac{1}{2}$ in. of the top surface, according to the recommendations of an interim study group.

The possibility of prestressing continuous pavements should not be overlooked. By creating a residual longitudinal compressive stress it is possible to eliminate cracking altogether, and to employ slabs substantially thinner than could otherwise be used. A test pavement designed for the state of Minnesota[1] used conventional No. 7 wire $\frac{7}{16}$-in.-diameter steel-strand cable to prestress a 1-mile section of pavement in one operation. Expansion joints were later sawed at 800-ft intervals. The economic success of pavement prestressing appears to depend upon stressing great lengths at once.

YIELD - LINE THEORY OF SLAB ANALYSIS

4.16 INTRODUCTION

Presently, the analysis for moments and shears of most reinforced-concrete slabs is based upon elastic theory, somewhat modified for inelastic redistribution. With idealized support conditions, and usually for only a uniformly distribution load, the critical forces and moments can be obtained for rectangular slabs.

[1] Minnesota to Prestress a Mile of Highway, *Civil Eng.*, vol. 36, no. 2, p. 92, February, 1966.

While these methods of moment analysis are based essentially on elastic behavior, the actual proportioning of slabs, as for beams and other members, is increasingly being performed by ultimate-strength methods which recognize the inelastic nature of deformations before failure. Although there is a certain inconsistency in combining elastic-moment analysis with ultimate-strength design of sections, this procedure is known to be safe and conservative. It can be shown that a structure analyzed and designed in this way will not fail at a lower load than anticipated (although it may possess substantial reserve strength relative to final collapse; see Art. 7.7).

In recent years, methods have been advanced for moment analysis of reinforced-concrete structures which are based on inelastic considerations, and which direct attention to the conditions that obtain in the structure just prior to failure (see Art. 7.7). In the case of slabs, this failure theory of structural analysis is known as *yield-line theory* (Refs. 4.10 and 4.11). It was first proposed by K. W. Johansen. A powerful tool in analysis, it permits the determination of failure moments in slabs of irregular as well as rectangular shapes for a variety of support conditions and loadings.

Figure 4.22a shows a simply supported, uniformly loaded, reinforced-concrete slab. For present purposes, it will be assumed to be underreinforced (as are most slabs), with $\rho < \rho_b$. The elastic distribution of moments is shown in Fig. 4.22b. As the load is gradually increased, and as the maximum applied moment becomes equal to the ultimate-moment capacity of the slab cross section, the tensile steel commences to yield along the transverse line of maximum moment.

Upon yielding, the curvature of the slab at the yielding section increases sharply, and deflection increases disproportionately. The elastic deformation of

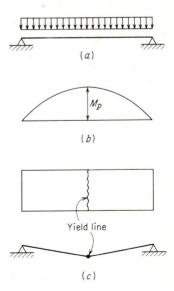

(a)

(b)

Yield line

(c)

Fig. 4.22 Simply supported, uniformly loaded slab.

the slab is of a small order of magnitude compared with the change in shape of the deflected structure due to the plastic deformation at the yield line. It is acceptable to assume that the slab portions between yield line and supports remain rigid, all the deformation taking place at the yield line, as shown in Fig. 4.22c.

The "hinge" which forms at the yield line rotates at essentially a constant moment. (This moment-rotation relation is shown graphically in Fig. 7.9a.) For all practical purposes, the restraining moment at the yielding hinge can be taken equal to the ultimate moment; that is, $M_p = M'_u$.

For a determinate slab such as that of Fig. 4.22, the formation of one yield line is tantamount to failure. A "mechanism" forms (the segments of the slab between the hinge and the supports are able to move without an increase in load), and gross deflection of the structure results.

Indeterminate structures, however, can maintain equilibrium even after the formation of one or more yield lines. The fixed-fixed slab of Fig. 4.23, for example, when loaded uniformly, will have an elastic distribution of moments as in (b). As the load is gradually increased (it is assumed for simplicity that the slab is equally reinforced for positive and negative moment), the more highly stressed sections at the supports commence yielding. Rotation of the end tangents occurs, but restraining moments of substantially constant amount M_p continue to act at the support lines. The load can be increased still further, until the moment at midspan becomes equal to the ultimate-

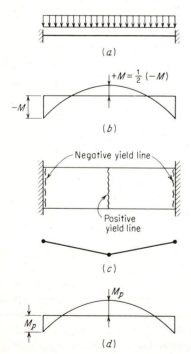

(a)

$+M = \frac{1}{2}(-M)$

$-M$

(b)

Negative yield line

Positive yield line

(c)

M_p

M_p

(d)

Fig. 4.23 Fixed-end, uniformly loaded slab.

moment capacity of the slab, and a third yield line forms as shown in (c). This converts the structure into a mechanism and results in collapse.

The moment diagram just before failure is shown in Fig. 4.23d. Note that the ratio of elastic positive to negative moments of 1:2 is no longer obtained. Owing to inelastic deformation, the ratio of these moments just before collapse is 1:1 for this particular structure. This is known as *inelastic redistribution of moments*.

Whether or not a concrete structure can sustain rotation and deformation such as to ensure full moment redistribution depends at least in part on the reinforcement ratio. Lightly reinforced members can generally undergo considerable rotation, while overreinforced members may fail by concrete crushing before much rotation occurs. For a more complete discussion of this important aspect of plastic or limit analysis of concrete structures, the reader is referred to Art. 7.7. It will be sufficient for the present to note that most slabs are relatively lightly reinforced, and that the required rotation capacity is normally available.

4.17 LOCATION OF YIELD LINES

The location and orientation of the yield line were evident in the case of the simple slab of Fig. 4.22. Similarly, for the one-way indeterminate slab of Fig. 4.23, the yield lines were easily established. For other cases it is helpful to have a set of guidelines for drawing yield lines and locating axes of rotation. When a slab is on the verge of collapse owing to the existence of a sufficient number of real or plastic hinges to form a mechanism, axes of rotation will be located along the lines of support or over point supports such as columns. The slab segments can be considered to rotate as rigid bodies in space about these axes of rotation. The yield line between any two adjacent slab segments is a straight line, being the intersection of two essentially plane surfaces. Since the yield line (as a line of intersection of two planes) contains all points common to these two planes, it must contain the point of intersection of the two axes of rotation, which is also common to the two planes. That is, the yield line (or yield line extended) must pass through the point of intersection of the axes of rotation of the two adjacent slab segments.

The terms *positive yield line* and *negative yield line* are used to distinguish between those associated with tension at the bottom and tension at the top of the slab, respectively.

Guidelines for establishing axes of rotation and yield lines are summarized as follows:

1. Yield lines are generally straight.
2. Axes of rotation generally lie along lines of support (the support line may be a real hinge, or it may establish the location of a yield line which acts as a plastic hinge).

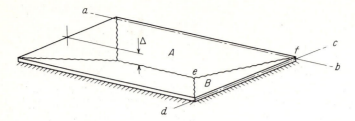

Fig. 4.24   Two-way slab on simple supports.

3. Axes of rotation pass over any columns.
4. A yield line passes through the intersection of the axes of rotation of adjacent slab segments.

In Fig. 4.24, which shows a slab simply supported along its four sides, rotation of slab segments A and B is about ab and cd, respectively. The yield line ef between these two segments is a straight line passing through f, the point of intersection of the axes of rotation.

Illustrations are given in Fig. 4.25 of the application of these guidelines to the establishment of yield-line locations and failure mechanisms for a number of slabs with various support conditions. Shown in (a) is a slab continuous over parallel supports. Axes of rotation are situated along the supports (negative yield lines) and near midspan, parallel to the supports (positive yield line). The particular location of the positive yield line in this case and the other cases of Fig. 4.25 depends upon the distribution of loading and the reinforcement of the slab. Methods for determining its location will be discussed later.

For the continuous slab on nonparallel supports, shown in (b), the midspan yield line (extended) must pass through the intersection of the axes of rotation over the supports. In (c) there are axes of rotation over all four simple supports. Positive yield lines form along the lines of intersection of the rotating segments of the slab. A rectangular two-way slab on simple supports is shown in (d). The diagonal yield lines must pass through the corners, while the central yield line is parallel to the two long sides (axes of rotation along opposite supports intersect at infinity in this case).

With this background, the reader should have no difficulty in applying the guidelines to the slabs of Fig. 4.25e to g to confirm the general pattern of yield lines shown.

4.18   EQUILIBRIUM METHOD OF ANALYSIS

Once the general pattern of yielding and rotation has been established by applying the guidelines of Art. 4.17, the specific location and orientation of the axes of rotation and the failure load for the slab can be established by either of two methods. The first of these is based on the equilibrium of the various

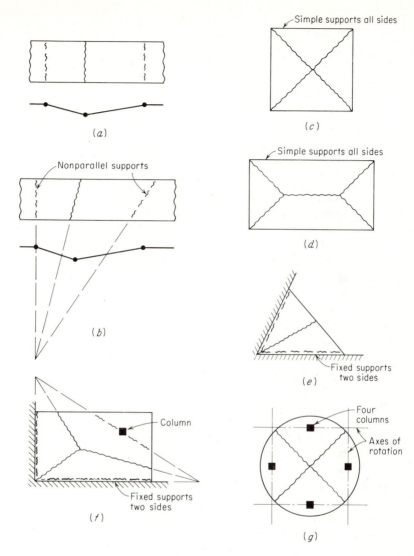

Fig. 4.25   Typical yield-line patterns.

segments of the slab. The second, to be discussed in Art. 4.19, makes use of the principle of virtual work.

The correct locations of axes of rotation and the collapse load of a slab can be found by considering the equilibrium of the slab segments. Each segment, studied as a free body, must be in equilibrium under the action of the applied loads, the moments along the yield lines, and the reactions or shear along the support lines. It is noted that, because the yield moments are principal moments, twisting moments are zero along the yield lines, and in most cases the shearing forces are also zero. Only the unit moment m generally is considered in writing equilibrium equations.

Example   The application of the equilibrium method will be demonstrated first with respect to the one-way, uniformly loaded, continuous slab of Fig. 4.26a. The slab has a 10-ft span and is reinforced so as to provide a resistance to positive bending of 5.0 ft-kips per ft through the span. In addition, negative steel over the supports provides moment capacities of 5.0 ft-kips per ft at A and 7.5 ft-kips per ft at C. Determine the ultimate load capacity of the slab.

The number of equilibrium equations required will depend upon the number of unknowns. One unknown is always the relation between the ultimate resisting moments of the slab and the load. Other unknowns are needed to define the locations of yield lines. In the present instance, one additional equation will suffice to define the distance of the yield line from the supports. Taking the left segment of the slab as a free body, and writing the equation for moment equilibrium about the left support line (see Fig. 4.26b),

$$(a)\quad \frac{wx^2}{2} - 10.0 = 0$$

Similarly, for the right slab segment,

$$(b)\quad \frac{w}{2}(10 - x)^2 - 12.5 = 0$$

Solving Eqs. (a) and (b) simultaneously for w and x results in

$$w = 0.89 \text{ kips per ft}^2 \qquad x = 4.75 \text{ ft}$$

If a slab is reinforced identically in orthogonal directions, the ultimate resisting moment is the same in these two directions, as it is along any other line, regardless of its direction. Such a slab is said to be *isotropically* reinforced.

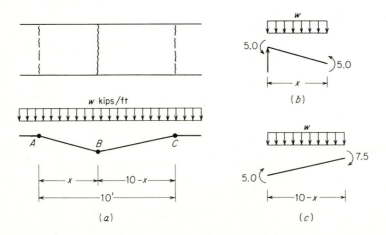

Fig. 4.26   Equilibrium method of analysis for one-way slab.

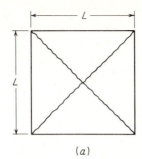

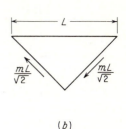

Fig. 4.27   (a)   (b)

If, however, the ultimate strengths are different in two perpendicular directions, the slab is called *orthogonally anisotropic*, or simply *orthotropic*. Only
isotropic slabs will be discussed in this section. Johansen has shown (Ref. 4.10)
that orthotropic slabs can be reduced to equivalent isotropic cases, for purposes
of analysis, by modifying the slab dimensions.

It is convenient in yield-line analysis to represent moments with vectors.
The standard convention, in which the moment acts in a clockwise direction
when viewed along the vector arrow, will be followed. Treatment of moments
as vector quantities will be illustrated by the following example.

Example   A square slab is simply supported along all sides and is to be isotropically reinforced. Determine the ultimate resisting moment m per linear foot required just to
sustain a uniformly distributed load of w psf.

Conditions of symmetry indicate the yield-line pattern shown in Fig. 4.27a.
Considering the moment equilibrium of any one of the identical slab segments about
its support (see Fig. 4.27b), one obtains

$$\frac{wL^2}{4}\frac{L}{6} - 2\frac{mL}{\sqrt{2}}\frac{1}{\sqrt{2}} = 0$$

$$m = \frac{wL^2}{24}$$

4.19 ANALYSIS BY VIRTUAL WORK

Alternative to the method of Art. 4.18 is a method of analysis using the principle of virtual work. Since the moments and loads are in equilibrium when
the yield-line pattern has formed, an infinitesimal increase in load will cause
the structure to deflect further. The external work done by the loads to cause a
small arbitrary virtual deflection must equal the internal work done as the slab
rotates at the yield lines to accommodate this deflection. The slab is therefore
given a virtual displacement, and the corresponding rotations at the various
yield lines may be calculated. By equating internal and external work, the
relation between the applied loads and the ultimate resisting moments of the
slab is obtained.

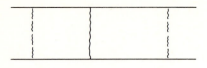

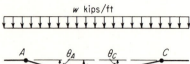

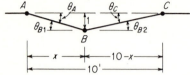

Fig. 4.28 Mechanism method of analysis for one-way slab.

Example Determine the ultimate-load capacity of the one-way, uniformly loaded, continuous slab of Fig. 4.28, using the method of virtual work. The resisting moments of the slab are 5.0, 5.0, and 7.5 ft-kips per ft at A, B, and C, respectively (Fig. 4.28).

A unit deflection is given to the slab at B. Then the external work done by the load is the sum of the loads times their displacements and is equal to

$$\frac{wx}{2} + \frac{w}{2}(10 - x)$$

The rotations at the hinges are calculated in terms of the unit deflection (Fig. 4.28) and are

$$\theta_a = \theta_{b1} = \frac{1}{x} \qquad \theta_{b2} = \theta_c = \frac{1}{10 - x}$$

The internal work is the sum of the moments times their corresponding rotation angles:

$$5 \times \frac{1}{x} \times 2 + 5 \times \frac{1}{10 - x} + 7.5 \times \frac{1}{10 - x}$$

Equating the external and internal work,

$$\frac{wx}{2} + 5w - \frac{wx}{2} = \frac{10}{x} + \frac{5}{10 - x} + \frac{7.5}{10 - x}$$

$$5w = \frac{10}{x} + \frac{25}{2(10 - x)}$$

$$w = \frac{2}{x} + \frac{5}{2(10 - x)}$$

To determine the minimum value of w, this expression is differentiated with respect to x and set equal to zero:

$$\frac{dw}{dx} = -\frac{2}{x^2} + \frac{5}{2(10 - x)^2} = 0$$

from which

$$x = 4.75 \text{ ft}$$

Substituting this value in the preceding expression for w, one obtains

$w = 0.89$ kips per ft^2

as before.

In many cases, particularly those with yield lines established by several unknown dimensions (such as Fig. 4.25f), direct solution by virtual work would become quite tedious. The ordinary derivatives in the above example would be replaced by several partial derivatives, producing a set of equations to be solved simultaneously. In such cases it is often more convenient to select an arbitrary succession of possible yield-line locations, solve the resulting mechanisms for the unknown load (or unknown moment), and determine the correct minimum load (or maximum moment) by trial.

Example The two-way slab of Fig. 4.29 is simply supported on all four sides and carries a uniformly distributed ultimate load of w psf. Determine the ultimate-moment resistance of the slab, which is to be isotropically reinforced.

Positive yield lines will form in the pattern shown in Fig. 4.29a, with the dimension a unknown. The correct dimension a will be such as to maximize the moment resistance required to support the load w. The values of a and m will be found by trial.

In Fig. 4.29a, the length of the diagonal yield is $\sqrt{25 + a^2}$. From similar triangles,

$$b = 5 \frac{\sqrt{25 + a^2}}{a} \qquad c = a \frac{\sqrt{25 + a^2}}{5}$$

Then the rotation of the plastic hinge at the diagonal yield line corresponding to a unit deflection at the center of the slab (see Fig. 4.29b) is

$$\theta_1 = \frac{1}{b} + \frac{1}{c} = \frac{a}{5\sqrt{25 + a^2}} + \frac{5}{a\sqrt{25 + a^2}} = \frac{1}{\sqrt{25 + a^2}}\left(\frac{a}{5} + \frac{5}{a}\right)$$

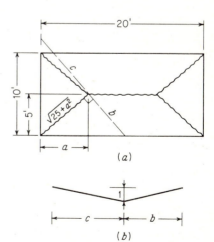

Fig. 4.29

(a)

(b)

The rotation of the yield line parallel to the long edges of the slab is

$$\theta_2 = \tfrac{1}{5} + \tfrac{1}{5} = 0.400$$

For a first trial, let $a = 6$ ft. Then the length of the diagonal yield line is

$$\sqrt{25 + 36} = 7.81 \text{ ft}$$

The rotation at the diagonal yield line is

$$\theta_1 = \frac{1}{7.81}\left(\frac{6}{5} + \frac{5}{6}\right) = 0.261$$

At the central yield line it is $\theta_2 = 0.400$. The internal work done as the incremental deflection is applied is

$$W_i = (m \times 7.81 \times 0.261 \times 4) + (m \times 8 \times 0.400) = 11.36m$$

The external work done during the same deflection is

$$W_e = (10 \times 6 \times \tfrac{1}{2}w \times \tfrac{1}{3} \times 2) + (8 \times 5w \times \tfrac{1}{2} \times 2)$$
$$+ (12 \times 5 \times \tfrac{1}{2}w \times \tfrac{1}{3} \times 2) = 80w$$

Equating W_i and W_e, one obtains

$$m = \frac{80w}{11.36} = 7.05w$$

Successive trials for different values of a result in the following data:

a	W_i	W_e	m
6.0	11.36m	80.0w	7.05w
6.5	11.08m	78.4w	7.08w
7.0	10.87m	76.6w	7.04w
7.5	10.69m	75.0w	7.02w

It is evident that the yield-line pattern defined by $a = 6.5$ ft is critical. The required resisting moment for the given slab is $7.08w$.

In common with the equilibrium method of analysis, the method of virtual work as applied to slabs is an "upper-bound" method in the sense

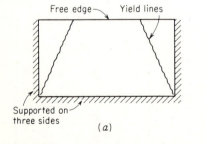

Free edge — Yield lines

Supported on three sides

(a)

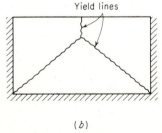

Yield lines

(b)

Fig. 4.30

that a yield pattern different from the one assumed may produce a lower failure load or, conversely, may require greater moment capacity to resist a given applied load. To illustrate, the rectangular slab of Fig. 4.30, supported along only three sides and free along the fourth, may fail by either of the two mechanisms shown. An analysis based on yield pattern *a* may indicate a minimum failure load higher than one based on pattern *b*. It is necessary to investigate all reasonable failure mechanisms in each case to establish that the correct solution has been found.

4.20 CORNER LEVERS

In the preceding discussion it has been assumed that yield lines enter the corners between the two intersecting sides. An alternative possibility is that the yield line forks before it reaches the corner, forming what is known as a *corner lever*, shown in Fig. 4.31*a*.

   If the corner is not held down, the triangular element *abc* will pivot about the axis *ab* and lift off the supports. If the corner is held down, a similar situation obtains, except that the line *ab* becomes a yield line. If cracking at the corners of such a slab is to be controlled, top steel, more or less perpendicular to the line *ab*, must be provided. The direction taken by the positive yield lines near the corner indicates the desirability of supplementary bottom-slab reinforcement at the corners, placed approximately parallel to the line *ab* (see Art. 4.7).

   Although yield-line patterns with corner levers are generally more critical than those without, they are often neglected in yield-line analysis. The analysis becomes considerably more complicated if the possibility of corner levers is introduced, and the error made by neglecting them is usually small.

   To illustrate, the uniformly loaded square slab of Art. 4.18, when analyzed for the assumed yield pattern of Fig. 4.27, required an ultimate-moment

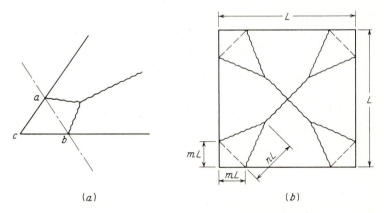

(*a*)   (*b*)

Fig. 4.31 Corner levers.

capacity of $wL^2/24$. The actual yield-line pattern at failure is probably as shown in Fig. 4.31b. Since two additional parameters m and n have necessarily been introduced to define the yield-line pattern, a total of three equations of equilibrium is now necessary. These equations are obtained by summing moments and vertical forces on the segments of the slab. Such an analysis results in a required moment of $wL^2/22$, an increase of about 9 per cent as compared with the results of an analysis neglecting corner levers.[1]

4.21 LIMITATIONS OF YIELD–LINE THEORY

The usefulness of yield-line theory is apparent from the preceding articles. In general, elastic solutions are available only for limited conditions, usually uniformly loaded rectangular plates. Even if available, they do not account for the effects of inelastic action. By means of yield-line analysis, a rational solution for the failure load may be found for slabs of any shape, supported in a variety of ways, and for concentrated loads as well as distributed and partially distributed loads. It is thus seen to be a powerful analytical tool for the structural engineer.

In applying yield-line analysis to slabs, it must be remembered that the analysis is predicated upon available rotation capacity at the yield lines. If the slab reinforcement happens to correspond closely to the elastic distribution of moments in the slab, little rotation is required. If, on the other hand, there is a marked difference, it may be that the required rotation will exceed the available rotation capacity, in which case the slab will fail prematurely. In general, slabs, being rather lightly reinforced, will have adequate rotation capacity to attain the ultimate loads predicted by yield-line analysis.

It should also be borne in mind that the yield-line analysis focuses attention on the moment capacity of the slab. It is presumed that earlier failure will not occur due to shear, bond, or other causes. In addition, yield-line theory gives no information on stresses, deflections, or severity of cracking under service-load conditions.

In spite of these limitations, yield-line analysis provides answers to problems of slab design which cannot be handled by other means, and so will undoubtedly assume a position of increasing importance in engineering practice.

REFERENCES

4.1. W. L. Gamble, M. A. Sozen, and C. P. Siess: Measured and Theoretical Bending Moments in Reinforced Concrete Floor Slabs, Univ. of Illinois, Dept. of Civil Engineering, Structural Research Series, No. 246, June, 1962.

4.2. D. Peabody, Jr.: Continuous Frame Analysis of Flat Slabs, *J. Boston Soc. Civil Engr.*, January, 1948.

[1] Analysis indicates (Ref. 4.11) that when the corner angle is acute, the error due to neglect of corner levers may be significant.

4.3. W. G. Corley, M. A. Sozen, and C. P. Siess: The Equivalent Frame Analysis for Reinforced Concrete Slabs, Univ. of Illinois, Dept. of Civil Engineering, Structural Research Series, No. 218, June, 1961.

4.4. W. G. Corley and N. M. Hawkins: Shearhead Reinforcement for Slabs. *ACI J.*, vol. 65, no. 10, pp. 811–824, October, 1968.

4.5. H. M. Westergaard: Stresses in Concrete Pavements Computed by Theoretical Analysis, *Public Roads*, April, 1926.

4.6. H. M. Westergaard: Analytical Tools for Judging Results of Structural Tests of Concrete Pavements, *Public Roads*, December, 1933.

4.7. G. Pickett and G. K. Ray: Influence Charts for Concrete Pavements, *Trans. ASCE*, vol. 116, pp. 49–73, 1951.

4.8. Thickness Design for Concrete Pavements, Portland Cement Association Concrete Information Bulletin, Chicago, 1966.

4.9. P. Rice: Design of Concrete Floors on Ground for Warehouse Loading, *J. ACI*, August, 1957.

4.10. E. Hognestad: Yield-line Theory for the Ultimate Flexural Strength of Reinforced Concrete Slabs, *J. ACI*, March, 1953.

4.11. L. L. Jones: "Ultimate Load Analysis of Reinforced and Prestressed Concrete Structures," John Wiley & Sons, Inc., New York, 1962.

5
Compression Members: Axial Compression Plus Bending

5.1 COMPRESSION MEMBERS, COLUMNS

Any member which carries an axial-compression load will be defined as a compression member, regardless of whether or not design calculations indicate simultaneous bending to be present. Correspondingly, in addition to the most common type, namely columns in the usual sense, compression members include arch ribs, rigid-frame members inclined or otherwise, compression members in trusses, shells or portions thereof which carry axial compression, and others. In this chapter the term column will often be used interchangeably with the term compression member, for brevity and also in conformity with general usage, but it should be understood that this does not limit application to columns in the literal sense.

Three types of reinforced-concrete compression members are in use:

1. Members reinforced with longitudinal bars and lateral ties.
2. Members reinforced with longitudinal bars and closely spaced spirals.
3. Composite compression members which are reinforced longitudinally with structural-steel shapes, pipe or tubing, with or without additional longitudinal bars.

259

Types 1 and 2 are by far the most common, and most of the discussion of this chapter will refer to them.

The ACI Code provides that the reinforcement ratio of the longitudinal steel shall be no less than 0.01 nor more than 0.08, and that at least six bars shall be used for spiral columns and at least four for tied columns.

5.2 COLUMNS REINFORCED WITH TIES OR SPIRALS

Figure 5.1 shows cross sections of the simplest kinds of columns, spiral-reinforced or provided with lateral ties. Other cross sections frequently found in buildings and bridges are shown in Fig. 5.2. In general, in members with large axial forces and small moments, longitudinal bars are spaced more or less uniformly around the perimeter (Fig. 5.2a to c). Contrariwise, when bending moments are large, much of the longitudinal steel is concentrated at the faces of largest compression or tension, i.e., at maximum distance from the axis of bending (Fig. 5.2d and e). In heavily loaded columns with large steel percentages, the result of large numbers of bars, each of them positioned and held individually by ties, is steel congestion in the forms and difficulties in placing the concrete. In such cases, bundled bars are frequently employed. Bundles consist of three or four bars tied in direct contact, wired or otherwise fastened together. These are usually placed in the corners. Tests have shown that adequately bundled bars act as one unit; i.e., they are detailed as if a bundle constituted a single round bar of area equal to the sum of the bundled bars.

Lateral reinforcement, in the form of individual relatively widely spaced ties or a continuous closely spaced spiral, serves several functions. For one, such

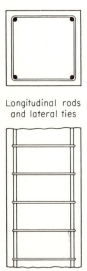

Longitudinal rods
and lateral ties

Fig. 5.1 Reinforced-concrete column.

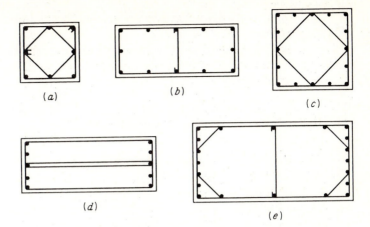

Fig. 5.2 Tie arrangements conforming to ACI Code.

reinforcement is needed to hold the longitudinal bars in position in the forms while the concrete is being placed. For this purpose, longitudinal and transverse steel are wired together to form cages, which are then moved into the forms and properly positioned prior to placing of the concrete. For another, transverse reinforcement is needed to prevent the highly stressed, slender longitudinal bars from buckling outward by bursting the thin concrete cover.

Closely spaced spirals evidently serve these two functions. Ties, which can be arranged and spaced in various ways, must be so designed that these two requirements are met. This means that the spacing must be sufficiently small to prevent buckling between ties and that, in any tie plane, a sufficient number of ties must be provided to position and hold all bars. On the other hand, in columns with many longitudinal bars, if the column section is crisscrossed by too many ties, they interfere with the placement of concrete in the forms. To achieve adequate tying, yet hold the number of ties to a minimum, the ACI Code gives the following rules for tie arrangement:

> All bars of tied columns shall be enclosed by *lateral ties*, at least No. 3 in size for longitudinal bars up to No. 10, and at least No. 4 in size for Nos. 11, 14, and 18 and bundled longitudinal bars. The spacing of the ties shall not exceed 16 diam. of longitudinal bars, 48 diam. of tie bars, nor the least dimension of the column. The ties shall be so arranged that every corner and alternate longitudinal bar shall have lateral support provided by the corner of a tie having an included angle of not more than 135°, and no bar shall be farther than 6 in. clear on either side from such a laterally supported bar. Welded wire fabric of equivalent area may be used instead of ties.

Where the bars are located around the periphery of a circle, complete circular ties may be used. Spirals, wires, or rods are specified to be of no less than $\frac{3}{8}$ in. in diameter, and the clear spacing between turns of the spiral is stipulated to be no more than 3 in. nor less than 1 in.

In view of this close spacing, the presence of a spiral affects both the ultimate load and the type of failure compared with an otherwise identical tied column.

The structural effect of a spiral is easily visualized by considering as a model a steel drum filled with sand (Fig. 5.3). When a load is placed on the sand, a lateral pressure is exerted by the sand on the drum, which causes hoop tension in the steel wall. The load on the sand can be increased until the hoop tension becomes large enough to burst the drum. The sand pile alone, if not confined in the drum, would have been able to support hardly any load. A cylindrical concrete column, to be sure, does have a definite strength without any lateral confinement. As it is being loaded, it shortens longitudinally and expands laterally, depending on Poisson's ratio. A closely spaced spiral confining the column counteracts the expansion, as did the steel drum in the model. This causes hoop tension in the spiral, while the carrying capacity of the confined concrete in the core is greatly increased. Failure occurs only when the spiral steel yields, which greatly reduces its confining effect, or when it fractures. It has been found that a given amount of steel per unit length of column, when in the form of a spiral, is at least twice as effective in adding to the carrying capacity as the same amount of steel used in the form of longitudinal bars.

A tied column fails at the load given by Eq. (2.5) as

$$P'_u = 0.85f'_c A_c + f_y A_s \tag{5.1}$$

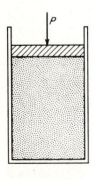

Fig. 5.3   Model for action of spiral.

Fig. 5.4   Failure of tied column.

At this load the concrete fails by crushing and shearing outward along inclined planes, and the longitudinal steel by buckling outward between ties (Fig. 5.4). In a spiral-reinforced column, when the same load is reached, the longitudinal steel and the concrete within the core are prevented from outward failing by the spiral. The concrete in the outer shell, however, not being so confined, does fail; i.e., the outer shell spalls off when the load P'_u is reached. It is at this stage that the confining action of the spiral takes effect, and if sizable spiral steel is provided, the load which will ultimately fail the column by causing the spiral to yield or burst can be significantly larger than that at which the shell spalled off.

In contrast to some foreign countries, it is reasoned in American practice that any excess capacity beyond the spalling load of the shell is wasted because the member, although not actually failed, would no longer be considered serviceable. For this reason the Code provides a minimum spiral reinforcement of such amount that its contribution to the carrying capacity is just slightly larger than that of the concrete in the shell. The situation is best understood from Fig. 5.5, which compares the performance of a tied column with that of a spiral column whose spalling load is equal to the ultimate load of the tied column. The failure of the tied column is abrupt and complete. This is true, to

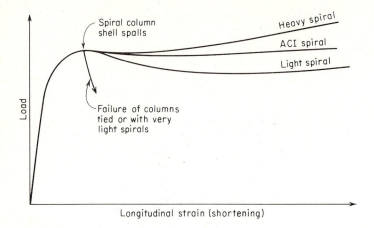

Fig. 5.5 Behavior of spiral and tied columns.

almost the same degree, of a spiral column with a spiral so light that its strength contribution is considerably less than the strength lost in the spalled shell. With a heavy spiral the reverse is true, and with considerable prior deformation, the spalled column would fail at a higher load. The "ACI spiral," its strength contribution about compensating for that lost in the spalled shell, hardly increases the ultimate load. However, by preventing instantaneous crushing of concrete and buckling of steel, it produces a more gradual and ductile failure, i.e., a tougher column.

To determine the right amount of spiral steel, one calculates

(a)   Strength contribution of shell $= 0.85f'_c(A_g - A_c)$

where A_g and A_c are, respectively, the gross and core concrete areas. On the other hand, since spiral steel is at least twice as effective as longitudinal steel, one has, conservatively,

(b)   Strength contribution of spiral $= 2\rho_s A_c f_{sy}$

Here f_{sy} is the spiral yield strength, and ρ_s is the ratio of volume of spiral reinforcement to volume of core, per unit length of column. Equating (a) to (b) and solving for the spiral ratio,

(c) $\rho_s = 0.425 \dfrac{(A_g - A_c)f'_c}{A_c f_{sy}}$

The ACI Code stipulates the minimum spiral ratio as

$$\rho_s = 0.45 \frac{(A_g/A_c - 1)f'_c}{f_{sy}} \tag{5.2}$$

which is seen to be but slightly larger than (c).

It follows that, when designed to the Code, two concentrically loaded columns, one tied and one with spiral but otherwise identical, will fail at about the same load, the former in a sudden and brittle manner, the latter gradually with prior shell spalling and with more ductile behavior. This advantage of the spiral column is much less pronounced if the load is applied with significant eccentricity or when bending from other sources is present simultaneously with axial load. For this reason the Code permits somewhat larger loads on spiral than on tied columns when loaded concentrically or with small eccentricities but reduces this difference for large eccentricities.

COMPRESSION PLUS BENDING

5.3 SAFETY PROVISIONS

It was pointed out in Art. 3.3 that adequate safety margins are established by applying overload factors to the design loads and strength-reduction factors to the theoretical ultimate strengths computed for members assumed to be perfect in workmanship and made of materials having exactly the specified strength. It was pointed out that the overload factors apply uniformly, while the strength-reduction factors are different for differently stressed members.

For members subject to compression or compression plus bending, the Code provides reduction factors

For tied members: $\phi = 0.70$
For spirally reinforced members: $\phi = 0.75$

The spread between these two values reflects the added safety furnished by the greater toughness of spirally reinforced columns. Special, more liberal ϕ values are provided for members with small axial loads and large moments; they are discussed in Art. 5.4a.

There are various reasons why the ϕ values for columns are considerably lower than those for flexure or shear (0.90 and 0.85, respectively). For one, the strength of underreinforced flexural members is not much affected by variations in concrete strength, since it depends primarily on the yield strength of the steel, while the strength of axially loaded members depends strongly on the concrete compression strength [compare, for example, Eqs. (2.5) and (2.27a)]. Because the cylinder strength of concrete under site conditions is less closely controlled than the yield strength of mill-produced steel, a larger occasional strength deficiency must be allowed for. This is particularly true for columns, in which concrete, being placed from the top down the long narrow form, is more subject to segregation than in horizontally cast beams. Moreover, electrical and other conduits are frequently located in building columns; this reduces their effective cross sections, often unbeknown to the designer, even though this is poor practice and restricted by the Code. Finally, the consequences of a column failure, say in a lower story, would be more catastrophic than that of a single beam in a floor system of the same building.

Correspondingly, then, the ultimate-strength values to be used for design are

$$P_u = \phi P'_u \tag{5.3a}$$
$$M_u = \phi M'_u \tag{5.3b}$$

where P'_u and M'_u are those simultaneous values of ultimate axial force and ultimate bending moment which are calculated to cause the "perfect" member to fail (see Art. 2.8b).

Additionally, the Code recognizes that no members are likely to be found which are purely axially loaded. Even if design calculations show a compression member to be free of any bending moments, imperfections of construction, such as slight beam eccentricities and deviations from straightness or verticality, will cause some unintentional bending moments. For this reason the Code specifies that the minimum load eccentricity for which a compression member must be designed is

For tied members: $\qquad\qquad e_{\min} = 0.10h$
For spirally reinforced members: $e_{\min} = 0.05h$

where h is the overall depth of a rectangular section or the outside diameter of a circular section, and $e = M/P$ is the eccentricity referred to the center line of the column. The difference between these values, again, represents a recognition of the greater toughness of spirally reinforced columns with zero or small eccentricities. Also, these minimum eccentricities are limited to no less than 1 in. for in situ construction, nor less than 0.6 in. for precast members under strict dimensional control. Evidently, if structural analysis shows eccentricities to be present which exceed these minima, the larger design eccentricities must be used in dimensioning the member.

5.4 RECTANGULAR COLUMNS

For a rectangular column reinforced by A_s and A'_s along the two faces parallel to the axis of bending, it was shown in Eqs. (2.66) and (2.67) that

$$P'_u = \frac{P_u}{\phi} = 0.85f'_c ab + A'_s f'_y - A_s f_s \tag{5.4}$$

$$P'_u e' = \frac{P_u e'}{\phi} = 0.85f'_c ab \left(d - \frac{a}{2} \right) + A'_s f'_y (d - d') \tag{5.5}$$

where the eccentricity e' in Eq. (5.5) is referred to the center of the tension steel. These equations hold for all but very small eccentricities; they assume that the steel A_s on the convex side will be in tension and that only part of the concrete section is in compression (see Fig. 2.31). For very small centroidal

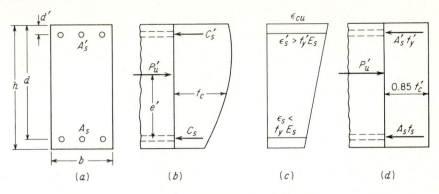

Fig. 5.6   Stresses and strains in eccentric column at failure.

eccentricities e, the distance to the neutral axis $c = a/k_1$ may exceed the depth h of the section. Then the entire cross-sectional area will be in compression, so that the stress in the steel A_s will also be compressive, though of magnitude smaller than f'_y. The corresponding distribution of strains and stresses is shown in Fig. 5.6. The stress distribution of Fig. 5.6b, which corresponds to the strain distribution of Fig. 5.6c, is approximated for purposes of calculation by the rectangular stress distribution of Fig. 5.6d. Summing forces, as well as moments, about the steel A_s, it is seen that

$$P'_u = \frac{P_u}{\phi} = 0.85f'_c bh + A'_s f'_y + A_s f_s \tag{5.4a}$$

$$P'_u e' = \frac{P_u e'}{\phi} = 0.85f'_c bh \left(d - \frac{h}{2}\right) + A'_s f'_y (d - d') \tag{5.5a}$$

In these equilibrium equations f_s is unknown, and therefore the additional condition of strain compatibility must be used for determining P'_u for a given eccentricity e'. This is discussed in detail in Art. 2.9b. Also, the direct numerical analysis without recourse to formulas, for large as well as small eccentricities, is illustrated in the Examples of Art. 2.9b.

If, now, the centroidal eccentricity e is reduced to zero, i.e., the column is loaded in concentric compression, then the strains of concrete and steel are the same everywhere. Assuming that the yield point of all steel in the section is the same, Eq. (5.4a) becomes

$$P'_0 = 0.85f'_c bh + A'_s f'_y + A_s f_y = 0.85f'_c A_c + A_{st} f_y \tag{5.6}$$

Here A_{st} is the total steel area in the section. It is seen that the ultimate load for an axially loaded column (that is, $M = 0$), computed as the limiting case of an eccentrically loaded member, is identical with Eq. (2.5), which was derived for concentric compression and is supported by extensive tests on axially loaded members.

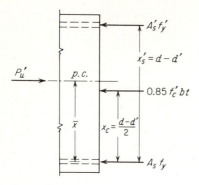

Fig. 5.7   Concentric load through plastic centroid.

[It should be noted that a concentrically loaded member is one whose eccentricity e about the axis of the cross section is zero. This eccentricity is different from e', used in Eqs. (5.5) and elsewhere, which is measured from the center of the steel A_s. For symmetrically reinforced sections the member is concentrically loaded when the load passes through the center of the section. For an unsymmetrically reinforced section, with $A'_s \neq A_s$, to be concentrically loaded so that Eq. (5.6) is true, the load must pass through a point known as the *plastic centroid*. This point is the location of the resultant of the three internal forces that make up Eq. (5.6). Its distance from the steel A_s can be found from Fig. 5.7 as

$$\bar{x} = \frac{0.85f'_c bhx_c + A'_s f'_y x'_s}{0.85f'_c bh + A'_s f'_y + A_s f_y}$$

Evidently, for reasons of equilibrium, the external load P must act along the same line as the resultant of all internal forces, i.e., must pass through the plastic centroid. In a symmetrical section the center and the plastic centroid coincide.]

The other limiting case of a member in combined compression plus bending is the situation in which the compression force has become zero, so that bending alone is present. It can be shown, as for the other extreme, that in this case (that is, $P = 0$) the general equations for compression plus bending result in an ultimate bending moment M'_0 equal to that of a simple flexural member with tension reinforcement A_s and compression reinforcement A'_s (see Art. 3.4b).

a. Interaction diagram for compression plus bending   The information of Art. 2.9, supplemented by the above discussion, shows the considerable complexity of behavior and of corresponding strength calculations for eccentrically loaded members. This behavior ranges from concentric compression ($P = P'_0, M = 0$), over the interval in which failure occurs by crushing of the concrete, through the balanced condition and the interval in which failure occurs by yielding of the steel, to the other extreme of simple bending ($M = M'_0$, $P = 0$). The situation is more easily visualized if the results of the corresponding calculations are graphically depicted in a so-called *interaction diagram*.

For a given cross section and reinforcement, an interaction diagram has the general shape of Fig. 5.8, plotted in terms of ultimate axial loads as ordinates and ultimate moments as abscissas. Moments and eccentricities are here referred to the plastic centroid (for symmetrical sections to the center line of the member), rather than to the center of the tension steel. Any point on the solid curve, such as point a, represents a pair of values P'_u and M'_u which, according to the ultimate-strength theory, will just fail the member.

For concentric compression ($M'_u = 0$) the curve starts at d with the strength P'_0 of a concentrically loaded member [Eq. (5.6)]. The portion db pertains to that range of small eccentricities in which failure is initiated by crushing of the concrete; the corresponding computations were discussed in Art. 2.9b. Point b represents the balanced condition; i.e., under the simultaneous action of the load P'_b [see Eq. (2.51)] and the corresponding moment M'_b, the concrete will reach its limiting strain (0.003) simultaneously with the tension steel reaching its yield stress f_y. The portion bc represents that range in which failure is initiated by yielding of the tension steel (see Art. 2.9b). Finally, the end point c refers to the moment capacity M'_0 in simple bending, i.e., when $P'_u = 0$. Any inclined line through the origin has a slope whose reciprocal represents the centroidal eccentricity corresponding to the particular combination of ultimate values P'_u and M'_u as shown; that is, $e = M'_u/P'_u$.

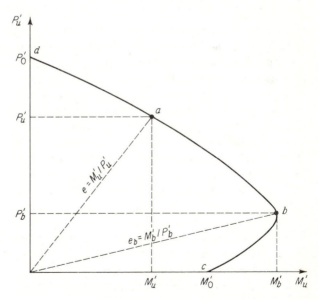

Fig. 5.8 Interaction diagram for compression plus bending, P'_u versus M'_u.

It will be observed that in the region bd of compression failure, the larger the axial load P'_u, the smaller is the moment M'_u which the section is able to sustain before failing. However, in the region bc of tension failures the reverse is true: the larger the axial load, the larger is the simultaneous moment capacity. This is easily understood. In region bd failure occurs through overstraining of the concrete. The larger the concrete compression strain caused by the axial load alone, the smaller is the margin of additional strain available for the added compression caused by bending. On the other hand, in region bc, yielding of the steel initiates failure. If the member is loaded in simple bending to the point at which yielding begins in the tension steel, and if an axial compression load is then added, the steel compression stresses caused by this load will superpose on the previous tension stresses. This reduces the total steel stress to a value below the yield strength. Consequently, an additional moment can now be sustained of such magnitude that the combination of the steel stress from the axial load and the increased moment again reaches the yield strength. Because the presence of axial compression increases the moment capacity as long as tension governs, the designer must make sure that only that amount of compression which is certain to be present is utilized in calculating the moment capacity for a given load combination.

b. ϕ factors for small axial loads and other Code requirements It was mentioned in Art. 5.3 that the Code stipulates ϕ values of 0.70 for tied columns and 0.75 for spiral-reinforced columns. It was noted, also, that these factors are lower than $\phi = 0.90$ for flexural members chiefly because the strength of members which fail in compression is more strongly affected by fluctuations in material strength than those which fail in tension. Referring to the interaction diagram of Fig. 5.8, compression governs over the range b–d, and hence, these lower ϕ values are appropriate for this range. However, in the range b–c, tension governs, the more so the smaller the axial force. Finally, when the axial force is zero, the member becomes an ordinary beam with moment capacity M'_0. The Code prescribes that beams must be so designed that they are underreinforced; hence, tension governs and $\phi = 0.90$. It follows that some transition is in order from the ϕ values of 0.70 or 0.75 for columns, when compression governs, to $\phi = 0.90$ for beams without axial load, when tension governs.

The Code therefore stipulates that for small values of the axial compression force P_u between $0.10f'_c A_g$ and zero, ϕ may be linearly increased from 0.70 for tied and 0.75 for spiral columns at $P_u = 0.10f'_c A_g$ to 0.90 at $P_u = 0$. In the rare cases where the balanced load P_b is smaller than $0.10f'_c A_g$, such increase of ϕ is to start at P_b rather than at $0.10f'_c A_g$. The effects of the safety provisions of the Code are schematically shown in Fig. 5.9. The solid curve $dabc$ is the same as in Fig. 5.8 and represents the actual carrying capacity, i.e., the combination of P'_u and M'_u, which will cause failure. The heavy dashed curve $DABC$ represents the reduced design strength according to the Code.

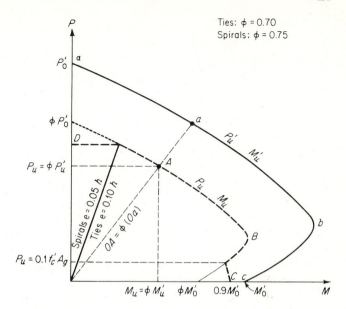

Fig. 5.9 Interaction diagram, design values P_u versus M_u by the ACI Code.

It is obtained by reducing the actual strengths by the capacity reduction factor ϕ for columns. As mentioned before, ϕ is 0.70 for tied columns and 0.75 for spiral columns. Thus, the actual strength represented by point a corresponds to the reduced design strength shown by point A, which is obtained by multiplying the length Oa by the reduction factor ϕ as shown. However, the Code also specifies minimum eccentricities of $0.10h$ for tied and $0.05h$ for spiral columns, but no smaller than 1.0 in. for cast-in-place nor 0.6 in. for precast members. Thus, any member with design eccentricity smaller than these values must be designed as if it had the indicated minimum eccentricity. This results in the horizontal cutoff beginning at point D. Lastly, for small axial loads P_u, the Code permits a linear increase of ϕ from ϕ at $0.10f_c'A_g$ to 0.90 for $P_u = 0$. This is shown at the lower end of the curve, beginning at point C.

c. Design aids In order to reduce the repetitive effort of dimensioning large numbers of columns, as they occur in any multistory building, a number of design aids are available in the form of tables or graphs. These are found in handbooks or special pamphlets published by the American Concrete Institute, the Concrete Reinforcing Steel Institute, and others. They cover the most common practical cases, i.e., symmetrically reinforced tied rectangular columns and circular and square spirally reinforced columns. Some of these incorporate the safety provisions of the Code, as represented by the dashed curve $DABC$ of Fig. 5.9.

Graphs 2 to 9 of the Appendix represent a set of such design aids. They give the actual, unreduced carrying capacities P'_u and M'_u corresponding to the solid curve *dabc* of Fig. 5.9. To obtain the design values P_u and M_u, the values read from the graphs must be multiplied by the applicable reduction factor ϕ, and attention must be paid to the prescribed minimum eccentricities, as discussed in section *b* above. Graphs 2 to 5 are drawn for rectangular columns with reinforcement along two opposite faces; Graphs 6 to 9 for spiral-reinforced circular columns. The graphs are valid for any concrete strength up to $f'_c = 4000$ psi. They are limited to this strength range because, for concretes of larger strength, the parameters α and β, which characterize the concrete stress distribution, and the factor β_1, which determines the equivalent rectangular stress block, are no longer constant but decrease with increasing f'_c. This was discussed in detail in Art. 2.4. The charts are drawn for $f_y = 60$ ksi. However, since $\mu = f_y/0.85f'_c$ is a main parameter of these charts, they can be used without substantial errors for other steel strengths, at least for preliminary dimensioning. The previously mentioned published design aids cover a wide range of steel and concrete strengths.

The graphs are seen to consist of interaction curves of the type just described. However, instead of plotting P'_u versus M'_u, convenient nondimensional parameters have been selected, that is, P'_u/bhf'_c versus $M'_u/bh^2f'_c$. Families of curves are drawn for various values of $\rho_t\mu$ which are used in most cases in conjunction with the family of radial lines representing different eccentricity ratios e/h. The use of these graphs and associated calculations is illustrated in the following examples.

Example   In a two-story structure an exterior column is to be designed for dead load of 56 kips, maximum live load 72 kips, dead-load moment 88 ft-kips, live-load moment 75 ft-kips. The minimum live load compatible with the full live-load moment is 36 kips, which obtains when no live load is placed on the roof but full live load is placed on the second floor. Design the column for $f'_c = 4000$ psi, $f_y = 60,000$ psi.

A cross section 12 × 20 in. will be tried, and the graphs used for the design.

(a) Design for full dead and live load. According to the safety provisions of the ACI Code (see Art. 3.3), the column must be designed for $P_u = 1.4 \times 56 + 1.7 \times 72 = 201$ kips and $M_u = 1.4 \times 88 + 1.7 \times 75 = 251$ ft-kips. Then, according to the Code, the required load and moment capacities are $P'_u = P_u/\phi = 201/0.70 = 287$ kips and $M'_u = M_u/\phi = 251/0.70 = 358$ ft-kips. For the selected column size, $e/h = (358 \times 12)/(287 \times 20) = 0.75$. To select the correct graph, with 2.5-in. cover from face to bar center, $\gamma/h = 15.0/20 = 0.75$. Hence, it is necessary to interpolate between Graphs 3 and 4 of the Appendix. To enter the graphs, $\kappa' = 287/(12 \times 20 \times 4) = 0.30$ and $\kappa'e/h = 0.30 \times 0.75 = 0.225$. Interpolating between the two graphs, one finds $\rho_t\mu = 0.40$. Since $\rho_t = A_s/bh$ and $\mu = f_y/0.85f'_c$, $A_s = 0.40 \times 12 \times 20 \times (0.85 \times \frac{4}{60}) = 5.45$ in.^{2} Four No. 9 bars and two No. 8 bars with an area of 5.57 in.^{2} will suffice, with the No. 9 bars placed in the four corners, and the No. 8 bars between them along the short sides. Often it will be considered better practice to select six No. 9 bars, giving 6.00 in.^{2} Even though this represents an excess of about 10 per cent over the required steel area, bar identification and placement are simplified and opportunities of error reduced.

(b) It is observed that in the graphs the point representing the given full axial load and full moment is located in the lower portion of the curves. In these portions the moment capacity is seen to decrease with decreasing axial force, in contrast to the upper portions, where the reverse is true. In other words, for the given data the column strength is controlled by tension. Therefore, its capacity must be checked additionally for maximum moment acting simultaneously with the minimum axial load consistent with the maximum moment. For this case, $M_u = 251$ ft-kips as before, but $P_u = 1.4 \times 56 + 1.7 \times 36 = 139.5$ ft-kips. Correspondingly, $P'_u = 139.5/0.70 = 199$ kips, and as before, $M'_u = 358$ ft-kips. Calculating κ' and $\kappa'e/h$ as before, interpolation between graphs now gives $\rho_t\mu = 0.46$ and $A_s = 6.25$ in.[2] Eight No. 8 bars furnish 6.28 in.[2], four each along the 12-in. faces. It is seen that the second load pattern, with full maximum moment and with minimum axial force consistent with this moment, governs the design. This would not be the case for members with large axial forces and relatively small moments, such as the lower-story columns of high-rise buildings.

d. Member with small axial load, tension reinforced   The majority of columns are symmetrical and symmetrically reinforced. Unsymmetrical members do occur, however. For these, design aids are not usually available, but they can be handled without difficulty by using the basic premises of ultimate-strength design. By way of example, in single-story rigid frames, such as portal frames, large bending moments often combine with relatively small axial loads. In such cases unsymmetrical reinforcement frequently presents the most economical solution. For illustration, such a member, reinforced for tension only, is analyzed herein.

For a member with tension reinforcement only, subject to a compression force with an eccentricity e' relative to the center of the tension steel, Eqs. (2.66) and (2.67) can be evaluated for the special case of $A'_s = 0$. In a manner similar to that used for deriving, from these equations, Eq. (2.74) for the special case of $A_s f_y = A'_s f'_y$, one obtains

$$P_u = \phi P'_u = \phi 0.85 f'_c bd \left[-\rho\mu + 1 - \frac{e'}{d} + \sqrt{\left(1 - \frac{e'}{d}\right)^2 + \frac{e'\rho\mu}{d}} \right]$$

when failure is initiated by yielding. Whether one chooses to use this formula or direct numerical calculations from general principles is a matter of individual preference.

Example   The member shown in Fig. 5.10 is identical with that analyzed for bending only in Art. 2.4f. It is now subject to a compression force with an eccentricity from the center line of $e = 55$ in. Determine the ultimate design force and moment.

With external and internal forces as shown, the two equilibrium equations are as follows: Taking moments about the center of the steel,

$$P'_u (55 + 10.5) = a \times 10 \times 0.85 \times 4000 \left(23 - \frac{a}{2} \right)$$

and equating internal and external longitudinal forces,

$$P'_u = a \times 10 \times 0.85 \times 4000 - 2.35 \times 60,000$$

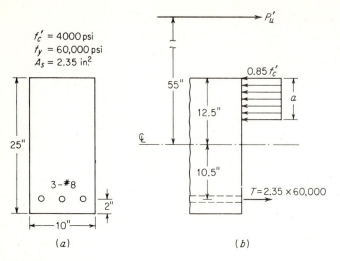

$f_c' = 4000\,\text{psi}$
$f_y = 60,000\,\text{psi}$
$A_s = 2.35\,\text{in}^2$

Fig. 5.10

These two simultaneous equations are solved either explicitly or by successive approximation (assuming a value of a and correcting). They give $a = 6.00$ in., $P_u' = 63,000$ lb, and consequently, $M_u' = 55 \times 63,000 = 3,470,000$ in.-lb. It is seen that P_u', and therefore P_u, are smaller than $0.1f_c'A_g = 0.1 \times 4000 \times 250 = 100,000$ lb. Hence, according to the ACI Code, a ϕ value between 0.7 and 0.9 is to be used, the precise value to be obtained by interpolation between 0.7 at $P_u = 0.1f_c'A_g$ and 0.9 at $P_u = 0$. While P_u' is known at this stage, ϕ, and therefore P_u, are not, so that ϕ can be determined only by trial. Thus, assuming $\phi = 0.8$ gives a trial value $P_u = 0.8 \times 63,000 = 50,300$ lb. Performing the interpolation as stipulated in the Code, $\phi = 0.7 + 0.2(100 - 50.3)/100 = 0.80$, which coincides with the assumed value. Hence, the design capacities are $P_u = 0.80 \times 63,000 = 50,300$ lb, and $M_u = 0.80 \times 3,470,000 = 2,770,000$ in.-lb.

In Art. 2.4f, the same member, but without axial force, was found to have an ultimate moment of $M_u' = 2,950,000$ in.-lb. Comparison with $M_u' = 3,470,000$ in.-lb, as calculated above, illustrates the previously discussed fact that for flexural members failing in tension, the presence of an axial force increases the moment capacity.

By the Code, the member without axial force has a design moment capacity $M_u = 0.90 \times 2,950,000 = 2,650,000$ in.-lb which, in spite of the difference in ϕ, is likewise smaller than M_u in the presence of P_u.

e. High-strength materials and contribution of intermediate bars   When large

bending moments are present, it is most economical to concentrate all or most of the steel along the outer faces parallel to the axis of bending. Such arrangements are shown in Fig. 5.2d and e. On the other hand, with small eccentricities so that axial compression is prevalent, and when a small cross section is desired, it is often advantageous to place the steel more uniformly around the perimeter, as in Fig. 5.2a to c. In this case, special attention must be paid to the intermediate bars, i.e., those which are not placed along the two faces that are most highly stressed. This is so because when the ultimate load is reached, the stresses in these intermediate bars are usually below the yield point, even though the bars along one or both extreme faces may be yielding. To analyze this situation, the distribution of strains along the cross section must be considered.

High-strength concrete and steel are effective in reducing cross sections of columns with small eccentricities. The highest-strength steel currently available has a yield point $f_y = 75$ ksi (see Table 1.2).

Example The column of Fig. 5.11a is loaded eccentrically, with $e = 9.0$ in.; $f'_c = 6.0$ ksi, and $f_y = 75$ ksi. Determine the magnitude of the design load and moment capacity.

Trial calculations indicate that the neutral axis is located 17 to 19 in. from the compression edge. A value of $c = 18$ in. is tentatively selected. When the concrete reaches its assumed failure strain of 0.003, the strain distribution is that shown in Fig. 5.11b. The strains at the locations of the four bar groups are found from similar triangles and are shown in the figure.

The yield strain corresponding to $f_y = 75$ ksi is $\epsilon_y = 75/29,000 = 0.00258$. Comparing this value with the strains at the four bar locations, it is seen that the bars

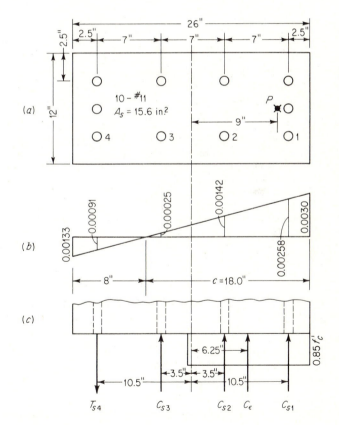

Fig. 5.11

of group 1 have just reached the yield strain (by coincidence), while those of all other bars are below yielding. Their stresses, then, are obtained by multiplying their strains by 29,000 ksi, resulting in the following values for the four groups:

$$f_{s1} = 75.0 \text{ ksi} \qquad f_{s2} = 41.2 \text{ ksi} \qquad f_{s3} = 7.3 \text{ ksi} \qquad f_{s4} = -26.4 \text{ ksi}$$

From the assumed value of $c = 18$ in., observing that for $f_c' = 6$ ksi, $\beta_1 = 0.85 - 2 \times 0.05 = 0.75$ [see Eq. (2.31)], one obtains the depth of the rectangular stress block $a = 0.75 \times 18.0 = 13.5$ in. Hence,

$$
\begin{aligned}
P_u' &= 0.85 f_c' ab + \Sigma f_{sn} A_{sn} = 0.85 \times 6 \times 13.5 \times 12 + 75 \times 4.68 + 41.2 \\
&\qquad\qquad\qquad\qquad\qquad\qquad\qquad\qquad \times 3.12 + 7.3 \times 3.12 - 26.4 \times 4.68 \\
&= 1205 \text{ kips} \\
M_u' &= 0.85 f_c' ab \times 6.25 + \Sigma f_{sn} A_{sn} z_n = 10{,}520 \text{ in.-kips}
\end{aligned}
$$

where z_n is the distance of bar group n from the center line of the section (see Fig. 5.11c).

Thus, the assumed location of the neutral axis, $c = 18$ in., results in an eccentricity of $e = 10{,}520/1205 = 8.75$ in., which differs by only 2.8 per cent from the given eccentricity of 9.0 in. No revision of the assumed axis location is needed, and the design capacities are

$$P_u = 0.7 \times 1205 = 845 \text{ kips} \qquad M_u = 0.7 \times 10{,}520 = 7400 \text{ in.-kips}$$

The following general observations can be made from this example:

1. Even with the relatively small eccentricity of about one-third of the depth of the section, only the bars of group 1 just barely reach their yield strain, and consequently their yield stress. All other bar groups are stressed far below their yield strength, which would also have been true for group 1 for a slightly larger eccentricity. It follows that the use of the more expensive high-strength steels is economical in symmetrically reinforced columns only for very small eccentricities, such as in the lower stories of high-rise buildings.
2. The contribution of the intermediate bars of groups 2 and 3 to both P_u and M_u is quite small because of their low stresses. Again, intermediate bars, except as they are needed to hold ties in place, are economical only for columns with very small eccentricities.

PROBLEM

Find the design capacities P_u and M_u for the same column if it is reinforced with four No. 18 bars, one in each corner, of yield strength 60 ksi, instead of the ten No. 11 high-strength bars used in the example. Considering that the price of high-strength reinforcement is about 25 per cent higher than that of regular-strength reinforcement, discuss the advantages or disadvantages of using, in columns with moderate eccentricities, (a) high-strength vs. regular-strength steel; (b) presence vs. absence of intermediate bars.

5.5 RECTANGULAR COLUMNS IN BIAXIAL BENDING

The methods discussed in Art. 5.4 permit rectangular columns to be designed or analyzed if bending is present about only one of the two principal axes. There are situations, and they are by no means exceptional, in which axial compression is accompanied by simultaneous bending about both principal axes of the section. Such is the case, for instance, in corner columns of tier buildings where beams and girders frame into the column in the directions of both walls and transfer their end moments into the column in two perpendicular planes. Similar situations can occur in interior columns, particularly in irregular column layouts, and in a variety of other structures.

The situation with respect to ultimate strength of biaxially loaded columns is shown schematically in Fig. 5.12. Let x and y denote the directions of the principal axes of the cross section. Then the curve denoted by M'_{ux0} represents the usual interaction curve if the column is bent about the x axis only; similarly, the curve M'_{uy0} represents bending about the y axis only. If bending is inclined so that it occurs about an axis subtending an angle α with the x axis, another interaction curve is obtained, as shown. For various angles α, these curves generate an interaction surface of the shape shown.

The horizontal plane in Fig. 5.12 is drawn for a given value of the axial force P'_u. If bending were only about the x or the y axis, the column, simultaneously with P'_u, could sustain the shown moments M'_{ux0} and M'_{uy0}, respectively. For bending inclined at an angle α, the corresponding moment capacity is M'_u. This moment has the components M'_{ux} and M'_{uy} in the two principal directions. These moments are of course smaller than M'_{ux0} and M'_{uy0}, respectively. Based on the usual propositions of strength design (rectangular stress block, ultimate concrete strain of 0.003, etc.), computer studies have been used to produce graphical or tabular design aids for biaxially bent columns (e.g., Ref. 5.2).

A simple, approximate design method has been developed by Bresler (Ref. 5.3) and has been satisfactorily verified by comparison with results of

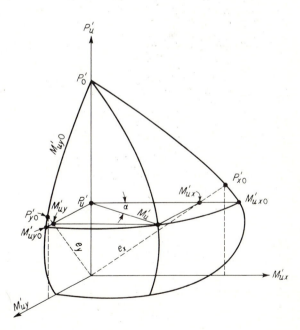

Fig. 5.12 Interaction surface for compression plus biaxial bending.

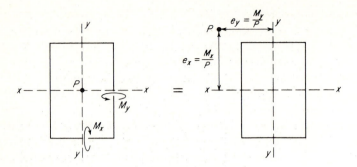

Fig. 5.13 Biaxial bending plus compression.

extensive tests and accurate computations (e.g., Ref. 5.4). Figure 5.13 shows a rectangular section under the simultaneous action of a concentric load P and the biaxial bending moments M_x and M_y. The resulting action is the same as when the load P is applied with eccentricities

$$e_x = \frac{M_x}{P} \quad \text{and} \quad e_y = \frac{M_y}{P}$$

about the centroidal axes, as shown.

Bresler's equation can then be written in the form

$$\frac{P'_u}{P'_{x0}} + \frac{P'_u}{P'_{y0}} - \frac{P'_u}{P'_0} = 1 \tag{5.7}$$

where P'_u = ultimate load for biaxial bending with eccentricities e_x and e_y
$\quad\quad P'_{x0}$ = ultimate load when only eccentricity e_x is present ($e_y = 0$)
$\quad\quad P'_{y0}$ = ultimate load when only eccentricity e_y is present ($e_x = 0$)
$\quad\quad P'_0$ = ultimate load for concentrically loaded column [Eq. (5.6)]

The values of these four loads are shown in Fig. 5.12, together with the inclined lines corresponding to eccentricities e_x and e_y. It is seen that the ultimate load for biaxial bending, P'_u, is of course smaller than either of the two loads for uniaxial bending, P'_{x0} and P'_{y0}.

The nature of Eq. (5.7) is easily illustrated. If, for instance, it were applied in the limit to a column bent only about the x axis, e_y would equal zero; therefore P'_{y0} would equal P'_0 by the definition of P'_{y0}. If $P'_{y0} = P'_0$ is substituted in Eq. (5.7), one finds $P'_u = P'_{x0}$, as one should. That is, when bending is present in one direction only, the equation gives the correct answer to the effect that the ultimate load is that for uniaxial bending in the particular direction. For another example, assume that $P'_u/P'_0 = 0.20$. This means that the eccentricities are relatively large, so that the ultimate load is a small fraction of that for the concentric column. Assume further that $P'_u/P'_{y0} = 0.40$. Substitution in Eq. (5.7) gives $P'_u/P'_{x0} = 0.80$. That is, the simultaneous bending about the y axis has reduced the ultimate load of this column by 20 per cent as compared with what it would have been had bending been present about the x axis only.

For use in design, the equation can be written in the alternative form

$$\frac{1}{P'_u} = \frac{1}{P'_{x0}} + \frac{1}{P'_{y0}} - \frac{1}{P'_0} \tag{5.8}$$

which is used as follows:

Given the size and reinforcement of the column and the load eccentricities e_x and e_y, one computes, by the methods of the preceding sections, the ultimate loads P'_{x0} and P'_{y0} for uniaxial bending around the x and y axes, respectively, and the ultimate load P'_0 for concentric loading. Then Eq. (5.8) permits one to compute $1/P'_u$, and therefrom, the calculated ultimate load P'_u for biaxial bending. The design ultimate load, as usual, is obtained as $P_u = \phi P'_u$.

Equations (5.7) and (5.8) are valid provided $P'_u \geq 0.10 P'_0$. They are not reliable where biaxial bending is prevalent and is accompanied by an axial force smaller than $P'_0/10$. In the case of strongly prevalent bending, failure is initiated by yielding of the steel, and the situation corresponds to the lowest tenth of the interaction diagram of Fig. 5.8. In this range it is safe and satisfactorily accurate to neglect the axial force entirely and to calculate the section for biaxial bending only. This is so because, as was discussed before and is evident from the shape of the interaction diagram, in the region of tension failure the addition of axial load increases the moment capacity, i.e., contributes to safety. If this contribution is neglected to simplify calculations, the result is on the conservative side.

Example A 12- × 20-in. column is reinforced with four No. 9 bars, one in each corner, giving $A_{st} = 4.0$ in.[2] The eccentricity about the major axis is $e_x = 17$ in., and about the minor axis $e_y = 3$ in. Calculate the design load for this column, when $f'_c = 4$ ksi, $f_y = 60$ ksi.

For use with the column charts the following quantities are needed: $\rho_t\mu = [4.0/(12 \times 20)][60/(0.85 \times 4)] = 0.294$; $(\gamma/h)_x = 15.0/20 = 0.75$; $(\gamma/h)_y = 7.0/12 = 0.585$; $e_x/h_x = 17/20 = 0.85$; $e_y/h_y = 3/12 = 0.25$. In the x direction, interpolating between the charts for $\gamma/h = 0.8$ and 0.7, one obtains $\kappa'_x = 0.205$; in the y direction, the chart for $\gamma/h = 0.60$ being close enough, $\kappa'_y = 0.605$. Hence, $P'_{x0} = 0.205 \times 4 \times 240 = 197$ kips, and $P'_{y0} = 0.605 \times 4 \times 240 = 580$ kips. Also, $P'_0 = 0.85 f'_c bd + A_{st} f_y = 0.85 \times 4 \times 12 \times 20 + 4 \times 60 = 1055$ kips. Then, from Eq. (5.8),

$$P'_u\left(\tfrac{1}{197} + \tfrac{1}{580} - \tfrac{1}{1055}\right) = 1$$

which gives

$$P'_u = 171 \text{ kips}$$

The carrying capacity to be used in design is then

$$P_u = \phi P'_u = 0.70 \times 171 = 120 \text{ kips} \tag{5.9}$$

One observes that, in this example, an eccentricity ratio e/h in the y direction amounting to almost 30 per cent of that in the x direction resulted in a capacity reduction of only 13 per cent (i.e. 171 versus 197 kips). This illustrates that the effect of minor bending moments in the direction perpendicular to the main moments is often negligibly small, which is some justification for the frequent practice in framed structures of neglecting the bending moments in the direction perpendicular to that of main framing. It follows that biaxial bending should be taken into account when estimated eccentricity ratios in the minor direction approach or exceed about 20 per cent of those in the major direction.

5.6 CIRCULAR COLUMNS, SPIRAL-REINFORCED

It was mentioned in Art. 5.2 that spiral-reinforced columns, when load eccentricities are small, show greater toughness, i.e., greater ductility, than tied columns, but that this difference fades out as the eccentricity is increased. For this reason, as was discussed in Art. 5.3, the Code provides a more favorable reduction factor $\phi = 0.75$ for spiral columns, as compared with $\phi = 0.70$ for tied columns. Also, the minimum stipulated design eccentricity for entirely or nearly axially loaded members is $e_{\min} = 0.05h$ for spiral-reinforced members, as compared with twice this value for tied members. It follows that spirally reinforced columns permit a somewhat more economical utilization of the materials, particularly for calculated eccentricities smaller than $0.05h$. Further advantages lie in the facts that spirals are available prefabricated, which may save labor in assembling column cages, and that the circular shape is frequently desired by the architect.

Figure 5.14 shows the cross section of a spiral-reinforced column. From six to ten and more longitudinal bars of equal size are provided for longitudinal reinforcement, depending on column diameter. The strain distribution at the instant at which the ultimate load is reached is shown in Fig. 5.14b. Bar groups 2 and 3 are seen to be strained to much smaller values than groups 1 and 4. The stresses in the four bar groups are easily found. For any of the bars with strains in excess of yield strain $\epsilon_y = f_y/E_s$, the stress at failure is evidently the yield stress of the bar. For those bars with smaller strains, the stress is found from $f_s = \epsilon_s E_s$.

One then has the internal forces shown in Fig. 5.14c. They must be in force and moment equilibrium with the external load P'_u. It will be noted that the situation is exactly the same as that discussed, for rectangular columns, in Art. 5.4e. Calculations can be carried out exactly as in the example of that article, except that for circular columns the concrete compression zone subject to the equivalent rectangular stress distribution has the shape of a segment of a circle, shown shaded in Fig. 5.14a.

Although the shape of the compression zone and the strain variation in the different groups of bars make longhand calculations awkward, no new principles are involved and computer solutions are easily developed.

Design or analysis of spirally reinforced columns is usually carried out by means of design aids, such as Graphs 6 to 9 of the Appendix. More elaborate and detailed tables and graphs are available, as mentioned in Art. 5.4c. In

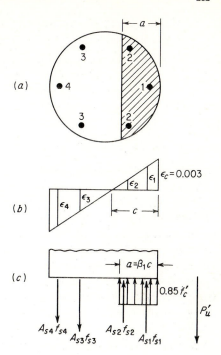

Fig. 5.14 Circular column, compression plus bending.

developing such design aids, the entire steel area is assumed to be arranged in a uniform, concentric ring, rather than being concentrated in the actual bar locations; this simplifies calculations without noticeably affecting results.

It should be noted that in order to qualify for the more favorable safety provisions for spiral columns, the steel ratio of the spiral must be at least equal to that given by Eq. (5.2) of Art. 5.2, for reasons discussed in that article.

5.7 SERVICE-LOAD DESIGN

It was pointed out in Arts. 3.1, 3.2, and 3.14 that for flexural members the Code permits design to be based on the stipulated service loads not multiplied by load factors, provided that appropriately reduced design strengths are used. In flexural members, as a concession to long tradition, this is achieved by basing design on assumed elastic material behavior and using elastic allowable stresses which are about half of the specified material strength. The results of such calculations are generally conservative and frequently uneconomical when compared with the more realistic modern methods of strength design which constitute the main design approach in the Code.

For columns, likewise, the Code permits service-load design as an alternative design method. Just as for beams, the method dispenses with the carefully graduated load factors and permits design to be made for the unfactored specified service loads, dead and live. The requisite safety margin is provided by specifying that for service-load design, the combined axial load and moment capacity shall be taken as 40 per cent of that computed by the strength design methods which have been discussed in the preceding articles. In other words, a flat safety factor of 2.5 replaces the combined load and ϕ factors of the principal design method of the Code. This service-load method, again a concession to long tradition, results in somewhat simpler "bookkeeping" in design calculations, achieved at the expense of reduced rationality, producing structures which have less consistent strengths in their various parts and which are generally less economical than when designed for strength.

PROBLEMS

1. A column 18 in. square is reinforced with four No. 11 bars; $f'_c = 4$ ksi, and $f_y = 60$ ksi. If the column is axially loaded (design eccentricity $e = 0$), find P_u.

2. The column of Prob. 1 carries a load with an eccentricity $e = 3.5$ in. Find P_u and M_u.

3. If the eccentricity of the column of Prob. 1 is $e = 11.0$ in., find P_u and M_u.

4. A column carries, axially, a dead load of 90 kips and a live load of 70 kips. Using $f'_c = 4$ ksi and $f_y = 60$ ksi, design (a) a square, tied column; (b) a circular, spiral column.

5. Redesign the columns of Prob. 4 if the live load is applied with an eccentricity $e = 6$ in.

6. Redesign the columns of Prob. 4 if the live load is applied with an eccentricity of 20 in.

7. For the situation of Prob. 6, design a rectangular tied column with ratio of long to short side of the order of 2. Compare economy, i.e., compare quantity of concrete and steel, with Prob. 6.

8. Design the columns of Prob. 4 and/or 5, 6, or 7 by the service-load design method, and compare results with those obtained by the strength design method.

9. Redesign the column of Prob. 7 for $f'_c = 4.5$ ksi and $f_y = 75$ ksi and compare results with Prob. 7. Are there reservations when using the charts of the Appendix for these specified steel and concrete strengths? If so, what are they? Are they likely to be serious for the given strength values?

SLENDER COLUMNS

5.8 GENERAL

A column is said to be slender if its cross-sectional dimensions are small as compared with its length. The degree of slenderness is generally expressed in terms of the slenderness ratio l/r, where l is the unsupported length of the member, and r the radius of gyration ($r = \sqrt{I/A}$) of its cross section. For square or circular members, the value of r is the same about any cross-sectional axis; for other shapes r is smallest about the minor principal axis, and it is generally this value which must be used in determining the slenderness ratio of a free-standing column.

It has long been known that a member of great slenderness will collapse under a smaller compression load than a stocky member of the same cross-sectional dimensions. This is so for the following reason: A stocky member, say with $l/r = 10$ (e.g., a square column of length equal to about three times its thickness h), when loaded in axial compression, will fail at the load given by Eq. (5.1), because at that load, both concrete and steel are stressed to their maximum carrying capacity and give way, respectively, by crushing and by yielding. If a member with the same cross section has a slenderness ratio $l/r = 100$ (e.g., a square column hinged at both ends and of length equal to about 30 times its thickness), it may fail under an axial load of one-half or less of that given by Eq. (5.1). In this case, collapse is caused by buckling, i.e., by lateral bending of the member, with consequent overstressing of steel and concrete by the bending stresses which superpose on the axial compression stresses.

With present-day high-strength materials and with improved methods of dimensioning members (such as ultimate-strength design), it is now possible, for a given value of axial load, with or without simultaneous bending, to design a much smaller cross section than in the recent past. Evidently, this makes for more slender members. It is chiefly because of this that rational and reliable design procedures for slender columns have become increasingly important. In this respect, the method first introduced in the 1971 edition of the ACI Code represents a great step ahead for two reasons: For one, the method is based on sound principles of structural mechanics and is extensively verified by tests so that it is both rational and reliable. For another, the method is basically the same, with due regard to difference in material behavior, as that used in steel-structure design according to the specifications of the American Institute of Steel Construction (AISC) and the American Iron and Steel Institute (AISI). In consequence, the design of slender columns has now been put on a common basis in this country, regardless of material.

5.9 SLENDERNESS EFFECTS

a. Concentrically loaded columns The basic information on the behavior of straight, concentrically loaded, slender columns was developed by Euler more than 200 years ago (Ref. 5.3). In generalized form, it states that such a member will fail by buckling at the critical load

$$P_{cr} = \frac{\pi^2 E_t I}{(kl)^2} \tag{5.10}$$

which, when both sides are divided by the cross-sectional area, gives the compression stress at which buckling occurs as

$$\left(\frac{P}{A}\right)_{cr} = \frac{\pi^2 E_t}{(kl/r)^2} \tag{5.11}$$

It is seen that the buckling stress or load decreases rapidly with increasing *slenderness ratio kl/r.*

For the simplest case of a column hinged at both ends and made of elastic material, E_t simply becomes Young's modulus and kl is equal to the actual length l of the column. At the load or stress given by Eq. (5.10) or (5.11), the originally straight member buckles into a half sine wave, as in Fig. 5.15a. In this bent configuration, bending moments Py act at any section such as a; y is the deflection at that section. These deflections continue to increase until the bending stress caused by the increasing moment, together with the original compression stress, overstresses and fails the member.

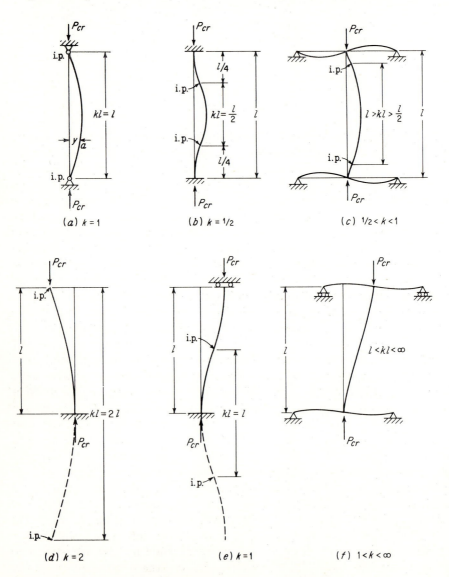

Fig. 5.15 Buckling and effective length of axially loaded columns.

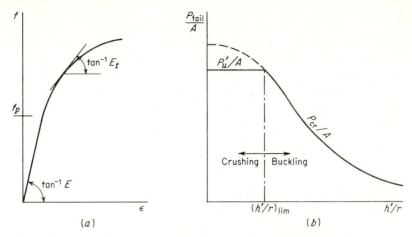

Fig. 5.16 Effect of slenderness on strength of axially loaded columns.

If the stress-strain curve of a short piece of the given member is of the shape of Fig. 5.16a, as it would be for reinforced-concrete columns, E_t is equal to Young's modulus, provided the buckling stress $(P/A)_{cr}$ is below the proportional limit f_p. If it is larger than f_p, buckling occurs in the inelastic range. In this case, in Eqs. (5.10) and (5.11), E_t is the tangent modulus, i.e., the slope of the tangent to the stress-strain curve. As the stress increases, E_t decreases. A plot of the buckling stress versus the slenderness ratio, a so-called column curve, therefore has the shape given in Fig. 5.16b, which shows the reduction in buckling strength with increasing slenderness. For very stocky columns one finds that the value of the buckling load or stress, calculated from Eq. (5.10) or (5.11), exceeds the direct crushing strength of the stocky column, given by Eq. (5.1). This is also shown in Fig. 5.16b. Correspondingly, there is a limiting slenderness ratio $(kl)/r_{lim}$. For values smaller than this, failure occurs by simple crushing, regardless of kl/r; for values larger than $(kl)/r_{lim}$, failure occurs by buckling, the buckling load or stress decreasing for greater slenderness.

If the member of Fig. 5.15a is fixed against rotation at both ends, it buckles in the shape of Fig. 5.15b, with inflection points (i.p.) as shown. The portion between inflection points is in precisely the same situation as the hinge-ended column of Fig. 5.15a, and thus the *effective length* kl of the fixed-fixed column, i.e., the distance between inflection points, is seen to be $kl = l/2$. Equation (5.10) shows that an elastic column, when fixed at both ends, will carry 4 times as much load as when hinged.

Columns in real structures are rarely either hinged or fixed, but rather have ends partially restrained against rotation by abutting members. This is schematically shown in Fig. 5.15c, from which it is seen that for such members the effective length kl, that is, the distance between inflection points, has a value between l and $l/2$. The precise value depends on the degree of end restraint, i.e., on the ratio r' of the rigidity EI/l of the column to the sum of rigidities EI/l of the restraining members at both ends. In the columns of

Fig. 5.15a to c it was assumed that one end was prevented from moving laterally relative to the other end, by horizontal bracing or otherwise. In this case it is seen that the effective length kl is always smaller than, or at most equal to, the real length l.

If a column is fixed at one end and entirely free at the other (cantilever column or flagpole), it buckles as shown in Fig. 5.15d. That is, the upper end moves laterally with respect to the lower, a kind of deformation known as sidesway. It buckles into a quarter of a sine wave and, therefore, is analogous to the upper half of the hinged column of Fig. 5.15a. The inflection points, one at the end of the actual column and the other at the imaginary extension of the sine wave, are seen to be a distance $2l$ apart, so that the effective length is $kl = 2l$.

If the column is rotationally fixed at both ends but one end can move laterally with respect to the other, it buckles as shown in Fig. 5.15e, with an effective length $kl = l$. If one compares this column, fixed at both ends but free to sidesway, with a fixed-fixed column which is braced against sidesway (Fig. 5.15b), one sees that the effective length of the former is twice that of the latter. By Eq. (5.10), this means that the buckling strength of an elastic fixed-fixed column which is free to sidesway is only one-quarter that of the same column when braced against sidesway. This is an illustration of the general fact that *compression members free to buckle in a sidesway mode are always considerably weaker than when braced against sidesway.*

Again, the ends of columns in actual structures are rarely either hinged, fixed, or entirely free, but are usually restrained by abutting members. If sidesway is not prevented, buckling occurs as in Fig. 5.15f, and the effective length, as before, depends on the degree of restraint r'. If the cross beams are very rigid as compared with the column, the case of Fig. 5.15e is approached, and kl is only slightly larger than l. On the other hand, if the restraining members are extremely flexible, a hinged condition is approached at both ends. Evidently, a column hinged at both ends and free to sidesway is unstable. It will simply topple, being unable to carry any load whatever.

In reinforced-concrete structures one is rarely concerned with single members, but rather with rigid frames of various configurations. The manner in which the described relationships affect the buckling behavior of frames is illustrated by the simple portal frame of Fig. 5.17, with loads applied concentrically to the columns. If sidesway is prevented, as indicated schematically by the brace of Fig. 5.17a, the buckling configuration will be as shown. The buckled shape of the column corresponds to that of Fig. 5.15c, except that the lower end is hinged. It is seen that the effective length kl is smaller than l. On the other hand, if no sidesway bracing is provided to an otherwise identical frame, buckling occurs as in Fig. 5.17b. The column is in a situation similar to that of Fig. 5.15d, upside down, except that the upper end is not fixed, but only partially restrained by the girder. It is seen that the effective length kl exceeds $2l$ by an amount depending on the degree of restraint. The buckling

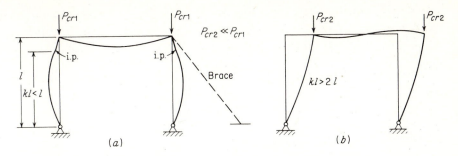

Fig. 5.17   Rigid-frame buckling. (*a*) Laterally braced; (*b*) unbraced.

strength depends on kl/r in the manner shown in Fig. 5.16*b*. In consequence, even though they are dimensionally identical, the unbraced frame will buckle at a radically smaller load than the braced frame.

In summary, the following can be noted:

The strength of concentrically loaded columns decreases with increasing slenderness ratio kl/r.

In columns which are *braced against sidesway* or which are parts of frames braced against sidesway, the effective length kl, that is, the distance between inflection points, falls between $l/2$ and l, depending on degree of end restraint.

The effective lengths of columns which are *not braced against sidesway*, or which are parts of frames not so braced, are always larger than l, the more so the smaller the end restraint. In consequence, the buckling load of a frame not braced against sidesway is always substantially smaller than that of the same frame when braced.

b. Compression plus bending   Most reinforced-concrete compression members are subject to simultaneous flexure, caused by transverse loads or by end moments due to continuity. The behavior of members subject to such combined loading also depends greatly on their slenderness.

Figure 5.18*a* shows such a member, often known as a *beam-column*, axially loaded by P and bent by equal end moments M_e. If no axial load were present, the moment M_0 in the member would be constant throughout and equal to the end moments M_e. This is shown in Fig. 5.18*b*. In this situation, i.e., in simple bending without axial compression, the member deflects as shown by the dashed curve of Fig. 5.18*a*, where y_0 represents the deflection at any point caused by bending only. When P is applied, the moment at any point increases by an amount equal to P times its lever arm. The increased moments cause additional deflections, so that the deflection curve under the simultaneous action of P and M_0 is the solid curve of Fig. 5.18*a*. At any point, then, the total moment is now

$$M = M_0 + Py \tag{5.12}$$

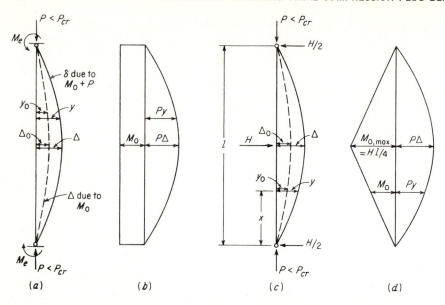

Fig. 5.18 Moments in slender members, compression plus bending, single curvature.

that is, the total moment consists of the moment M_0 which acts in the absence of P and the additional moment caused by P, equal to P times the deflection.

A similar situation is shown in Fig. 5.18c, where bending is caused by the transverse load H. When P is absent, the moment at any point x is $M_0 = Hx/2$, with a maximum at midspan equal to $Hl/4$. The corresponding M_0 diagram is shown in Fig. 5.18c. When P is applied, additional moments Py are caused again, distributed as shown, and the total moment at any point in the beam-column consists of the same two parts as in Eq. (5.12).

The deflections y of elastic beam-columns of the type shown in Fig. 5.18 can be calculated from the deflections y_0 (i.e., from the deflections of the corresponding beam without axial load), using the following expression (see, for example, Ref. 5.5):

$$y = y_0 \frac{1}{1 - P/P_{cr}} \tag{5.13}$$

If P and P_{cr} in Eq. (5.13) are divided by the area A, and $(P/A)_{cr}$ is replaced with the expression given in Eq. (5.11), then Eq. (5.13) can be rewritten in the form

$$y = y_0 \frac{1}{1 - P(kl/r)^2/A\pi^2 E} \tag{5.14}$$

Let Δ be the deflection at the point of maximum moment M_{max}, as in Fig. 5.18. Then, from Eqs. (5.12) and (5.13),

$$M_{max} = M_0 + P\Delta = M_0 + P\Delta_0 \frac{1}{P/P_{cr}} \tag{5.15}$$

It can be shown (Ref. 5.6) that Eq. (5.15) can be written

$$M_{max} = M_0 \frac{1 + \psi P/P_{cr}}{1 - P/P_{cr}} \qquad (5.16)$$

where ψ is a coefficient which depends on type of loading and varies between about ± 0.20 for most practical cases. Considering that P/P_{cr} is always significantly smaller than 1, it is seen that the second term in the numerator of Eq. (5.16) becomes quite small compared with 1. Neglecting this term, one obtains the simplified design equation

$$M_{max} = M_0 \frac{1}{1 - P/P_{cr}} \qquad (5.17)$$

where $1/(1 - P/P_{cr})$ is known as a moment magnification factor, which reflects the amount by which the beam moment M_0 is magnified by the presence of a simultaneous axial force P.

Since P_{cr} decreases with increasing slenderness ratio, it is seen from Eq. (5.17) that the moment M in the member increases with increasing slenderness ratio kl/r. The situation is shown schematically in Fig. 5.19. It indicates that, for a given transverse loading, i.e., a given value of M_0, an axial force P causes a larger additional moment in a slender member than in a stocky member.

In the two members in Fig. 5.18, the largest moment caused by P, namely, $P\Delta$, adds directly to the maximum value of M_0, for example,

$$M_0 = \frac{Hl}{4}$$

in Fig. 5.18d. As P increases, the maximum moment at midspan increases at a rate faster than that of P in the manner given by Eqs. (5.12) and (5.17) and shown in Fig. 5.20. The member will fail when the simultaneous values of P and M become equal to P'_u and M'_u, the ultimate strength of the cross section at the location of maximum moment.

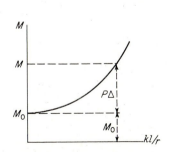

Fig. 5.19 Effect of slenderness on column moments.

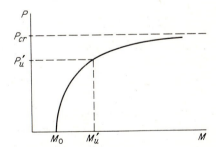

Fig. 5.20 Effect of axial load on column moments.

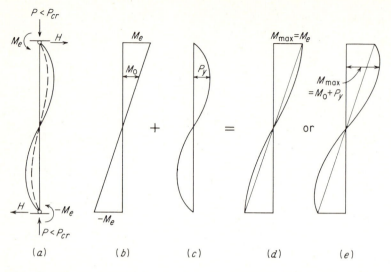

$P < P_{cr}$

M_e

H

M_e

M_0

P_y

$M_{max} = M_e$

M_{max} $= M_0 + P_y$

$+$

$=$

or

H

$-M_e$

$P < P_{cr}$

$-M_e$

(a)   (b)   (c)   (d)   (e)

Fig. 5.21   Moments in slender members, compression plus bending, double curvature.

This direct addition of the maximum moment caused by P to the maximum moment caused by transverse load, clearly the most unfavorable situation, does not obtain for all types of deformations. For instance, the member of Fig. 5.21a, with equal and opposite end moments, has the M_0 diagram of Fig. 5.21b. The deflections caused by M_0 alone are again magnified when an axial load P is applied. In this case, these deflections under simultaneous bending and compression can be approximated by (Ref. 5.5)

$$y = y_0 \frac{1}{1 - P/4P_{cr}} \tag{5.18}$$

By comparison with Eq. (5.13), it is seen that the deflection magnification here is much smaller.

The additional moments Py caused by axial load are distributed as shown in Fig. 5.21c. While the M_0 moments are largest at the ends, the Py moments are seen to be largest some distance from the ends. Depending on their relative magnitudes, the total moments $M = M_0 + Py$ are distributed as in either Fig. 5.21d or e. In the former case, the maximum moment continues to act at the end and to be equal to M_e; the presence of the axial force, then, does not result in any increase in the maximum moment. Alternatively, in the case of Fig. 5.21e, the maximum moment is located some distance from the end; at that location M_0 is significantly smaller than its maximum value M_e, and for this reason the added moment Py increases the maximum moment to a value only moderately greater than M_e.

Comparing Figs. 5.18 and 5.21, one can generalize as follows: The beam moment M_0 will be magnified most strongly when the location where M_0 is largest coincides with that where the design deflection y is largest. This obtains for members bent into single curvature by symmetrical loads or equal end moments. If the two end moments of Fig. 5.18a are unequal but of the same sign, i.e., producing single curvature, M_0 will still be strongly magnified, though not quite as much as for equal end moments. On the other hand, as evident from Fig. 5.21, there will be little or possibly no magnification if the end moments are of opposite sign and produce an inflection point along the member.

It can be shown (Ref. 5.6) that the way in which moment magnification depends on the relative magnitude of the two end moments (as in Figs. 5.18a and 5.21) can be expressed by a modification of Eq. (5.14) as follows:

$$M_{max} = M_0 \frac{C_m}{1 - P/P_{cr}} \tag{5.19}$$

where

$$C_m = 0.6 + 0.4 \frac{M_1}{M_2} \geq 0.4 \tag{5.20}$$

Here M_1 is the numerically smaller and M_2 the numerically larger of the two end moments; hence, by definition, $M_0 = M_1$. The fraction M_1/M_2 is defined as positive if the end moments produce single curvature, and negative if they produce double curvature. It is seen that when $M_1 = M_2$, as in Fig. 5.18a, $C_m = 1$, so that Eq. (5.19) becomes Eq. (5.17), as it should. It is to be noted that Eq. (5.20) applies only to members braced against sidesway. As will become apparent from the discussion which follows, for members not braced against sidesway, maximum moment magnification usually occurs, that is, $C_m = 1$.

Members which are braced against sidesway include columns that are parts of structures in which sidesway is prevented in one of various ways: by walls or partitions sufficiently strong and rigid in their own planes effectively to prevent horizontal displacement; by special bracing in vertical planes, such as wind bracing; in tier buildings by designing the utility core to resist horizontal loads and furnish bracing to the frames; or by bracing the frame against some other essentially immovable support.

If no such bracing is provided, *sidesway can occur only for the entire frame simultaneously*, not for individual columns in the frame. If this is the case, the combined effect of bending and axial load is somewhat different from that in braced columns. For illustration, consider the simple portal frame of Fig. 5.22a subject to a horizontal load H, such as a wind load, and compression forces P, such as from gravity loads. The moments M_0 caused by H alone, in the absence of P, are shown in Fig. 5.22b; the corresponding deformation of the frame is given in dashed curves. When P is added, additional moments are caused which

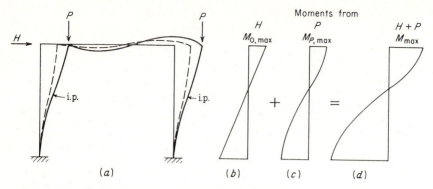

Fig. 5.22   Fixed portal frame, laterally unbraced.

result in the magnified deformations shown in solid curves and in the moment diagram of Fig. 5.22c. It is seen that the maximum values of M_0, both positive and negative, and the maximum values of the additional moments M_P of the same sign occur at the same locations, namely, at the ends of the columns. They are therefore fully additive, leading to a large moment magnification. In contrast, if the frame of Fig. 5.22 is laterally braced and vertically loaded, Fig. 5.23 shows that the maximum values of the two different moments occur in different locations; the moment magnification, if any, is therefore much smaller, as correctly expressed by C_m. For the portal hinged at the bottom, as in Fig. 5.24, the discussion of Fig. 5.22 holds: the maximum moments of both kinds are fully additive at the column tops. The difference between the cases of Figs. 5.22, and 5.24, therefore, lies not in the presence or absence of an inflection point, but only in the greater stiffness of the columns fixed at the bottom. This results in the fact that their effective length kl is smaller than that in the hinged situation.

     It should be noted that the moments which cause a frame to sidesway need not be caused by horizontal loads as in Figs. 5.22 and 5.24. Asymmetries,

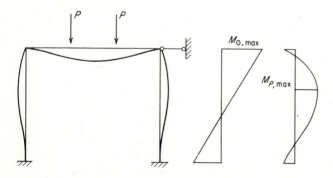

Fig. 5.23   Fixed portal frame, laterally braced.

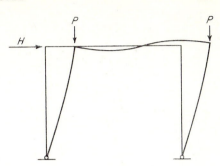

Fig. 5.24 Hinged portal frame, laterally unbraced.

either of frame configuration or vertical loading or both, also result in sidesway displacements. In this case the presence of axial column loads results in the same deflection and moment magnification as before.

In summary, it can be stated as follows:

In flexural members the presence of axial compression causes additional deflections and additional moments Py. Other things being equal, the additional moments increase with increasing slenderness ratio kl/r.

In *members braced against sidesway* and bent in single curvature, the maxima of both types of moments, M_0 and Py, occur at the same or at nearby locations and are fully additive; this leads to large moment magnifications. If the M_0 moments result in double curvature (i.e., in the occurrence of an inflection point), the opposite is true, and less or no moment magnification occurs.

In *members of frames not braced against sidesway*, the maximum moments of both kinds, M_0 and Py, almost always occur at the same locations, the ends of the column; they are fully additive, regardless of the presence or absence of an inflection point. Here, too, other things being equal, the additional deflections and the corresponding moments increase with increasing kl/r.

This discussion is a simplified presentation of a fairly complex subject. The provisions of the ACI Code regarding slender columns are based on the behavior and the corresponding equations which have just been presented. They take account, in an approximate manner, of the additional complexities which arise from the fact that concrete is not an elastic material, that tension cracking changes the moment of inertia of a member, and that under sustained load, creep increases the short-time deflections and, thereby, the moments caused by these deflections.

5.10 SLENDER REINFORCED-CONCRETE COLUMNS—ACI CODE

A slender reinforced-concrete column reaches the limit of its strength when the combination of P and M at the most highly stressed section will make that section fail. In general, P is substantially constant along the length of the member. This means that the column approaches failure when at the most highly stressed section the axial force P combines with a moment $M = M_{max}$,

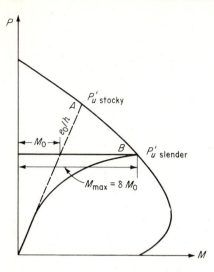

Fig. 5.25 Effect of slenderness on carrying capacity.

as given by Eq. (5.19), so that this combination becomes equal to P'_u and M'_u, which will fail the section. This is easily visualized by means of Fig. 5.25.

For a column of given cross section, Fig. 5.25 presents a typical interaction curve. For simplicity, suppose the column is bent in single curvature with equal eccentricities at both ends. For this eccentricity, the strength of the cross section is given by point A on the interaction curve. If the column is sufficiently stocky so that the moment magnification is negligibly small, the $P'_{u, stocky}$ at point A represents the member strength of the column under the simultaneous moment $M'_{u, stocky} = e_0 P'_{u, stocky}$.

On the other hand, if the same column is sufficiently slender, significant moment magnification will occur with increasing P (cf. Fig. 5.20). Then the moment at the most highly stressed section is M_{max}, as given by Eq. (5.19) (with $C_m = 1$ because of equal end eccentricities). The solid curve in Fig. 5.25 shows the nonlinear increase of M_{max} as P increases. The point where this curve intersects the interaction curve, i.e., point B, defines the member strength $P'_{u, slender}$ of the column, combined with the simultaneously applied end moments $M_0 = e_0 P'_{u, slender}$. If end moments are unequal, the factor C_m is to be included, as discussed in Art. 5.10.

Correspondingly, the Code specifies that axial loads and end moments in columns should be determined by a conventional elastic-frame analysis (see Chap. 8). The member is then to be designed for that axial load and a simultaneous magnified column moment

$$M_c = \frac{M_2 C_m}{1 - P'_u / P_{cr}}$$

in accord with Eq. (5.19). In terms of the ultimate design load $P_u = \phi P'_u$, the

Code equation is written

$$M_c = \delta M_2 \tag{5.21}$$

where the moment magnification factor (see Fig. 5.25) is

$$\delta = \frac{C_m}{1 - P_u/(\phi P_{cr})} \tag{5.22}$$

The critical load P_{cr}, in accord with Eq. (5.10), is given as

$$P_{cr} = \frac{\pi^2 EI}{(kl_u)^2} \tag{5.23}$$

where l_u is defined as the unsupported length of the compression member, and $C_m = 0.6 + 0.4M_1/M_2 \geq 0.4$ is as previously given, in Eq. (5.20).

   In homogeneous, elastic members, such as steel columns, the rigidity EI is easily obtained from Young's modulus and the usual moment of inertia. Reinforced-concrete columns, however, are nonhomogeneous, since they consist of both steel and concrete. And while steel is substantially elastic, concrete is not, and is, in addition, subject to creep and to cracking if tension occurs on the convex side of the column. All these factors affect the effective rigidity EI of a reinforced-concrete member. It is possible by computer methods to calculate fairly realistic effective rigidities, taking account of these factors. Even these calculations are no more accurate than the assumptions on which they are based, which, at this writing, are still not entirely certain. On the basis of elaborate studies, both analytical and experimental (Ref. 5.7), the Code permits EI to be determined by either of the following equations:

$$EI = \frac{E_c I_g/5 + E_s I_s}{1 + \beta_d} \tag{5.24}$$

or the simpler expression

$$EI = \frac{E_c I_g/2.5}{1 + \beta_d} \tag{5.25}$$

where E_c = modulus of elasticity of concrete, psi [see Eqs. (1.1) and (1.2)]
   I_g = moment of inertia of gross section of column, in.[4]
   E_s = modulus of elasticity of steel, 29,000,000 psi
   I_s = moment of inertia of reinforcement about centroidal axis of cross section, in.[4]
   β_d = ratio of maximum design dead-load moment to maximum design total-load moment, always positive

   The factor β_d accounts approximately for the effect of creep. That is, the larger the moments caused by sustained dead loads, the larger are the creep deformations and corresponding curvatures. Consequently, the larger the sustained loads relative to the temporary loads, the smaller is the effective rigidity, as correctly reflected in Eqs. (5.24) and (5.25).

Both Eqs. (5.24) and (5.25) are conservative lower limits for large numbers of investigated actual members (Ref. 5.7). The simpler, but more conservative, Eq. (5.25) is not unreasonable for lightly reinforced members, but greatly underestimates the effect of reinforcement for more heavily reinforced members, i.e., for the range of higher ρ values. Equation (5.24) is more reliable for the entire range of ρ and definitely preferable for medium and high ρ values.

An accurate determination of the effective length factor k is essential in connection with Eqs. (5.21) to (5.23). In Art. 5.9, it was shown that for frames braced against sidesway, k varies from $\frac{1}{2}$ to 1, while for laterally unbraced frames, it varies from 1 to ∞, depending on the degree of rotational restraints at both ends. This was illustrated in Fig. 5.15. For a rigid frame, it is seen that this degree of rotational restraint depends on whether the rigidities of the beams framing into the column at top and bottom are large or small compared with the rigidity of the column itself. An approximate but generally satisfactory way of determining k is by means of the alignment charts of Fig. 5.27 (Ref. 5.6). The charts are based on isolating the given column plus all members framing into it at top and bottom, as shown in Fig. 5.26. Then the degree of end restraint at each end is $\psi = \Sigma(EI/l \text{ of columns})/\Sigma(EI/l \text{ of floor members})$. Only floor members which are in a plane at either end of the column are to be included.

It is seen that k must be known before a column in a frame can be dimensioned. Yet k depends on the rigidity EI/l of the members to be dimensioned, as well as on that of the abutting members. Thus, the dimensioning process necessarily involves iteration; i.e., one assumes member sizes, calculates rigidities and corresponding k values, and then calculates more accurate member sizes on the basis of these k values until assumed and final member sizes

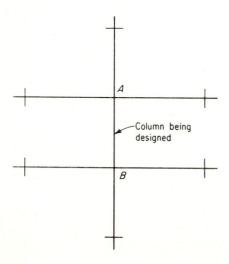

Fig. 5.26

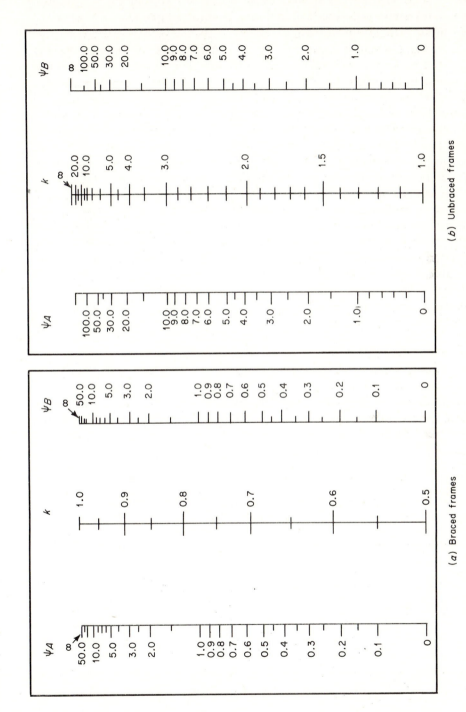

(a) Braced frames

(b) Unbraced frames

Fig. 5.27 Alignment charts for effective length factors k. ψ = ratio of $\Sigma EI/l$ of compression members to $\Sigma EI/l$ of flexural members in a plane at one end of a compression member. k = effective length factor. (*From Ref. 5.8.*)

coincide or are satisfactorily close. The question of determining the correct rigidities EI/l for both beams and columns is not a simple one. In the near-failure state for which dimensioning is done, the beams will be cracked on the tension side and the columns may or may not show tension cracking, depending on the eccentricity ratio e/h. The Code specifies that in determining k, due consideration shall be given to the effects on the rigidities of reinforcement and of cracking.

As a first approximation in the design iteration, when only a rough estimate of the member sizes has been made, it is sufficiently accurate to take one-half of the gross moment of inertia of the uncracked concrete section for the beams, and the full gross moment of inertia for the columns. This reflects the fact that the reduction of rigidity because of cracking is generally larger for beams than for columns. Once improved member sizes have been established on this basis, a more accurate determination is called for. Comparisons of a computer study with results of laboratory tests on simple frames have shown that the following procedure leads to satisfactorily accurate results: Calculate $(EI/l)_{\text{beam}}$ on the basis of the moment of inertia I_{ct} of the cracked transformed section of the beams, taking the average between support and span sections if these sections are different, and calculate $(EI/l)_{\text{col}}$ from Eq. (5.24), taking the creep factor $\beta_d = 0$.

Once the end restraint coefficients $\psi = \Sigma(EI/l)_{\text{col}}/\Sigma(EI/l)_{\text{beam}}$ at both ends of a column are determined, the effective length factor is obtained from chart a or b of Fig. 5.27, depending on whether the member is laterally braced or not. To establish whether or not a member is laterally braced, the following conditions should be considered:

Where there is more than one column in a given frame or story of a building, sidesway can occur only by simultaneous lateral motion of all columns of the particular story. Hence, for bracing to be effective, it must be sufficiently rigid to brace all columns of that story when they are fully loaded. Bracing generally consists of shear walls, utility cores, or other wall-like components. Bracing which holds the upper ends of all columns fully immovable relative to the lower ends is, of course, impossible of achievement. The function of adequate lateral bracing, therefore, is to limit lateral deflections of one end relative to the other to insignificant and inconsequential amounts. Reference 5.8 suggests that "a compression member braced against sidesway is a member in a story in which the bracing members [shear walls, etc.] have a total stiffness to resist lateral movement of that story, which is at least six times the sum of the lateral stiffness of all columns in that story."

Because sidesway can occur only for all columns of a story simultaneously, rather than for any individual column, the Code specifies that "in frames not braced against sidesway, the value δ shall be computed for the entire story, assuming all columns to be (fully) loaded." Correspondingly, in Eq. (5.22), P_u and P_{cr} shall be taken as ΣP_u and ΣP_{cr} for all columns in the story.

This procedure takes care of sidesway buckling. However, if it is ascertained in this manner that adequate safety against sidesway buckling exists in a given story, it is still possible for one or another of the individual columns to be so slender that its capacity is inadequate, even though it is fully braced against sidesway by the other columns. To safeguard against this possibility in laterally unbraced frames, it is necessary to check each column for two values of δ: (1) the value obtained for sidesway of the entire story, as above defined, which value is the same for all columns in that story, and (2) the value obtained for the individual column, assuming it to be laterally braced by the other columns.

The procedure of designing slender columns is inevitably fairly lengthy, particularly since it generally involves a trial-and-error procedure. At the same time, studies of existing structures have shown that a large fraction of all actual building columns is sufficiently stocky so that slenderness effects reduce their capacities only by a few per cent (Ref. 5.7). To permit the designer to dispense with these procedures, the Code provides upper limits of slenderness below which the effects of slenderness are insignificant and may be neglected.

These limits are adjusted so as to result in a maximum unaccounted reduction of column capacity of no more than 5 per cent. The corresponding Code provision reads: "For compression members braced against sidesway, the effects of slenderness may be neglected when kl_u/r is less than $34 - 12M_1/M_2$ and for compression members not braced against sidesway when kl_u/r is less than 22, l_u being the unsupported, free length of the column." The radius of gyration r for rectangular columns may be taken as $0.30h$, where h is the overall cross-sectional dimension in the direction in which stability is being considered. For circular members, it may be taken as 0.25 times the diameter. For other shapes, r may be computed for the gross concrete section.

This discussion of the design of slender columns has dealt with all the main features of this important problem. To save space, some of the minor points which apply only in rare cases have been omitted. For these and for more complete information, the reader is referred to the Code, to Ref. 5.8, and particularly to Ref. 5.7, which also contains an extensive bibliography.

Example   Part of a three-story frame is shown in Fig. 5.28. The frame is not braced against sidesway. Gravity loads govern; i.e., wind loading is not critical. The columns of all three stories are identical in dimensions and reinforcement. Likewise, the floor beams at levels A and B are the same. For these beams $I_b = I_{ct}$ has been calculated, resulting in $I_b/l = 43.3$ in.[3] Design the member AB as a square column, if it is loaded as follows: dead loads result in $P_d = 46$ kips, $M_d = 92$ kip-ft; live loads result in $P_l = 94$ kips, $M_l = 230$ kip-ft. Use $f'_c = 4$ ksi, $f_y = 60$ ksi. (The majority of examples in this chapter are calculated for those materials merely because Graphs 2 to 9 of the Appendix are drawn for these values. Published design aids for a wide variety of materials combinations are, or will shortly become, available.)

According to the Code, the design strength to be furnished by the column is

$P_u = 1.4 \times 46 + 1.7 \times 94 = 224$ kips
$M_u = 1.4 \times 92 + 1.7 \times 230 = 520$ kip-ft

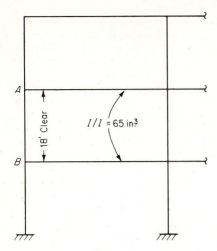

Fig. 5.28

Preliminary calculations indicate a 22- × 22-in. column with approximately $\rho_t = 0.03$. For these dimensions, $I_g = 22^4/12 = 19{,}500$ in.^{4}; $I_s = 2 \times 0.015 \times 22^2 \times 8.5^2 = 1050$ in.^{4} $E_c = 57{,}000 \sqrt{4000} = 3.6 \times 10^6$ psi. $E_s = 29 \times 10^6$ psi. Substitution of these values in Eq. (5.24) with $\beta_d = 0$ yields for the columns $EI_c = 4.45 \times 10^{10}$ lb-in.^{2}

Thus, for determining the effective length factor k for the columns, $EI_c/l = 4.45 \times 10^{10}/(12 \times 18) = 2.08 \times 10^8$ lb-in.^{2}, and for the beams, $EI_b/l = 3.6 \times 10^6 \times 43.3 = 1.56 \times 10^8$ lb-in.^{2} At both points A and B, one beam frames into two columns. Hence, at both points, the rigidity ratio is $\psi = 2 \times 2.08 \times 10^8/(1.56 \times 10^8) = 2.67$. Then, from Fig. 5.27$b$, the effective length factor $k = 1.75$.

According to the Code, in laterally unbraced columns, slenderness effects can be neglected if $kl/r = 22$. With $r = 0.3 \times 22 = 6.6$ in., $kl/r = 1.75 \times 18 \times 12/6.6 = 57$, so that here slenderness effects must be included in the design calculations.

Then, from Eq. (5.23),

$$P_{cr} = \frac{\pi^2 \times 4.45 \times 10^{10}}{(1.75 \times 18 \times 12)^2} = 3.07 \times 10^6 \text{ lb}$$

With $P_u = 2.24 \times 10^5$ lb and $C_m = 1.0$ as for laterally unbraced frames, the moment magnification factor from Eq. (5.22) is

$$\delta = \frac{1}{1 - (2.24 \times 10^5)/(0.7 \times 3.07 \times 10^6)} = 1.12$$

Consequently, the moment for which the column is to be designed according to Eq. (5.21) is

$$M_c = 1.12 \times 520 = 581 \text{ kip-ft}$$

together with $P_u = 224$ kips.

For use with the column design graphs, the eccentricity ratio $e/h = 581 \times 12/224 \times 22 = 1.42$ and $P_u/\phi f_c' bh = 224{,}000/(0.7 \times 4000 \times 22^2) = 0.165$. With $\gamma/h = 17/22 = 0.775$, Graph 3 for $\gamma/h = 0.8$ is sufficiently accurate. Entering the graph with the above values, one obtains $\rho_t \mu = 0.50$ and $\rho_t = 0.50 \times 0.85 \times 4/60 = 0.0283$. This is sufficiently close to the assumed value of $\rho_t = 0.03$ so that no further approximation is needed.

Hence, for the 22- $\times$ 22-in. column, the required steel area $A_t = 0.0283 \times 22^2 = 13.7$ in.^{2} which is furnished by ten No. 11 bars with 15.6 in.^{2} or by six No. 10 plus four No. 11 bars giving 13.84 in.^{2} Half of these are to be placed in each of the two opposite faces.

Note: In Arts. 5.10 and 5.11, it was pointed out that sidesway can occur only for all columns in a story, but not for any one single column alone. Consequently, in this example, for calculating δ, ΣP_u and ΣP_{cr} should have been used, the sums to extend to all columns in that story. In addition, the given individual column should have been checked also as if laterally braced. The former was not done because it would have unduly increased the length of the example without illustrating any new principles. The additional check is not necessary in this case because it can be seen by inspection that the given column is safe as a braced member provided the other columns in the story are of approximately the same size, as they would be in a normal, simple three-story building.

REFERENCES

5.1. A. H. Mattock, L. B. Kriz, and E. Hognestad: Rectangular Concrete Stress Distribution in Ultimate Strength Design, *J. ACI*, vol. 57, p. 875, 1961.

5.2. A. L. Parme, J. M. Nieves, and A. Gouwens: Capacity of Rectangular Columns Subject to Biaxial Bending, *J. ACI*, vol. 63, p. 911, 1966.

5.3. B. Bresler: Design Criteria for Reinforced Concrete Columns under Axial Load and Biaxial Bending, *J. ACI*, vol. 57, p. 481, 1960; discussion p. 1612.

5.4. L. N. Ramamurthy: Investigation of Ultimate Strength of Square and Rectangular Columns under Biaxial Eccentric Loads, Symposium on Reinforced Concrete Columns, *ACI. Publ.* SP-13, 1966.

5.5. S. P. Timoshenko and J. M. Gere: "Theory of Elastic Stability," 3d ed., McGraw-Hill Book Company, New York, 1969.

5.6. B. G. Johnston (ed.): "Guide to Design Criteria for Metal Compression Members," 2d ed., John Wiley & Sons, Inc., New York, 1966.

5.7. J. G. MacGregor, J. E. Breen, and E. O. Pfrang: Design of Slender Concrete Columns, *J. ACI,* vol. 67, p. 6, 1970.

5.8. Commentary to Building Code Requirements for Reinforced Concrete, Report of ACI Committee 318, 1971.

6
Footings

6.1 TYPES AND FUNCTION OF SUBSTRUCTURES

The substructure, or foundation, is that part of a structure which is usually placed below the surface of the ground and which transmits the load to the underlying soil or rock. All soils compress noticeably when loaded and cause the supported structure to settle. The two essential requirements in the design of foundations are that the total settlement of the structure shall be limited to a tolerably small amount and that differential settlement of the various parts of the structure shall be eliminated as nearly as possible. With respect to possible structural damage, the elimination of differential settlement, i.e., different amounts of settlement within the same structure, is even more important than limitations on uniform overall settlement.

To limit settlements as indicated, it is necessary (1) to transmit the load of the structure to a soil stratum of sufficient strength and (2) to spread the load over a sufficiently large area of that stratum to minimize bearing pressure. If adequate soil is not found immediately below the structure, it becomes necessary to use deep foundations such as piles or caissons to transmit the load to deeper, firmer layers. If satisfactory soil directly underlies the structure, it is

303

merely necessary to spread the load, by footings or other means. Such substructures are known as *spread* foundations, and it is mainly this type which will be discussed. Information on the more special types of deep foundations can be found in texts on foundation engineering, e.g., Ref. 6.1.

6.2 TYPES OF SPREAD FOUNDATIONS

Footings generally can be classified as wall and column footings. The horizontal outlines of the most common types are given in Fig. 6.1. A wall footing is simply a strip of reinforced concrete, wider than the wall, which distributes its pressure. Single-column footings are usually square, sometimes rectangular, and represent the simplest and most economical type. Their use under exterior columns meets with difficulties if property rights prevent the use of footings projecting beyond the exterior walls. In this case combined footings or strap footings are used which enable one to design a footing which will not project beyond the wall column. Combined footings under two or more columns are also used under closely spaced, heavily loaded interior columns where single footings, if they were provided, would completely or nearly merge.

Such individual or combined column footings are the most frequently used types of spread foundations on soils of reasonable bearing capacity. If the soil is weak and/or column loads are great, the required footing areas become so large as to be uneconomical. In this case, unless a deep foundation is called for by soil conditions, a mat or raft foundation is resorted to. This consists of a solid reinforced-concrete slab which extends under the entire building and which, consequently, distributes the load of the structure over the maximum available area. Such a foundation, in view of its own rigidity, also minimizes differential settlement. It consists, in its simplest form, of a concrete slab reinforced in both directions. A form which provides more rigidity and, at the same time, is often more economical consists of an inverted beam-and-girder floor. Girders are located in the column lines in one direction, with beams in the other, mostly at closer intervals. If the columns are arranged in a square pattern, girders are equally spaced in both directions, and the slab is

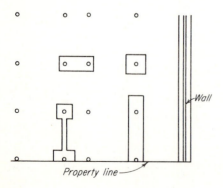
Property line

Fig. 6.1 Types of footings.

provided with two-way reinforcement. Inverted flat slabs, with capitals at the bottoms of the columns, are also used for mat foundations.

6.3 FACTORS AFFECTING THE DESIGN OF CONCRETE FOOTINGS

In ordinary constructions the load on a wall or column is transmitted vertically to the footing, which in turn is supported by the upward pressure of the soil on which it rests. If the load is symmetrical with respect to the bearing area, the bearing pressure is assumed to be uniformly distributed (Fig. 6.2a). It is known that this is only approximately true. Under footings resting on coarse-grained soils the pressure is larger at the center of the footing and decreases toward the perimeter (Fig. 6.2b). This is so because the individual grains in such soils are somewhat mobile, so that the soil located close to the perimeter can shift very slightly outward in the direction of lower soil stresses. In contrast, in clay soils pressures are higher near the edge than at the center of the footing, since in such soils the load produces a shear resistance around the perimeter which adds to the upward pressure (Fig. 6.2c). It is customary to disregard these nonuniformities (1) because their numerical amount is uncertain and highly variable, depending on type of soil, and (2) because the influence of these irregularities on the magnitudes of bending moments and shearing forces in the footing is relatively small.

On compressible soils footings should be loaded concentrically to avoid tilting which would result if bearing pressures are significantly larger under one side of the footing than under the opposite side. This means that single footings should be placed concentrically under the columns, and wall footings concentrically under the walls, and that for combined footings the centroid of the footings area should coincide with the resultant of the column loads. Eccentrically loaded footings can be used on highly compacted soils and on rock. It follows that one should count on rotational restraint of the column by a single footing only when such favorable soil conditions are present and when the footing is designed both for the column load and the restraining moment. Even then, less than full fixity should be assumed, except for footings on rock.

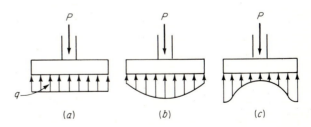

Fig. 6.2   Bearing pressure distribution (a) as assumed; (b) actual, for granular soils; (c) actual, for cohesive soils.

The accurate determination of stresses, particularly in single-column footings, is not practical, since they represent relatively massive blocks which cantilever from the column in all four directions. Under uniform upward pressure they deform in a bowl shape, a fact which would greatly complicate an accurate stress analysis. For this reason present procedures for the design of such footings are based almost entirely on the results of two extensive experimental investigations, both carried out at the University of Illinois (Refs. 6.2 and 6.3). These tests have later been reevaluated, particularly in the light of newer concepts of strength in shear and diagonal tension (Refs. 6.4 and 6.5).

6.4 LOADS, BEARING PRESSURE, FOOTING SIZE

Allowable bearing pressures are established from principles of soil mechanics, on the basis of load tests and other experimental determinations (see, for example, Refs. 6.1 and 6.6). Allowable bearing pressures q_a under service loads usually are based on a safety factor of 2.5 to 3.0 against exceeding the ultimate bearing capacity of the particular soil and to keep settlements within tolerable limits. Many local building codes contain allowable bearing pressures for the types of soils and soil conditions found in the particular locality.

For concentrically loaded footings the required area is determined from

$$A_{\text{req}} = \frac{D + L}{q_a} \tag{6.1}$$

In addition, most codes permit a 33 per cent increase in allowable pressure when effects of wind, W, or earthquake, E, are included, in which case

$$A_{\text{req}} = \frac{D + L + W}{1.33q_a} \text{ or } \frac{D + L + E}{1.33q_a} \tag{6.2}$$

A_{req}, the required footing area, is the larger of those determined by Eq. (6.1) or (6.2). The loads in the numerators of Eqs. (6.1) and (6.2) must be calculated at the level of the base of the footing, i.e., at the contact plane between soil and footing. This means that the weight of the footing and surcharge (i.e., fill and possible liquid pressure on top of the footing) must be included. Wind loads and other lateral loads cause a tendency to overturning. In checking for overturning of a foundation, only those live loads which contribute to overturning should be included, and dead loads which stabilize against overturning should be multiplied by 0.9. A safety factor of at least 1.5 should be maintained against overturning, unless otherwise specified by the local building code (Ref. 6.7).

A footing is eccentrically loaded if the supported column is not concentric with the footing area or if the column transmits at its juncture with the footing not only a vertical load but also a bending moment. In either case the load effects at the footing base can be represented by the vertical load P and a

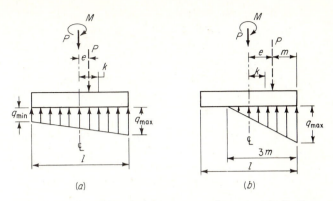

Fig. 6.3 Assumed bearing pressures under eccentric footings.

bending moment M. The resulting bearing pressures are again assumed to be linearly distributed. As long as the resulting eccentricity $e = M/P$ does not exceed the kern distance k of the footing area, the usual flexure formula

$$q_{\substack{max \\ min}} = \frac{P}{A} \pm \frac{Mc}{I} \qquad (6.3)$$

permits the determination of the bearing pressures at the two extreme edges, as shown in Fig. 6.3a. The footing area is found by trial and error from the condition $q_{max} \leq q_a$. If the eccentricity falls outside the kern, Eq. (6.3) gives a negative value (tension) for q along one edge of the footing. Because no tension can be transmitted at the contact area between soil and footing, Eq. (6.3) is no longer valid and bearing pressures are distributed as in Fig. 6.3b. For rectangular footings of size $l \times b$ the maximum pressure can be found from

$$q_{max} = \frac{2P}{3bm} \qquad (6.4)$$

which, again, must be no larger than the allowable pressure q_a. For nonrectangular footing areas of various configurations, kern distances and other aids for calculating bearing pressures can be found in Refs. 6.1 and 6.7 and elsewhere.

Once the required footing area has been determined, the footing must then be designed to develop the necessary strength to resist all moments, shears, and other internal actions caused by the applied loads. For this purpose, the load factors of the Code apply to footings as to all other structural components. Correspondingly, for strength design the footing is dimensioned for the effects of the following external loads (see Art. 3.3):

$$U = 1.4D + 1.7L \qquad (3.1a)$$

or if wind effects are to be included,

$$U = 0.75(1.4D + 1.7L + 1.7W) \tag{3.1b}$$

In seismic zones earthquake forces E must be considered, in which case $1.1E$ shall be substituted for W in Eq. (3.1b). Equation (3.1c), that is,

$$U = 0.9D + 1.3W \tag{3.1c}$$

will hardly ever govern the strength design of a footing. However, lateral earth pressure Q may on occasion affect footing design, in which case

$$U = 1.4D + 1.7L + 1.7Q \tag{3.1d}$$

For horizontal pressures F of liquids, such as groundwater, $1.4F$ shall be substituted for $1.7Q$ in Eq. (3.1d). Vertical liquid pressures are to be added to dead loads; i.e., the load factor 1.4 applies to them.

These factored loads must be counteracted and equilibrated by corresponding bearing pressures of the soil. Consequently, once the footing area is determined, the bearing pressures are recalculated for the factored loads for purposes of strength computations. These are fictitious pressures which are needed only to produce the required ultimate strength of the footing. To distinguish them from the actual pressures q under service loads, the design pressures which equilibrate the factored loads U will be designated q_u.

6.5 REINFORCED-CONCRETE WALL FOOTINGS

The simple principles of beam action apply to wall footings with only minor modifications. Figure 6.4 shows a wall footing with the forces acting on it. If bending moments were computed from these forces, the maximum moment would be found to occur at the middle of the width. Actually, the very large rigidity of the wall modifies this situation, and the tests cited in Art. 6.3 show that for footings under concrete walls it is satisfactory to compute the moment at the face of the wall (section 1-1). Tension cracks in these tests formed at the locations shown in Fig. 6.4, i.e., under the face of the wall rather than in the middle. For footings supporting masonry walls, the maximum moment is

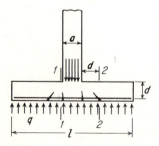

Fig. 6.4 Wall footing.

computed midway between the middle and the face of the wall, because masonry is generally less rigid than concrete. The maximum bending moment in footings under concrete walls is therefore given by

$$M_u = \tfrac{1}{8}q_u(b - a)^2 \tag{6.5}$$

For determining shear stresses, the vertical shear force is computed on section 2-2, located, as in beams, at a distance d from the face of the wall. Thus,

$$V_u = q_u \left(\frac{b - a}{2} - d \right) \tag{6.6}$$

The calculation of development length is based on the section of maximum moment, i.e., section 1-1.

Example: design of wall footing A 16-in. concrete wall supports a dead load $D = 14$ kips per ft and a live load $L = 10$ kips per ft. The allowable bearing pressure is $q_a = 4.5$ kips per ft^2 at the level of the bottom of the footing, which is 4 ft below grade. Design a footing for this wall using 3000-psi concrete and Grade 40 steel.

Assuming a 12-in.-thick footing, the footing weight per square foot is 150 lb per ft^2, and the weight of the 3-ft fill on top of the footing is $3 \times 100 = 300$ lb per ft^2. Consequently, the portion of the allowable bearing pressure which is available or effective for carrying the wall load is

$q_e = 4500 - (150 + 300) = 4{,}050$ lb/ft^2

The required width of the footing is therefore $b = 24{,}000/4{,}050 = 5.93$ ft. A 6-ft-wide footing will be assumed.

The bearing pressure for strength design of the footing, caused by the factored loads, is

$$q_u = \frac{1.4 \times 14 + 1.7 \times 10}{6} \times 10^3 = 6100 \text{ lb per ft}^2$$

From this, moment and shear for strength design are

$M_u = \tfrac{1}{8} \times 6100(6 - 1.33)^2 \times 12 = 199{,}500$ in.-lb per ft

and assuming $d = 9$ in., the shear at section 2-2 is

$V_u = 6100[\tfrac{1}{2}(6 - 1.33) - \tfrac{9}{12}] = 9700$ lb per ft

Shear usually governs the depth of footings, particularly since the use of shear reinforcements in footings is generally avoided as uneconomical. With $v_u = 2\sqrt{3000} = 110$ psi, the required depth becomes

$d = 9700/(0.85 \times 110 \times 12) = 8.7$ in.

Since the ACI Code calls for a 3-in. clear cover of bars, a 12-in.-thick footing will be selected, giving $d = 8.5$ in. This is sufficiently close to the assumed values that the calculations need not be revised.

To determine the required steel area, $M_u/(\phi bd^2) = 199{,}500/(0.9 \times 12 \times 8.5^2) = 256$ is used to enter Graph 1b of the Appendix. For this value, the curve 40/3 gives the steel ratio $\rho = 0.0067$. The required steel area is then $A_s = 0.0067 \times 8.5 \times 12 = 0.68$ in.^{2} per ft. Number 7 bars, 10 in. on centers, furnish $A_s = 0.72$ in.^{2} per ft. With the area of a single bar $A_b = 0.60$ in.^{2}, the required development length according to Table 3.2 is $l_d = 0.04 \times 0.60 \times 40{,}000/\sqrt{3000} = 17.5$ in. This length is to be furnished from section 1-1 outward. The length of each bar, if end cover is 3 in., is $72 - 6 = 66$ in., and the actual development length from section 1-1 to the nearby end is $\frac{1}{2}(66 - 16) = 25$ in., which is more than the required development length.

Longitudinal shrinkage and temperature reinforcement, according to the Code, must be at least $0.002 \times 12 \times 12 = 0.29$ in.^{2} per ft. Number 4 bars on 10-in. centers will furnish 0.29 in.^{2} per ft.

6.6 SINGLE–COLUMN FOOTINGS: GENERAL INFORMATION

In plan, single-column footings are usually square. Rectangular footings are used if space restrictions dictate this choice or if the supported columns are of strongly elongated rectangular cross section. In the simplest form, they consist of a single slab (Fig. 6.5a). Another type is that of Fig. 6.5b, where a pedestal or cap is interposed between the column and the footing slab; the pedestal provides for a more favorable transfer of load and in many cases is required in order to provide the necessary development length for dowels. This form is also known as a *stepped* footing. All parts of a stepped footing must be poured in a single pour, in order to provide monolithic action. Sometimes sloped footings such as those in Fig. 6.5c are used. They require less concrete than stepped footings, but the additional labor necessary to produce the sloping surfaces (formwork, etc.) usually makes stepped footings more economical. In general, single-slab footings (Fig. 6.5a) are most economical for thicknesses up to 3 ft.

Single-column footings represent, as it were, cantilevers projecting out from the column in both directions and loaded upward by the soil pressure. Corresponding tension stresses are caused in both these directions at the bottom surface. Such footings are therefore reinforced by two layers of steel, perpendicular to each other and parallel to the edges.

The required bearing area is obtained by dividing the total load, including the weight of the footing, by the selected bearing pressure. Weights of footings, at this stage, must be estimated and amount usually to 4 to 8 per cent of the column load, the former value applying to the stronger types of soils.

In computing bending moments and shears, only the upward pressure q_u which is caused by the factored column loads is considered. The weight of the

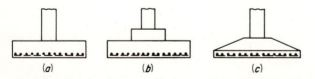

| (a) | (b) | (c) |

Fig. 6.5 Types of single-column footings.

footing proper does not cause moments or shears, just as, obviously, no moments or shears are present in a book lying flat on a table.

6.7 DESIGN OF SINGLE FOOTINGS

a. Shear   Once the required footing area A_{req} has been established from the allowable bearing pressure q_a and the most unfavorable combination of service loads, including weight of footing and overlying fill (and such surcharge as may be present), the thickness h of the footing must be determined. In single footings the effective depth d is mostly governed by shear. Since such footings are subject to two-way action, i.e., they bend in both major directions, their performance in shear is much like that of flat slabs in the vicinity of columns (see Art. 4.8). However, in contrast to two-way floor and roof slabs, it is generally not economical in footings to use shear reinforcement. For this reason only the design of footings in which all shear is carried by the concrete will be discussed here. For the rare cases where the thickness is restricted so that shear reinforcement must be used, the information in Art. 4.8 about slabs applies also to footings.

Two different types of shear strength are distinguished in footings: two-way or punching shear, and one-way or beam shear.

A column supported by the slab of Fig. 6.6 tends to punch through that slab because of the shear stresses which act in the footing around the perimeter of the column. At the same time the concentrated compression stresses from the column spread out into the footing so that the concrete adjacent to the column is in vertical or slightly inclined compression, in addition to shear. In consequence, if failure occurs, the fracture takes the form of the truncated pyramid shown in Fig. 6.6 (or of a truncated cone for a round column), with sides sloping outward at an angle approaching 45°. The average shear stress in the concrete which fails in this manner can be taken as that acting on vertical planes laid through the footing around the column on a perimeter a distance $d/2$ from the faces of the column (vertical section through $abcd$ on Fig. 6.7).

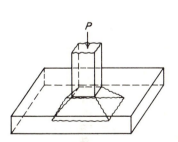

Fig. 6.6   Punching-shear failure in single footing.

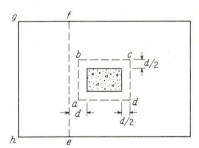

Fig. 6.7   Critical sections for shear.

Thus, the nominal shear stress in the concrete in this perimeter section is

$$v_{u1} = \frac{V_{u1}}{b_0 d} \qquad (6.7a)$$

where b_0 = length of perimeter of critical section (abcd in Fig. 6.7)

V_{u1} = total shear force on critical perimeter section = q_u times total footing area minus area within perimeter abcd

For design purposes, according to the ACI Code, the nominal shear stress to be used is

$$v_{u1} = \frac{V_{u1}}{\phi b_0 d} \qquad (6.7b)$$

The concrete subject to this shear stresss v_{u1} is also in vertical compression from the stresses spreading out from the column, and in horizontal compression in both major directions because of the biaxial bending moments in the footing. This triaxiality of stress increases the shear strength of the concrete (see Art. 1.9). Tests of footings and of flat slabs have shown, correspondingly, that for punching-type failures the shear stress computed on the critical perimeter area is larger than in one-way action (e.g., beams) and is approximately equal to $4\sqrt{f_c'}$. Accordingly, the Code specifies that v_{u1} given by Eq. (6.7b) shall not exceed this value, or

$$v_{u1} = \frac{V_{u1}}{\phi b_0 d} \leq 4\sqrt{f_c'} \qquad (6.8)$$

Shear failure can also occur as in beams or one-way slabs at a section a distance d from the face of the column (such as section ef in Fig. 6.6). The shear stress on that section is

$$v_{u2} = \frac{V_{u2}}{bd} \qquad (6.9a)$$

where b = width of footing at distance d from face of column (ef in Fig. 6.7)

V_{u2} = total shear force on that section = q_u times footing area outside that section (area efgh in Fig. 6.7)

and, for design purposes, according to the Code,

$$v_{u2} = \frac{V_{u2}}{\phi bd} \qquad (6.9b)$$

Since this type of one-way-shear action is the same as in beams or one-way slabs, the same shear stress v_c as given by Eqs. (3.24a) and (3.24b) will cause cracking in footings as in beams. (In footing design, the simpler and more conservative formula is generally used, that is, $v_c = 2\sqrt{f_c'}$.) Accordingly, the

Code also specifies

$$v_{u2} = \frac{V_{u2}}{\phi bd} \le v_c \tag{6.10}$$

where $\phi = 0.85$ in both Eqs. (6.8) and (6.10).

The required depth d is the larger of the values calculated from Eqs. (6.8) and (6.10). In square footings Eq. (6.8) usually governs; in strongly elongated footings, Eq. (6.10).

b. Bearing—transfer of stress at base of column When a column rests on a footing or pedestal, it transfers its load to only a part of the total area of the supporting member. The adjacent footing concrete provides lateral support to the directly loaded part of the concrete. This causes triaxial compression stresses that increase the strength of the concrete which is loaded directly under the column. Based on tests, the Code provides that when the supporting area is wider than the loaded area on all sides,

$$\text{Maximum bearing stress} = 0.85\phi f_c' \sqrt{\frac{A_2}{A_1}} \le 2 \times 0.85\phi f_c' \tag{6.11}$$

For bearing on concrete, $\phi = 0.70$, f_c' is the cylinder strength of the footing concrete which occasionally may be less than that of the column, A_1 is the loaded area, and A_2 is the maximum area of the portion of the supporting surface which is geometrically similar to and concentric with the loaded area. The meaning of A_2 for a somewhat unusual case is illustrated by Fig. 6.8. (For the rare cases of bearing pressure on sloped or stepped footings, see Sec. 10.14.4 of the Code and Ref. 6.8.)

All axial forces and bending moments which act at the bottom section of a column must be transferred to the footing at the bearing surface by compression in the concrete and by reinforcement. If the maximum bearing stress as given by Eq. (6.11) is exceeded, the excess must be provided for by reinforcement. This may be done either by extending the column bars into the footing or by providing dowels which are embedded in the footing and project above it. In the latter case the column bars merely rest on the footing and, in most cases, are tied to dowels. This results in a simpler construction procedure

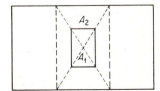

Fig. 6.8

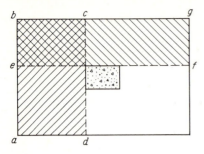

Fig. 6.9 Critical sections for bending and bond.

than extending the column bars into the footing.[1] To assure a minimum of tying between column and footing, the Code requires that the minimum area of reinforcement which crosses the bearing surface (dowels or column bars) be 0.5 per cent of the column area and that no less than four such bars be provided. The length of the dowels or bars of diameter d_b must be sufficient on both sides of the bearing surface to provide the required development length for compression bars, that is, $l_d = 0.02 f_y d_b / \sqrt{f_c'} \le 0.003 f_y d_b$ (see Art. 3.6).

The two largest bar sizes, Nos. 14 and 18, are frequently used in columns with large axial forces. Under normal circumstances the Code specifically prohibits the lap splicing of these bars because tests have shown that welded splices or other positive connections are necessary to fully develop these heavy bars. However, a specific exception is made for dowels for Nos. 14 and 18 column bars. Based on long-standing successful use, the Code now permits these heavy bars to be spliced to dowels of lesser diameter (i.e., No. 11 or smaller) provided the dowels have a development length into the column corresponding to that of the column bar (i.e., No. 14 or 18, as the case may be) and into the footing as prescribed for the particular dowel size (i.e., No. 11 or smaller, as the case may be).

c. Bending moments, reinforcement, and bond If a vertical section is passed through a footing, the bending moment which is caused in the section by the net upward soil pressure (i.e., column load divided by bearing area) is obtained from simple statics. Figure 6.9 shows such a section cd located along the face of the column. The bending moment about cd is that caused by the upward pressure q_u on the area to one side of the section, i.e., the area $abcd$. The reinforcement perpendicular to that section, i.e., the bars running in the long direction, are calculated from this bending moment. Likewise, the moment about section ef is caused by the pressure q_u on area $befg$, and the reinforcement in the short direction, i.e., perpendicular to ef, is calculated for this bending

---

[1] In past Code editions it was required that at least one dowel be provided for each column bar and that the total area of dowels be not less than that of the column bars. The 1971 edition of the Code is more liberal in requiring that dowels merely provide for the excess of the load which cannot be resisted by concrete in direct bearing.

moment. In footings which support reinforced-concrete columns, these critical sections for bending are located at the faces of the column, as shown.

In footings supporting steel columns, the sections ab and ef are located, not at the edge of the steel base plate, but halfway between the edge of the column and that of the steel base plate.

In footings with *pedestals*, the width resisting compression in sections cd and ef is that of the pedestal; the corresponding depth is the sum of the thicknesses of pedestal and footing. Further sections parallel to cd and ef are passed at the edge of the pedestal, and the moments determined in the same manner, to check the strength at locations in which the depth is that of the footing only.

For footings with relatively small pedestals, the latter are often discounted in moment and shear computation, and bending is checking at the face of the column, with width and depth equal to that of the footing proper.

In *square footings*, the reinforcement is uniformly distributed over the width of the footing in each of the two layers; i.e., the spacing of the bars is constant. The moments for which the two layers are designed are the same. However, the effective depth d for the upper layer is less by 1 diam than that of the lower layer. Consequently, the required A_s is larger for the upper layer. Instead of using different spacings or different bar diameters in each of the two layers, it is customary to determine A_s for the upper layer and to use the same arrangement of reinforcement for the lower layer.

In *rectangular footings*, the reinforcement in the long direction is again uniformly distributed over the pertinent (shorter) width. In locating the bars in the short direction, one has to consider that the support provided to the footing by the column is concentrated near the middle. Consequently, the curvature of the footing is sharpest, i.e., the moment per foot largest, immediately under the column, and it decreases in the long direction with increasing distance from the column. For this reason a larger steel area per longitudinal foot is needed in the central portion than near the far ends of the footing. The ACI Code provides, therefore, that:

> In the short direction that portion of the reinforcement determined by Eq. (6.12) shall be uniformly distributed across a bandwidth B centered with respect to the column and having a width equal to the length of the short side of the footing. The remainder of the reinforcement shall be uniformly distributed in the outer portion of the footing.

$$\frac{\text{Reinforcement in bandwidth } (B)}{\text{Total reinforcement in short direct}} = \frac{2}{\beta + 1} \qquad (6.12)$$

In Eq. (6.12), β is the ratio of the long side to the short side of the footing.

The minimum reinforcement ratio of $200/f_y$ prescribed by the Code to prevent precipitous fracture (see footnote p. 127) applies in footings separately to each of the two directions.

The critical sections for development length of footing bars are the same as those for bending. Development length may also have to be checked at all vertical planes in which changes of section or of reinforcement occur, as at the edges of pedestals or where part of the reinforcement may be terminated.

Example: design of a square footing A column 18 in. square, with $f_c' = 4$ ksi, reinforced with eight No. 8 bars of $f_y = 50$ ksi, supports a dead load of 225 kips and a live load of 175 kips. The allowable soil pressure q_a is 5 ksf. Design a square footing with base 5 ft below grade, $f_c' = 4$ ksi, $f_y = 50$ ksi.

Since the space between the bottom of the footing and the surface will be occupied partly by concrete and partly by soil (fill), an average unit weight of 125 lb per ft³ will be assumed. The pressure of this material at the 5-ft depth is $5 \times 125 = 625$ psf, leaving a bearing pressure of $q_e = 5000 - 625 = 4375$ psf available to carry the column service load. Hence, the required footing area $A_{req} = (225 + 175)/4.375 = 91.5$ ft². A base 9 ft 6 in. square is selected, furnishing a footing area of 90.3 ft², which differs from the required area by about 1 per cent.

For strength design, the upward pressure caused by the factored column loads is $q_u = (1.4 \times 225 + 1.7 \times 175)/9.5^2 = 6.80$ ksf.

The footing depth in square footings is usually determined from the two-way or punching shear on the critical perimeter $abcd$ of Fig. 6.10. Trial calculations suggest $d = 19$ in. Hence, the length of the critical perimeter is

$$b_0 = 4(18 + d) = 148 \text{ in.}$$

The shear force acting on this perimeter, being equal to the total upward pressure minus that acting within the perimeter $abcd$, is

$$V_{u1} = 6.80[9.5^2 - (37/12)^2] = 550 \text{ kips}$$

The corresponding design shear stress [see Eq. (6.8)]

$$v_{u1} = 550,000/(0.85 \times 148 \times 19) = 231 \text{ psi}$$

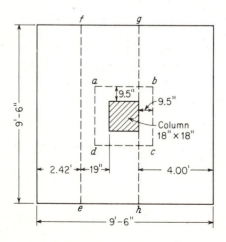

Fig. 6.10

is smaller than $4 \sqrt{f_c'} = 4 \sqrt{4000} = 253$ psi by about 10 per cent. Hence, the selected value of $d = 19$ in. will now be checked for one-way or beam shear on section ef of Fig. 6.10. The total shear force acting on that section is

$$V_{u2} = 6.80 \times 2.42 \times 9.50 = 156 \text{ kips}$$

and the nominal shear stress

$$v_{u2} = 156{,}000/(0.85 \times 9.5 \times 12 \times 19) = 85 \text{ psi}$$

This is less than $v_c = 2 \sqrt{4000} = 126$ psi, so that the selected 19-in. depth satisfies both shear requirements.

The bending moment on section gh of Fig. 6.10 is

$$M_u = 6.80 \times 9.5(4.0^2/2)12 = 6200 \text{ in.-kips}$$

The depth required for shear being greatly in excess of that required for bending, the steel ratio will be low and the corresponding depth of the rectangular stress block small. Assuming it to be $a = 2$ in., the required steel area is

$$A_s = \frac{6200}{0.9 \times 50(19 - 1)} = 7.65 \text{ in.}^2$$

Checking the minimum steel ratio by Code, $\rho_{min} = 200/50{,}000 = 0.004$ gives the required minimum reinforcement $A_s = \rho_{min}bd = 0.004(12 \times 9.5)19 = 8.66$ in.^{2} This governs because it is larger than the 7.65 in.^{2} calculated for bending. Eleven No. 8 bars furnishing 8.64 in.^{2} will be used in each direction. The required development length beyond section gh is

$$l_d = 0.04 \times 0.79 \times 50{,}000/\sqrt{4000} = 25.0 \text{ in.}$$

which is more than adequately met by the actual length of bars beyond section gh, namely, $48 - 3 = 45$ in.

Checking for bearing of the column on the footing, the bearing stress produced by the factored loads is $613{,}000/18^2 = 1890$ psi. It is easily verified that of the two requirements of Eq. (6.11), the second governs; i.e., the maximum bearing stress to be carried by concrete alone is $2 \times 0.85 \times 0.7 \times 4000 = 4760$ psi. This being greatly in excess of the actual bearing stress, 1890 psi, the entire column load can be transferred by concrete alone. To provide proper connection between column and footing, it is therefore only necessary to supply the minimum dowel area by Code, that is, 0.5 per cent of the gross column area. Hence, the minimum dowel area is $0.005 \times 18^2 = 1.62$ in.^{2} This can be supplied by four No. 6 bars. It is felt, however, that at least the four corner bars of the column should be spliced to dowels of the same diameter as those of the bars. Correspondingly, four No. 8 dowels will be furnished, one in each corner as shown in Fig. 6.11. The minimum development length of the dowels, above and below the top surface of the footing, is $l_d = 0.02 \times 0.79 \times 50{,}000/\sqrt{4000} = 15.8$ in., which is easily accommodated in a footing with $d = 19$ in.

For concrete in contact with ground, a minimum cover of 3 in. is required for corrosion protection. With $d = 19.0$ in. measured from the top of the footing to the center of the upper layer of bars, the total thickness of the footing which is required in order to provide 3 in. clear cover for the lower steel layer is

$$h = 19 + 1.5 \times 1.00 + 3 = 23.5 \text{ in.}$$

The footing, with 24 in. thickness, is shown in Fig. 6.11.

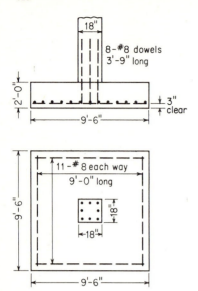

Fig. 6.11

6.8 COMBINED FOOTINGS

Spread footings which support more than one column or wall are known as *combined footings*. They can be divided into two categories: those which support two columns, and those which support more than two, and generally large numbers, of columns.

Examples of the first type, i.e., two-column footings, are shown in Fig. 6.1. In buildings where the allowable soil pressure is large enough so that single footings are adequate for most columns, two-column footings are seen to become necessary in two situations: (1) if columns are so close to the property line that single-column footings cannot be made without projecting beyond that line, and (2) if some adjacent columns are so close to each other that their footings would merge. Both situations are shown in Fig. 6.1.

When the bearing capacity of the subsoil is low so that large bearing areas become necessary, individual footings are replaced by *continuous strip footings* which support more than two columns, and usually all columns in a row. Mostly, such strips are arranged in both directions, in which case a *grid foundation* is obtained as shown in Fig. 6.12. Such a grid foundation can be made to develop a much larger bearing area much more economically than can be done by single footings, because the individual strips of the grid foundation represent continuous beams whose moments are much smaller than the cantilever moments in large single footings which project far out from the column in all four directions.

For still lower bearing capacities, the strips are made to merge, resulting in a mat foundation as shown in Fig. 6.13. That is, the foundation consists of a solid reinforced-concrete slab under the entire building. In structural

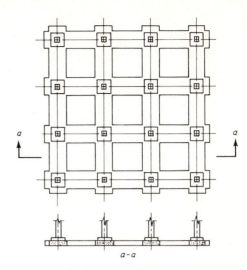

Fig. 6.12 Grid foundation. (*Adapted from Ref*. 6.10.)

action such a mat is very similar to a flat slab or a flat plate, upside down, i.e., loaded upward by the bearing pressure and downward by the concentrated column reactions. The mat foundation evidently develops the maximum available bearing area under the building. If the soil's capacity is so low that even this large bearing area is insufficient, some form of deep foundation, such as piles or caissons, must be used. These are discussed in texts on foundation design and fall outside the scope of the present volume.

Grid and mat foundations may be designed with the column pedestals as shown in Figs. 6.12 and 6.13, or without them, depending on whether or not they are necessary for shear strength and development length of dowels.

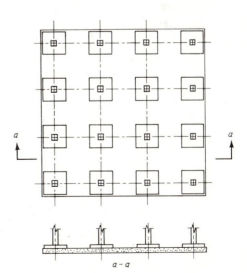

Fig. 6.13 Mat foundation. (*Adapted from Ref*. 6.10.)

   Apart from developing large bearing areas, another advantage of grid and mat foundations is that their continuity and rigidity help in reducing differential settlements of individual columns relative to each other, which may otherwise be caused by local variations in quality of subsoil, or other causes. For this purpose continuous spread foundations are frequently used in situations where the superstructure or the type of occupancy makes for unusual sensitivity to differential settlement.

6.9 TWO–COLUMN FOOTINGS

It is desirable to design combined footings so that the centroid of the footing area coincides with the resultant of the two column loads. This produces uniform bearing pressure over the entire area and forestalls a tendency for the footing to tilt. In plan, such footings are rectangular, trapezoidal, or T-shaped,

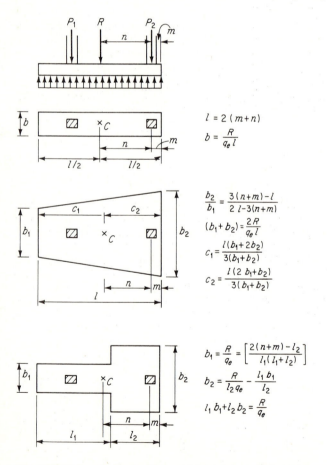

For the rectangular footing:

$$l = 2\,(m+n)$$

$$b = \frac{R}{q_e\,l}$$

For the trapezoidal footing:

$$\frac{b_2}{b_1} = \frac{3\,(n+m)-l}{2\,l-3\,(n+m)}$$

$$(b_1+b_2) = \frac{2R}{q_e\,l}$$

$$c_1 = \frac{l\,(b_1+2b_2)}{3\,(b_1+b_2)}$$

$$c_2 = \frac{l\,(2\,b_1+b_2)}{3\,(b_1+b_2)}$$

For the T-shaped footing:

$$b_1 = \frac{R}{q_e}\left[\frac{2\,(n+m)-l_2}{l_1\,(l_1+l_2)}\right]$$

$$b_2 = \frac{R}{l_2\,q_e} - \frac{l_1\,b_1}{l_2}$$

$$l_1\,b_1+l_2\,b_2 = \frac{R}{q_e}$$

Fig. 6.14 Two-column footings. (*Adapted from Ref. 6.7.*)

the details of the shape being arranged to produce coincidence of centroid and resultant. The simple relationships of Fig. 6.14 facilitate the determination of the shape of the bearing area (from Ref. 6.7). In general, the distances m and n are given, the former being the distance from the center of the exterior column to the property line, and the latter the distance from that column to the resultant of both column loads.

Another expedient which is used if a single footing cannot be centered under an exterior column is to place the exterior column footing eccentrically and to connect it with the nearest interior-column footing by a beam or strap. This strap, being counterweighted by the interior column load, resists the tilting tendency of the eccentric exterior footings and equalizes the pressure under it. Such foundations are known as *strap, cantilever, or connected footings*.

The two examples which follow demonstrate some of the peculiarities of the design of two-column footings.

Example 1: Design of a combined footing supporting one exterior and one interior column An exterior 24- × 18-in. column with $D = 170$ kips, $L = 130$ kips and an interior 24- × 24-in. column with $D = 250$ kips, $L = 200$ kips are to be supported on a combined rectangular footing whose outer end cannot protrude beyond the outer face of the exterior column (see Fig. 6.1). The distance center to center of columns is 18 ft 0 in., and the allowable bearing pressure of the soil is 6000 psf. The bottom of the footing is 6 ft below grade, and a surcharge of 100 psf is specified on the surface. Design the footing for $f'_c = 3000$ psi, $f_y = 60{,}000$ psi.

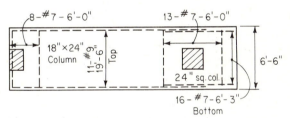

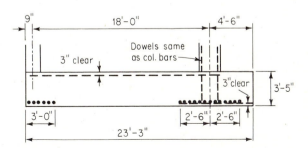

Fig. 6.15

The space between the bottom of footing and the surface will be occupied partly by concrete (footing, concrete floor) and partly by backfill. An average unit weight of 125 lb/ft³ can be assumed. Hence, the effective portion of the allowable bearing pressure which is available for carrying the column loads is $q_e = q_a -$ (weight of fill and concrete + surcharge) $= 6000 - (6 \times 125 + 100) = 5150$ psf. Then the required area $A_{req} =$ sum of column loads/$q_e = 750/5.15 = 145.5$ ft². The resultant of the column loads is located from the center of the exterior column a distance $450 \times 18/750 = 10.8$ ft. Hence, the length of the footing must be $2(10.8 + 0.75) = 23.1$ ft. A length of 23 ft 3 in. is selected. The required width is then $145.5/23.25 = 6.3$ ft. A width of 6 ft 6 in. is selected (see Fig. 6.15).

Longitudinally, the footing represents an upward-loaded beam spanning between columns and cantilevering beyond the interior column. Since this beam is considerably wider than the columns, the column loads are distributed crosswise by transverse beams, one under each column. In the present relatively narrow and long footing it will be found that the required minimum depth for the transverse beams is smaller than is required for the footing in the longitudinal direction. These "beams," therefore, are not really distinct members, but merely represent transverse strips in the main body of the footing, so reinforced that they are capable of resisting the transverse bending moments and the corresponding shears. It then becomes necessary to decide how large the effective width of this transverse beam can be assumed to be. Obviously, the strip directly under the column does not deflect independently and is strengthened by the adjacent parts of the footing. The effective width of the transverse beams is therefore evidently larger than that of the column. In the absence of definite rules for this case, or of research results on which to base such rules, the authors recommend conservatively that the load be assumed to spread outward from the column into the footing at a slope of 2 vertical to 1 horizontal. This means that the effective width of the transverse beam is assumed to be equal to the width of the column plus $d/2$ on either side of the column, d being the effective depth of the footing.

Strength design in longitudinal direction. The net upward pressure caused by the factored column loads is

$$q_u = \frac{1.4(170 + 250) + 1.7(130 + 200)}{23.25 \times 6.5} = 7.60 \text{ kips/ft}^2$$

Then the net upward pressure per linear foot in the longitudinal direction is $7.60 \times 6.5 = 49.4$ kips/ft. The maximum negative moment between the columns occurs at the section of zero shear. Let x be the distance from the outer edge of the exterior column to this section. Then (see Fig. 6.16)

$$V = 49,400x - 459,000 = 0$$

results in $x = 9.30$ ft. The moment at this section is

$$M = [49,400 \times 9.30^2/2 - 459,000(9.30 - 0.75)]12 = -21,400,000 \text{ in.-lb}$$

The moment at the right edge of the interior column is

$$M = 49,400(3.5^2/2)12 = 3,630,000 \text{ in.-lb}$$

and the details of the moment diagram are as shown in Fig. 6.16. Try $d = 37.5$ in.

From the shear diagram of Fig. 6.16 it is seen that the critical section for flexural shear is at a distance d to the left of the left face of the interior column. At that point the shear force is

$$V = 418,000 - (37.5/12)49,400 = 264,000 \text{ lb}$$

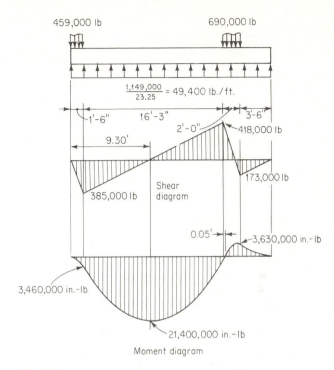

Fig. 6.16

and the nominal shear stress is

$$v_u = 264{,}000/(0.85 \times 78 \times 37.5) = 106 \text{ psi}$$

The allowable shear stress is $v_c = 2\sqrt{3000} = 110$ psi, indicating that $d = 37.5$ in. is adequate.

Additionally, as in single footings, punching shear should be checked on a perimeter section a distance $d/2$ around the column, for which the allowable shear stress $v_c = 4\sqrt{3000} = 220$ psi. Of the two columns, the exterior one with a three-sided perimeter a distance $d/2$ from the column is more critical in regard to this "punching shear." The perimeter is

$$b_0 = 2\left(1.5 + \frac{37.5/12}{2}\right) + \left(2.0 + \frac{37.5}{12}\right) = 11.24 \text{ ft}$$

and the shear force, being the column load minus the soil pressure within the perimeter, is

$$V = 459{,}000 - 3.06 \times 5.12(7600) = 340{,}000 \text{ lb}$$

Consequently, the nominal shear stress on the perimeter section is

$$v_u = \frac{340{,}000}{0.85 \times 11.24 \times 12 \times 37.5} = 79.0 \text{ psi}$$

considerably less than the allowable value.

With $d = 37.5$ in., and with 3.5-in. insulation from the center of the bars to the top surface of the footing, the total thickness is 41 in.

To determine the required steel area, $M_u/(\phi bd^2) = 21{,}400{,}000/(0.9 \times 78 \times 37.5^2) = 217$ is used to enter Graph 1b of the Appendix. For this value, the curve 60/3 gives the steel ratio $\rho = 0.0037$. The required steel area is $A_s = 0.0037 \times 37.5 \times 78 = 10.8$ in.^{2} Eleven No. 9 bars furnish 11.00 in.^{2} The required development length is $l_d = 1.4 \times 0.04 \times 1.00 \times 60{,}000/\sqrt{3000} = 61.3$ in. $= 5.1$ ft. From Fig. 6.16, the distance from the point of maximum moment to the nearer left end of the bars is seen to be $9.30 - \frac{3}{12} = 9.05$ ft, much larger than the required minimum development length. The selected reinforcement is therefore adequate for both bending and bond.

For the portion of the longitudinal beam which cantilevers beyond the interior column, the minimum required steel area controls. $\rho_{min} = 200/60{,}000 = 0.0033$. $A_s = 0.0033 \times 78 \times 37.5 = 9.75$ in.^{2} Sixteen No. 7 bars, with $A_s = 9.62$ in.^{2}, are selected, and their development length computed as above, but for bottom bars is found adequate.

Design of transverse beam under interior column. The width of the transverse beam under the interior column can now be established as previously suggested and is $24 + 2(d/2) = 24 + 2 \times 18.75 = 61.5$ in. The net upward load per linear foot of the transverse beam is $690{,}000/6.5 = 106{,}000$ lb per ft. The moment at the edge of the interior column is

$$M = 106{,}000(2.25^2/2)12 = 3{,}220{,}000 \text{ in.-lb}$$

Since the transverse bars are placed on top of the longitudinal bars (see Fig. 6.15), the actual value of d furnished is $37.5 - 1.0 = 36.5$ in. The minimum required steel area controls; i.e.,

$$A_s = (200/60{,}000) \times 61.5 \times 36.5 = 7.48 \text{ in.}^2$$

Thirteen No. 7 bars are selected and placed within the 61.5 in. effective width of the transverse beam.

Punching shear at the perimeter a distance $d/2$ from the column has been checked before. The critical section for regular flexural shear, at a distance d from the face of the column, lies beyond the edge of the footing, and therefore no further check on shear is needed.

The design of the transverse beam under the exterior column is the same as the design of that under the interior column, except that the effective width is 36.75 in. The details of the calculations are not shown. It will be easily checked that eight No. 7 bars, placed within the 36.75-in. effective width, satisfy all requirements. Design details are shown in Fig. 6.15.

Example 2: Design of a strap footing In a strap or connected footing, the exterior footing is placed eccentrically under its column so that it does not project beyond the property line. Such an eccentric position would result in a strongly uneven distribution of bearing pressure which could lead to tilting or even tipping of the footing. To counteract this eccentricity, the footing is connected by a beam or strap to the nearest interior footing.

Both footings are so proportioned that under service load the pressure under each of them is uniform and the same under both footings. To achieve this, it is necessary, as in other combined footings, that the centroid of the combined area for the two footings coincide with the resultant of the column loads. The resulting forces are shown schematically in Fig. 6.17. They consist of the loads P_e and P_i of the exterior and interior columns, respectively, and of the net upward pressure q, which is uniform and equal under both footings. The resultants R_e and R_i of these upward

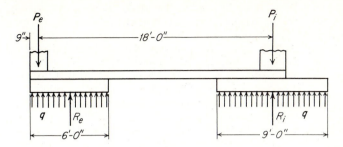

Fig. 6.17

pressures are also shown. Since the interior footing is concentric with the interio,
column, R_i and P_i are colinear. This is not the case for the exterior forces R_e and P_er
where the resulting couple just balances the effect of the eccentricity of the column
relative to the center of the footing. The strap proper is generally constructed so that
it will not bear on the soil. This can be achieved by providing formwork not only for
the sides, but also for the bottom face, and withdrawing it before backfilling.

To illustrate this design, the columns of Example 1 will now be supported on a
strap footing. Its general shape, plus dimensions as determined only subsequently by
calculations, is seen in Fig. 6.18. With an allowable bearing pressure of $q_a = 6.0$ kips
per ft^2 and a depth of 6 ft to the bottom of the footing as before, the bearing pressure

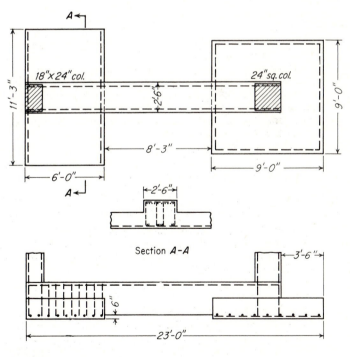

Fig. 6.18

available for carrying the external loads applied to the footings is $q_e = 5.15$ kips per ft², as in Example 1. These external loads, for the strap footing, consist of the column loads and of the weight plus fill and surcharge of that part of the strap which is located between the footings. (The portion of the strap located directly on top of the footing displaces a corresponding amount of fill and therefore is already accounted for in the determination of the available bearing pressure q.) Assuming the bottom of the strap to be 6 in. above the bottom of the footings to prevent bearing on soil, the total depth to grade is 5.5 ft. Estimating the strap width to be 2.5 ft, its estimated weight plus fill and surcharge is $2.5 \times 5.5 \times 0.125 + 0.100 \times 2.5 = 2$ kips per ft. Estimating the gap between footings to be 8 ft, the total weight of the strap is 16 kips. Hence, for purposes of determining the required footing area, 8 kips will be added to the dead load of each column. The required total area of both footings is then $(750 + 16)/5.15 = 149$ ft². The distance of the resultant of the two column loads plus the strap load from the axis of the exterior column, sufficiently accurately, is $458 \times \frac{18}{766} = 10.75$ ft, or 11.50 ft from the outer edge, almost identical with that calculated for Example 1. Trial calculations show that a rectangular footing 6 ft 0 in. $\times$ 11 ft 3 in. under the exterior column and a square footing 9×9 ft under the interior column have a combined area of 149 ft² and a distance from the outer edge to the centroid of the combined areas of $(6 \times 11.25 \times 3 + 9 \times 9 \times 18.75)/149 = 11.55$ ft, which is almost exactly equal to the previously calculated distance to the resultant of the external forces.

For strength calculations, the bearing pressure caused by the factored external loads, including that of the strap with its fill and surcharge, is

$$q_u = [1.4(170 + 250 + 16) + 1.7(130 + 200)]/149 = 7.8 \text{ kips per ft}^2$$

Design of footings. The exterior footing performs exactly like a wall footing of 6-ft length. Even though the column is located at its edge, the balancing action of the strap was so arranged as to result in uniform bearing pressure, the downward load being transmitted to the footing uniformly by the strap. Hence, the design is carried out exactly as for a wall footing (see Art. 6.5).

The interior footing, even though it merges in part with the strap, can safely be designed as an independent, square single-column footing (see Art. 6.7). The main difference is that, because of the presence of the strap, punching shear cannot occur along the truncated pyramid surface of Fig. 6.6. For this reason, two-way or punching shear, according to Eq. (6.7), should be checked along a perimeter section located at a distance $d/2$ outward from the longitudinal edges of the strap and from the free face of the column, d being the effective depth of the footing. Flexural or one-way shear, as usual, is checked at a section a distance d from the face of the column.

Design of strap. Even though the strap is in fact monolithic with the interior footing, the effect on the strap of the soil pressure under this footing can safely be neglected, because the footing has been designed to withstand the entire upward pressure, as if the strap were absent. In contrast, the exterior footing having been designed as a wall footing which receives its load from the strap, the upward pressure from the wall footing becomes a load to be resisted by the strap. With this simplification of the actually somewhat more complex situation, the strap represents a single-span beam loaded upward by the bearing pressure under the exterior footing and supported by downward reactions at the center lines of the two columns (Fig. 6.19). A width of 30 in. is selected. For a column width of 24 in. this permits beam and column bars to be placed without interference where the two members meet and allows the column forms to be supported on the top surface of the strap. The maximum moment, as determined by equating the shear force to zero, occurs close to the inner edge of the exterior footing. Shear forces are large in the vicinity of the exterior column. The foot-

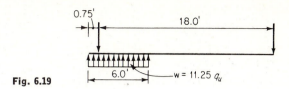

Fig. 6.19

ing is drawn approximately to scale in Fig. 6.18, which also shows the general arrangement of the reinforcement in footings and strap.

PROBLEMS

1. Complete the strength design of Example 2 above and determine all dimensions and reinforcement. Compare total volume of concrete of the strap footing of Example 2 with that of the rectangular combined footing of Example 1. It will be found that the strap footing is significantly more economical; this economy would increase with increasing distance between columns.

2. Assuming for the columns of Example 1 that because of extraneous obstructions the combined footing cannot project beyond the edge of the interior column, design a trapezoidal footing to accommodate this situation.

3. For the conditions of Prob. 2, design a T-shaped footing.

6.10 CONTINUOUS STRIP, GRID, AND MAT FOUNDATIONS

As mentioned in Art. 6.8, in the case of heavily loaded columns, particularly if they are to be supported on relatively weak or uneven soils, continuous foundations are resorted to. These may consist of a continuous strip footing supporting all columns in a given row, or more often of two sets of such strip footings intersecting at right angles so that they form one continuous *grid foundation* (Fig. 6.12). For even larger loads or weaker soils the strips are made to merge, resulting in a *mat foundation* (Fig. 6.13).

For the design of such continuous foundations it is essential that reasonably realistic assumptions be made regarding the distribution of bearing pressures which act as upward loads on the foundation. For compressible soils it can be assumed in first approximation that the deformation or settlement of the soil at a given location and the bearing pressure at that location are proportional to each other. Now, if columns are spaced at moderate distances and if the strip, grid, or mat foundation is very rigid, the settlements in all portions of the foundation will be substantially the same. This means that the bearing pressure, also known as *subgrade reaction*, will be the same provided that the centroid of the foundation coincides with the resultant of the loads. If they do not coincide, then for such rigid foundations the subgrade reaction can be assumed as linear and determined from statics, in the same manner as was discussed for single footings (see Fig. 6.3). In this case all loads, the downward column loads as well as the upward bearing pressures, are known. Hence, moments and shear forces in the foundation can be found by statics alone.

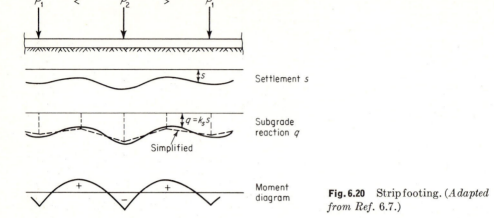

Settlement s

Subgrade
reaction q

Moment
diagram

Fig. 6.20 Strip footing. (*Adapted from Ref. 6.7.*)

Once these are determined, the design of strip and grid foundations is similar to that of inverted continuous beams and that of mat foundations to that of inverted flat slabs or plates.

On the other hand, if the foundation is relatively flexible and the column spacing large, settlements will no longer be uniform or linear. For one thing, the more heavily loaded columns will cause larger settlements, and thereby larger subgrade reactions, than the lighter ones. Also, since the continuous strip or slab midway between columns will deflect upward relative to the nearby columns, this means that the soil settlement, and thereby the subgrade reaction, will be smaller midway between columns than directly at the columns. This is shown schematically in Fig. 6.20. In this case the subgrade reaction can no longer be assumed as uniform. A reasonably accurate but fairly complex analysis can then be made using the theory of beams on elastic foundations (Ref. 6.9).

A simplified procedure has been developed which covers the most frequent situations of strip and grid foundations (Refs. 6.10 and 6.11). The method first defines the conditions under which a foundation can be regarded as rigid so that uniform or overall linear distribution of subgrade reactions can be assumed. This is the case when the average of two adjacent span lengths in a continuous strip does not exceed $1.75/\lambda$, provided also that the adjacent span and column loads do not differ by more than 20 per cent of the larger value. Here

$$\lambda = \sqrt[4]{\frac{k_s b}{4 E_c I}} \tag{6.13}$$

where $k_s = S k_s'$
$k_s' =$ coefficient of subgrade reaction as defined in soils mechanics, basically the force per unit area required to produce a unit settlement, in kips/ft³

b = width of footing, ft

E_c = modulus of elasticity of concrete, kips/ft^2

I = moment of inertia of footing, ft^4

S = shape factor, being $[(b + 1)/2b]^2$ for granular soils such as sands, and $(n + 0.5)/1.5n$ for cohesive soils such as clays, n being the ratio of longer to shorter side of the strip

If the average of two adjacent spans exceeds $1.75/\lambda$, the foundation is regarded as flexible. Provided that adjacent spans and column loads differ by no more than 20 per cent, the complex curvilinear distribution of subgrade reaction can be replaced by a set of equivalent trapezoidally distributed reactions, which are also shown on Fig. 6.20. References 6.10 and 6.11 contain fairly simple equations for determining the intensities p of the equivalent pressures under the columns and at the middle of the spans, and also give equations for the positive and negative moments caused by these equivalent subgrade reactions. With this information, the design of continuous strip and grid footings proceeds similarly to that of footings under two columns (Art. 6.10).

Mat foundations, likewise, require different approaches, depending on whether they can be classified as rigid or flexible. As in strip footings, if the column spacing is less than $1/\lambda$, the structure may be regarded as rigid, the soil pressure can be assumed as uniformly or linearly distributed, and the design is based on statics. On the other hand, when the foundation is considered flexible as above defined, and if the variation of adjacent column loads and spans is not greater than 20 per cent, the same simplified procedure as for strip and grid foundations can be applied to mat foundations. The mat is divided into two sets of mutually perpendicular strip footings of width equal to the distance between midspans, and the distribution of bearing pressures and bending moments is carried out for each strip, as explained before. Once moments are determined, the mat in essence is treated the same as a flat slab or plate, with the reinforcement allocated between column and middle strips as in these slab structures (see Chap. 4).

This approach is feasible only when columns are located in a regular rectangular grid pattern. When a mat which can be regarded as rigid supports columns at random locations, the subgrade reactions can still be taken as uniform or as linearly distributed and the mat analyzed by statics. If it is a flexible mat which supports such randomly located columns, the design is based on the theory of plates on elastic foundation. An outline of this procedure is found in Ref. 6.11.

6.11 FOOTINGS ON PILES

If the bearing capacity of the upper soil layers is insufficient for a spread foundation, but firmer strata are available at greater depth, piles are used to transfer the loads to these deeper strata. Piles are generally arranged in groups or clusters, one under each column. The group is capped by a spread footing or cap which distributes the column load to all piles in the group. These pile caps are in most ways very similar to footings on soil, except for two features. For one, reactions on caps act as concentrated loads at the individual piles, rather than as distributed pressures. For another, if the total of all pile reactions in a cluster is divided by the area of the footing to obtain an equivalent uniform pressure (for purposes of comparison only), it is found that this equivalent pressure is considerably higher in pile caps than for spread footings. This means that moments, and particularly shears, are also correspondingly larger, which requires greater footing depths than for a spread footing of similar horizontal dimensions. In order to spread the load evenly to all piles, it is in any event advisable to provide ample rigidity, i.e., depth for pile caps.

Allowable bearing capacities of piles R_a are obtained from soil exploration, pile-driving energy, and test loadings, and their determination is not within the scope of the present book. As in spread footings, the effective portion of R_a available to resist the unfactored column loads is the allowable pile reaction less the weight of footing, backfill, and surcharge per pile. That is,

$$R_e = R_a - W_f \qquad (6.14)$$

where W_f is the total weight of footing, fill, and surcharge divided into the number of piles.

Once the available or effective pile reaction R_e is determined, the number of piles in a concentrically loaded cluster is the integer next larger than

$$n = \frac{D + L}{R_e}$$

As far as the effects of wind, earthquake moments at the foot of the columns, and safety against overturning are concerned, design considerations are the same as in Art. 6.4 on spread footings. These effects generally produce an eccentrically loaded pile cluster in which different piles carry different loads. The number and location of piles in such a cluster is determined by successive approximation from the condition that the load on the most heavily loaded pile shall not exceed the allowable pile reaction R_a. Assuming a linear distribution of pile loads due to bending, the maximum pile reaction is

$$R_{\max} = \frac{P}{n} + \frac{M}{I_{pg}/c} \qquad (6.15)$$

where P is the maximum load (including weight of cap, backfill, etc.) and M the moment to be resisted by the pile group, both referred to the bottom of the cap; I_{pg} is the moment of inertia of the entire pile group about the centroidal axis about which bending occurs; and c is the distance from that axis to the extreme pile. $I_{pg} = \sum_{1}^{n} (1 \times y_n^2)$; that is, it is the moment of inertia of n piles, each counting as one unit and located a distance y_n from the described centroidal axis.

Piles are generally arranged in tight patterns, which minimizes the cost of the caps, but they cannot be placed closer than conditions of driving and of undisturbed carrying capacity will permit. A spacing of about 3 times the butt (top) diameter of the pile but no less than 2 ft 6 in. is customary. Commonly, piles with allowable reactions of 30 to 70 tons are spaced at 3 ft 0 in. (Ref. 6.7).

The *strength design* of footings on piles is similar to that of single-column footings. One approach is to design the cap for the pile reactions calculated for the factored column loads. For a concentrically loaded cluster this would give $R_u = (1.4D + 1.7L)/n$. However, since the number of piles was taken as the next larger integral according to Eq. (6.15), determining R_u in this manner can lead to a design where the strength of the cap is less than the capacity of the pile group. It is therefore recommended (Ref. 6.7) that the pile reaction for strength design be taken as

$$R_u = R_e \times \text{avg load factor} \qquad (6.16)$$

where the average load factor $= (1.4D + 1.7L)/(D + L)$. In this manner the cap is designed to be capable of developing the full allowable capacity of the pile group. Details of a typical pile cap are shown in Fig. 6.21.

As in single-column spread footings, the depth of the pile cap is usually governed by shear. In this regard both punching or two-way shear and flexural

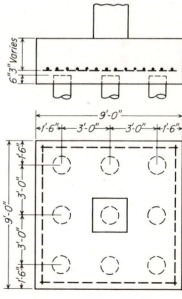

Fig. 6.21 Typical single-column footing on piles (pile cap).

TYPICAL PILE CAP

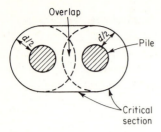

Fig. 6.22 Critical section for punching shear, closely spaced piles. (*Adapted from Ref. 6.8.*)

or one-way shear need to be considered, similarly as in Art. 6.7a. The critical sections are the same as given in that article. The difference is that shears on caps are caused by concentrated pile reactions rather than by distributed bearing pressures. This poses the question of how to calculate shear if the critical section intersects the circumference of one or more piles. For this case the ACI Code accounts for the fact that pile reaction is not really a point load, but is distributed over the pile bearing area. Correspondingly, for piles with diameters d_p, it stipulates as follows:

> In computing the external shear on any section, the entire reaction from any pile whose center is located $d_p/2$ or more outside the section shall be assumed as producing shear on the section. The reaction from any pile whose center is located $d_p/2$ or more inside the section shall be assumed as producing no shear on the section. For intermediate positions, the portion of the pile reaction assumed as producing shear on this section shall be based on straight-line interpolation between full value at $d_p/2$ outside the section and zero at $d_p/2$ inside the section.

In addition to checking two-way and one-way shear as just discussed, punching shear must also be investigated for the individual pile. Particularly in caps on a small number of heavily loaded piles, it is this possibility of a pile punching upward through the cap which may govern the required depth. The critical perimeter for this action, again, is located at a distance $d/2$ outside the upper edge of the pile. However, for relatively deep caps and closely spaced piles, critical perimeters around adjacent piles may overlap. In this case, fracture, if any, would undoubtedly occur along an outward-slanting surface around both adjacent piles. For such situations the critical perimeter is so located that its length is a minimum, as shown for two adjacent piles in Fig. 6.22.

6.12 SERVICE-LOAD DESIGN OF FOOTINGS

As was pointed out in Arts. 3.1 and 3.14, in addition to strength design, which is the main recommended design method, the Code permits dimensioning to be carried out by an alternative method, generally known as service-load design. In this method no load factors are used; i.e., the structure is calculated as being loaded by the stipulated service loads. Likewise, no ϕ factors are used; that is, $\phi = 1$ is taken throughout. Safety is provided by stipulating allowable stresses, which are of the order of one-half of the stresses used in strength design. Details of the allowable stresses and other features of this method are found in Art. 3.14.

This design method can also be applied to footings, pile caps, and other foundations. The main difference is that instead of the bearing pressure q_u or pile reaction R_u used for strength design, calculations are made for the smaller corresponding values q_e and R_e. These have been used, it will be recalled, for determining footing sizes or number of piles, respectively. In connection with these smaller external forces, the reduced allowable stresses are employed, as mentioned before. In general, particularly when higher-strength materials are used, service-load design will lead to somewhat less economical structures.

REFERENCES

6.1. W. C. Tang: "Foundation Design," Prentice Hall, Inc., Englewood Cliffs, N.J., 1962.
6.2. A. N. Talbot: Reinforced Concrete Wall Footings and Column Footings, *Univ. Illinois Eng. Exp. Sta. Bull.* 67, 1913.
6.3. F. E. Richart: Reinforced Concrete Wall and Column Footings, *J. ACI*, vol. 45, pp. 97 and 237, 1948.
6.4. E. Hognestad: Shearing Strength of Reinforced Column Footings, *J. ACI*, vol. 50, p. 189, 1953.
6.5. Shear and Diagonal Tension, part 3, Slabs and Footings, Report of ACI-ASCE Committee 326, *J. ACI*, vol. 59, p. 353, 1962.
6.6. K. Terzaghi and R. E. Peck: "Soil Mechanics in Engineering Practice," 2d ed., John Wiley & Sons, Inc., New York, 1968.
6.7. F. Kramrisch: Footings, chap. 8 in M. Fintel (ed.), "Handbook for Concrete Engineering," Van Nostrand–Reinhold Publishing Company, New York, 1973.
6.8. Commentary to Building Code Requirements for Reinforced Concrete, ACI Standard 318–71, 1971.
6.9. M. Heteneyi: "Beams on Elastic Foundations," University of Michigan Press, Ann Arbor, Mich., 1946.
6.10. F. Kramrisch and P. Rogers: Simplified Design of Combined Footings, *J. Soil Mech. Div. Proc. ASCE*, vol. 87, no. SM5, 1961.
6.11. Suggested Design Procedures for Combined Footings and Mats, Report of ACI Committee 436, *J. ACI*, vol. 63, p. 1041, 1966.

7
Analysis of Continuous Beams and Frames

7.1 CONTINUITY OF REINFORCED-CONCRETE STRUCTURES

The individual members which compose a steel or timber structure are fabricated or cut separately and joined together by rivets, bolts, welds, or nails. Unless the joints are specially designed for rigidity, they are too flexible to transfer moments of significant magnitude from one member to another. In contrast, in reinforced-concrete structures, as much of the concrete as is practical is poured in one single operation. Reinforcing steel is not terminated at the ends of a member, but is extended through the joints into adjacent members. At construction joints, special care is taken to bond the new concrete to the old by carefully cleaning the latter, by extending the reinforcement through the joint, and by other means. Even in precast-concrete construction, connections often provide for transfer of moment as well as shear and thrust. As a result, reinforced-concrete structures usually represent monolithic or continuous units. A load applied at one location causes deformation and stress at all other locations. This characteristic behavior is shown in Fig. 7.1.

With simple joints, as provided in steel construction by many types of riveted and bolted connections, only the beam 3-4 would bend in the frame of Fig. 7.1a, while all other members would remain essentially straight. With

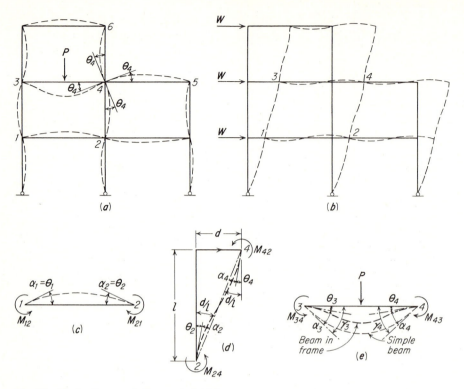

Fig. 7.1 Deflections and slopes of rigid frames.

rigid joints, such as in a reinforced-concrete frame, the distortion caused by a load on one single member is seen to spread to all other members of the frame, although the magnitude of deformation decreases with increasing distance from the loaded member. Consequently, a beam such as 1-2 is subjected to end rotations α_1 and α_2 and accompanying curvature as shown in Fig. 7.1c. Since bending moment is proportional to curvature, this beam and all other members are subject to bending moment, even though they carry no transverse loads.

If horizontal forces, such as wind, seismic, or blast forces, act on the frame, it deforms, as shown in Fig. 7.1b. Here too, all members of the frame distort, even though the forces act only on the left side; the amount of distortion is seen to be about the same for all corresponding members, regardless of their distance from points of loading, in contrast to the case of vertical loading. A member such as 2-4, even though it carries no transverse load, will experience deformations and associated bending moment, as shown in Fig. 7.1d. It is seen that, in addition to joint rotations α_2 and α_4, one end is displaced horizontally with respect to the other by the amount d, resulting in an additional rotation d/l. If a member carries transverse load too, as does beam 3-4, the end rotations γ caused by the load must be superimposed on those due to continuity at the joints (Fig. 7.1e). The final end slopes θ of each individual member are seen to be the sums of these individual contributions α, d/l, and γ.

In statically determinate structures, such as simple beams, the deflected shape and the moments and shears depend only on the type and magnitude of the loads and the dimensions of the member. In contrast, inspection of Fig. 7.1 shows that in statically indeterminate frames, the deflection curve of any member, in addition to depending on the loads, depends on the end slopes θ, whose magnitudes in turn depend on the distortion of adjacent, rigidly connected members. In particular, for a rigid joint such as joint 4 (Fig. 7.1a), the slopes θ of all abutting members must be the same and equal to the rotation of the joint. For a correct design of such frames, it is evidently necessary to determine moments, shears, and thrusts with due consideration of the effect of continuity at the joints.

7.2 PLACEMENT OF LOADS

The individual members of a structural frame must be designed for the worst combination of loads which can reasonably be expected to occur during its useful life. Internal moments, shears, and thrusts are brought about by the combined effect of dead and live loads. While the former are constant, live loads such as floor loads from human occupancy can be placed in various ways, some of which will result in larger effects than others.

In Fig. 7.2a, only span CD is loaded by live load. The distortions of the various frame members are seen to be largest in, and immediately adjacent to, the loaded span and to decrease rapidly with increasing distance from the load. Since bending moments are proportional to curvatures, the moments in more remote members are correspondingly smaller than those in, or close to, the loaded span. However, the loading of Fig. 7.2a does not produce the maximum possible positive moment in CD. In fact, if additional live load were placed on span AB, this span would bend down, BC would bend up, and CD itself would bend down in the same manner, although to a lesser degree, as it is bent by its own load. Hence, the positive moment in CD is increased if AB, and by the same reasoning EF, are loaded simultaneously. By expanding the same reasoning to the other members of the frame, it is easily seen that the "checkerboard pattern" of live load of Fig. 7.2b produces the largest possible positive moments not only in CD, but in all loaded spans. Hence, two such checkerboard patterns are required to obtain the maximum positive moments in all spans.

In addition to maximum span moments, it is often necessary to investigate minimum span moments. Dead load, acting as it does on all spans, usually produces only positive span moments. However, live load, placed as in Fig. 7.2a, and even more so in Fig. 7.2b, is seen to bend the unloaded spans upward,

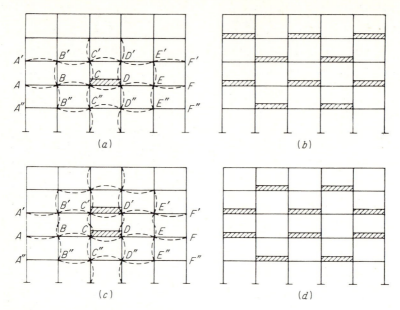

Fig. 7.2

i.e., to produce negative moments in the span. If these negative live-load moments are larger than the generally positive dead-load moments, a given girder, depending on load position, may be subject at one time to positive span moments and at another to negative span moments. It must be designed to withstand both types of moments; i.e., it must be furnished with tensile steel at both top and bottom. Thus, the loading of Fig. 7.2b, in addition to giving maximum span moments in the loaded spans, gives minimum span moments in the unloaded spans.

Maximum negative moments at the supports of the girders are obtained, on the other hand, if loads are placed on the two spans adjacent to the particular support and in a corresponding pattern on the more remote girders. A separate loading scheme of this type is then required for each support for which maximum negative moments are to be computed.

In each column, the largest moments occur at the top or bottom. While the loading of Fig. 7.2c results in large moments at the ends of columns CC' and DD', the reader can easily convince himself that these moments are further augmented if additional loads are placed as shown in Fig. 7.2d.

It is seen from this brief discussion that, in order to calculate the maximum possible moments at all critical points of a frame, live load must be placed in a great variety of different schemes. In most practical cases, however, consideration of the relative magnitude of effects will permit limitation of analysis to a small number of significant cases.

7.3 ANALYSIS OF SUBFRAMES

Considering the complexity of many practical building frames, and the need to account for the possibility of alternative loadings, there is evidently a need to simplify. This can be done by means of certain approximations which allow the determination of moments with reasonable accuracy while reducing substantially the amount of computation.

Numerous trial computations have shown that, for building frames with reasonably regular outline, not involving unusual asymmetry of loading or shape, the influence of sidesway caused by vertical loads can be neglected. In that case, moments due to vertical loads are determined with sufficient accuracy by dividing the entire frame into simpler subframes. Each of these consists of one continuous beam, plus the top and bottom columns framing into that particular beam. Placing the live loads on the beam in the most unfavorable manner permits sufficiently accurate determination of all beam moments, as well as the moments at the top ends of the bottom columns and the bottom ends of the top columns. For this partial structure, the far ends of the columns are considered fixed, except for such first-floor or basement columns where soil and foundation conditions dictate the assumption of hinged ends. Such an approach is explicitly permitted by the ACI Code, which specifies for floor and roof members:

> (1) The live load may be considered to be applied only to the floor or roof under consideration, and the far ends of the columns may be assumed as fixed.
>
> (2) Consideration may be limited to combinations of dead loads on all spans with full live load on two adjacent spans (for negative support moments) and with full live load on alternate spans (for positive span moments).

When investigating the maximum negative moment at any joint, negligible error will result if the joints second removed in each direction are considered to be completely fixed. Similarly, in determining maximum or minimum span moments, the joints at the far ends of the adjacent spans may be considered fixed. Thus, individual portions of a frame of many members may be investigated separately.

In regard to columns, the Code indicates:

> Columns shall be designed to resist the axial forces from loads on all floors and the maximum bending due to design loads on a single adjacent span of the floor under consideration. Account shall also be taken of the loading condition giving the maximum ratio of bending moment to axial load. In building frames, particular atten-

tion shall be given to the effect of unbalanced floor loads on both exterior and interior columns and of eccentric loading due to other causes. In computing moments in columns due to gravity loading, the far ends of columns which are monolithic with the structure may be considered fixed.

7.4 METHODS OF ANALYSIS

A number of methods have been developed over the years for the analysis of continuous beams and frames. The so-called "classical" methods, such as application of the theorem of three moments, the method of least work (Castigliano's second theorem), and the general method of consistent deformation, have proved useful mainly in the analysis of continuous beams having few spans, or of very simple frames. For the more complicated cases usually met in practice, such methods prove to be exceedingly tedious, and alternative approaches are preferred. For many years the closely related methods of slope deflection and moment distribution provided the basic analytical tools for the analysis of indeterminate concrete beams and frames. In offices which have access to high-speed digital computers, these have been supplanted largely by matrix methods of analysis. Where computer facilities are not available, moment distribution is still the most common method. Approximate methods of analysis, based either on an assumed shape of the deformed structure or on moment coefficients, provide a means for rapid estimation of internal forces and moments. Such estimates are useful in preliminary design and in checking more exact solutions and, in the case of structures of minor importance, may serve as the basis for final design.

In view of the number of excellent texts now available treating methods of analysis (e.g., Refs. 7.1 to 7.5), the present discussion will be confined to an evaluation of the usefulness of several of the more important of these, with particular reference to the analysis of reinforced-concrete structures. Certain idealizations and approximations which facilitate the solution in practical cases will be described in more detail.

a. Slope deflection The method of slope deflection was developed independently by Bendixen in Germany in 1914 and by Maney in the United States in 1915. The method entails writing two equations for each member of a continuous frame, one at each end, expressing the end moment as the sum of four contributions: (1) the restraining moment associated with an assumed fixed-end condition for the loaded span, (2) the moment associated with rotation of the tangent to the elastic curve at the near end of the member, (3) the moment associated with rotation of the tangent at the far end of the member, and (4) the moment associated with translation of one end of the member with respect to the other. These equations are related through application of requirements of equilibrium and compatibility at the joints. A set of simultaneous, linear algebraic equations results for the entire structure in which the structural displacements are the unknowns. Solution for these displacements permits the calculation of all internal forces and moments (Refs. 7.1 and 7.2).

The method is well suited to solving continuous beams, provided there are not very many spans. Its usefulness is extended through modifications which take advantage of symmetry and antisymmetry, and of hinged-end support conditions where they exist. However, for multistory and multibay frames in which there are a large number of members and joints, and which will, in general, involve translation as well as rotation of these joints, the effort required to solve the correspondingly large number of simultaneous equations will be prohibitive. Other methods of analysis are more attractive.

b. Moment distribution In 1932 Hardy Cross developed the method of moment distribution to solve problems in frame analysis which involve many unknown joint displacements. For the next three decades, moment distribution provided the standard means in engineering offices for the analysis of indeterminate frames. Even now, it serves as the basic analytical tool when computer facilities are not available.

The moment-distribution method (Refs. 7.1 and 7.2) can be regarded as an iterative solution of the slope-deflection equations. Starting with fixed-end moments for each member, these are modified in a series of cycles, each converging on the precise final result, to account for rotation and translation of the joints. The resulting series may be terminated whenever one reaches the degree of accuracy required. After obtaining member end moments, all member stress resultants can be obtained by use of the laws of statics.

It has been found by comparative analyses that, except in unusual cases, building-frame moments found by modifying fixed-end moments by only two cycles of moment distribution will be sufficiently accurate for design purposes.

c. Matrix analysis The introduction of matrix methods of analysis to the structural engineering profession in the early 1950's, coupled with the increasing availability of high-speed digital computers, has produced changes in practice which can only be described as revolutionary. Use of matrix theory makes it possible to reduce the detailed numerical operations required in the analysis of an indeterminate structure to systematic processes of matrix manipulation which can be performed automatically and rapidly by computer. Such methods permit the rapid solution of problems involving large numbers of unknowns. As a consequence, less reliance is placed on special techniques limited to certain types of problems, and powerful methods of general applicability have emerged, such as the matrix displacement method and the matrix force method (Refs. 7.3 to 7.5). By such means, an "exact" determination of moments, shears, and thrusts throughout an entire building frame may be obtained quickly and at small expense. Account may be taken of such factors as rotational restraint provided by members perpendicular to the plane of a frame. A large number of alternative loadings may be considered. Provided that computer facilities are available, highly refined analyses are possible at lower cost than for approximate analyses previously employed.

Some engineers prefer to write individual programs for structural analysis, particularly suited to their own needs, making use of standard programming languages such as FORTRAN or PL/I. In many offices, standard, general-purpose programs are used which are applicable to a broad range of problems. Several such programs are currently available. They use problem-oriented languages developed specifically for structural engineering applications. The engineer can communicate with the computer in his own language, to describe the type, makeup, and arrangement of the structure, the loads, and the kind of results (e.g., displacements, member forces, reactions) requested. One example of such a problem-oriented language is STRESS (Ref. 7.6), for the analysis of two- and three-dimensional framed structures composed of linear members. STRESS has been designed for use even with small "office model" computers having limited storage capacity. More sophisticated and powerful general-purpose programs such as STRUDL (Ref. 7.7) require larger machines.

d. Approximate analysis In spite of the development of refined methods for the analysis of beams and frames, increasing attention is being paid to various approximate methods of analysis (Ref. 7.8). There are several reasons for this. Prior to performing a complete analysis of an indeterminate structure, it is necessary to estimate the proportions of its members in order to know their relative stiffness upon which the analysis depends (see Art. 7.6). These dimensions may be obtained on the basis of approximate analysis. Also, even with the availability of computers, most engineers find it desirable to make a rough check of results, using approximate means, to detect gross errors. Further, for structures of minor importance, it is often satisfactory to design on the basis of results obtained by rough calculation. For these reasons, many engineers at some stage in the design process estimate the values of moments, shears, and thrusts at critical locations, using approximate sketches of the structure deflected by its loads.

Provided that points of inflection (locations in members at which the bending moment is zero and there is a reversal of curvature of the elastic curve) can be located accurately, the stress resultants for a framed structure usually may be found on the basis of static equilibrium alone. Each portion of the structure must be in equilibrium under the application of its external loads and the internal stress resultants.

For the fixed-end beam of Fig. 7.3a, for example, the points of inflection under uniformly distributed load are known to be located $0.211l$ from the ends of the span. Since the moment at these points is zero, imaginary hinges may be placed there without modifying the member behavior. The individual segments between hinges can be analyzed by statics, as shown in Fig. 7.3b.

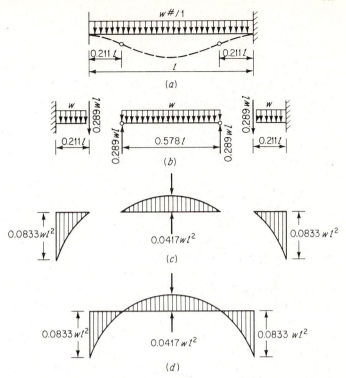

Fig. 7.3 Analysis of fixed-end beam by locating inflection points.

Starting with the center segment, shears equal to $0.289wl$ must act at the hinges. These, together with the transverse load, produce a midspan moment of $0.0417wl^2$. Proceeding next to the outer segments, a downward load is applied at the hinge representing the shear from the center segment. This, together with the applied load, produces support moments of $0.0833wl^2$. Note that, for this example, since the correct position of the inflection points was known at the start, the resulting moment diagram of Fig. 7.3c agrees exactly with the true moment diagram for a fixed-end beam shown in Fig. 7.3d. In more practical cases inflection points must be estimated, and the results obtained will only approximate the true values.

The use of approximate analysis in determining stress resultants in frames is illustrated by Fig. 7.4. Figure 7.4a shows the geometry and loading of a two-member rigid frame. In Fig. 7.4b an exaggerated sketch of the probable deflected shape is given, together with the estimated location of points of inflection. On this basis, the central portion of the girder is analyzed by statics, as shown in Fig. 7.4d, to obtain girder shears at the inflection points of 7 kips, acting with a thrust P (still not determined). Similarly, the requirements of statics applied to the outer segments of the girder in Fig. 7.4c and e give vertical shears of 11 and 13 kips at B and C, respectively, and end moments

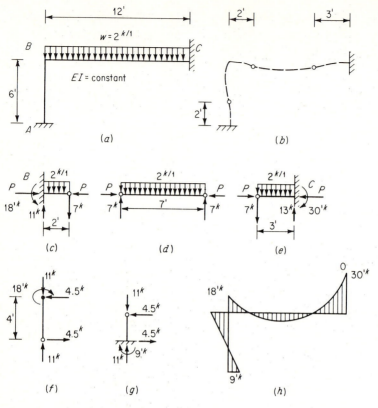

Fig. 7.4 Approximate analysis of rigid frame.

of 18 and 30 ft-kips at the same locations. Proceeding then to the upper segment of the column, shown in Fig. 7.4f, with known thrust of 11 kips and top
moment of 18 ft-kips acting, a horizontal shear of 4.5 kips at the inflection
point is required for equilibrium. Finally, static analysis of the lower part
of the column indicates a requirement of 9 ft-kips moment at A, as shown in
Fig. 7.4g. The value of P equal to 4.5 kips is obtained by summing horizontal
forces at joint B.

The moment diagram resulting from approximate analysis is shown in
Fig. 7.4h. For comparison, an exact analysis of the frame indicates member
end moments of 8 ft-kips at A, 16 ft-kips at B, and 28 ft-kips at C. The results
of the approximate analysis would be satisfactory for design in many cases;
if a more exact analysis is to be made, a valuable check is available on the
magnitude of results.

A specialization of the approximate method described, known as the
portal method, is commonly used to estimate the effects of sidesway due to
lateral forces acting on multistory building frames. For such frames, it is
usual to assume that horizontal loads are applied at the joints only. If this

is true, moments in all members vary linearly and, except in hinged members, have opposite signs at the ends. In addition, the points of zero moment are located reasonably close to the midpoint of each member.

For a simple rectangular portal frame having three members, the shear forces are the same in both legs and are each equal to half the external horizontal load. If one of the legs is more rigid than the other, it would require a larger horizontal force to displace it horizontally the same amount as the more flexible leg. Consequently, the portion of the total shear resisted by the stiffer column is larger than that of the more flexible column.

In multistory building frames, moments and forces in the girders and columns of each individual story are distributed in substantially the same manner as just discussed for single-story frames. The portal method of computing approximate moments, shears, and axial forces from horizontal loads is therefore based on the following three simple propositions:

1. The total horizontal shear in all columns of a given story is equal and opposite to the sum of all horizontal loads acting above that story.
2. The horizontal shear is the same in both exterior columns; the horizontal shear in each interior column is twice that in an exterior column.
3. The inflection points of all members, columns and girders, are located midway between joints.

Although the last of these propositions is commonly applied to all columns, including those of the bottom floor, the authors prefer to deal with the latter separately, depending on conditions of foundation. If the actual conditions are such as practically to prevent rotation (foundation on rock, massive pile foundations, etc.), the inflection points of the bottom columns are above midpoint and may be assumed to be at a distance $2h/3$ from the bottom. If little resistance is offered to rotation, such as for relatively small footings on compressible soil, the inflection point is located closer to the bottom and may be assumed to be at a distance $h/3$ from the bottom, or even lower. (In ideal hinges the inflection point is at the hinge, i.e., at the very bottom.) Since shears and corresponding moments are largest in the bottom story, a judicious evaluation of foundation conditions as they affect the location of inflection points is of considerable importance.

The first of the three cited propositions follows from the requirement that horizontal forces be in equilibrium at any level. The second takes account of the fact that in building frames interior columns are generally more rigid than exterior ones because (1) the larger axial loads require larger cross section and (2) exterior columns are restrained from joint rotation only by one abutting girder, while interior columns are so restrained by two such members. The third proposition is very nearly true because, except for the top and bottom columns and, to a minor degree, for the exterior girders, each member in a building frame is restrained about equally at both ends. For this reason it deflects under horizontal loads in an antisymmetrical manner, with the inflection point at midlength.

The actual computations in this method are extremely simple. Once column shears are determined from propositions 1 and 2, and inflection points located from proposition 3, all moments, shears, and forces are simply computed by statics. The process is illustrated in Fig. 7.5a.

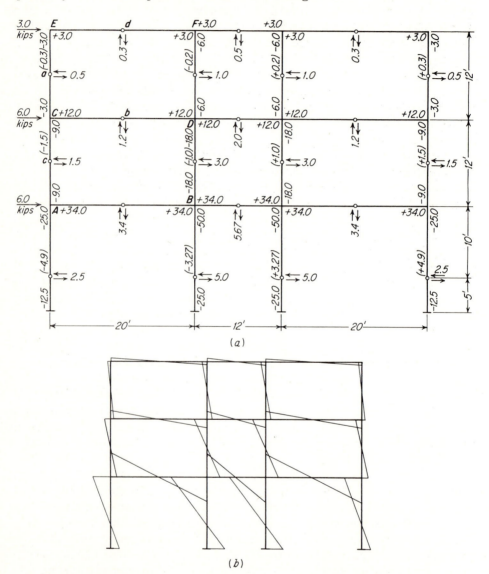

Fig. 7.5 Portal method for determining moments from wind load in building frame.

Consider joints C and D. The total shear in the second story is $3 + 6 = 9$ kips. According to proposition 2, the shear in each exterior column is $9/6 = 1.5$ kips, and in each interior column $2 \times 1.5 = 3.0$ kips. The shears in the other floors, obtained in the same manner, act at the hinges as shown. Consider the equilibrium of the rigid structure between hinges a, b, and c; the column moments, 3.0 and 9.0, respectively, are obtained directly by multiplying the shears by their lever arms, 6 ft. The girder moment at C, to produce equilibrium, is equal and opposite to the sum of the column moments. The shear in the girder is obtained by recognizing that its moment (i.e., shear times half the girder span) must be equal to the girder moment at C. Hence, this shear is $12.0/10 = 1.2$ kips. The moment at end D is equal to that at C, since the inflection point is at midspan. At D, column moments are computed in the same manner from the known column shears and lever arms. The sum of the two girder moments, to produce equilibrium, must be equal and opposite to the sum of the two column moments, from which the girder moment to the right of C is $18.0 + 6.0 - 12.0 = 12.0$. Axial forces in the columns also follow from statics. Thus, for the rigid body aEd, a vertical shear of 0.3 kip is seen to act upward at d. To equilibrate it, a tensile force of -0.3 kip is required in the column CE. In the rigid body abc an upward shear of 1.2 kips at b is added to the previous upward tension of 0.3 kip at a. To equilibrate these two forces, a tension force of -1.5 kips is required in column AC. If the equilibrium of all other partial structures between hinges is considered in a similar manner, all moments, forces, and shears are rapidly determined.

In the present case relatively flexible foundations were assumed, and the location of the lowermost inflection points was estimated to be at $h/3$ from the bottom. The general character of the resulting moment distribution is shown in Fig. 7.5b.

e. ACI moment coefficients   The ACI Code includes moment and shear coefficients which may be used for the analysis of buildings of usual types of construction, spans, and story heights. They are reprinted in Table 7.1. The ACI coefficients were derived with due consideration of a maximum allowable ratio of live to dead load (3:1); maximum allowable span difference (the larger of two adjacent spans not to exceed the shorter by more than 20 per cent); the fact that reinforced-concrete beams are never simply supported, but either rest on supports of considerable width, such as walls, or are built monolithically with columns; and other factors. Since all these influences are considered, the ACI coefficients are necessarily quite conservative, so that actual, accurate moments in any particular design are likely to be considerably smaller than indicated. Consequently, in many reinforced-concrete structures, significant economy can be effected by making a more precise analysis. This becomes mandatory for beams and slabs with spans differing by more than 20 per cent or sustaining loads which are not uniformly distributed.

Table 7.1   Moment and shear values using ACI coefficients*

Positive moment:

  End spans:

    If discontinuous end is unrestrained . . . . $\frac{1}{11}wl_n^2$

    If discontinuous end is integral with the support . . . . $\frac{1}{14}wl_n^2$

  Interior spans . . . . $\frac{1}{16}wl_n^2$

Negative moment at exterior face of first interior support:

  Two spans . . . . $\frac{1}{9}wl_n^2$

  More than two spans . . . . $\frac{1}{10}wl_n^2$

Negative moment at other faces of interior supports . . . . $\frac{1}{11}wl_n^2$

Negative moment at face of all supports for (1) slabs with spans not exceeding 10 ft,
  and (2) beams and girders where ratio of sum of column stiffnesses to beam stiff-
  ness exceeds 8 at each end of the span . . . . $\frac{1}{12}wl_n^2$

Negative moment at interior faces of exterior supports for members built integrally
  with their supports:

    Where the support is a spandrel beam or girder . . . . $\frac{1}{24}wl_n^2$

    Where the support is a column . . . . $\frac{1}{16}wl_n^2$

Shear in end members at first interior support . . . . $1.15\dfrac{wl_n}{2}$

Shear at all other supports . . . . $\dfrac{wl_n}{2}$

---

* w = total load per unit length of beam or per unit area of slab.

  l_n = clear span for positive moment and shear and the average of the two adjacent clear
      spans for negative moment.

7.5   IDEALIZATION OF THE STRUCTURE

It is seldom possible for the engineer to analyze an actual complex redundant structure. Almost without exception, certain idealizations must be made in devising an analytical model, so that the analysis will be practically possible. Thus, three-dimensional members are represented by straight lines, generally coincident with the actual centroidal axis. Supports are idealized as rollers, hinges, or rigid joints. Loads actually distributed over a finite area are assumed to be point loads. In three-dimensional framed structures, analysis is commonly confined to plane frames, each of which is assumed to act independently, even though they are actually linked and interact.

    In the idealization of reinforced-concrete frames, certain questions require special comment. The most important of these pertain to effective span lengths, effective moments of inertia, and conditions of support.

a. Effective span length   In elastic frame analysis, a structure is usually represented by a simple line diagram, based dimensionally on the center-line distances between columns and between floor beams. Actually, the depths of beams and the widths of columns (in the plane of the frame) amount to sizable fractions of the respective lengths of these members; their clear lengths are therefore considerably smaller than their center-line distances between joints.

It is evident that the usual assumption in frame analysis that the members are prismatic, with constant moment of inertia between center lines, is not strictly correct. A beam intersecting a column may be prismatic up to the column face, but from that point to the column center line it has a greatly increased depth, with a moment of inertia which could be considered infinite compared with that of the remainder of the span. A similar variation in width and moment of inertia is obtained for the columns. Thus, to be strictly correct, the actual variation in member depth should be considered in the analysis. Qualitatively, this would increase beam support moments somewhat, and decrease span moments. In addition, it is apparent that the critical section for design for negative bending would be at the face of the support, and not at the center line, since for all practical purposes an unlimited effective depth is obtained in the beam across the width of the support.

It will be observed that, in the case of columns, the moment curve is not very steep, so that the difference between center-line moment and the moment at the top or bottom face of the beam is small and can in most cases be disregarded. However, the moment diagram for the beam is usually quite steep in the region of the support, and there will be a substantial difference between support center-line moment and face moment. If the former were used in proportioning the member, an unnecessarily large section would result. It is desirable, then, to reduce support moments found by elastic analysis to account for the finite width of the supports.

In Fig. 7.6, the change in moment between the support center line and the support face will be equal to the area under the shear diagram between those two points. For knife-edge supports, this shear area is seen to be very nearly equal to $VaL/2$. Actually, however, the reaction is distributed in some unknown way across the width of the support. This will have the effect of modifying the shear diagram as shown by the dashed line; it has been proposed that the reduced area be taken equal to $VaL/3$. The fact that the reaction is distributed will modify the moment diagram as well as the shear diagram, causing a slight rounding of the negative-moment peak, as shown in the figure, and the reduction of $VaL/3$ is properly applied to the moment diagram after the peak has been rounded. This will give nearly the same face moment as would be obtained by deducting the amount $VaL/2$ from the peak moment.

Another effect is present, however: the modification of the moment diagram due to the increased moment of inertia of the beam at the column. This effect is similar to that of a haunch, and it will mean slightly increased negative moment and slightly decreased positive moment. For ordinary values of the ratio a, this shift in the moment curve will be of the order of $VaL/6$. Thus, it is convenient simply to deduct the amount $VaL/3$ from the unrounded peak moment obtained from elastic analysis. This allows for (1) the actual rounding of the shear diagram and the negative-moment peak due to the dis-

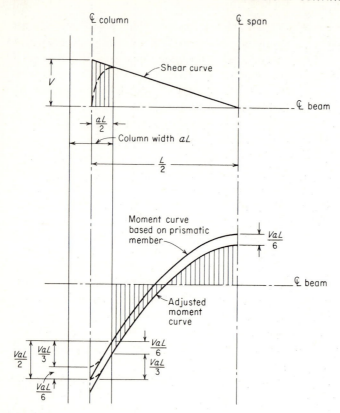

Fig. 7.6

tributed reaction and (2) the downward shift of the moment curve due to the haunch effect at the supports. The consistent reduction in positive moment of $VaL/6$ is illustrated in Fig. 7.6.

In connection with moment reductions, it should be noted that there are certain conditions of support for which no reduction in negative moment is justified. For example, when a continuous beam is carried by a girder of approximately the same depth, the negative moment in the beam at the center line of the girder should be used in designing the negative reinforcing steel.

b. Moments of inertia   The design of flexural members is based on the cracked section, i.e., on the supposition that the part of the concrete which is in tension is ineffective. It would seem, therefore, that moments of inertia to be used for frame analysis should be determined in the same manner.

In most cases in frame analysis, however, it is only the ratio of stiffnesses which influences the result, and not the absolute values of the stiffnesses. Stiffness ratios are but little affected by different assumptions (e.g., use of cracked section vs. gross section) in computing moment of inertia, provided there is consistency for all members. In recognition of this fact, it is generally convenient and sufficiently accurate to base stiffness calculations for frame analysis on the full concrete cross section. The contribution of the reinforcement is usually neglected, which compensates to some extent for the neglect of the influence of cracks. There may be some justification for including the transformed area of the steel in computing moments of inertia of columns, because often the column stress will be entirely compressive, with no cracking to compensate for the neglect of the steel. In addition, steel percentages are considerably higher for columns than for beams. However, it is the usual practice to compute moments of inertia of columns just as for beams, considering only the cross section of the concrete.

For T beams, allowance should be made for the effect of the flange on member stiffness. It seems reasonable to assume the same effective width of flange for computing moment of inertia as is used in stress computations.

A problem arises in continuous T-beam construction, in that for flexural design the effective cross section is rectangular in the negative-moment regions. However, it is not unusual to provide compressive reinforcement in the negative-moment regions for such construction. This increases the stiffness over that of the plain rectangular concrete section. Consequently, there is some justification for the common practice of computing stiffnesses of such members on the basis of the gross cross section of the concrete, including flanges, and using this stiffness throughout the entire span.

It is somewhat laborious to compute, in every case, the neutral axis of the effective T section and to find the moment of inertia about that axis of the unsymmetrical section. In the usual case it is satisfactory to estimate the moment of inertia of the unsymmetrical section under the assumption that it will be some multiple of that of the rectangular cross section of web width b' and total height h. For sections of usual proportions (flange width about 4 to 6 times web width, and flange thickness from 0.2 to 0.4 times total depth), this multiplier will be very close to 2.0.

c. Conditions of support   For purposes of analysis, many structures can be divided into a number of two-dimensional frames. Even for such cases, however, there are situations in which it is impossible to predict with accuracy what the conditions of restraint might be at the ends of a span; yet moments are frequently affected to a considerable degree by the choice made. In many other cases, it is necessary to recognize that structures may be three-dimensional. The rotational restraint at a joint may be influenced or even governed by the characteristics of members framing into that joint at right angles. Adjacent members or frames parallel to the one under primary consideration may likewise influence its performance.

If floor beams are cast monolithically with reinforced-concrete walls (frequently the case when first-floor beams are carried on foundation walls), the moment of inertia of the wall about an axis parallel to its face may be so large that the beam end could be considered completely fixed for all practical purposes. If the wall is relatively thin, or the beam particularly massive, the moment of inertia of each should be calculated, that of the wall being equal to $bt^3/12$, where t is the wall thickness and b the wall width tributary to one beam (i.e., center to center of adjacent panels).

If the outer ends of concrete beams rest on masonry walls, as is sometimes the case, an assumption of zero rotational restraint (i.e., hinged support) is probably closest to the actual case.

For columns supported on relatively small footings, which in turn rest on compressible soil, a hinged end is generally assumed, since such soils offer but little resistance to rotation of the footing. If, on the other hand, the footings rest on solid rock, or if a cluster of piles is used with their upper portion encased by a concrete cap, the effect is to provide almost complete fixity for the supported column, and this should be assumed in the analysis. Columns supported by a continuous foundation mat should likewise be assumed fixed at their lower ends.

If members framing into a joint in a direction perpendicular to the plane of the frame under analysis have sufficient torsional stiffness, and if their far ends can be considered fixed or nearly so, their effect on joint rigidity should be included in the computations. The torsional stiffness of a member of length L is given by the expression GJ/L, where G is the shear modulus of elasticity of concrete (approximately equal to $E_c/2.2$), and J is the torsional stiffness factor of the member. For beams of rectangular cross section, or of sections made up of rectangular elements, J can be taken equal to $\Sigma(hb^3/3 - b^4/5)$, in which h and b are the cross-sectional dimensions of each rectangular element, b being the lesser dimension in each case. In moment distribution, when the effect of torsional rigidity is included, it is important that the absolute flexural stiffness $4EI/L$ be used, rather than relative I/L values.

A common situation in beam-and-girder floors and concrete-joist floors is illustrated in Fig. 7.7. The sketch shows a beam-and-girder floor system in

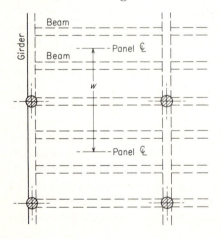

Fig. 7.7

which longitudinal beams are placed at the third points of each bay, supported by transverse girders, in addition to the longitudinal beams supported directly by the columns. If the transverse girders are quite stiff, it is apparent that the flexural stiffness of all beams in the width w should be balanced against the stiffness of one set of columns in the longitudinal bent. If, on the other hand, the girders have little torsional stiffness, there would be ample justification for making two separate longitudinal analyses, one for the beams supported directly by the columns, in which the rotational resistance of the columns would be considered, and a second for the beams framing into the girders, in which case simple supports would be assumed. Probably, it would generally be sufficiently accurate to consider the girders very stiff torsionally and to add directly the stiffness of all beams tributary to a single column. This has the added advantage that all longitudinal beams will have the same cross-sectional dimensions and the same reinforcing steel, which will greatly facilitate construction.

7.6 PRELIMINARY DESIGN

In making an elastic analysis of a structural framework, it is necessary to know at the outset the cross-sectional dimensions of the members, so that moments of inertia and stiffnesses can be calculated. Yet the determination of these same cross-sectional dimensions is the precise purpose of the elastic analysis. Obviously, a preliminary estimate of member sizes must be one of the first steps in the analysis. Subsequently, with the results of the analysis at hand, members are proportioned, and the resulting dimensions compared with those previously assumed. If necessary, the assumed section properties are modified, and the analysis is repeated. Since the procedure may become quite laborious, it is obviously advantageous to make the best possible original estimate of member sizes, in the hope of avoiding repetition of the analysis.

In this connection, it is worth repeating that in the ordinary frame analysis one is concerned with relative stiffnesses only, not the absolute stiffnesses. If, in the original estimate of member sizes, the sizes of all beams and columns are overestimated or underestimated by about the same amount, correction of these estimated sizes after the first analysis will have little or no effect on the relative stiffnesses. Consequently, no revision of the analysis would be required. If, on the other hand, a nonuniform error in estimation is made, and relative stiffnesses differ from assumed values by more than about 30 per cent, a new analysis should be made.

The experienced designer can estimate beam and column sizes with surprising accuracy. Those with little or no experience must rely on trial calculations or arbitrary rules, modified to suit particular situations. In building frames, beam sizes are usually governed by the negative moments and the shears at the supports, where their effective section is rectangular. Moments can be approximated by the fixed-end moments for the particular span, or by using the ACI moment coefficients. In most cases, shears will not differ greatly from simple beam shears. Alternatively, many designers prefer to estimate the depth of beams at about $\frac{3}{4}$ in. per ft of span, with the width about one-half the depth. Obviously, these dimensions are subject to modification, depending on the type and magnitude of the loads, methods of design, and material strength.

Column sizes are governed primarily by axial loads, which can be estimated quickly, although the presence of moments in the columns is cause for some increase of the area as determined by axial loads. For interior columns, in which unbalanced moments will not be large, a 10 per cent increase may be sufficient, while for exterior columns, particularly for upper stories, an increase of 50 per cent in area may be appropriate. In deciding on these estimated increases, the following factors should be considered: Moments are larger in exterior than in interior columns, since, in the latter, dead-load moments from adjacent spans will largely balance, in contrast to the case in exterior columns. In addition, the influence of moments, as compared with that of axial loads, is larger in upper-floor than in lower-floor columns, because the moments are usually of about the same magnitude, while the axial loads are larger in the latter than in the former. Judicious consideration of factors such as these will enable a designer to produce a reasonably accurate preliminary design, which in most cases will permit a satisfactory analysis to be made on the first trial.

7.7 LIMIT ANALYSIS OF CONTINUOUS BEAMS AND FRAMES

Since it is known that concrete does not respond elastically to loads of more than about one-half the ultimate, there is a certain inconsistency in designing reinforced cross sections by ultimate-strength methods (taking into account inelastic behavior) when the moments for which those sections are being designed have been found by elastic analysis. Although this presently accepted procedure by which elastic analysis is coupled with inelastic design of sections is not consistent, it is safe and conservative. It can be shown that a frame so analyzed and designed will not fail at a lower load than anticipated. On the other hand, it is known that an indeterminate beam or frame normally will not fail when the ultimate-moment capacity of just one critical section is reached. After formation of *plastic hinges* at the more highly stressed sections, substantial changes may occur in the ratio of moments at critical sections as loads are further increased before collapse of the structure takes place. Recognition of such redistribution of moments is important in that it permits a more realistic appraisal of the strength of the structure, thus leading to improved economy. In addition, it permits the designer to modify, within limits, the moment diagrams for which members are to be designed. Certain sections may be deliberately underreinforced, provided that corresponding adjustments are made elsewhere in the span. Such arbitrary reassignment of moments enables the designer to reduce the congestion of reinforcement which often occurs in high-moment areas, such as at the junction of girders with columns.

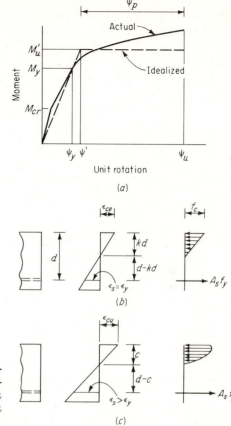

Fig. 7.8 Plastic-hinge characteristics in a reinforced-concrete member. (*a*) Typical moment-rotation diagram; (*b*) strains and stresses at start of yielding; (*c*) strains and stresses at incipient failure.

a. Plastic hinges and collapse mechanisms If a short segment of a reinforced-concrete beam is subjected to a bending moment, curvature of the beam axis will result, and there will be a corresponding rotation of one face of the segment with respect to the other. It is convenient to express this in terms of an angular change ψ per unit length of the member. A representative relation between applied moment M and resulting unit rotation ψ for an underreinforced beam segment is shown in Fig. 7.8*a*.

The diagram is linear up to the cracking moment M_{cr}, after which a straight line of somewhat flatter slope is obtained. At the moment which initiates yielding, M_y, the unit rotation starts to increase disproportionately. Further increase in applied moment causes extensive inelastic rotation, until eventual collapse is obtained when the rotation exceeds the ultimate capacity ψ_u.

The yield moment M_y is easily calculated on the basis of the elastic stress distribution shown in Fig. 7.8*b*.

$$M_y = A_s f_y \left(d - \frac{kd}{3} \right)$$

The conventional ultimate-moment capacity M_u' (Fig. 7.8c), given by the expression

$$M_u' = A_s f_y \left(d - \frac{k_1 c}{2} \right)$$

is always somewhat above M_y due to the slight increase in the internal lever arm as extensive yielding occurs and as the concrete stress distribution changes shape. Some increase in resisting moment beyond M_u' is almost always observed, due largely to strain-hardening of the reinforcement.

For purposes of limit analysis, the M-ψ curve is usually idealized, as shown by the dashed line. The slope of the elastic portion of the curve is obtained with satisfactory accuracy using the moment of inertia of the cracked transformed section. After the calculated ultimate moment M_u' is reached, continued plastic rotation is assumed to occur with no change in applied moment. The elastic curve of the beam will show an abrupt change in slope at such a section. The beam behaves as if there were a hinge at that point. However, the hinge will not be "friction-free," but will have a fairly constant resistance to rotation.

If such a plastic hinge forms in a determinate structure, as shown in Fig. 7.9, uncontrolled deflection takes place, and the structure will collapse. The resulting system is referred to as a *mechanism*, in analogy to linkage systems in mechanics. Generalizing, one can say that a statically determinate system requires the formation of only one plastic hinge in order to become a mechanism.

This is not so for indeterminate structures. In this case, stability may be maintained even though hinges have formed at several cross sections. The formation of such hinges in indeterminate structures permits a redistribution of moments within the beam or frame. It will be assumed for simplicity that the indeterminate beam of Fig. 7.10a is symmetrically reinforced, so that the negative bending capacity is the same as the positive. Let the load P be increased gradually until the elastic moment at the fixed support, $\frac{3}{16}PL$, is just equal to the plastic moment capacity of the section M_u. This load is

$$(a) \quad P = P_{el}' = \frac{16}{3} \frac{M_u'}{L} = 5.33 \frac{M_u'}{L}$$

At this load, the positive moment under the load is $\frac{5}{32}PL$, as shown in Fig. 7.10b. The beam still responds elastically everywhere but at the left sup-

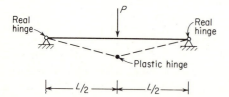

Real hinge

Real hinge

Plastic hinge

$L/2$ $L/2$

Fig. 7.9

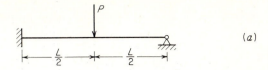

(a)

(b)

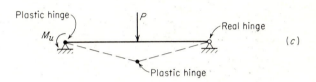

(c)

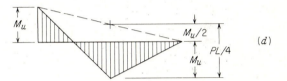

(d)

Fig. 7.10

port. At that point the actual fixed support can be replaced for purposes of analysis with a plastic hinge offering a known resisting moment M'_u. Because a redundant reaction has been replaced by a known moment, the beam is now determinate.

The load can be increased further until the moment under the load also becomes equal to M'_u, at which load the second hinge forms. The structure is converted into a mechanism, as shown in Fig. 7.10c, and collapse occurs. The moment diagram at collapse load is shown in Fig. 7.10d.

The magnitude of load causing collapse is easily calculated from the geometry of Fig. 7.10d.

$$M'_u + \frac{M'_u}{2} = \frac{PL}{4}$$

from which

$$(b) \quad P = P'_u = \frac{6M'_u}{L}$$

By comparison of Eqs. (b) and (a), it is evident that an increase in P of 12.5 per cent is possible, beyond that load which caused the formation of the first plastic hinge, before the beam will actually collapse. Due to the formation of plastic hinges, a redistribution of moments has occurred such that, at failure, the ratio between positive moment and negative moment is equal to that assumed in reinforcing the structure.

b. Rotation requirement   It may be evident that there is a direct relation between the amount of redistribution desired and the amount of inelastic rotation at the critical sections of a beam required to produce the desired redistribution. In general, the greater the modification of the elastic-moment ratio, the greater the required rotation capacity to accomplish that change. To illustrate, if the beam of Fig. 7.10a had been reinforced according to the elastic-moment diagram of Fig. 7.10b, no inelastic-rotation capacity at all would be required. The beam would, at least in theory, yield simultaneously at the left support and at midspan. On the other hand, if the reinforcement at the left support had been deliberately reduced (and the midspan reinforcement correspondingly increased), inelastic rotation at the support would be required before the strength at midspan could be realized.

The amount of rotation required at plastic hinges for any assumed moment diagram can be found by considering the requirements of compatibility. The member must be bent, under the combined effects of elastic moment and plastic hinges, so that the correct boundary conditions are satisfied at the supports. Usually, zero support deflection is to be maintained. The moment-area and conjugate-beam principles are useful in quantitative determination of rotation requirements (Ref. 7.9). In deflection calculations, it is convenient to assume that plastic hinging occurs at a point, rather than being distributed over a finite *hinging length*, as is actually the case. Consequently, in loading the conjugate beam with unit rotations, plastic hinges are represented as concentrated loads.

Calculation of rotation requirements will be illustrated by the two-span continuous beam of Fig. 7.11a. The elastic-moment diagram resulting from a single concentrated load is shown in Fig. 7.11b. The moment at support B is 0.096Pl, while that under the load is 0.182Pl. If the deflection of the beam at support C were calculated using the unit rotations equal to M/EI, based on this elastic-moment diagram, a zero result would be obtained.

Figure 7.11c shows an alternative, statically admissible moment diagram which was obtained by arbitrarily increasing the support moment from 0.096Pl to 0.150Pl and making the statically consistent reduction in span moment from 0.182Pl to 0.150Pl. If the beam deflection at C were calculated using this moment diagram as a basis, a nonzero value would be obtained. This indicates the necessity for inelastic rotation at one or more points to maintain geometric compatibility at the right support.

If the beam were reinforced according to Fig. 7.11c, increasing loads would produce the first plastic hinge at D, where the beam has been deliberately made understrength. Continued loading would eventually result in formation of the second plastic hinge at B, creating a mechanism and leading to collapse of the structure.

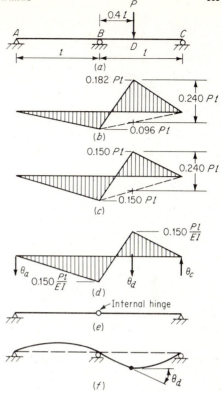

Fig. 7.11 Moment redistribution in two-span beam. (*a*) Loaded beam; (*b*) elastic moments; (*c*) modified moments; (*d*) M/EI loads; (*e*) conjugate beam; (*f*) deflection curve.

Limit analysis requires calculation of rotation at all plastic hinges up to, but not including, the last hinge which triggers actual collapse. Figure 7.11*d* shows the M/EI load to be imposed on the conjugate beam of Fig. 7.11*e*. Also shown is the concentrated angle change θ_d, which is to be evaluated. Starting with the left span, taking moments of the M/EI loads about the internal hinge of the conjugate beam at B, one obtains the left reaction of the conjugate beam (equal to the slope of the real beam):

$$\theta_a = 0.025 \frac{Pl^2}{EI}$$

With that reaction known, moments are taken about the support C of the conjugate beam and set equal to zero to obtain

$$\theta_d = 0.060 \frac{Pl^2}{EI}$$

This represents the necessary discontinuity in the slope of the elastic curve shown in Fig. 7.11*f* to restore the beam to zero deflection at the right support. The beam must be capable of developing at least that amount of plastic rotation if the modified moment diagram assumed in Fig. 7.11*c* is to be valid.

c. Rotation capacity   The capacity of concrete structures to absorb inelastic rotations at plastic-hinge locations is not unlimited. The designer adopting limit analysis in concrete must calculate not only the amount of rotation required at critical sections to achieve the degree of moment redistribution he has assumed, but also the rotation capacity of the members at those sections to ensure that it is adequate.

Curvature at initiation of yielding is easily calculated from the elastic strain distribution shown in Fig. 7.8b.

$$(a) \quad \psi_y = \frac{\epsilon_y}{d(1 - k)}$$

in which the ratio k establishing the depth of the elastic neutral axis is found from Eq. (3.41). The ultimate unit rotation may be obtained from the geometry of Fig. 7.8c.

$$(b) \quad \psi_u = \frac{\epsilon_{cu}}{c}$$

Although it is customary in flexural-strength analysis to adopt $\epsilon_{cu} = 0.003$, for purposes of limit analysis a more refined value is desirable. Extensive experimental studies (Ref. 7.10) indicate that the ultimate-strain capacity of concrete in the negative-moment region of continuous beams is strongly influenced by the support width b, by the moment-shear ratio M/V, and by the presence of additional reinforcement in the form of compression steel and "binding" steel (i.e., web reinforcement). The last parameter is conveniently introduced by means of a steel ratio ρ'', defined as the ratio of the volume of one stirrup plus its tributary compressive-steel volume to the concrete volume tributary to one stirrup. On the basis of empirical studies, the ultimate flexural strain at a plastic hinge is

$$(c) \quad \epsilon_{cu} = 0.003 + 0.02b\,\frac{V}{M} + 0.20\rho''$$

Based on Eqs. (a) to (c), the inelastic unit rotation for the idealized relation shown in Fig. 7.8a is

$$(d) \quad \psi_p = \psi_u - \psi_y \times \frac{M'_u}{M_y}$$

This plastic rotation is not confined to one cross section, but is distributed over a finite length referred to as the *hinging length*. The experimental studies upon which Eq. (c) is based measured strains and rotations in a length equal to the effective depth d of the test members. Consequently, ϵ_{cu} is an *average* value of ultimate strain over a finite length, and ψ_p, given by Eq. (d), is an *average* value of unit rotation. The total inelastic rotation θ_p can be found by multiplying the average unit rotation by the hinging length.

$$(e) \quad \theta_p = \left(\psi_u - \psi_y \times \frac{M'_u}{M_y} \right) l_p$$

On the basis of current evidence, it appears that the hinging length l_p in support regions, on either side of the support, can be approximated by the expression

$$(f) \quad l_p = 0.5d + 0.05z$$

in which z is the distance from the point of maximum moment to the nearest point of zero moment.

d. Moment redistribution under the ACI Code   Full utilization of the plastic capacity of reinforced-concrete beams and frames requires an extensive analysis of all possible mechanisms and an investigation of rotation requirements and capacities at all proposed hinge locations. The increase of design time may not be justified by the limited gains obtained. On the other hand, a restricted amount of redistribution of elastic moments can safely be made without complete analysis, yet may be sufficient to obtain most of the advantages of limit analysis.

A limited amount of redistribution is permitted under the ACI Code, depending upon a rough measure of available ductility, without explicit calculation of rotation requirements and capacities. The ratio ρ/ρ_b, or in the case of doubly reinforced members, $(\rho - \rho')/\rho_b$, is used as an indicator of rotation capacity, where ρ_b is the balanced steel ratio, given by Eq. (3.2). For singly reinforced members with $\rho = \rho_b$, experiments indicate almost no rotation capacity, since the concrete strain is nearly equal to ϵ_{cu} when steel yielding is initiated. Similarly, in a doubly reinforced member, when $(\rho - \rho') = \rho_b$, very little rotation will occur after yielding before the concrete crushes. However, when ρ or $(\rho - \rho')$ is low, extensive rotation is usually possible. Accordingly, the ACI Code provision reads:

> Except where approximate values for bending moments are used, the negative moments calculated by elastic theory at the supports of continuous flexural members for any assumed loading arrangement may each be increased or decreased by not more than $20 \left(1 - \dfrac{\rho - \rho'}{\rho_b} \right)$ per cent. These modified negative moments shall be used for calculation of the moments at sections within the spans. Such an adjustment shall be made only when the section at which the moment is reduced is so designed that ρ or $\rho - \rho'$ is equal to or less than $0.50\rho_b$.

Redistribution for steel ratios above $0.50\rho_b$ is conservatively prohibited.

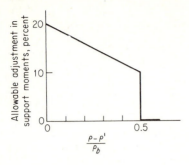

Fig. 7.12 Allowable moment redistribution under the ACI Code.

Code provisions are shown graphically in Fig. 7.12. To illustrate their application, and to demonstrate the advantage of moment redistribution when alternative loadings are involved, consider the concrete beam of Fig. 7.13. A three-span continuous beam is shown, with dead load of 1 kip per ft and live load of 2 kips per ft. To obtain maximum moments at all critical design sections, it is necessary to consider three alternative loadings. Case a, with live and dead load over exterior spans and dead load only over interior span, will produce maximum positive moment in the exterior spans. Case b, with dead load on exterior spans and dead and live load on the interior span, will produce the maximum positive moment in the interior span. The maximum negative moment over the interior support is obtained by placing dead and live load on the two adjacent spans and dead load only on the far exterior span, as shown in case c.

It will be assumed for simplicity that a 20 per cent adjustment of support moments is permitted throughout, provided span moments are modified accordingly. An overall reduction in design moments through the entire three-span beam may be possible. Case a, for example, produces an elastic maximum span moment in the exterior spans of 109 ft-kips. Corresponding to this is an elastic negative moment of 82 ft-kips at the interior support. Adjusting the support moment upward by 20 per cent, one obtains a negative moment of 98 ft-kips, which results in a downward adjustment of the span moment to 101 ft-kips.

Now consider case b: By a similar redistribution of moments, a reduced middle-span moment of 57 ft-kips is obtained through increase of the support moment from 78 to 93 ft-kips.

The moment obtained at the first interior support for loading case c can be adjusted in the reverse direction; i.e., the support moment is decreased by 20 per cent to 107 ft-kips. To avoid increasing the controlling span moment of the interior span, the right interior support moment is adjusted upward by 20 per cent to 80 ft-kips. The positive moments in the left exterior span and in the interior span corresponding to these modified support moments are 96 and 56 ft-kips, respectively.

It will be observed that the reduction obtained for the span moments in cases a and b were achieved at the expense of increasing the moment at the first

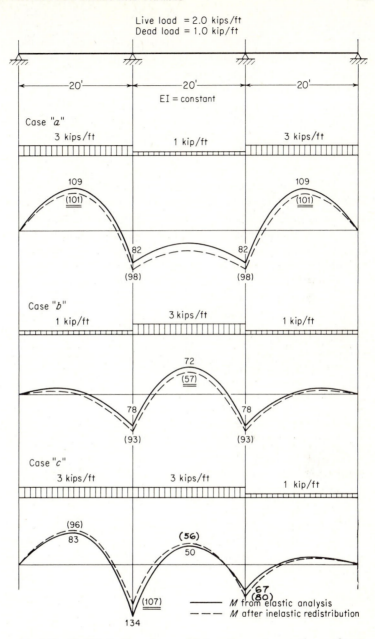

Fig. 7.13   Redistribution of moments in a three-span continuous beam.

interior support. However, the increased support moment in each case was less than the moment for which that support would have to be designed based on the loading c, which produced maximum support moment. Similarly, the reduction in support moment in case c was taken at the expense of an increase in span moments in the two adjacent spans. But in each case, the increased span moments were less than the maximum span moments obtained for other loading conditions. The final design moments at all critical sections are underlined in Fig. 7.13. It may be seen, then, that the net result is a reduction in design moments over the entire beam. This modification of moments does not mean a reduction in safety factor below that implied in code safety provisions; rather, it means a reduction of the *excess* strength which would otherwise be present in the structure because of the actual redistribution of moments that would occur before failure. It reflects the fact that the maximum design moments are obtained from alternative load patterns, which could not exist concurrently. The end result is a more realistic appraisal of the actual collapse load of the indeterminate structure.

REFERENCES

7.1. J. S. Kinney: "Indeterminate Structural Analysis," Addison-Wesley Publishing Company, Inc., Reading, Mass., 1957.
7.2. C. H. Norris and J. B. Wilbur: "Elementary Structural Analysis," 2d ed., McGraw-Hill Book Company, New York, 1960.
7.3. H. C. Martin: "Introduction to Matrix Methods of Structural Analysis," McGraw-Hill Book Company, New York, 1966.
7.4. M. F. Rubinstein: "Matrix Computer Analysis of Structures," Prentice-Hall, Inc., Englewood Cliffs, N.J., 1966.
7.5. H. I. Laursen: "Structural Analysis," McGraw-Hill Book Company, New York, 1969.
7.6. S. J. Fenves: "Computer Methods in Civil Engineering," Prentice-Hall, Inc., Englewood Cliffs, N.J., 1967.
7.7. ICES-STRUDL-II Engineering User's Manual, Massachusetts Institute of Technology, Dept. of Civil Engineering, Cambridge, Mass., 1968.
7.8. J. R. Benjamin: "Statically Indeterminate Structures," McGraw-Hill Book Company, New York, 1959.
7.9. G. C. Ernst: A Brief for Limit Design, *Trans. ASCE*, vol. 121, pp. 605–632, 1956.
7.10. A. H. Mattock: Rotation Capacity of Hinging Regions in Reinforced Concrete Frames, Proceedings of International Symposium on Flexural Mechanics of Reinforced Concrete, Miami, Fla., 1964, ACI Publ. SP-12.

8
Reinforced-Concrete Buildings

8.1 INTRODUCTION

In the conceptual stages of the design of reinforced-concrete structures, it should be borne in mind that reinforced concrete is a material with its own unique features and characteristics. To use concrete in structural systems which are patterned after articulated structures of steel or timber betrays a lack of design imagination and denies the monolithic nature of concrete construction. From a practical point of view, such imitative design will usually fail to realize the true potential of the material and will often result in highly uneconomical structures. The most successful designs take full advantage of the special features of concrete construction. Two fine examples of good building design in concrete are shown in Figs. 8.1 and 8.2.

The advantages of reinforced-concrete construction for buildings are many, but the more significant points are summarized as follows:

Versatility of form Usually placed in the structure in the fluid state, the material is readily adaptable to a wide variety of architectural and functional requirements.

Fig. 8.1 Toronto City Hall, combining reinforced, precast, and prestressed concrete. (*Architect-engineers: Viljo Revell, John B. Parkin Assoc.*)

Durability With proper concrete protection of the steel reinforcement, the structure will have long life even under highly adverse climatic or environmental conditions.

Fire resistance With proper protection for the reinforcement, a reinforced-concrete structure provides the maximum in fire protection.

Fig. 8.2 Lake Point Tower in Chicago. With a height of 645 ft, having 70 stories, it is presently the world's tallest reinforced-concrete building.

Speed of construction In terms of the entire period from the date of approval of the contract drawings to the date of completion, a concrete building may often be completed in less time than a steel structure. Although the field erection of a steel building is more rapid, this phase must necessarily be preceded by prefabrication of all parts in the shop.

Cost   In many cases the first cost of a concrete structure is less than that of a comparable steel structure. In almost every case, maintenance costs are less.

Availability of labor and material   It is always possible to make use of local sources of labor, and in many inaccessible areas a nearby source of good aggregate can be found, so that only the cement and reinforcement need be brought in from a remote source.

8.2 FLOOR AND ROOF LOADS

The minimum live loads for which the floors and roof of any building should be designed are usually specified in the building code that governs at the site of construction. There are local municipal codes, state codes, and regional codes, as well as "model" codes proposed for general usage. Representative values of minimum live loads to be used in the design of a wide variety of buildings are found in Building Code Requirements for Minimum Design Loads in Buildings and Other Structures (Ref. 8.1), a portion of which is reprinted in Table 8.1.

In addition to these uniformly distributed loads, it is recommended, as an alternative to the uniform load, that floors be designed to support safely certain concentrated loads, if these produce a greater stress. For example, office floors are to be designed to carry a load of 2000 lb distributed over an area $2\frac{1}{2}$ ft square, to allow for the weight of a safe or other heavy equipment, and stair treads must safely support a 300-lb load applied on the center of the tread. The specified uniform and concentrated loads are assumed to include adequate allowance for ordinary impact loading. Certain reductions are permitted in live load of 100 psf or less for members supporting 150 ft² of floor or more, on the premise that it is not likely that the entire area would be fully loaded at one time. The reduction specified is at the rate of 0.08 per cent per square foot of area supported by the member, except that no reduction is permitted in places of public assembly. The reduction is not to exceed 60 per cent nor the percentage

$$100\,\frac{D + L}{4.33L}$$

in which D is the dead load per square foot of area supported by the member, and L is the live load. No reduction is permitted for loads over 100 psf (characteristic of warehouses and factories, in which cases it is not unlikely that the entire area may be loaded at one time), except that column loads may be reduced 20 per cent.

The specified minimum live loads cannot always be used. The type of occupancy should be considered, and the probable loads computed as accurately as possible. Warehouses for heavy storage may be designed for loads as high as 500 psf or more; unusually heavy operations in manufacturing build-

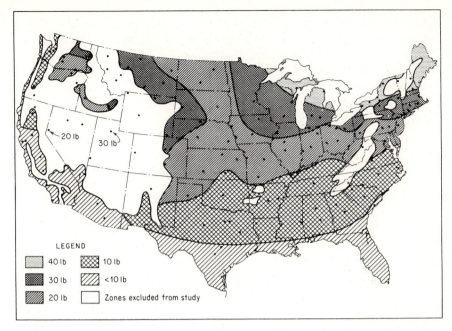

Fig. 8.3 Estimated weight of seasonal snowpack equaled or exceeded one year in ten. (*From Ref. 8.1, courtesy USA Standards Institute.*)

ings may require a large increase in the 125-psf value specified in Table 8.1; special provision must be made for all definitely located heavy concentrated loads.

In all cases, the dead weight of the floor must be included in the total design load. When plastered ceilings are specified, an additional allowance of about 10 psf is usual. Floor fill and finish floors, together with suspended loads such as piping and lighting fixtures, also must be included. If the locations of partitions are not definitely established, an extra allowance of from 10 to 20 psf may be in order.

Snow loads for the design of building roofs are likewise generally specified by local codes. They are preferably based on statistical studies, over an extended period, of the local depth and density of snowfall. Such a set of data has resulted in the snow-load map shown in Fig. 8.3. A minimum roof load of 20 psf is specified to provide for construction and repair loads and to ensure reasonable stiffness. More detailed information on snow loads is usually available for any particular locality.

8.3 HORIZONTAL LOADS

Horizontal loads on building frames are caused primarily by wind pressure. In addition, earthquake shocks produce horizontal sway, which results in inertia

**Table 8.1   Minimum uniformly distributed live loads
(from Ref. 8.1, courtesy of USA Standards Institute)**

Occupancy or use	Live load, psf	Occupancy or use	Live load, psf
Apartments (see Residential)		Residential:	
Armories and drill rooms	150	Multifamily houses:	
Assembly halls and other places of assembly:		Private apartments	40
		Public rooms	100
Fixed seats	60	Corridors	60
Movable seats	100	Dwellings:	
Balcony (exterior)	100	First floor	40
Bowling alleys, poolrooms, and similar recreational areas	75	Second floor and habitable attics	30
		Uninhabitable attics	20
Corridors:		Hotels:	
First floor	100	Guest rooms	40
Other floors, same as occupancy served except as indicated		Public rooms	100
		Corridors serving public rooms	100
Dance halls	100	Public corridors	60
Dining rooms and restaurants	100	Private corridors	40
Dwellings (see Residential)		Reviewing stands and bleachers	100
Garages (passenger cars)	100	Schools:	
Floors shall be designed to carry 150 per cent of the maximum wheel load anywhere on the floor.		Classrooms	40
		Corridors	100
		Sidewalks, vehicular driveways, and yards subject to trucking	250
Grandstands (see Reviewing stands)		Skating rinks	100
Gymnasiums, main floors, and balconies	100	Stairs, fire escapes, and exitways	100
Hospitals:		Storage warehouse:	
Operating rooms	60	Light	125
Private rooms	40	Heavy	250
Wards	40	Stores:	
Hotels (see Residential)		Retail:	
Libraries:		First floor, rooms	100
Reading rooms	60	Upper floors	75
Stack rooms	150	Wholesale	125
Manufacturing	125	Theaters:	
Marquees	75	Aisles, corridors, and lobbies	100
Office buildings:		Orchestra floors	60
Offices	80	Balconies	60
Lobbies	100	Stage floors	150
Penal institutions:		Yards and terraces, pedestrians	100
Cell blocks	40		
Corridors	100		

forces acting horizontally on the structure. It was found by extensive observation that these horizontal forces, rather than those induced by the simultaneous vertical motion of the ground, are chiefly responsible for structural earthquake damage. For this reason building codes in localities with high earthquake incidence require that structures be designed to resist corresponding horizontal forces.

The accurate estimation of wind and earthquake loads is a very difficult matter. An excellent summary of current thinking and improved methods for predicting the magnitude of such loads is found in Ref. 8.2. Much useful information pertaining specifically to earthquake effects is found in Ref. 8.3. Most building codes presently specify design wind pressures per square foot of vertical wall surface. Depending upon locality, these vary from about 10 to 50 psf. In dealing with forces due to seismic action, present codes generally specify that buildings in zones of high earthquake probability be designed to resist horizontal forces which are estimated as stipulated percentages of the total load of the buildings above the given floor. These are of the order of 10 per cent of that load.

For wind pressures, it is customary to compute the load on a joint from the area of exterior wall surface pertaining to it, i.e., from the rectangle outlined by vertical center lines between adjacent frames and by horizontal center lines between floor levels. Earthquake forces are distributed to individual joints in similar ways. These distributions are not very exact, particularly if the external forces do not act in the planes of the frames and if the outline of the structure is irregular in plan or elevation. The actual distribution of wind forces, in particular, is frequently very different from the assumed uniform pressure on the surface of the windward walls.

Buildings resist horizontal loads by different means, depending upon their type and form. In older buildings, with massive bearing walls, such forces are resisted by these walls acting as cantilever beams (loaded in the plane of the wall). Similar action is obtained in modern reinforced-concrete buildings containing integral shear walls. Many modern frame buildings include only light-weight curtain walls, nonstructural in nature, which cannot contribute in a significant way to the lateral resistance of a structure. In such cases, horizontal loads must be resisted by the frame itself. In all cases, floor and roof slabs serve the important function of transmitting horizontal loads to the joints of the frame or to shear-resistant vertical walls. They act as relatively rigid diaphragms, tending to equalize the lateral deflection of all frames or walls at a given level.

For combinations of dead and live load plus wind or earthquake loads, it must be recognized that the coincidence of full horizontal force with the most unfavorably distributed live load is extremely unlikely. Also, the rarity and temporary nature of the maximum horizontal loads are likely to result in less damage than if those same forces were of great frequency and extended duration. Accordingly, most building codes treat wind and earthquake stresses as somewhat secondary effects, in comparison with the main effects of dead and live load. Under the ACI Code, for example, the usual load factors are reduced by 25 per cent when wind or earthquake loads are combined with dead and live loads.

A high degree of uncertainty prevails regarding the magnitude as well as the distribution of wind and earthquake forces. It is useless, in analysis, to aim at an accuracy greater than that of the load data and other assumptions. It follows that highly refined analysis for horizontal load effects is often not justified, and the use of the approximate methods described in Art. 7.4d is perfectly satisfactory.

REINFORCED-CONCRETE FLOORS AND ROOFS

8.4 TYPES OF CONSTRUCTION

The types of concrete floor and roof systems are so numerous as to defy concise classification. In steel construction the designer is usually limited to the utilization of structural plates and shapes which have been standardized in form and size by the relatively few producers in the field. In reinforced concrete, on the other hand, the engineer has more control over the form of the structural components of a building. In addition, many small producers of reinforced-concrete structural elements and construction accessories can compete profitably in this field, since plant and equipment requirements are not excessive. This has resulted in the development of a wide variety of concrete systems. Only the more common types can be mentioned in this text.

In general, the commonly used reinforced-concrete floor and roof systems can be classified as follows:

1. One-way reinforcing systems (in which the main reinforcement in each structural element runs in one direction)
   a. One-way solid slabs
      (1) Slab supported on monolithic concrete beams and girders
      (2) Slab supported on steel beams (shear connectors may be used to provide for composite action)
      (3) Slab with light-gage steel decking which serves as form and reinforcement for the concrete
   b. One-way ribbed slabs (also known as concrete joist floors)
   c. Precast systems, including precast slab or beams or both
2. Two-way reinforcing systems (in which the main reinforcement in at least one of the structural elements runs in two directions)
   a. Two-way solid slabs
      (1) Slab on monolithic concrete beams
      (2) Slab on steel beams
   b. Two-way ribbed slab on concrete or steel beam supports

    c. Beamless slabs
        (1) Flat slab with drop panels or column capitals or both
        (2) Flat plate with no drop panels or column capitals (may be precast
            and lifted to final elevation)
        (3) Ribbed slab

Each of these types is described in the following articles.

8.5 MONOLITHIC BEAM–AND–GIRDER CONSTRUCTION

A solid-slab beam-and-girder floor consists of a series of parallel beams supported at their extremities by girders, which in turn frame into concrete
columns placed at more or less regular intervals over the entire floor area. This
framework is covered by a one-way reinforced-concrete slab, the load from
which is transmitted first to the beams and then to the girders and columns.
The beams are usually spaced so that they come at the midpoints, at the third
points, or at the quarter points of the girders, as shown in Fig. 8.4. The arrangement of beams and spacing of columns should be determined by economical and
practical considerations. These will be affected by the use to which the building
is to be put, the size and shape of the ground area, and the load which must be
carried. A comparison of a number of trial designs and estimates should be
made if the size of the building warrants, and the most satisfactory arrangement selected. Since the slabs, beams, and girders are built monolithically,
the beams and girders are designed as T beams, and advantage is taken of
continuity.
    Beam-and-girder floors, as they are usually called, are adapted to any
loads and to any spans that might be encountered in ordinary building con-

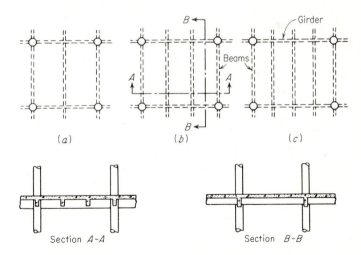

Fig. 8.4 Framing of beam-and-girder floors.

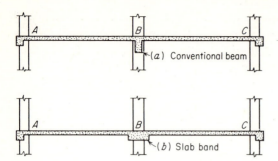

Fig. 8.5 Comparison of sections through conventional and slab-band floors.

struction. The normal maximum spread in live-load values is from 40 to 400 psf, and the normal range in column spacings is from 16 to 32 ft. A complete design of a typical beam-and-girder floor panel is given in Arts. 8.23 to 8.26.

In normal beam-and-girder construction the depth of a beam is about twice its stem width. For light loads, however, it may be economical to omit the intermediate beams and deep column-line girders entirely and to have the one-way slab supported by wide, shallow beams which are centered on the column lines and which frame directly into the columns. The interior beam becomes little more than a thickened portion of the slab; this form of construction, in fact, is commonly referred to as *slab-band* construction. There are a number of advantages associated with its use. In the slab span, a haunch is created which aids in resisting the relatively high negative moments that will be present over the interior support. This will reduce both the positive slab moment and the critical negative slab moment at the face of the haunch. Although beam steel will be increased because of the reduced effective depth as compared with a beam of normal proportions, cost comparisons have shown that this may often be outweighed by savings in slab steel. Other advantages include a reduced depth of construction, permitting a reduction in overall height of the building, and greater flexibility in locating columns. Formwork is simplified because of the reduction in the number of framing members. In Fig. 8.5 are shown comparative sections through conventional and slab-band floors.

8.6 COMPOSITE CONSTRUCTION WITH STEEL BEAMS

One-way reinforced-concrete slabs are also frequently used in structural-steel framed buildings. Spacing of the steel beams is usually about 6 to 8 ft. Often the slab is poured so that its underside is flush with the underside of the top flange of the beam, as shown in Fig. 8.6a. This simplifies construction to a certain extent, since the plywood slab form can be wedged between the upper and lower flanges of the steel beam. Of even greater significance is the fact that positive bracing against lateral buckling is provided for the compressive flange of the beam, whereas if the bottom of the slab were flush with the top of the

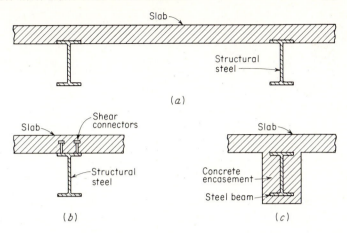

Fig. 8.6 Slab floors with steel beams.

steel, one would have to rely upon adhesion between the concrete and steel to prevent lateral movement.

Increasingly in recent years buildings are being designed for composite action. Shear connectors are welded to the top of the steel beam and are embedded in the concrete slab, as shown in Fig. 8.6b; they prevent relative movement between the concrete and steel in the direction of the beam axis. The entire cross section, consisting of the concrete slab and the steel beam, then acts to resist longitudinal flexural stress. Shallower sections can be used which are nevertheless stiffer than their deeper, noncomposite counterparts. The design of composite steel-concrete members is discussed in detail in Chap. 12.

If it is necessary to fireproof a steel frame building, a form may be built around the beam, and the steel encased in concrete as shown in Fig. 8.6c. Although some composite action undoubtedly exists, this may not be reliable for various reasons, such as the presence of a paint film between the concrete and steel, and the contribution of the concrete is usually disregarded. As an alternative to the solid encasement of the steel, it may be wrapped with expanded-metal lath or wire mesh, which is then troweled or sprayed with a fireproofing vermiculite plaster, affording a substantial reduction in dead load.

8.7 COMPOSITE CONSTRUCTION WITH LIGHT–GAGE STEEL DECKING

In recent years considerable use has been made of light-gage steel deck covered with a concrete slab for floors and roofs in steel-frame office and apartment buildings. An example of this type of construction is shown in Fig. 8.7. The steel deck in this case serves a double purpose: (1) it provides a form for the wet concrete, eliminating the necessity for temporary wood forms; and (2) it provides tensile reinforcement for the hardened slab. Often a light welded

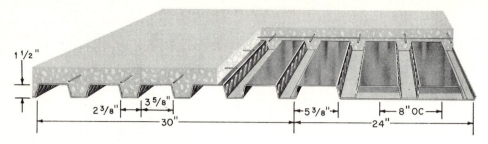

Fig. 8.7 Composite concrete slab with light-gage steel-deck reinforcing. (*Courtesy Inland-Ryerson Construction Products Co.*)

wire mesh is used in conjunction with the steel deck, to reinforce for shrinkage and temperature stresses in the direction perpendicular to the deck ribs. In one case, transverse wires are welded to the ridges of corrugated sheet steel, reinforcing shrinkage and temperatures stress as well as improving bond between the panels and the concrete. In some installations, additional negative steel bars are placed over supports; in other cases, the temperature mesh, although not adequate to reinforce for full negative moment, will at least control the cracking which will occur over supports.

The design of steel-deck-reinforced concrete slabs follows the same principles as that of any one-way slab. The steel is assumed to be concentrated at the level of the centroid of the deck cross section. The question of bond between the deck and the concrete becomes of particular importance, and precautions must be taken to ensure that the panels are free of oil film or dirt prior to pouring the slab.

8.8 ONE-WAY RIBBED SLABS

A ribbed floor consists of a series of small, closely spaced reinforced-concrete T beams, framing into girders, which in turn frame into the supporting columns. The T beams (called *joists* or *ribs*) are formed by creating void spaces in what otherwise would be a solid slab. The voids may be formed using wood or specially waterproofed cardboard, but usually special steel pans are used. The girders which support the joists are built as regular T beams.

Since the strength of concrete in tension is small and is commonly neglected in design, elimination of much of the tension concrete in a slab by the use of pan forms results in a saving of weight with little alteration in the structural characteristics of the slab. Ribbed floors are economical for buildings such as apartment houses, hotels, and hospitals, where the live loads are fairly small and the spans comparatively long. They are not suitable for heavy construction, such as in warehouses, printing plants, and heavy manufacturing buildings.

Steel pan forms are available in depths of 6, 8, 10, 12, and 14 in. They are tapered in cross section, as shown in Fig. 8.8, to facilitate removal and

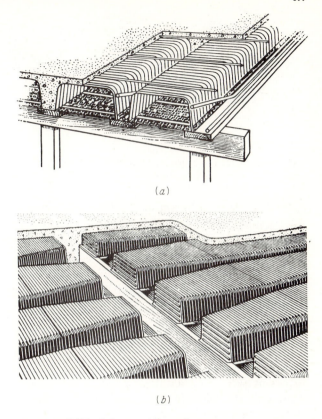

(a)

(b)

Fig. 8.8   Ribbed floors with steel pans.

are usually provided with a bottom width of 20 in., although 10- and 30-in. pans are available. Any width of joist may be obtained by varying the width of the soffit form. Figure 8.8a and b shows steel pan forms in place on wooden falsework. The tapered end pans shown in Fig. 8.8b are used where it is desired to obtain a wider joist width near the end supports, because of high shear or negative bending moment.

Concrete is placed on the soffit boards in the space between rows of steel pans, and above the pans to form a slab 2 in. or more thick.[1] After the concrete has hardened, the steel pans are generally removed for reuse. The reinforcement of the joists usually consists of two bars, one bent and one straight. The slab is reinforced, primarily for temperature and shrinkage stresses, with wire mesh or small bars placed at right angles to the joists; the area of the reinforcement is usually about 0.18 per cent of the cross-sectional area of the slab.

[1] The ACI Code specifies a minimum slab thickness of 2 in., but not less than one-twelfth the clear distance between joists.

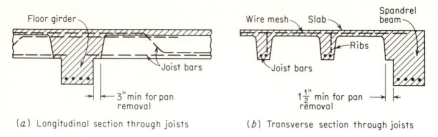

(a) Longitudinal section through joists   (b) Transverse section through joists

Fig. 8.9 Ribbed-floor cross sections.

The girders which support the joists are rectangular beams or T beams with a maximum flange thickness equal to the total floor thickness, as shown in Fig. 8.9. The rows of pans are stopped at the proper distance from the stem of the girder. End caps are used at each end of each row of pans to form the sides of the girder flange. Figure 8.9 shows transverse and longitudinal cross sections of a typical ribbed-slab floor. A complete ribbed-slab floor design is given in Arts. 8.27 to 8.30.

8.9 PRECAST–CONCRETE SYSTEMS

Precast concrete has accounted for a significant portion of building construction in recent years. Although in many cases prestressing is combined with precasting, this is not always so. Many manufacturers produce standard precast members which utilize ordinary reinforcing bars.

Precast-concrete units may be produced as stock items, adaptable for use under a wide variety of conditions and for many jobs, or they may be produced to a specific shape for a specific use. In the first category are single- and double-T sections, channel slabs, and various types of hollow-core rectangular members, as well as standard I sections. In other cases, special members may be designed for use as exterior facades, grillages, etc.

Precast units are designed by standard reinforced-concrete theory. In some cases, a relation of the usual requirements for concrete cover around bars may be justified when the units are made under carefully controlled factory methods. Higher-strength concrete is more easily obtained under these conditions also. In all cases, special attention must be given to details at connections, so that the required forces may be transferred, and so that the resulting structure has the necessary resistance to lateral forces.

Precast construction is treated in detail in Chap. 9.

8.10 TWO–WAY SLABS WITH EDGE BEAMS

Two-way solid slabs supported by beams on all four sides, with main reinforcement running in two directions, have been discussed in detail in Chap. 4.

The perimeter beams may be either concrete, poured monolithically with the slab, or structural steel, often encased in concrete. Such two-way systems are suitable for intermediate and heavy loads on spans up to about 30 ft. The range of usefulness of two-way beam-supported slabs corresponds generally to that of beamless slabs with drop panels and column capitals, described below. The latter are often preferred because of the complete elimination of projecting beam webs below the slab.

8.11 BEAMLESS SLABS WITH DROP PANELS OR COLUMN CAPITALS

By suitably proportioning and reinforcing the slab, it is possible to eliminate supporting beams altogether. This type of construction, described in detail in Chap. 4, is often referred to as *flat-slab* construction (although almost all slabs dealt with in ordinary practice are flat). The slab is supported directly on columns. Figure 8.10 shows a typical flat-slab floor. Note the thickened portion of the slab, termed a *drop panel*, at the columns, and the enlarged sections at the tops of the columns, called *column capitals*. Both the drop panels and column capitals serve the double purpose of reducing shear stress

Fig. 8.10   Flat-slab garage floor. (*Courtesy Portland Cement Assoc.*)

in the slab near the column supports and providing greater effective depth for negative bending moment.

In general, flat-slab construction is economical for live loads of 100 psf or more and for spans up to about 30 ft. For lighter loads, such as are found in apartment houses, hotels, and office buildings, some form of ribbed-floor construction will usually be less expensive than a flat-slab floor, although in recent years flat slabs have been used economically for a wide range of loading. For spans longer than about 30 ft, beams and girders are desirable because of improved stiffness.

8.12 FLAT-PLATE CONSTRUCTION

A flat-plate floor is essentially a flat-slab floor with the drop panels and column capitals omitted, so that a floor slab of uniform thickness is carried directly by prismatic columns. Flat-plate floors have been found economical and otherwise advantageous for such uses as apartment buildings, as shown in Fig. 8.11, where the spans are moderate and the loads relatively light. Construction depth for each floor is held to the absolute minimum, with resultant savings

Fig. 8.11 Flat-plate floor construction. (*Courtesy Portland Cement Assoc.*)

Fig. 8.12 Student dormitory at Clemson College, South Carolina. Lift-slab construction used with flat-plate floors and roof.

in the overall height of the building. The smooth underside of the slab can be painted directly and left exposed for ceiling, or plaster can be applied to the concrete.

Certain problems are inherent in flat-plate construction. One of these is the fact that shear stresses in the concrete may be very high near the columns.

Often special reinforcement is required in this area in the form of steel-beam sections or extra bar reinforcement (see Art. 4.8b). Problems have arisen with excessive longtime deflections of the relatively thin slabs in some installations of this type.

Many flat-plate buildings are constructed by the lift-slab method, shown in Fig. 8.12. A casting bed (often doubling as the ground-floor slab) is poured, steel columns are erected and braced, and at ground level successive slabs, which will later become the upper floors, are cast. A membrane or sprayed parting agent is laid down between successive pours so that each slab can be lifted in its turn. Jacks placed atop the columns are connected to threaded rods extending down the faces of the columns and connecting, in turn, to lifting collars embedded in the slabs, as shown in Fig. 4.14d. When a slab is in its final position, the collar is welded to the column. Lifting collars such as that shown in Fig. 4.14d, in addition to providing anchorage for the lifting rods, serve to increase the effective size of the support for the slab, and consequently to improve the shear strength of the slab.

8.13 TWO-WAY RIBBED SLABS

As in one-way slab systems, the dead weight of two-way slabs can be reduced considerably by creating void spaces in what would otherwise be a solid slab. For the most part, the concrete removed is in tension and ineffective, so that the lighter floor will have virtually the same structural characteristics as the corresponding solid floor. Voids are usually formed using dome-shaped steel forms (Fig. 8.13), which are removed after the concrete slab has hardened. The lower flange of each dome contacts that of the adjacent dome, so that the concrete is poured entirely against a metal surface, resulting in good finished appearance of the slab. A wafflelike appearance, which has in many instances been used to architectural advantage, is imparted to the underside of the floor or roof (see Fig. 8.14). Because of their appearance, two-way ribbed slabs are often termed grid slabs or waffle slabs.

Void spaces formed by dome-shaped inserts may be incorporated with beam-supported two-way slabs, as well as with beamless flat slabs and flat

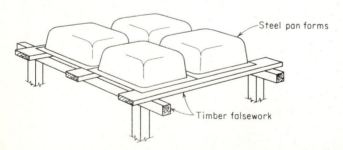

Fig. 8.13 Forms for two-way ribbed slab.

Fig. 8.14 Regency House Apartments, San Antonio, Texas. Cantilevered grid slab used with column-line beams.

plates. In the latter case, dome forms are commonly omitted near the columns, providing a solid slab better able to resist negative bending moment and high shear in that region.

8.14 FLOOR SURFACES

Many types of wearing surfaces are used on concrete floors. Concrete topping $\frac{3}{4}$ to 2 in. thick is frequently used. Such topping may be either monolithic or bonded. Monolithic finishes should be placed within 45 min after the base slab is poured; bonded finishes are placed on a fully hardened base slab. To obtain an effective bond in the latter case, the base slab should be rough. The value of epoxy cement to improve the bond between new and old concrete is well established. Mix proportions for the topping surface vary, and quite often an artificial hardening agent is added.[1]

[1] See Portland Cement Association bulletins "Surface Treatments for Concrete Floors" and "Specifications for Heavy-duty Floor Concrete Floor Finish" for more complete coverage of this important topic.

A wooden floor may be provided for, if desired, by embedding beveled nailing strips or "sleepers," usually 2 by 4's laid flat, in a layer of cinder concrete on top of the main slab. Small vitrified-clay flat tiles embedded in a 2-in. layer of portland cement mortar also provide a satisfactory floor surface. Asphalt or linoleum tile placed over a smooth cement base is frequently used.

The durability of any wearing surface depends in a large measure upon the method of placing the surface. If not properly constructed, the concrete finish might spall, the wood floor "dry-rot," the tiles curl, and the linoleum crack. To ensure the maximum degree of serviceability from a given type of floor surface, a special study of methods which have proved successful for laying that particular type of floor should be made.

8.15 CONCRETE ROOFS

The design of roofs is similar to the design of floors. In addition to the structural requirements, however, the roofs must be impervious to the passage of water, provide for adequate drainage, and furnish protection against condensation.

To provide adequate drainage, the roof slab may be pitched slightly, or a filling of some light material such as cinder concrete, covered with a suitable roofing material, may be used, the thickness of the filling being varied so as to give the required slope to the roof surface. The amount of slope required for drainage will depend upon the smoothness of the exposed surface. A value of $\frac{1}{8}$ in. per ft might be used with a surface of hard tile. Felt and gravel roofs should have a pitch of at least $\frac{1}{4}$ in. per ft. Some form of flashing is required along the parapets to prevent the drainage from seeping into the building at the edges of the roof slab.

Condensation may best be guarded against by proper ventilation and insulation. The form of insulation to be used will depend upon the particular class of building under consideration.

Imperviousness may best be provided for by the application of some form of separate roof covering, such as multiple layers of treated felt cemented together and to the slab by means of coal-tar pitch or asphalt, and overlaid with gravel, vitrified tile embedded in asphalt, or any of the commercial types of built-up roofings. Tin, corrugated iron, or copper roofings are sometimes placed on reinforced-concrete buildings but are usually more expensive than other types of coverings.

WALLS AND PARTITIONS

8.16 PANEL AND CURTAIN WALLS

As a general rule, the exterior walls of a reinforced-concrete building are supported at each floor by the skeleton framework, their only function being to enclose the building. Such walls are called *panel* walls. They may be made of concrete (often precast), cinder concrete block, brick, tile blocks, or insulated metal panels. The latter may be faced with aluminum, stainless steel, or a porcelain-enamel finish over steel, backed by 1 or 2 in. of insulating material and an inner surface sheathing. The thickness of each of these types of panel walls will vary according to the material, type of construction, climatological conditions, and the building requirements governing the particular locality in which the construction takes place.

The pressure of the wind is usually the only load that is considered in determining the structural thickness of a wall panel, although in some cases exterior walls are used as diaphragms to transmit forces caused by horizontal loads down to the building foundations.

Curtain walls are similar to panel walls except that they are not supported at each story by the frame of the building, but are self-supporting. However, they are often anchored to the building frame at each floor to provide lateral support.

8.17 BEARING WALLS

A bearing wall may be defined as one which carries any vertical load in addition to its own weight. Such walls may be constructed of stone masonry, brick, concrete block, or reinforced concrete. Occasional projections or pilasters add to the strength of the wall and are often used at points of load concentration. In small commercial buildings, bearing walls may be used with economy and expediency. In larger commercial and manufacturing buildings, when the element of time is an important factor, the delay necessary for the erection of the bearing wall and the attending increased cost of construction often dictate the use of some other arrangement.

Bearing walls may be of either single or double thickness, the advantage of the latter type being that the air space between the walls renders the interior of the building less liable to temperature variations and makes the wall itself more nearly moistureproof. On account of the greater gross thickness of the double wall, such construction reduces the available floor space. This feature is often sufficient in itself to warrant the selection of the solid wall unless the factors of condensation and temperature are of great importance. Cavity wall construction is usually limited to a total height of about 40 ft.

The thickness of bearing walls varies with the height. The New York City Building Code requires that, for bearing walls made of solid masonry units (brick, stone, concrete brick, etc.) and not more than 75 ft in height, the thickness of the uppermost 55 ft of height shall be at least 12 in., and the thickness below this shall be at least 16 in., except that for walls in buildings of one, two, and three stories the top-story thickness in each case may be 8 in. In computing wall thicknesses, a maximum story height of 13 ft shall be assumed.

According to the ACI Code, the load capacity of a wall is given by

$$P_u = 0.55\phi f_c' A_g \left[1 - \left(\frac{l_c}{40h} \right)^2 \right]$$

(8.1)

where A_g = gross area of section, in.²
  l_c = vertical distance between supports, in.
  h = thickness of wall, in.
  ϕ = 0.70

In the case of concentrated loads, the length of the wall to be considered as effective for each shall not exceed the center-to-center distance between loads, nor shall it exceed the width of the bearing plus 4 times the wall thickness. Reinforced-concrete bearing walls shall have a thickness of at least 1/25 the unsupported height or width, whichever is shorter. Reinforced-concrete bearing walls of buildings shall be not less than 6 in. thick for the uppermost 15 ft of their height; for each successive 25 ft downward, or fraction thereof, the minimum thickness shall be increased 1 in.

The area of horizontal reinforcement shall be not less than 0.0025 and that of vertical reinforcement not less than 0.0015 times the area of the concrete. These values may be reduced to 0.0020 and 0.0012, respectively, if the reinforcement is not larger than $\frac{5}{8}$ in. in diameter and consists of either welded wire fabric or deformed bars with f_y of 60 ksi or greater.

8.18 BASEMENT WALLS

In determining the thickness of basement walls, the lateral pressure of the earth, if any, must be considered in addition to other structural features. If it is part of a bearing wall, the lower portion may be designed either as a slab supported by the basement and first floors or as a retaining wall, depending upon the type of construction. If columns and wall beams are available for support, each basement wall panel of reinforced concrete may be designed to resist the earth pressure as a simple slab reinforced in either one or two directions.

A minimum thickness of 8 in. is generally specified for reinforced-concrete basement walls. In wet ground a minimum thickness of 12 in. should be used. In any case, the thickness cannot be less than that of the wall above.

Care should be taken to brace a basement wall thoroughly from the inside if the earth is backfilled before the wall has obtained sufficient strength to resist the lateral pressure without such assistance or if it is placed before the first-floor slab is in position.

8.19 PARTITION WALLS

Interior walls used for the purpose of subdividing the floor area may be made of cinder block, brick, precast concrete, metal lath and plaster, clay tile, or metal. The type of wall selected will depend upon the fire resistance required, flexibility of rearrangement, ease with which electrical conduits, plumbing, etc., can be accommodated, and architectural requirements.

STAIRWAYS

8.20 TYPES OF CONCRETE STAIRS

The simplest form of reinforced-concrete stairway consists of an inclined slab supported at the ends upon beams, with steps formed upon its upper surface. Such a stair slab is usually designed as a simple slab with a span equal to the horizontal distance between supports. This method of design requires steel to be placed only in the direction of the length of the slab. Transverse steel, usually one bar to each tread, is used only to assist in distribution of the load and to provide temperature reinforcement. It sometimes becomes necessary to include a platform slab at one or both ends of the inclined slab. Many successful designs made as outlined above for the simple inclined slab indicate that the effect of the angle that occurs in a slab of this type can safely be disregarded, although reinforcement should be carefully detailed.

It is advisable to keep the unsupported span of a stair slab reasonably short. If no break occurs in the flight between floors, intermediate beams, supported by the structural framework of the building, may be employed. If the stair between floors is divided into two or more flights, beams as described above may be used to support the intermediate landing, with these in turn supported as above for the long straight flight, or the intermediate slab may be suspended from a beam at the upper floor level by means of rod hangers. Where conditions permit, the intermediate slab may be supported directly by the exterior walls of the building.

8.21 BUILDING-CODE REQUIREMENTS

The required number of stairways, and many of the details, are governed to a large extent by the provisions of the governing building code. Among other things, these provisions stipulate the maximum distance from the most remote point in the floor area to the stairway, the minimum width of stairway, the maximum height of any straight flight, the maximum height (or rise) of a single step, the minimum distance (the run) between the vertical faces of two consecutive steps, and the required relation between the rise and the run to give safety and convenience in climbing.

In most codes the minimum width of any stair slab and the minimum dimension of any landing are about 44 in. The maximum rise of a stair step is usually specified as about $7\frac{3}{4}$ in., and the minimum run or tread width, exclusive of nosing, $9\frac{1}{2}$ in. A rise of less than $6\frac{1}{2}$ in. is not considered generally satisfactory. In order to give a satisfactory and comfortable ratio of rise to run, various rules have been adopted. One requires that for steps without nosings, the sum of the rise and run shall be $17\frac{1}{2}$ in., but the rise shall not be less than $6\frac{1}{2}$ or more than $7\frac{3}{4}$ in.

The maximum height of a straight flight between landings is generally 12 ft, except for stairways serving as exits from places of assembly, where a maximum of 8 ft is normally specified.

The number of stairways is governed by the width of the stair slab, the number of probable occupants on each floor, and the dimensions of the floor area. One typical code specification is that the distance from any point in an open floor area to the nearest stairway or exit shall not exceed 100 ft, that the corresponding distance along corridors in a particular area shall not exceed 125 ft, and that the combined width of all stairways in any story shall be such that the stairs may accommodate at one time the total number of persons occupying the largest floor area served by such stairs above the floor area under consideration on the basis of one person for each 22 in. of stair width and $1\frac{1}{2}$ treads on the stairs, and one person for each $3\frac{1}{2}$ ft² floor area on the landings and halls within the stairway enclosure.

In all buildings over 40 ft in height, and in all mercantile buildings regardless of height, the required stairways must be completely enclosed by fireproof partitions, and at least one stairway must continue to the roof. An open ornamental stairway can be used from the main entrance floor to the floor next above, provided it is not the only stairway.

8.22 CONSTRUCTION DETAILS

The usual practice is to construct the stairways after the main structural framework of the building has been completed, in which event recesses should be left in the beams to support the stair slab, and dowels should be provided to furnish the necessary anchorage. Occasionally, however, the stairs are poured at the same time as the floors, in which event negative-moment reinforcement should be furnished over the supports of the stair slab, as in any continuous-beam construction. The steps are usually poured monolithically with the slab, but they may be molded after the main slab is in place. In the latter instance, provision must be made for securing the step to the slab.

DESIGN OF A BEAM-AND-GIRDER FLOOR

8.23 DATA AND SPECIFICATIONS

To illustrate the application of the principles of reinforced-concrete design to the design of a concrete floor of the slab–beam-and-girder type, typical portions of the second floor of the building shown in Figs. 8.16 and 8.17 will be designed. The building is to sustain a live load of 210 psf. The ACI moment coefficients of Table 7.1 are applicable and will be used for this example, although improved economy would undoubtedly result from an elastic analysis of the building frame. The outside faces of all the exterior columns will be kept in the same plane so that the pilaster will have smooth faces, and all the wall beams and girders will be placed flush with the outer face of the pilasters so that the entire outer walls, except for the windows, will be in one plane. A $1\frac{1}{2}$-in. bonded wearing surface, which will add to the dead load but will not be considered part of the effective depth, will be placed on the slab. Steel reinforcement with a yield stress $f_y = 40,000$ psi will be used with concrete of strength $f_c' = 3000$ psi. Economic considerations lead to the choice of a maximum steel ratio of $0.40\rho_b$, or 0.0148, in the present case.

For purposes of illustration, most of the computations for this design will be presented on computation sheets of the type that are typical in design office practice.

8.24 DESIGN OF THE SLABS

The computations for the design of the floor slabs $S1$ and $S2$ are shown in Computation Sheet 8.1. The clear spans were obtained from the assumption that the floor beams will be 8 in. in width. In computing the negative moments at the faces of supports, the average of the two adjacent clear spans was used. A negative-moment coefficient at the interior face of the exterior support of $\frac{1}{24}$ was taken in accordance with Table 7.1. Reinforcing steel is placed with centers 1 in. from the top or bottom of the slab.

For the present slab, which is very shallow and in which only about one-half the steel is to be bent, the approximate bend points and cutoff points of Fig. 3.16 are satisfactory.

For slab $S1$:
    Positive steel exterior, bend at $7 \times 12/7 = 12$ in. from support face
    Positive steel interior, bend at $7 \times 12/5 = 17$ in. from support face
    Negative steel exterior, cut at $7 \times 12/5 = 17$ in. from support face
    Negative steel interior, cut at $7 \times 12/3 = 28$ in. from support face
For slab $S2$:
    Positive steel, bend at $6.33 \times 12/4 = 19$ in. from support face
    Negative steel, cut at $6.33 \times 12/3 = 25$ in. from support face

All details of the slab are shown in Fig. 8.16 on page 397.

8.25 DESIGN OF THE BEAMS

The computations for the design of beams $B1$ and $B2$ are shown in Computation Sheets 8.2 and 8.3. The clear spans for the beams were obtained using the same assumptions made in the design of the slab, with the further assumption that the beams will be framed into girders 12 in. in width. The beams framing

Computation Sheet 8.1   Beam-and-girder floor

Design of floor slab:

Live load	$210 \times 1.7 = 358$ psf
Dead load (assume 4″ slab $+ 1\frac{1}{2}$″ finish)	$69 \times 1.4 = 96$ psf
	Total $= 454$ psf

Clear span for $S1 = 7'0''$
Clear span for $S2 = 6'4''$ } Avg clear span $= 6'8''$

S1

Center:

$M = +\frac{1}{14}(454)(7.0)^2(12) = 19,000$ in.-lb

At continuous end:

$M = -\frac{1}{12}(454)(6.67)^2(12) = -20,200$ in.-lb

At wall end:

$M = -\frac{1}{24}(454)(7.0)^2(12) = -11,050$ in.-lb

From Graph 1 in Appendix:

$\dfrac{M_u}{\phi b d^2} = 525$

$d^2 = \dfrac{20,200}{0.9 \times 12 \times 525} = 3.56$ in.^{2}

$d = 1.89$ in.

Use a minimum d of 3 in.

Slab thickness = 4 in. as assumed.

Center:

$\dfrac{M_u}{\phi b d^2} = \dfrac{19,000}{0.9 \times 12 \times 9} = 196$

From Graph 1:

$\rho = 0.0051$

$A_s = 0.0051 \times 12 \times 3 = 0.18$ in.^{2}

At continuous end:

$\dfrac{M_u}{\phi b d^2} = \dfrac{20,200}{0.9 \times 12 \times 9} = 208$

$\rho = 0.0054$

$A_s = 0.0054 \times 12 \times 3 = 0.19$ in.^{2}

At wall end:

$\dfrac{M_u}{\phi b d^2} = \dfrac{11,050}{0.90 \times 12 \times 9} = 114$

$\rho = 0.0030$

$A_s = 0.0030 \times 12 \times 3 = 0.11$ in.^{2}

S2

Center:

$M = +\frac{1}{16}(454)(6.33)^2(12) = 13,600$ in.-lb

At first interior support:

$M = -\frac{1}{12}(454)(6.67)^2(12) = 20,200$ in.-lb

At all other interior supports:

$M = -\frac{1}{12}(454)(6.33)^2(12) = 18,100$ in.-lb

Use 4-in. slab thickness throughout.

Center:

$\dfrac{M_u}{\phi b d^2} = \dfrac{13,600}{0.9 \times 12 \times 9} = 140$

$\rho = 0.0036$

$A_s = 0.0036 \times 12 \times 3 = 0.13$ in.^{2}

At first interior support:

$\dfrac{M_u}{\phi b d^2} = \dfrac{20,200}{0.9 \times 12 \times 9} = 208$

$\rho = 0.0054$

$A_s = 0.0054 \times 12 \times 3 = 0.19$ in.^{2}

At other interior supports:

$\dfrac{M_u}{\phi b d^2} = \dfrac{18,100}{0.9 \times 12 \times 9} = 186$

$\rho = 0.0048$

$A_s = 0.0048 \times 12 \times 3 = 0.17$ in.^{2}

Outer support	S1 center	First interior support	S2 center	Other supports
0.11	0.18	0.19	0.13	0.17
#3 straight at 14″	#3 straight at 14″	#3 bent at 7″	#3 bent at 16″	#3 bent at 8″
#3 bent at 14″	#3 bent at 14″		#3 straight at 16″	
$A_s = 0.19$	$A_s = 0.19$	$A_s = 0.18$	$A_s = 0.17$	$A_s = 0.17$

Temperature steel

$A_s = 0.0018 \times 12 \times 4 = 0.086$ Use #3 at 15″.

into girders were designed, since these beams have clear spans slightly longer than those framing into columns; however, to simplify the construction, all beams are of the same dimensions. Beams will be proportioned for a maximum steel ratio of $0.40\rho_b = 0.0148$. While a portion of the bottom steel will be bent up in each beam, stirrups will be provided to take all the diagonal tensile stresses.

Computation Sheet 8.2   Beam-and-girder floor

Design of beams:

Uniform load per foot from floor $\quad 7 \times 454 = 3175$ plf

Dead load of stem (assumed) $\qquad 225 \times 1.4 = \quad 315$ plf

$$\text{Total} = 3490 \text{ plf}$$

Clear span for $B1 = 22'6''$
Clear span for $B2 = 22'0''$ } Avg clear span $= 22'3''$

B1	B2
Center:	Center:
$M = +\frac{1}{14}(3490)(22.5)^2(12)$	$M = +\frac{1}{16}(3490)(22)^2(12)$
$\qquad = 1,520,000$ in.-lb	$\qquad = 1,269,000$ in.-lb
At continuous end:	At first interior support:
$M = -\frac{1}{10}(3490)(22.25)^2(12)$	$M = -\frac{1}{11}(3490)(22.25)^2(12)$
$\qquad = 2,080,000$ in.-lb	$\qquad = 1,866,000$ in.-lb
At wall ends:	At all other interior supports:
$M = -\frac{1}{24}(3490)(22.5)^2(12)$	$M = -\frac{1}{11}(3490)(22)^2(12)$
$\qquad = 885,000$ in.-lb	$\qquad = 1,842,000$ in.-lb
$V_{max} = \frac{1}{2}(1.15)(3490)(22.5) = 45,200$ lb	$V_{max} = \frac{1}{2}(3490)(22) = 38,400$ lb
At continuous end:	Beam dimensions will be kept the same
$\rho = 0.0148$	throughout.
$\dfrac{M_u}{\phi bd^2} = 525$	At first interior support:
$bd^2 = \dfrac{2,080,000}{0.9 \times 525} = 4410$	$\dfrac{M_u}{\phi bd^2} = \dfrac{1,866,000}{0.9 \times 4800} = 433$
Use $b = 12$ in., $d = 20$ in.	$\rho = 0.12$
$\dfrac{M_u}{\phi bd^2} = \dfrac{2,080,000}{0.90 \times 4800} = 482$	$A_s = 0.012 \times 240 = 2.88$ in.^{2}
$\rho = 0.0135$	
$A_s = 0.0135 \times 12 \times 20 = 3.24$ in.^{2}	
At wall end:	At all other interior supports:
$\dfrac{M_u}{\phi bd^2} = \dfrac{885,000}{0.9 \times 4800} = 206$	$\dfrac{M_u}{\phi bd^2} = \dfrac{1,842,000}{0.9 \times 4800} = 428$
$\rho = 0.0054$	$\rho = 0.0118$
$A_s = 0.0054 \times 240 = 1.30$ in.^{2}	$A_s = 0.0118 \times 240 = 2.84$ in.^{2}
Center:	Center:
$A_s = \dfrac{1,520,000}{0.9 \times 40,000(20-2)} = 2.35$ in.^{2}	$A_s = \dfrac{1,269,000}{0.9 \times 40,000(20-2)} = 1.96$ in.^{2}
$a = \dfrac{2.35 \times 40}{0.85 \times 3 \times 67.5} = 0.58$ in.	$a = 0.48$
Revised $A_s = 2.35 \times 18/19.71 = 2.14$ in.^{2}	Revised $A_s = 1.96 \times 18/19.76 = 1.79$ in.^{2}

Computation Sheet 8.3   Beam-and-girder floor

<table>
<tr><td style="text-align:center">*B*1</td><td style="text-align:center">*B*2</td></tr>
</table>

*B*1

Shear (use #3U stirrups)
At wall end:

V = 39,200 lb

$$v = \frac{39,200}{0.85 \times 240} = 192 \text{ psi}$$

$v_c = 2\sqrt{3000} = 110 \text{ psi}$

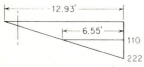

$s_{max} = 20/2 = 10 \text{ in.}$

$$s_{max} = \frac{0.22 \times 40,000}{50 \times 12} = 14.7 \text{ in.}$$

At distance d:

$v_u - v_c = 82 \times 3.15/4.82 = 54 \text{ psi}$

$$s = \frac{0.22 \times 40,000}{54 \times 12} = 13.6 \text{ in.} > s_{max}$$

Space 1 at 5″, 9 at 10″.
At continuous end:

At distance d:

$v_u - v_c = 112 \times 4.88/6.55 = 84 \text{ psi}$

$$s = \frac{0.22 \times 40,000}{84 \times 12} = 8.8 \text{ in.}$$

Space 1 at 4″, 5 at 8″, 7 at 10″.

*B*2

At either end:

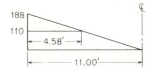

At distance d:

$v_u - v_c = 78 \times 2.91/4.58 = 50 \text{ psi}$

$$s = \frac{0.22 \times 40,000}{50 \times 12} = 14.7 \text{ in.} > s_{max}$$

Space 1 at 5″, 9 at 10″.

On the basis of Computation Sheets 8.2 and 8.3, the positive steel requirement of 2.14 in.² for *B*1 will be provided using four No. 7 bars ($A_s = 2.41$). Two of these will be bent up at each end. At the wall end of *B*1, the required area of 1.30 in.² will be met by adding one No. 5 bar to the two No. 7 bent bars ($A_s = 1.51$ in.²). In beam *B*2, the positive steel requirement of 1.79 in.² will be provided by four No. 6 bars ($A_s = 1.77$ in.²), two of which are bent up at each end. At the first interior support, the total requirement is 3.24 in.² Two No. 7 straight bars will be added to the two No. 6 and two No. 7 bent bars ($A_s = 3.29$ in.²). At typical interior supports, where the steel requirement is 2.84 in.², two No. 7 straight bars are added to the two No. 6 bars bent up from each side ($A_s = 2.97$ in.²). Bar bend and cutoff points will be found with the aid of Graph 11 of the Appendix and Fig. 3.15. For No. 7 bars, the greater of

$12d_b$ or d is 20 in.; the development length is 18 in., and $l_d + 20 = 38$ in. for bottom bars, while $1.4l_d + 20 = 45$ in. for top bars. For No. 6 bars, the greater of $12d_b$ or d is also 20 in.; this plus the development length of 13 in. gives $l_d + 20 = 33$ in. for bottom bars and $1.4l_d + 20 = 38$ in. for top bars. Similarly, for No. 5 bars, $l_d + 20 = 30$ in. for bottom bars and $1.4l_d + 20 = 34$ in. for top bars. Tentatively,

$B1$:

Two No. 7 bars can be bent up at

$$0.23 \times 22.5 \times 12 - 10 = 52 \text{ in.}$$

from the face of either support. The remaining two No. 7 bars are continued 6 in. into the support faces. At the wall end, the No. 5 straight bar is continued a distance d beyond the point of inflection and is cut off at

$$0.12 \times 22.5 \times 12 + 20 = 52 \text{ in.}$$

from the support face. At the continuous end, straight bars are cut off at

$$0.28 \times 22.5 \times 12 + 20 = 96 \text{ in.}$$

from the support face.

$B2$:

Two No. 6 bars are bent up at

$$0.25 \times 22 \times 12 - 10 = 56 \text{ in.}$$

from the support faces. The remaining two No. 6 bars are continued 6 in. into the face of the supports. At the first interior support, the four No. 7 straight bars are cut off at

$$0.24 \times 22.25 \times 12 + 20 = 84 \text{ in.}$$

beyond the support face. At all other interior supports, the two No. 7 and two No. 6 straight bars are cut off at

$$0.24 \times 22 \times 12 + 20 = 84 \text{ in.}$$

from the support face.

A check of requirements of development length past theoretical bend and cutoff points for both bottom and top bars confirms satisfaction of all requirements of Fig. 3.15.

8.26 DESIGN OF THE GIRDERS

A portion of the sample computations for the design of the girders is shown in Computation Sheets 8.4 and 8.5. The clear spans here were obtained by making use of all previous assumptions and further assuming that the exterior columns supporting the second floor have a depth of 14 in., measured from the face of

Computation Sheet 8.4   Beam-and-girder floor

Design of girders:
 Load from floor beams = 76,800 lb
 Live load on girder   358 plf
 Dead load of floor finish on girder   26 plf
 Dead load of girder (assumed)   530 × 1.4 =   745 plf
                                          Total = 1129 plf

$$\text{Clear span of } G1 = 20'2''$$
$$\text{Clear span of } G2 = 19'8''\Big\} \text{ Avg clear span used} = 19'11''$$

Simple-span maximum moment
$$\tfrac{1}{8}(1129)(19.92)^2(12) = \quad 672{,}000 \text{ in.-lb}$$
$$76{,}800(6.58)(12) = 6{,}020{,}000 \text{ in.-lb}$$
$$M_s = \overline{6{,}692{,}000} \text{ in.-lb}$$

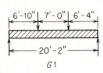

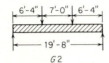

G1

Center:
$$M = +8/14M_s = 3{,}820{,}000 \text{ in.-lb}$$
At continuous end:
$$M = -8/10M_s = -5{,}350{,}000 \text{ in.-lb}$$
At wall end:
$$M = -8/16M_s = -3{,}340{,}000 \text{ in.-lb}$$
$$V_{\max} = 1.15\left(76{,}800 + 1129 \times \frac{20.2}{2}\right)$$
$$= 101{,}000 \text{ lb}$$

$\rho_{\max} = 0.0148$
At continuous end:
$$\frac{M_u}{\phi b d^2} = 525$$
$$bd^2 = \frac{5{,}350{,}000}{0.9 \times 525} = 11{,}320$$
Use $b = 16$ in., $d = 26.5$ in.
 $A_s = 0.0148 \times 423 = 6.27 \text{ in.}^2$
At wall end:
$$\frac{M_u}{\phi b d^2} = \frac{3{,}340{,}000}{0.9 \times 11{,}320} = 328$$
$\rho = 0.0088$
 $A_s = 0.0088 \times 423 = 3.72 \text{ in.}^2$
At center ($b = 60.5$ in.):
$$A_s = \frac{3{,}820{,}000}{0.9 \times 40{,}000(26.5 - 2)} = 4.33 \text{ in.}^2$$
$$a = \frac{4.33 \times 40}{0.85 \times 3 \times 60.5} = 1.12 \text{ in.}$$
Revised $A_s = 4.33 \times 24.50/25.94 = 4.10 \text{ in.}^2$

G2

Center:
$$M = +8/16M_s = 3{,}340{,}000 \text{ in.-lb}$$
At ends:
$$M = -8/11M_s = -4{,}860{,}000 \text{ in.-lb}$$

$$V_{\max} = 76{,}800 + 1129 \times \frac{19.67}{2}$$
$$= 87{,}500 \text{ lb}$$
Girder dimensions will be kept the same throughout.

At supports:
$$\frac{M_u}{\phi b d^2} = \frac{4{,}860{,}000}{0.90 \times 11{,}320} = 477$$
$\rho = 0.0133$
 $A_s = 0.0133 \times 423 = 5.63 \text{ in.}^2$

At center:
$$A_s = \frac{3{,}340{,}000}{0.9 \times 40{,}000(26.5 - 2)} = 3.78 \text{ in.}^2$$
$$a = \frac{3.78 \times 40}{0.85 \times 3 \times 60.5} = 0.975 \text{ in.}$$
Revised $A_s = 3.78 \times 24.50/26.01 = 3.56 \text{ in.}$

the wall, and that the interior first-floor columns are 16 in. in diameter. The average clear span was used in computing moments caused by the uniformly distributed load, and the average distance between beams and adjacent columns in $G1$ for computing moments caused by concentrated loads. These assumptions are justifiable since they produce moments slightly larger than those which actually exist in either span. The inherently approximate nature of calculations of this sort does not warrant closer refinement in computations.

From Computation Sheets 8.4 and 8.5 the positive steel-area requirement for $G1$ of 4.10 in.² will be provided by two No. 9 straight bars and two No. 10 bent bars ($A_s = 4.53$ in.²). At the wall end of $G1$, the requirement of 3.72 in.² is met by adding one No. 10 straight bar to the two No. 10 bent bars ($A_s = 3.79$ in.²). In girder $G2$, the positive requirement of 3.56 in.² calls for two No. 8 straight bars and two No. 9 bent bars ($A_s = 3.57$ in.²). At the first

Computation Sheet 8.5   Beam-and-girder floor

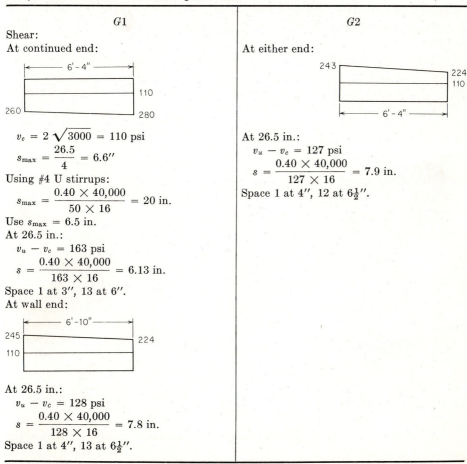

G1

Shear:

At continued end:

$$v_c = 2 \sqrt{3000} = 110 \text{ psi}$$

$$s_{max} = \frac{26.5}{4} = 6.6''$$

Using #4 U stirrups:

$$s_{max} = \frac{0.40 \times 40,000}{50 \times 16} = 20 \text{ in.}$$

Use $s_{max} = 6.5$ in.

At 26.5 in.:

$$v_u - v_c = 163 \text{ psi}$$

$$s = \frac{0.40 \times 40,000}{163 \times 16} = 6.13 \text{ in.}$$

Space 1 at 3″, 13 at 6″.

At wall end:

At 26.5 in.:

$$v_u - v_c = 128 \text{ psi}$$

$$s = \frac{0.40 \times 40,000}{128 \times 16} = 7.8 \text{ in.}$$

Space 1 at 4″, 13 at $6\frac{1}{2}$″.

G2

At either end:

At 26.5 in.:

$$v_u - v_c = 127 \text{ psi}$$

$$s = \frac{0.40 \times 40,000}{127 \times 16} = 7.9 \text{ in.}$$

Space 1 at 4″, 12 at $6\frac{1}{2}$″.

interior support, adding two No. 9 straight bars to the two No. 10 and two No. 9 bent bars provides an area of 6.53 in.², very close to the required 6.27 in.² At the typical interior supports, two No. 9 straight bars are added to the two No. 9 bent bars from each adjacent span, providing 6.00 in.², slightly in excess of the required 5.63 in.²

The points at which the bars may be bent up are determined by assuming that the positive moment is zero at a distance of $\frac{2}{3}L/3$ from the concentrated load to the center line of the support, measured toward the support, and that the moment diagram is a straight line between the maximum and zero values. The latter assumption is in error only because of the uniformly distributed weight of the girder; this is very small in comparison with the large concentrated loads. The two No. 9 bars in $G2$ may be bent up at a distance of

$$\frac{2.00}{3.57} \times \frac{2}{3} \times \frac{21 \times 12}{3} = 31 \text{ in.}$$

to the left of the first concentrated load. They will be extended 6 in. beyond that point to meet, very nearly, the ACI requirement for $d/2$ extension while still meeting requirements for negative bending.

A similar assumption for the negative-moment variation can be made; the point of zero negative moment is therefore $\frac{2}{3}L/3$ from the center line of the support. The overlapping of the positive- and negative-moment diagrams which results from these assumptions provides for any variation in the moments that might result from unequal placing of the live load on adjoining spans. The two No. 9 bars at the first interior support of $G2$ may be bent down at a distance of

$$\frac{2.00}{6.53} \times \frac{2}{3} \times \frac{21 \times 12}{3} = 17 \text{ in.}$$

from the support. Actual bend points are located to satisfy these requirements and are shown in Fig. 8.15. The remaining top bars are carried a distance

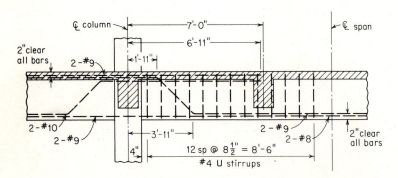

Fig. 8.15 Details of girder.

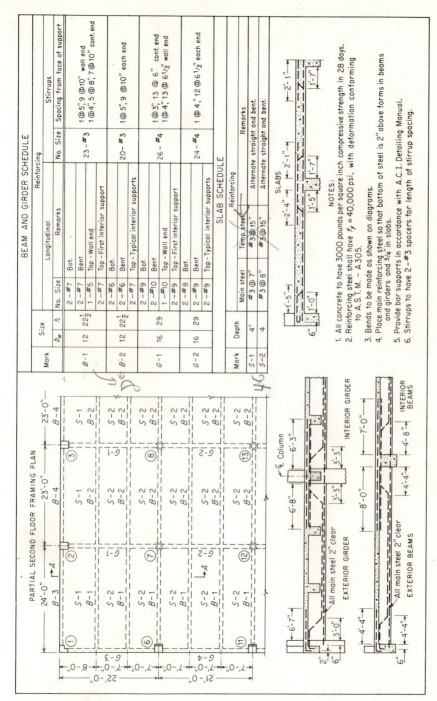

Fig. 8.16 Engineering drawing of beam-and-girder floor.

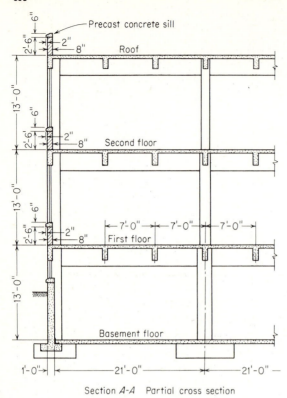

Section *A-A* Partial cross section   **Fig. 8.17**

d past the extreme position of the point of inflection and are cut off

$$\frac{2}{3} \times \frac{21 \times 12}{3} + 26.5 = 83 \text{ in.}$$

from the support as shown. Development length requirements are met throughout the member.

Vertical No. 4 U stirrups are provided to accommodate all the diagonal tensile stresses, in neglect of the contribution of the bent bars. With reference to the shear diagrams of Computation Sheet 8.5, the excess shear is very nearly constant from the support face to the center line of the concentrated load. In each case, the spacing required at the critical section a distance d from the support face is used through the entire region in which stirrups are required.

Complete engineering details for the portion of the building designed above are shown in Figs. 8.16 and 8.17.

DESIGN OF A RIBBED FLOOR WITH STEEL PAN FORMS

8.27 DEAD LOADS

The weight of a ribbed floor can be computed from the known concrete dimensions. The general shape of the cross section of the steel forms must be known,

Table 8.2   Weight of concrete in ribbed floors with steel pan fillers

Depth of joist below slab, in.	Thickness of slab, in.	Average weight of floor, psf
6	2	42
	$2\frac{1}{2}$	48
8	2	48
	$2\frac{1}{2}$	54
10	2	55
	$2\frac{1}{2}$	61
12	2	61
	$2\frac{1}{2}$	67
	3	73
14	2	69
	$2\frac{1}{2}$	75
	3	81

since the taper of the sides and the chamfering of the upper corners affect the volume of concrete in the ribs or joists. The cross section can be obtained from the manufacturers' catalogues. The average values in Table 8.2 can be used for joists (or ribs) 5 in. wide at the bottom and 25 in. on centers.

The values given in this table do not include the weight of an extra floor finish or a plastered ceiling below the floor. The former can be computed from the specified thickness of finish or type of floor surface. In computing design loads, an allowance of 10 psf is usually made for a plastered ceiling. An additional allowance of from 10 to 20 psf is made for the weight of partitions not definitely located on the architect's plans or where there is a possibility of future rearrangement of partitions. The latter condition is very likely to occur in buildings of the types to which ribbed-floor construction is adapted. When definitely located partitions are parallel with the joists, a joist thicker than the normal 5 in. is usually placed under these partitions, or if the partition is located between two joists, both these joists are made thicker than the others, and the slab thickness between them is increased. The increased slab thickness is obtained by using shallower pans in the one row or by lowering the pans in that row. When definitely located partitions are perpendicular to the joists, the partition weight is considered as a concentrated load in designing the joists. Transverse bridging ribs 4 in. wide and the same depth as the joists are commonly used at the center of span of less than 24 ft and at the third points of a longer span to stiffen the joists and distribute concentrated floor loads to adjacent joists.

8.28   DATA FOR DESIGN

A typical interior panel of a hotel floor is to be built as a ribbed floor, using removable steel pan forms. The joists are to be supported at the ends on

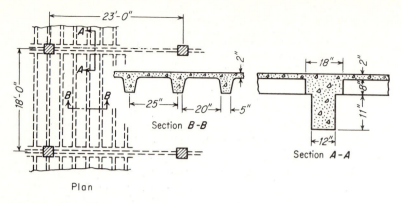

Fig. 8.18 Details of ribbed floor with removable steel pans.

concrete girders, as shown in Fig. 8.18. The span of the girders is 23 ft 0 in., and the distance center to center of girders is 18 ft 0 in. The live load is 60 psf, and allowances for dead loads other than the weight of the concrete are to be made as follows: partitions, 15 psf; plastered ceiling, 10 psf; wood floor, with sleepers in cinder-concrete fill, 18 psf. A 4000-psi concrete is to be used in the floor, with bars for which $f_y = 60{,}000$ psi. It is required to design the panel.

8.29 DESIGN OF JOISTS

Computation of moments and shears The joists are designed as regular continuous T beams, with a flange width equal to the distance center to center of joists. To compute the weight of the floor, the depth of the joists below the slab and the thickness of the slab must be assumed. A combination of 8-in. joists and 2-in. slab will be tried. The weight of the concrete per square foot of floor is 48 psf, and the total design dead load is

$$(15 + 10 + 18 + 48)(1.4) = 128 \text{ psf}$$

The design live load is $60 \times 1.7 = 102$ psf, and the total design load is $128 + 102 = 230$ psf. With 5-in. joists spaced 25 in. on centers, each joist supports $25/12 = 2.08$ ft² of foot of joist, and the total factored load on each joist is

$$230 \times 2.08 = 478 \text{ plf}$$

The panel is fully continuous, and the effective span for moment is equal to the clear distance between the stems of the girders. If the width of the girder stems is assumed as 12 in., the joist span is 17 ft. The maximum negative moment at the face of the support is

$$M = \tfrac{1}{11} \times 478 \times 17^2 \times 12 = 151{,}000 \text{ in.-lb}$$

and the maximum positive moment at the center of the span is

$$M = \tfrac{1}{16} \times 478 \times 17^2 \times 12 = 104,000 \text{ in.-lb}$$

The shear at the face of the supporting girder is

$$V_u = 478 \times \tfrac{17}{2} = 4070 \text{ lb}$$

Design for shear It is rather difficult to place stirrups in narrow joists, such as are used in this form of construction; hence the shearing unit stress should be kept below the value permitted on the concrete web. The ACI Code permits a 10 per cent increase in nominal shear stress carried by the concrete in joists, above the usual value. Accordingly, in the present case,

$$v_c = 1.1 \times 2 \sqrt{4000} = 140 \text{ psi}$$

The effective shearing width of a joist is taken as the width at the bottom of the joist; the increased section due to the side slope of the pans is neglected. If necessary, tapered pans can be used at the ends of each row to increase the shearing area, or deeper joists can be used to serve the same purpose. At the critical section for shear, a distance d (estimated as 8 in.) from the face of supports, the shear is

$$V_u = 4070 - 478 \times \tfrac{8}{12} = 3750 \text{ lb}$$

The minimum web area is

$$b_w d = \frac{V_u}{\phi v_u} = \frac{3750}{0.85 \times 140} = 31.6 \text{ in.}^2$$

If straight end pans are used, with $b_w = 5$ in., the minimum effective depth is 6.33 in. With 8-in. joists under a 2-in. slab, the effective depth is $10 - 1\tfrac{1}{8} = 8.87$ in. This will be satisfactory and will allow for a minimum clear insulation under the bars of about $\tfrac{3}{4}$ in.

If the problem of deciding whether to use tapered end pans or deeper pans throughout arises, it should be remembered that the steel area required decreases with an increase in the depth of the joists, and both the cost of the steel and the cost of the concrete should be considered. The cost of the forms can be omitted from the comparison, because this cost does not vary materially.

Computation of steel area At the face of the girder stem, the negative joist moment is 151,000 in.-lb. Then

$$\frac{M_u}{\phi b_w d^2} = \frac{151,000}{0.9 \times 5 \times 8.87^2} = 427$$

From Graph 1 of the Appendix, $\rho = 0.0076$. Then

$$A_s = \rho b_w d = 0.0076 \times 5 \times 8.87 = 0.33 \text{ in.}^2$$

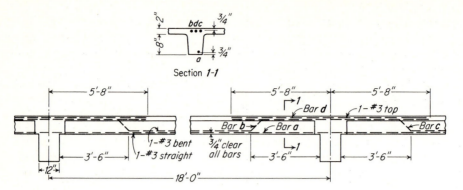

Fig. 8.19 Joist details.

At midspan, assuming a stress-block depth equal to the slab thickness,

$$\text{Trial } A_s = \frac{M_u}{\phi f_y(d - t/2)} = \frac{104,000}{0.9 \times 60,000 \times 7.87} = 0.25 \text{ in.}^2$$

The corresponding stress-block depth is

$$a = A_s f_y / 0.85 f_c' b = 0.25 \times 60/(0.85 \times 4 \times 25) = 0.177 \text{ in.}$$

The revised lever arm is 8.78 in., and

$$A_s = 0.25 \times 7.87/8.78 = 0.22 \text{ in.}^2$$

Arrangement of reinforcement As shown in Fig. 8.19, two No. 3 bars
($A_s = 0.22$ in.²) will be used for positive-moment reinforcement. Bending up
one bar in each span and adding one No. 3 straight top bar will provide sufficient steel ($A_s = 0.33$ in.²) to meet the negative-moment requirements. In
accordance with Graph 11 of the Appendix, the positive steel will be bent up at
3.50 ft from the support face and, together with the added straight bars, will
be continued a distance of 5.14 ft beyond the far face of the supports. The
remaining positive steel will be carried 6 in. into the face of supports. Adequate
development length is provided for both top and bottom steel.

In addition to the main joist reinforcement, temperature steel is placed
in the slab at right angles to the joists, in the amount of 0.0018 times the gross
area of the concrete slab between ribs. Quite often welded wire mesh is used
to reinforce the slab.

8.30 DESIGN OF GIRDER

Computation of loads The loads which are brought to the girder from the
joists are theoretically concentrated at the points at which the joists frame
into the girder. They are spaced so closely, however, that they may be con-

sidered as uniformly distributed throughout the length of the girder without materially affecting the maximum moment in the girder. Each pair of adjoining joists, one on either side of the girder, transmits a load to the girder equal to twice the end shear of one joist, or $2 \times 4070 = 8140$ lb. Since the joists are 25 in., or 2.08 ft, on centers, the uniform load per foot of girder is $8140/2.08 = 3910$ plf. To this must be added the weight of the girder and the floor and ceiling loads directly over and under the girder. With total girder depth assumed to be 23 in. and width 12 in., the weight of the girder is

$$12 \times 23 \times 150/144 = 290 \text{ plf}$$

The weights of partitions, ceiling, and fill over the 12-in. girder width add 43 plf. The factored dead load is then

$$(290 + 43)(1.4) = 468 \text{ plf}$$

The additional live load directly over the girder is

$$60 \times 1.7 = 102 \text{ plf}$$

and the total design load is

$$3910 + 468 + 102 = 4480 \text{ plf}$$

Computation of moment and shear   The girder is continuous at both ends, so that with 18-in. columns the effective span is

$$23 - \tfrac{18}{12} = 21.5 \text{ ft}$$

The maximum negative moment at the face of the support is

$$M = \tfrac{1}{11} \times 4480 \times 21.5^2 \times 12 = 2{,}260{,}000 \text{ in.-lb}$$

and the maximum positive moment at the center of the span is

$$M = \tfrac{1}{16} \times 4480 \times 21.5^2 \times 12 = 1{,}560{,}000 \text{ in.-lb}$$

The maximum shear is

$$V = 4480 \times 21.5/2 = 48{,}200 \text{ lb}$$

Design at supports   A maximum steel ratio of

$$0.40\rho_b = 0.0115$$

will be adopted. Then, from Graph 1 of the Appendix,

$$\frac{M_u}{\phi b d^2} = 620$$

from which

$$bd^2 = \frac{M_u}{620\phi} = \frac{2{,}260{,}000}{620 \times 0.9} = 4050 \text{ in.}^3$$

For a girder width of 12 in., the required effective depth is 18.4 in. An effective depth of 18.5 in. will be used, providing an overall girder depth of 21 in. No revision in the assumed dead load is required. The total tension steel required is

$$A_s = 0.0115 \times 222 = 2.55 \text{ in.}^2$$

Design at center   The girder is a T beam, but because of the comparatively thick flange, the neutral axis will undoubtedly be in the flange, and the girder can be designed for moment as a rectangular beam with a width equal to the width of the flange. Accordingly,

$$\frac{M_u}{\phi b d^2} = \frac{1,560,000}{0.9 \times 18 \times 18.5^2} = 281$$

From Graph 1 of the Appendix, $\rho = 0.0048$, and

$$A_s = 0.0048 \times 18 \times 18.5 = 1.60 \text{ in.}^2$$

Checking the location of the neutral axis,

$$a = \frac{1.60 \times 60}{0.85 \times 4 \times 18} = 1.57 \text{ in.}$$

well within the depth of the flange as assumed.

The positive-moment requirement of 1.60 in.² will be provided using two No. 6 and one No. 8 bars ($A_s = 1.67$ in.²). The single No. 8 bar will be bent up 4.00 ft from the face of the column, in accordance with Graph 11. At the support, the required area of 2.59 in.² is closely met by adding a single No. 9 straight bar to the two bent-up No. 8 bars ($A_s = 2.57$ in.²). The bent-up bars and the straight bar will be continued 7.25 ft beyond the face of the column, dictated by the need for a development length of $1.4L_d + d = 71$ in., beyond the point of peak stress for the No. 9 bar, occurring 16 in. past the face of the support due to bending down of one No. 8 bar. Requirements for development length elsewhere are easily met.

Design of web reinforcement   Resistance to diagonal tension will be provided using vertical stirrups, in neglect of the contribution of the bent No. 8 bars. The computations are the same as for any uniformly loaded, continuous beam; a typical example is given in Art. 8.25.

TYPICAL ENGINEERING DRAWINGS

Design information is conveyed to the builder mainly by engineering drawings. Their preparation is therefore a matter of utmost importance, and they should be checked by the designer to ensure that concrete dimensions as well as reinforcement quantities and placement are as he has determined by his calculations.

Engineering drawings for buildings usually consist of a plan view of each floor showing overall dimensions and locating the main structural elements, cross-sectional views through typical members, and beam and slab schedules which give detailed information on the reinforcement in tabular form. Sectional views are usually drawn to a larger scale than the plan, and serve to locate the steel and establish bend points, as well as to define the shape of the construction. Usually, a separate drawing is included which gives, in the form of sections and schedules, details of columns and footings.

It is a wise precaution to include, on each drawing, the material strengths used in the design of the structure, as well as the service live load on which the calculations were based.

An illustration has already been given (Fig. 8.16) of an engineering drawing for a beam-and-slab floor. Figures 8.20, 8.21, and 8.22 (pp. 406–411) show, respectively, typical engineering drawings for a one-way ribbed floor, a flat-plate floor, and a column schedule.

REFERENCES

8.1. Building Code Requirements for Minimum Design Loads in Buildings and Other Structures, USASI A-59.1, USA Standards Institute, New York, 1971.

8.2. William McGuire, "Steel Structures," Prentice-Hall, Inc., Englewood Cliffs, N.J., 1968.

8.3. Recommended Lateral Force Requirements and Commentary, Report by Seismology Committee, Structural Engineers Association of California, San Francisco, 1968.

8.4. Manual of Standard Practice for Detailing Reinforced Concrete Structures, 5th ed., ACI Publ. R315-65, American Concrete Institute, Detroit, 1970.

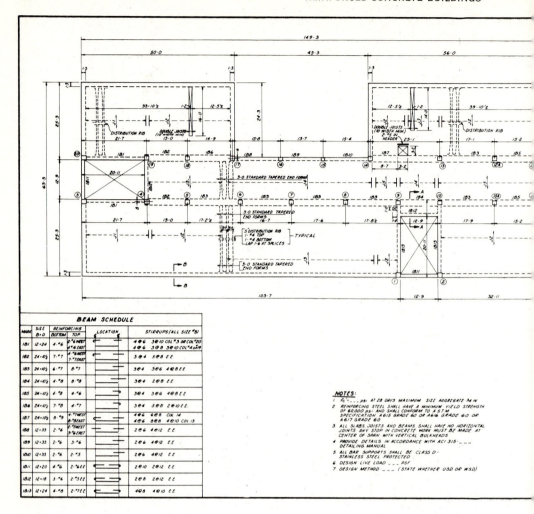

Fig. 8.20   Engineering drawing for one-way ribbed floor.

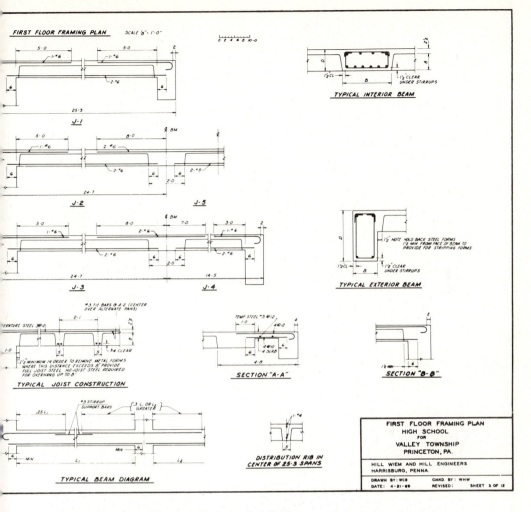

FIRST FLOOR FRAMING PLAN   SCALE ⅛"=1'-0"

TYPICAL INTERIOR BEAM

J-1

J-2   J-5

TYPICAL EXTERIOR BEAM

J-3   J-4

TYPICAL JOIST CONSTRUCTION

SECTION "A-A"

SECTION "B-B"

TYPICAL BEAM DIAGRAM

DISTRIBUTION RIB IN
CENTER OF 25'-3 SPANS

FIRST FLOOR FRAMING PLAN
HIGH SCHOOL
FOR
VALLEY TOWNSHIP
PRINCETON, PA.

HILL WIEM AND HILL ENGINEERS
HARRISBURG, PENNA.

DRAWN BY: WCB   CHKD. BY: WHW
DATE: 4-21-69   REVISED: SHEET 3 OF 12

(Courtesy American Concrete Institute, Ref. 8.4.)

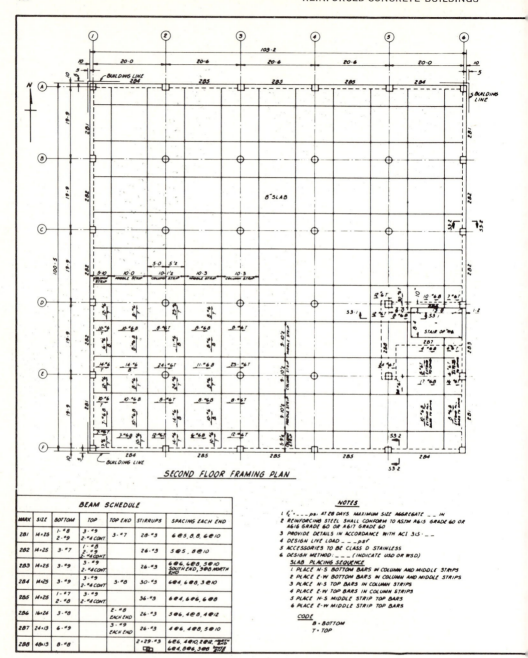

SECOND FLOOR FRAMING PLAN

		BEAM SCHEDULE				
MARK	SIZE	BOTTOM	TOP	TOP END	STIRRUPS	SPACING EACH END
2B1	14×25	1-#8 2-#9	3-#9 2-#4 CONT.	3-#7	28-#3	6@5, 8, 8, 6@10
2B2	14×25	3-#7	1-#8 2-#9 2-#4 CONT.		26-#3	5@5, 8@10
2B3	14×25	3-#9	3-#9 2-#4 CONT.		26-#3	6@6, 6@8, 5@10 SOUTH END, 9@8 NORTH END
2B4	14×25	3-#9	3-#9 2-#4 CONT.	3-#8	30-#3	6@4, 6@8, 3@10
2B5	14×25	1-#7 2-#8	3-#9 2-#4 CONT.		36-#3	6@4, 6@6, 6@8
2B6	16×24	3-#8		2-#8 EACH END	26-#3	5@6, 4@8, 4@12
2B7	24×13	6-#9		3-#9 EACH END	26-#3	4@6, 4@8, 5@10
2B8	48×13	8-#8		2-29-#3		6@6, 4@10, 2@12, ... 6@4, 8@6, 3@8 ...

Fig. 8.21 Engineering drawing for flat-plate floor.

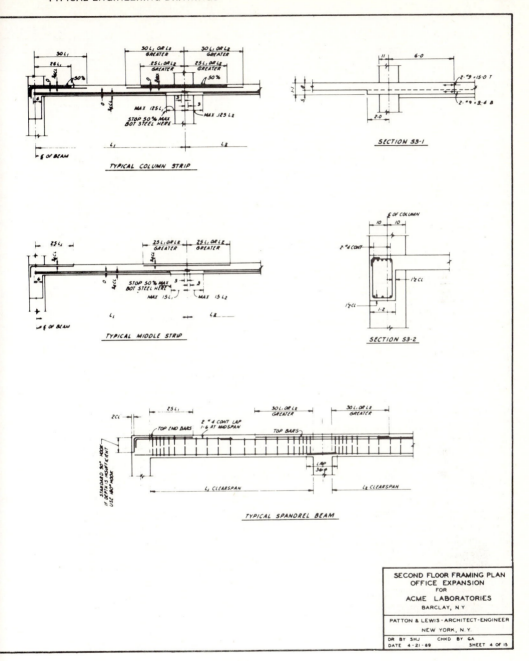

TYPICAL COLUMN STRIP

SECTION S3-1

TYPICAL MIDDLE STRIP

SECTION S3-2

TYPICAL SPANDREL BEAM

SECOND FLOOR FRAMING PLAN
OFFICE EXPANSION
FOR
ACME LABORATORIES
BARCLAY, N Y

PATTON & LEWIS · ARCHITECT - ENGINEER
NEW YORK, N.Y.

DR BY SHJ   CHKD BY GA
DATE 4-21-69   SHEET 4 OF 15

(*Courtesy American Concrete Institute, Ref. 8.4.*)

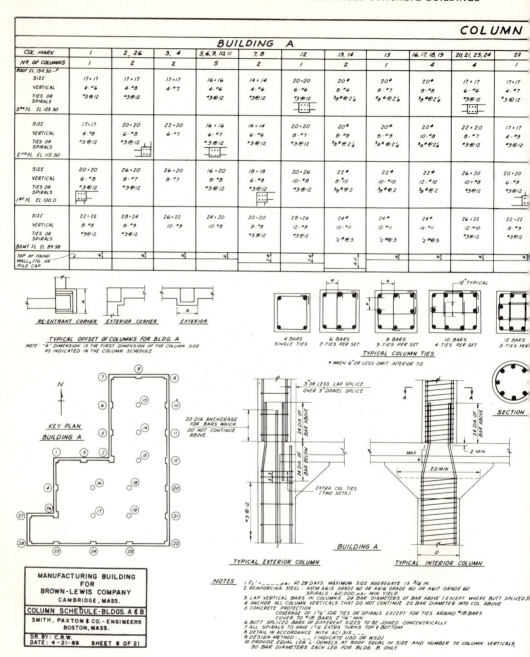

Fig. 8.22   Engineering drawing for columns.

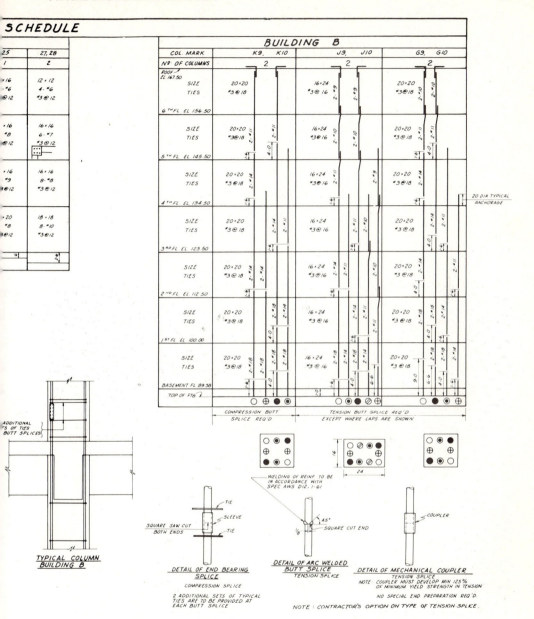

(*Courtesy American Concrete Institute, Ref. 8.4.*)

9

Precast-concrete Construction

9.1 GENERAL

From the early sixties to the present, building costs have increased at a considerably faster rate than that of most industrial products. This is true not only for the United States but for most industrialized nations. One of the main reasons for these high building costs is the large amount of site labor involved in traditional construction processes. Even disregarding cost, the demand for skilled on-site building labor is outrunning supply and will increasingly do so in most industrial and many developing nations. This trend can be slowed or halted only through greater industrialization of construction, that is, the maximum replacement of site labor by factory-production methods.

Answering this need, *precast-concrete construction* has developed rapidly and continues to grow in importance. Industrialization is achieved by mass production of repetitive and often standardized units: columns, beams, floor and roof elements, wall panels, etc. These are produced in precasting yards under factory conditions. On large jobs precasting yards are sometimes constructed on or adjacent to the site. More frequently, these yards are stationary regional enterprises which supply precast members to sizable areas within reasonable shipping distances, of the order of 200 miles. Advantages of precast

413

construction are less labor per unit because of mechanized series production; use of unskilled local labor, in contrast to skilled mobile construction labor; shorter construction time because site labor mostly involves only foundations and the connecting of the precast units; better quality control and higher concrete strength (up to $f' = 7500$ psi) achievable under factory conditions; and greater independence of construction from weather and season. Disadvantages are the greater cost of transporting prefab units as compared with transporting materials, and the additional technical problems and costs of site connections of precast elements.

Precast construction is used in all major types of structures: industrial buildings, residential and office buildings, halls of sizable span, bridges, etc. Precast members frequently are prestressed in the casting yard. In the context of the present chapter it is irrelevant whether a precast member is also prestressed (see Chap. 11). Discussion is focused on types of precast members and precast structures and on methods of connection; these are essentially independent of whether the desired strength of the member was achieved with ordinary reinforcement or by prestressing.

9.2 TYPES OF PRECAST MEMBERS AND ASSEMBLIES

A number of types of precast units have established themselves. Though not formally standardized at this time, they are widely available, with minor local variations. At the same time, the precasting process is sufficiently adaptable so that special shapes, developed for particular projects, can be produced economically, provided the number of repetitive units is sufficiently large. This is particularly important for exterior wall panels which permit a wide variety of architectural treatments.

Wall panels are made in a considerable variety of shapes, depending on architectural requirements. The most frequent four shapes are shown in Fig. 9.1. These units are produced in one- to four-story-high sections and up to 8 ft in width. They are used either as curtain walls attached to columns and beams or as bearing walls. To improve thermal insulation, sandwich panels are used which consist of an insulation core (e.g., foam glass, glass fiber, or expanded plastics) between two layers of normal or lightweight concrete. The two layers must be adequately interconnected through the core to act as one unit. A variety of surface finishes can be produced through the use of special exposed aggregates or of colored cement, sometimes employed in combination.

Stresses in wall panels are frequently more severe in handling and during erection than in the finished structure, and design must provide for these temporary conditions. Also, control of cracking is of greater importance in wall panels than in other precast units, for appearance more than for safety. To control cracking, the maximum tensile stress in the concrete, calculated by straight-line theory, should not exceed the modulus of rupture of the particular concrete with an adequate margin of safety. The Code specifies

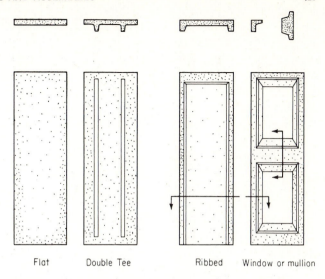

Flat   Double Tee   Ribbed   Window or mullion

Fig. 9.1   Precast wall panels. (*Adapted from Ref.* 9.1.)

the modulus of rupture as $f_r = 7.5\sqrt{f'_c}$ for ordinary concrete, multiplied by 0.75 for all-lightweight aggregate or 0.85 for sand-lightweight concrete. A safety factor of 1.65 for normal-weight concrete and 2.0 for lightweight concrete is recommended in Ref. 9.1, which also presents suggestions for special crack-control reinforcement. A wealth of further information on precast wall panels is found in Refs. 9.2 to 9.5.

Roof and floor elements are made in a wide variety of shapes adapted to specific conditions, such as span lengths, magnitude of loads, desired fire ratings, appearance, etc. Figure 9.2 shows typical examples of the most common shapes, arranged in approximate order of increasing span length, even though the spans covered by the various configurations overlap widely.

Flat slabs (Fig. 9.2a) are usually 4 in. thick, although they are used as thin as $2\frac{1}{2}$ in. when continuous over several spans, and are produced in widths of 4 to 8 ft and in lengths up to 36 ft. Depending on magnitude of loads and on deflection limitations, they are used over roof and floor spans ranging from 8 to about 22 ft. For lower weight and better insulation and to cover longer spans, *hollow planks* (Fig. 9.2b) of a variety of shapes are used. Some of these are made by extrusion in special machines. Depths range from 4 to about 8 in., widths from 2 to 4 ft. Again depending on load and deflection requirements, they are used on roof spans from about 16 to 34 ft, and on 12- to 26-ft floor spans, which can be augmented to about 30 ft if a 2-in. topping is applied to act monolithically with the hollow plank.

For longer spans *double tees* (Fig. 9.2c) are the most widely used shapes. Usual depths are from 14 to 22 in. They are generally used on roof spans up to about 60 ft. When used as floor members, a concrete topping of at least 2 in.

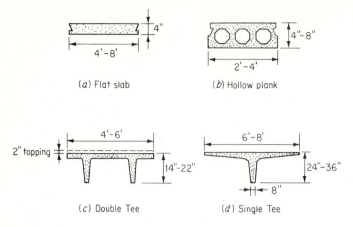

Fig. 9.2 Precast floor and roof elements. (*Adapted in part from Ref. 9.1.*)

is usually applied to act monolithically with the precast members for spans up to about 50 ft, depending on load and deflection requirements. Finally, *single tees* are available in dimensions shown in Fig. 9.2d, mostly used for roof spans up to 100 ft and more.

In all these units, the member itself or its flange constitutes the roof or floor slab. If the floor or roof proper is made of other material (plywood, gypsum, plank, etc.), it can be supported on *precast joists* in a variety of shapes for spans from about 15 to 60 ft.

For floor and roof units of reinforced rather than prestressed concrete with spans up to 35 ft, Ref. 9.6 contains valuable and authoritative information on design (including a method of designing by test rather than analysis), details, fabrication and erection, materials, and quality control. Even though parts of this information are geared to the 1963 edition of the Code and need some updating, this does not impair the utility of most of the material. The document does not apply to prestressed-concrete floor and roof units.

The shape of *precast beams* depends chiefly on the manner of framing. If floor and roof members are supported on top of the beams, these are mostly rectangular in shape (Fig. 9.3a). To reduce total depth of floor and roof construction, the tops of beams are often made flush with the top surface of the floor elements. To provide bearing, the beams are then constructed as ledger beams (Fig. 9.3b) or L beams (Fig. 9.3c). Although these shapes pertain to building construction, precast beams or girders are also frequently used in highway bridges. As an example Fig. 9.3d shows one of the various AASHO bridge girders, so named because they were developed by the American Association of State Highway Officials (AASHO).

If *precast columns* of single-story height are used so that the beams rest on top of the columns, simple prismatic columns are employed which are

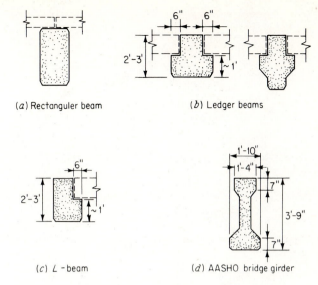

(a) Rectanguler beam   (b) Ledger beams

(c) L-beam   (d) AASHO bridge girder

Fig. 9.3   Precast beams and girders.

available in sizes from about 12×12 to 24×24 in. (Fig. 9.4a). In this case the beams are usually made continuous over the columns. Alternatively, in multistory construction, the columns can be made continuous for up to about four stories. In this case corbels are frequently used to provide bearing for the beams, as shown in Fig. 9.4b. Occasionally, T columns are used for direct support of double-tee floor members without the use of intermediate beams (Fig. 9.4c).

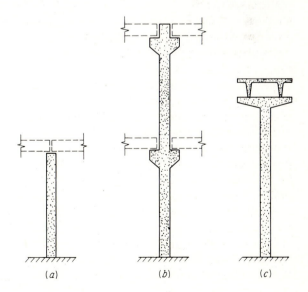

Fig. 9.4   Precast columns.   (a)   (b)   (c)

Fig. 9.5 Two-story garage over railroad tracks. (*Courtesy Engineering News-Record, Sept. 30, 1971, copyright McGraw-Hill, Inc. All rights reserved.*)

Figures 9.5 to 9.11 illustrate some of the many ways in which precast members have been used. Figure 9.5 shows a two-story parking garage under construction, erected over railroad tracks. A single-tee member is seen being placed on one of the two-story columns which are furnished with corbels to support floor members and curtain wall panels. Figure 9.6, from the same project, shows T columns supporting double-tee members to produce a major area over railroad tracks for two-story town houses, for garages, and for recreational use. Figure 9.7 demonstrates that unusual architectural designs can be realized in precast concrete, as in this all-precast administration building. Wall panels are utilized to produce a curved facade. Wedge-shaped repetitive floor units span freely from the exterior facade to the interior curved beam and column framework. In the insurance building of Fig. 9.8, 99′4″ precast girders span between exterior walls supported on four points each and provide six floors of office space entirely free of interior supports. The convention headquarters of Fig. 9.9 combines cast-in-place frames and floor slabs with precast double-tee roof beams and precast wall panels of special design. Figure 9.10 shows a 21-story hotel under construction, which, except for the service units, consists entirely of box-shaped, room-sized modules completely prefabricated and stacked on top of each other. Abroad, such precast modules, with plumbing, wiring, and heating preinstalled, are widely used for multistory apartment buildings, as an alternative to making similar apartment structures in precast wall, roof, and floor panels, which are more easily shipped but less easily erected than box-shaped modules.

Finally, Fig. 9.11 shows an example of the frequent combined use of structural steel with precast concrete. In this case the framing of an eight-story

Fig. 9.6 Construction of multipurpose area over railroad tracks. (*Courtesy Engineering News-Record, Sept. 30, 1971, copyright McGraw-Hill, Inc. All rights reserved.*)

Fig. 9.7 All-precast administration building. (*Courtesy Portland Cement Assoc.*)

motor inn was done in bolted structural steel, while precast-concrete floor and
roof planks and precast wall panels were used for all other main structural
components. This construction is economical for 6- to 12-story buildings, where
it saves both in cost and in construction time. It is one example of the increasingly important combined use of various structural materials and methods.

9.3 CONNECTIONS

Cast-in-place reinforced-concrete structures by their very nature tend to be
monolithic and continuous. Connections, in the sense of joining two hitherto
separate pieces, rarely occur in that type of construction. Precast structures,
on the other hand, resemble steel construction in that the final structure consists
of large numbers of prefabricated elements which are connected on the site
to form the finished structure. In both types of construction, such connections
can be detailed to transmit gravity forces only, or gravity and horizontal forces,
or moments in addition to these forces. In the latter case a continuous structure
is obtained much as in cast-in-place construction, and connections which
achieve such continuity by appropriate use of special hardware, rebars, and

concrete to transmit all tension, compression, and shear stresses are sometimes called "hard" connections. In contrast, connections which transmit reactions in one direction only, analogous to rockers or rollers in steel structures, but permit a limited amount of motion to relieve other forces, such as horizontal reaction components, are sometimes known as "soft" connections (Ref. 9.8). In almost all precast connections, bearing plates are used to ensure distribution and reasonable uniformity of bearing pressures. If these bearing plates are of steel and the plates of the two connected members are suitably joined by welding or otherwise, a hard connection is obtained in the sense that at least horizontal, as well as vertical, forces are transmitted. On the other hand, bearing plates are in use which permit relief of horizontal forces by limited amounts of motion, such as by use of Teflon pads, which effectively eliminate bearing friction, or of

Fig. 9.8 99'4" precast girders for column-free interior. (*Courtesy Portland Cement Assoc.*)

Fig. 9.9 Precast roof and wall panels combined with cast-in-place frames and floor slabs. (*Courtesy Portland Cement Assoc.*)

pads with low shear rigidity (elastomeric or rubber pads of various types). These transmit gravity loads but permit sizable horizontal deformations.

Precast-concrete structures are subject to dimensional changes from creep, shrinkage, and relaxation of prestress in addition to temperature, while in steel structures only temperature changes produce dimensional variations. In the earlier development of precast construction there was a tendency to use soft connections extensively in order to permit these dimensional changes to occur without causing restraint forces in the members, and particularly in the connections. More recent experience with extensive use of soft connections indicates that the resulting structures tend to show insufficient stability against lateral forces, such as high wind and particularly earthquake effects (Refs. 9.11 and 9.12). Therefore those with long experience with precast structures tend to advocate the use of hard connections, resulting in a high degree of continuity (Ref. 9.8, among others). This trend is also reflected in recent design aids (e.g., Ref. 9.1) and runs parallel to the trend toward continuity in steel structures by means of welding or high-tensile bolts. In connections of this sort, provision must be made for resistance to the restraint forces which are caused by the above-described volume changes (see, for example, Ref. 9.1).

Bearing stresses on plain concrete are limited by the Code to $0.85\phi f_c'$, except when the supporting area is wider on all sides than the loaded area A_1. In such a case this value of the permissible bearing stress may be multiplied by $\sqrt{A_2/A_1}$, where A_2 is the maximum portion of the supporting surface, which is geometrically similar to and concentric with the loading area. Reference 1 contains somewhat more restrictive bearing-stress recommendations. Higher bearing stresses may be used if special confinement reinforcement is provided. Illustrations of such reinforcement are discussed in connection with corbels and with beam-end connections, below.

Column-base connections are generally accomplished by means of steel base plates which are anchored into the precast column. Anchor bolts with two nuts each are provided in the footing or other supporting member (Fig. 9.12). The column is set on the lower nuts of the anchors. In this condition the entire temporary erection load is carried by the anchors, unless relieved by shims. When the column is positioned on the anchors, grout is placed under the anchor bolts to provide uniform bearing. If a flush appearance is desired, the anchors can be recessed in notches at the four corners of the column, which are subsequently dry-packed with grout (see, for example, Ref. 9.8).

Fig. 9.10   Precast room-size modules for 21-story hotel. (*Courtesy Portland Cement Assoc.*)

Fig. 9.11 Steel framing combined with precast floor plank and wall panels (not shown) for eight-story hotel. (*Bethlehem Steel Co. and Slingerland & Booss, Construction Engineers.*)

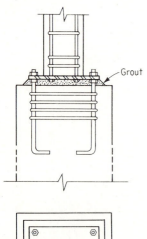

Grout

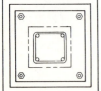

Fig. 9.12 Column-base connection.

Corbels and brackets are widely used in precast construction for supporting beams at columns or walls, but are also found in cast-in-place reinforced-concrete structures. Bearing plates are used to transfer beam reactions to corbels; their outer edge should be 2 in. or more from the outer edge of the corbel to prevent cracking of the corner. As shown in Fig. 9.13a, the beam reaction consists either of the vertical force V_u only or of an additional horizontal force N_u. The first situation obtains if bearing pads are used which relieve horizontal forces either by slipping or by shear deformation (e.g., Teflon pads or elastomeric or rubber pads), the second if steel bearing plates are welded to anchored plates or angles in both corbel and beam. The horizontal forces arise from volume changes as caused by shrinkage, temperature, and creep; unless relieved by appropriate pads, they should be provided for in design (see Ref. 9.1). The vertical reaction V_u should be assumed to act at the outer third point of the bearing plate. N_u should be determined by analysis, but shall not be taken as less than $0.20V_u$. It is regarded as live load with a load factor of 1.7.

The structural performance of a corbel or bracket is fairly complex, but may be approximately visualized from Fig. 9.13a. A correctly designed corbel, when overloaded, develops a crack, as shown. The applied forces V_u and N_u

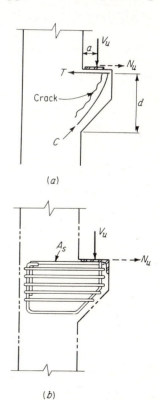

Fig. 9.13   Corbel. (*Adapted in part from Ref. 9.1.*)   (b)

(if present) are then equilibrated in trusslike fashion by the tension force T developed in the main tensile reinforcement and the compression force C which develops in the concrete compression strut outside the crack. To counteract cracking, either of the kind shown in Fig. 9.13a or closer to the root section, horizontal stirrups must be placed in the upper two-thirds of the corbel with a total area not less than one-half that of the tension reinforcement A_s, the latter being calculated as usual for the moment $a \times V_u$ at the root section of the corbel with depth d. The Code provides that the shear stress on the same section V_u/bd shall not exceed ϕv_u, where $\phi = 0.85$ and

$$v_u = \left(6.5 - 5.1 \sqrt{\frac{N_u}{V_u}}\right)(1 - 0.5a/d)\left\{1 + \left[64 + 160 \sqrt{\left(\frac{N_u}{V_u}\right)^3}\right]\rho\right\} \sqrt{f_c'}$$

$$(9.1)$$

It is stipulated that $\rho = A_s/bd$ shall not be less than $0.04 \times f_c'/f_y$. If it exceeds $0.13f_c'/f_y$, the latter value shall be substituted in Eq. (9.1). This and additional minor provisions in Sec. 11.14 of the Code have been developed from extensive tests (Ref. 9.7). The reinforcement of a corbel is shown in Fig. 9.13b. It is essential that the main tension reinforcement near the top face be anchored as close to the outer edge as possible. This can be done by welding to a corner angle as shown or, in the absence of such an angle, by welding to a cross bar.

Beam-end connections must be detailed to forestall trouble from high-stress concentrations near points of bearing. In these as in some other connections, the reinforcement necessary for the integrity of the member near the connection is determined on the basis of the so-called *shear-friction theory*. This method applies wherever direct shear is likely to cause failure, in contrast to the usual situation in beams and girders where shear is merely a measure for that diagonal tension which in fact gives rise to cracking.

In the *shear-friction method* one assumes a crack to occur along a shear plane in an unfavorable location. Any such crack must be traversed by well-

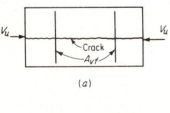

(a)

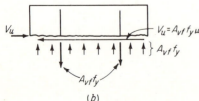

(b)   **Fig. 9.14**

anchored reinforcement (Fig. 9.14a). Once the crack is formed, sliding along the crack is counteracted by the interlocking of the two rough faces, a phenomenon known as shear friction even though this interlock is somewhat different from true friction. The normal force or pressure necessary to develop friction is developed in the reinforcement and is caused by the roughness of the crack, which tends to separate the two faces as sliding occurs. This causes tension in the steel which crosses the crack (Fig. 9.14b) and sets up an equal and opposite pressure between the crack faces. It is well documented by tests (Refs. 9.9 and 9.10) that the full yield stress f_y is developed in the reinforcement, causing a total pressure $A_{vf}f_y$, where A_{vf} is the shear-friction reinforcement, i.e., the steel area crossing the crack. The frictional force which resists sliding is then

$$V_u = \mu A_{vf}f_y \qquad (9.2a)$$

or the nominal shear stress

$$v_u = \mu \rho f_y \qquad (9.2b)$$

where μ is the friction coefficient. The Code specifies

$$\mu = \begin{cases} 1.4 & \text{for concrete cast monolithically} \\ 1.0 & \text{for concrete placed against hardened concrete} \\ 0.7 & \text{for concrete placed against as-rolled structural steel} \end{cases}$$

It also sets an upper limit of v_u, namely, $0.2f'_c$ or 800 psi, whichever is less. Thus, the shear-friction reinforcement is determined from

$$A_{vf} = \frac{V_u}{\phi \mu f_y} \qquad (9.3)$$

where $\phi = 0.85$.

The shear-friction theory is used for a variety of connections of precast members. For instance, corbels for which a/d equals 1 or less may be designed by this approach in lieu of Eq. (9.1). Likewise, the local reinforcement needed in *end connections of beams and girders* is determined in this manner. Figure 9.15 shows such a connection for a beam supported on a narrow bearing plate. The corner could spall off along a crack, shown in Fig. 9.15a. To counteract this, shear-friction reinforcement A_{vf} is provided as shown, and is anchored by welding it to a corner angle. Neglecting the small inclination of the bars to the horizontal, A_{vf} is directly determined by Eq. (9.3). The maximum tension force in this reinforcement, $T_u = A_{vf}f_y$, could cause a different crack, of the shape shown in Fig. 9.15b. To counteract this tendency, hoop reinforcement A_h is provided as shown and is calculated from $A_h = T_u/\mu f_y$. (Division by ϕ is not indicated here because ϕ was already used in determining A_{vf}, and thereby T_u.) Other examples of connections designed by the shear-friction method can be found in Refs. 9.1, 9.8, and 9.9.

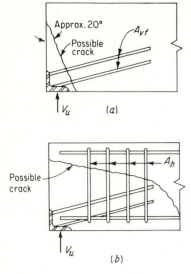

Fig. 9.15 Auxiliary beam-end reinforcement. (*Adapted from Ref. 9.1.*)

Cast-in-place concrete is frequently used in conjunction with precast concrete to provide continuity and monolothic action. This is done in either of two ways: (1) Small amounts of concrete or grout are placed at *connections of two adjacent beams or girders* after negative reinforcement over supports has been spliced by welding to provide continuity of adjacent spans. In Fig. 9.16 such splicing is done by welding the protruding negative steel of adjacent ledger beams to a splice angle. (2) Larger amounts of cast-in-place concrete are frequently used in *composite action* with precast members, such as *floor-*

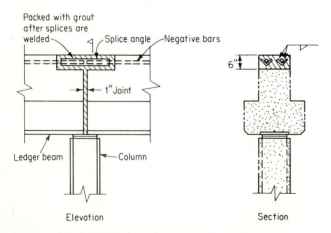

Fig. 9.16 Continuity connection of ledger beams. (*Adapted from Design Handbook for Standard Products, Concrete Materials, Inc., Charlotte, N.C., 1969.*)

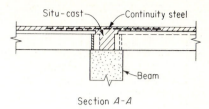

Section *A-A*

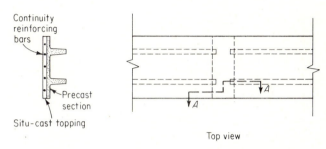

Top view

Fig. 9.17 Continuity connection of double-tee beams. (*Adapted from Ref.* 9.13.)

topping concrete placed over double tees (Ref. 9.13). Here likewise, continuity is developed between adjacent precast members by placing negative steel in the cast-in-place concrete, over the supports between adjacent members, as shown in Fig. 9.17.

This brief discussion of precast-concrete construction in general, and of connections in particular, is in no way meant to be comprehensive. It is merely intended to indicate by examples the peculiarities of precast construction and the manner in which it is used. The increasing importance of industrialization of construction and of systems building is bound to make precasting an ever more important method of building.

REFERENCES

9.1. P.C.I. Design Handbook: Precast and Prestressed Concrete, Prestressed Concrete Institute, Chicago, 1971.
9.2. Symposium on Precast Concrete Wall Panels, ACI Publ. SP-11, 1965.
9.3. Quality Standards and Tests for Precast Concrete Wall Panels, Report of ACI Committee 533, *J. ACI*, vol. 66, p. 270, 1969.
9.4. Selection and Use of Materials for Concrete Wall Panels, Report by ACI Committee 533, *J. ACI*, vol. 66, p. 814, 1969.
9.5. Design of Precast Concrete Wall Panels, Report of ACI Committee 533, *J. ACI*, vol. 68, p. 504, 1971.
9.6. Recommended Practice for Manufactured Reinforced Concrete Floor and Roof Units, Report of ACI-ASCE Committee 512, *J. ACI*, vol. 63, p. 625, 1966.
9.7. L. B. Kriz and C. R. Raths: Connections in Precast Concrete Structures: Strength of Corbels, *J. Prestressed Concrete Inst.*, vol. 10, p. 16, 1965.

9.8. P. W. Birkeland and H. W. Birkeland: Connections in Precast Concrete Construction, *J. ACI*, vol. 63, p. 345, 1966.

9.9. R. F. Mast: Auxiliary Reinforcement in Concrete Connections, *Proc. ASCE*, vol. 94, no. ST6, p. 1485, 1968.

9.10. J. F. Hofbeck, I. O. Ibrahim, and A. H. Mattock: Shear Transfer in Reinforced Concrete, *J. ACI*, vol. 66, p. 119, 1969.

9.11. G. V. Berg and J. L. Stratta: "Anchorage and the Alaska Earthquake," American Iron and Steel Institute, New York, 1964.

9.12. W. E. Kunze, J. A. Sbarounis, and J. E. Amrheim: The March 27 Alaskan Earthquake: Effects on Structures in Anchorage, *J. ACI*, vol. 62, p. 635, 1965.

9.13. F. S. Rostasy: Connections in Precast Concrete Structures: Continuity in Double-T Floor Construction, *J. Prestressed Concrete Inst.*, vol. 7, no. 4, p. 18, 1962.

10
Arch and Shell Roofs

10.1 INTRODUCTION

The search for structures which cover long spans and large areas without intermediate support, and which do so using a minimum of material, has long occupied the structural engineering profession. If a flat plane surface is not necessary to meet the functional requirements of a structure, then singly or doubly curved members such as arches and thin shells offer important advantages. Not only are they architecturally appealing; they permit substantial saving of material when compared with designs using slabs, beams, and columns. These nonlinear, nonplanar systems owe their material economy to a unique capacity to resist loads primarily by direct stress, rather than by flexural and transverse shear stress.

A beam is one of the least efficient of structural members. Commonly, only one cross section of a beam is subject to its maximum design moment, and consequently, if the member is prismatic, only one cross section of the beam is working at its full capacity. This situation is somewhat improved in continuous structures, in which two sections may be simultaneously subjected to the maximum moment, and by inelastic-moment redistribution (see Art. 7.7), which further utilizes excess strength. Still, it is impossible for the beam

431

to compete with the efficiency of the simple compression strut in which the strength of every cross section along the member axis is fully utilized over its entire width and depth. This supports the choice, if design constraints permit, of a structural form in which direct stress (in the case of concrete, compressive stress) is dominant over flexural and transverse shear stress. The degree to which direct stress is dominant is closely related to the structural efficiency of the design.

The use of arches is the oldest structural method for bridging spans too large for straight beams. Arches and vaults were perfected chiefly by the Romans for their bridges, aqueducts, and large public buildings more than two thousand years ago. They have been in wide use ever since, constructed of masonry, concrete, steel, and timber. Arches may be regarded as linear members, in common with beams and columns, in the sense that the cross-sectional dimensions are small in comparison with the length. In the case of an arch, however, the member axis is curved in such a way that the line of action of the thrust resulting from loads coincides very nearly with the axis of the member. In this way bending moments and transverse shears are minimized; for particular loadings they may be eliminated altogether.

Shells are often referred to as surface structures because both the length and width of the structural unit are large relative to its thickness. They may be curved in one direction in the form of a cylinder, or doubly curved to form a dome or a saddle-shaped surface. Their economy over long spans results from their ability to translate applied loads into "membrane" thrusts and shears acting in a plane tangent to the surface at any point. By this means bending and twisting moments, and shears transverse to the surface, are reduced or eliminated. Moreover, in the case of shells, significant overall economy results from the fact that the main structural unit serves a space-enclosing function as well.

10.2 ARCHES

Modern arches are of three types: the fixed arch, which is rigidly connected to the abutments, the two-hinged arch, which is supported at each end by a hinge resting on the abutment, and the three-hinged arch, which has an additional hinge at the crown. Reinforced-concrete arches are mostly of the first two types.

The structural action and advantage of arches are easily recognized by considering a two-hinged arch. Figure 10.1a shows a uniformly loaded simple beam with $M_{max} = wL^2/8$, $V_{max} = wL/2$, and axial force $N = 0$. If this beam is curved, as in Fig. 10.1b, M_{max} is the same as before. The shear at the support, however, is decreased to $V_{max} = (wL/2) \cos \theta$; in addition, an axial force is now present in the member which, at the supports, equals $N = (wL/2) \sin \theta$. This curved beam tends to straighten under load since, owing to bending, the bottom fibers are stretched and the top fibers compressed. This action increases

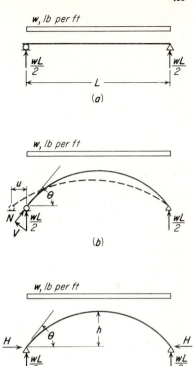

Fig. 10.1

the span, as shown in Fig. 10.1b. To maintain the original span, as in the two-hinged arch with immovable supports (Fig. 10.1c), a horizontal reaction H must be applied. In this two-hinged arch

$$M_{\max} = \frac{wL^2}{8} - Hh$$

$$V_{\max} = \frac{wL}{2} \cos \theta - H \sin \theta$$

$$N = \frac{wL}{2} \sin \theta + H \cos \theta$$

It is seen that, in comparison with the simple beam, the bending moment and the shear force in the arch are greatly decreased. On the other hand, an axial compression force N, not present in the beam, acts in the arch. It is exactly this state of affairs which is desirable in reinforced concrete, since concrete is relatively weak in tension and shear but strong in compression.

The choice of two-hinged or fixed arches depends on the particular conditions of design. Fixed arches require immovable abutments which prevent not only vertical and horizontal motion, but also rotation of the support. For roof arches resting on columns or walls, it is obviously impossible to fix the

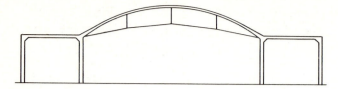

Fig. 10.2 Two-hinged roof arch with cambered tie rod.

supports against rotation, since the rigidity of the supporting structure is not sufficient for this purpose. Two-hinged arches, on the other hand, are insensitive to rotations and to moderate differential settlements of the abutments, which, for this reason, can often be made less substantial than for fixed arches. Tie rods are often provided in two-hinged arches to resist the horizontal reaction H; hence the supports have to resist vertical reactions only, with a corresponding further decrease in their cost. Tied two-hinged arches are obviously advantageous in roof structures of the type shown in Fig. 10.2, where large horizontal reactions would require much heavier columns and frames. Large hangars and halls of the type of Fig. 10.3 have been built both with fixed arches and with tied two-hinged arches; in the latter case tie rods were located below the floor. Effective hinges can be constructed at reasonable cost, provided that the reactions are not excessively large. They may be architecturally undesirable, however, and represent a maintenance problem in some cases. A well-considered decision on the type of arch to be used in a given design depends, therefore, on a balanced evaluation of factors such as those just enumerated.

Fig. 10.3 Hangar at Chicago Municipal Airport. Fixed arches spaced at 28 ft $11\frac{1}{2}$ in.; clear span 257 ft; height at center 60 ft; thickness of concrete shell between arches from $3\frac{1}{2}$ to 6 in. (*Ammann and Whitney.*)

As was noted in the general discussion of arch action (Fig. 10.1), arch ribs are subject to compression and bending, and usually only small shear. They are therefore designed and analyzed in the same manner as eccentrically compressed columns, one layer of steel being located near the top surface and another near the bottom. It is economical to keep steel percentages small, on the order of 1 to 2 per cent. Lateral ties for the reinforcement are used in arch ribs as in columns.

10.3 THEORY OF SYMMETRICAL, TWO-HINGED ARCHES

The two-hinged arch (Fig. 10.1c) differs from a curved simple beam (Fig. 10.1b) in that horizontally immovable hinges or tie rods prevent the spread of supports. The forces in such an arch can be analyzed by introducing a fictitious roller support at one of the hinges, which converts the arch into a simple beam. The horizontal displacement u can then be determined. To convert this fictitious beam into the actual arch, a horizontal force H must be introduced of sufficient magnitude to produce a displacement equal and opposite to u. In other words, this force brings the fictitious roller support back to its original position as it actually is in the arch.

To determine the horizontal displacement u, take a small element ds (Fig. 10.4a) of the arch, and consider the effect of the deformation of this one element on the deformation of the arch as a whole. Thus all parts of the arch, with the exception of this one element, are considered rigid. The contribution of the deformation of the element ds to the total displacement can be determined as follows: A bending moment M, considered constant over the small length ds, will cause the left cross section to rotate with respect to the right one by an amount $d\phi$, which can be determined by the moment-area principle, i.e., from the fact that the rotation of one section with respect to another is equal to the area of the moment curve between them divided by EI. Hence,

$$d\phi = \frac{M\,ds}{EI} \tag{10.1}$$

The entire left portion of the arch, rigidly attached as it is to the cross section, will likewise rotate through an angle $d\phi$, as represented by the dotted position of the axis in Fig. 10.4b. Consequently, the left end of the arch describes an arc $r\,d\phi$. It will be noted that triangle ABC is similar to triangle CDE. Hence, from simple proportions, $du/r\,d\phi = y/r$, and the horizontal displacement is

$$du = y\,d\phi = \frac{My\,ds}{EI} \tag{10.1a}$$

This is the contribution made to the total displacement u by the angular deformation due to bending of the single element ds. To obtain u itself, it

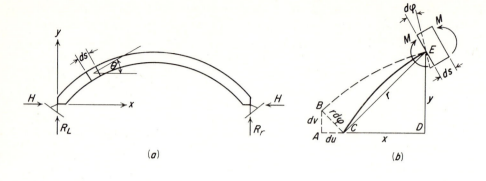

(a)   (b)

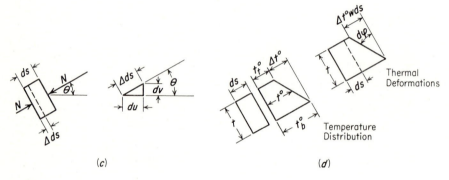

(c)   (d)

Fig. 10.4

is only necessary to sum these contributions for all elements ds along the arch, or

$$u = \int_0^s \frac{My\,ds}{EI} \tag{10.2}$$

Let M_0 represent the bending moment due to any external arch loads caused in the fictitious simple curved beam, as in Fig. 10.1b. These moments, of course, are the same as those in the straight beam (Fig. 10.1a). Then the outward displacement due to these "simple beam moments" is

$$u_{M_0} = \int_0^s \frac{M_0 y\,ds}{EI} \tag{10.2a}$$

The sign convention is the same as in beam analysis; i.e., moments are considered positive if they produce tension in the bottom fiber. In addition to M_0, the horizontal reaction H causes a moment at section x which is equal to

$$M_H = -Hy$$

and the inward displacement caused by these moments is

$$u_{M_H} = -H \int_0^s \frac{y^2 \, ds}{EI} \tag{10.2b}$$

Since, actually, no displacement occurs at the immovable hinge,

$$u_{M_0} = -u_{M_H}$$

from which one obtains the magnitude of the horizontal reaction

$$H = -\frac{\int_0^s (M_0 y \, ds/EI)}{\int_0^s (y^2 \, ds/EI)} \tag{10.3}$$

If displacements u were caused exclusively by bending, Eq. (10.3) would furnish all the information necessary for analyzing a two-hinged arch. Since such an arch, as is seen from Fig. 10.1, has only one statically indeterminate reaction H, it is now possible to find moments, shears, and axial forces at any point in the arch from equilibrium conditions only, because the previous unknown force H is given by Eq. (10.3). However, other factors in addition to bending moments may contribute significantly to the magnitude of u, and must therefore be considered in determining H.

In addition to bending moments, shear forces and axial thrusts act on element ds. Deformations due to shear in arches are even smaller, relatively, than in beams, and can therefore be neglected. Those caused by axial compression forces N, on the other hand, may affect the magnitude of H significantly and must be considered.

Consider an element ds with a concentric force N on both faces, as shown in Fig. 10.4c. The axial shortening Δds caused by the normal force N_0 is $N_0 \, ds/EA$. N_0 is the component, parallel to the arch axis at the element ds, of all external forces and reactions on the fictitious curved beam to the right or left of the element ds. The component of Δds in the direction of u is

$$du = \Delta ds \cos \theta \tag{10.4}$$

Hence,

$$du_{N_0} = \frac{N_0 \cos \theta \, ds}{EA}$$

and the total displacement due to N_0 is

$$u_{N_0} = -\int_0^s \frac{N_0 \cos \theta \, ds}{EA} \tag{10.5}$$

The minus sign indicates that this displacement is inward, in contrast to the outward displacement u_{M_0}. On the other hand, the horizontal reaction H results in an axial compression, at any point of the arch, of magnitude $H \cos \theta$; thus the inward displacement caused by this action is

$$u_{N_H} = -H \int_0^s \frac{\cos^2 \theta \; ds}{EA} \tag{10.6}$$

Temperature changes which may occur in roof arches produce further deformation of the arch. A uniform temperature increase of $t°$ causes the arch element ds to increase in length by the amount

$$\Delta \; ds = \omega t° \; ds$$

where ω is the temperature coefficient of expansion. Substitution of this expression in Eq. (10.4) and integration over the entire arch give the total support displacement due to a uniform temperature increase:

$$u_t = \omega t° \int_0^s \cos \theta \; ds = \omega t° l \tag{10.7}$$

Another factor which results in inward displacement u is the inevitable shrinkage of concrete. It has been shown, however, that the unfavorable effect of shrinkage is greatly decreased by the longtime plastic flow of concrete, which allows the structure to adapt itself to those stresses which are caused by change of shape, such as shrinkage, rather than by external loads. It has been proposed that the limited effect of shrinkage can be accounted for by an equivalent drop in temperature of 15°F added to the actual seasonal temperature drop for which the arch is designed. The temperature coefficient may be taken as $\omega = 5.5 \times 10^{-6}$ per degree Fahrenheit.

In addition to the uniform temperature change just considered, a temperature differential may exist between the inside and outside surfaces in roof arches, particularly if, as in the hangar of Fig. 10.3, the roof slab or shell is placed at middepth of the arch ribs. The total temperature changes and corresponding deformations may then be represented as in Fig. 10.4d. The amount by which the bottom fiber ds expands more than the top fiber is $\Delta t° \; \omega \; ds$. It is seen that this results in

$$d\phi = \frac{\Delta t° \; \omega \; ds}{t} \tag{10.8}$$

If this value is substituted in the left equality of Eq. (10.1a), and the result integrated over the arch, the total displacement caused by the temperature gradient becomes

$$u_{\Delta t°} = \omega \; \Delta t° \int_0^s \frac{y \; ds}{t} \tag{10.9}$$

These are all the types of displacements that occur in an arch with immovable hinges. In arches with tie rods, hinges are not immovable, since the force H in the tie rod causes it to stretch. For a tie rod of area A_{tr} and modulus E_{tr}, this stretch amounts to

$$u_{tr} = H \frac{l}{E_{tr} A_{tr}} \tag{10.10}$$

For a tied arch, the sum of all other displacements must be equal to the stretch of the tie; i.e.,

$$u_{M_0} + u_{M_H} + u_{N_0} + u_{N_H} + u_{t^\circ} + u_{\Delta t^\circ} = u_{tr}$$

If the proper values are substituted in the equation, the unknown horizontal force H may be found.

$$H = \frac{\displaystyle\int_0^s \frac{M_0 y \, ds}{I} - \int_0^s \frac{N_0 \cos \theta \, ds}{A} + E\omega \left(t^\circ l + \Delta t^\circ \int_0^s \frac{y \, ds}{t} \right)}{\displaystyle\int_0^s \frac{y^2 \, ds}{I} + \int_0^s \frac{\cos^2 \theta \, ds}{A} + \frac{EI}{E_{tr} A_{tr}}} \tag{10.11}$$

This lengthy expression can usually be simplified by omitting terms which are zero or negligibly small, depending on the particular condition. Thus, for an arch without tie rod, the last term in the denominator drops out, since $u_{tr} = 0$. If no temperature gradient can occur between inside and outside surfaces, such as in roof arches located below a well-insulated surface, the term containing Δt° is zero. Furthermore, in many cases the displacements u_{N_0} and u_{N_H} caused by axial compression are very small compared with those caused by bending moments (u_{M_0} and u_{M_H}). For a parabolic arch, Timoshenko (Ref. 10.1) has computed the amount of error obtained by neglecting the influence of axial compression in determining H. His results are given in the table below, in which t_c is the thickness at the crown.

h/l	$\frac{1}{12}$			$\frac{1}{8}$			$\frac{1}{4}$		
t_c/l	$\frac{1}{10}$	$\frac{1}{20}$	$\frac{1}{30}$	$\frac{1}{10}$	$\frac{1}{20}$	$\frac{1}{30}$	$\frac{1}{10}$	$\frac{1}{20}$	$\frac{1}{30}$
Error, %	17.1	5.1	2.4	8.4	2.2	1.0	1.8	0.4	0.2

Since the usual accuracy of engineering designs is hardly greater than 5 per cent on account of uncertainties of loading, materials properties, and various other factors, the influence of axial compression can be neglected in all but very flat or unusually thick arches. Normally then, the second term in both the numerator and denominator can be dropped as being negligibly small. Hence, for an arch of usual dimensions without tie rod and without temperature gradient, H can be determined with sufficient accuracy from

$$H = \frac{\displaystyle\int_0^s (M_0 y \, ds/I) + E\omega t^\circ l}{\displaystyle\int_0^s (y^2 \, ds/I)} \tag{10.11a}$$

The determination of the integrals in the equation for H is simple, provided that M_0, N_0, I, A, h, and y can be expressed as functions of s and that these functions can be integrated. Ordinarily this is possible for only a few very simple cases in which the arch axis is either a parabola or a circular arc. In most other cases it becomes necessary to approximate these integrals with a finite sum of numerical terms. The details of this procedure will be given after the theory of fixed arches is developed.

10.4 THEORY OF SYMMETRICAL FIXED ARCHES

The fixed arch shown in Fig. 10.5a has a total of six reactions, i.e., a vertical and a horizontal reaction and a moment at each support. There are, then, three redundant reactions to be determined. If no horizontal loads are present, the condition that the sum of horizontal forces be zero results in $H_l = H_r$.

For purposes of analysis the arch is cut through the crown, and each half arch is regarded as a statically determinate cantilever fixed at its support. In the plane of separation, in general, a horizontal and a vertical force H_c and V_c and a bending moment M_c will act from one half onto the other. These three quantities are taken as the redundants. They are shown, in Fig. 10.5b, acting in the direction and sense which will be assumed as positive in the following derivations. In any other cross section of the structure, M, N, and V will be assumed positive, as shown in Fig. 10.5b. If M_c, H_c, and V_c are known, M, N, and V at any point can be determined from statics.

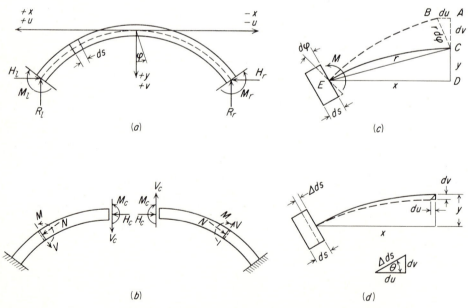

Fig. 10.5

The crown sections of the separate halves, through the action of loads, temperature, and other influences, will be displaced from their original positions. In general, there will be a horizontal displacement u, a vertical displacement v, and a rotation of the normal to the crown section ϕ. The assumed positive directions of these displacements are shown in Fig. 10.5a. If the structure in fact consisted of two independent cantilevers, u, v, and ϕ would, in general, be different for the two halves. Since, actually, the arch is one continuous structure, these displacements must be equal in magnitude and sign. Consequently,

$$u_l = u_r \qquad v_l = v_r \qquad \phi_l = \phi_r \tag{10.12}$$

where the subscripts l and r refer to the left and right half arch, respectively. These three equations permit the determination of the redundants H_c, V_c, and M_c. In other words, in the arch these three quantities must be of such magnitude as to join the left and the right cantilevers into one continuous structure.

Isolating an element ds of the arch (Fig. 10.5c), one can determine the contributions du, dv, and $d\phi$ made to the total displacements u, v, and ϕ by the deformations of this single element. The bending moment M causes a rotation $d\phi$ which is determined by Eq. (10.1). The part of the arch between the section and the crown, considered rigid and represented by its axis in Fig. 10.5c, rotates through the same angle $d\phi$, which, consequently, also represents the rotation of the crown section caused by the deformation of the element ds. Furthermore, since the triangles ABC and CDE are similar, it is seen that

$$\frac{du}{r\,d\phi} = \frac{y}{r} \qquad \text{and} \qquad \frac{dv}{r\,d\phi} = \frac{x}{r}$$

from which

$$du = y\,d\phi \qquad dv = x\,d\phi \tag{10.13}$$

Substituting $d\phi$ from Eq. (10.1), one obtains

$$du = \pm\frac{My\,ds}{EI} \qquad dv = \mp\frac{Mx\,ds}{EI} \qquad d\phi = \pm\frac{M\,ds}{EI} \tag{10.13a}$$

In these and the following expressions, if two signs are shown, the upper refers to the left half arch, and the lower to the right.

In addition to the angular rotation $d\phi$, the elements ds will be uniformly compressed or stretched by normal forces N, temperature variations, and shrinkage. If $\Delta\,ds$ is the change of ds due to any of these actions, it is seen, from Fig. 10.5d, that

$$du = \Delta\,ds\cos\theta \qquad dv = \Delta\,ds\sin\theta \tag{10.14}$$

An axial force N results in $\Delta\,ds = N\,ds/EA$, and hence in horizontal and vertical displacements

$$du = \frac{N \, ds \, \cos \theta}{EA} \qquad dv = \frac{N \, ds \, \sin \theta}{EA} \tag{10.14a}$$

At any point of the arch, x, y, the moment M, and the axial force N are the resultant effects of the three crown quantities M_c, H_c, and V_c and of the external loads acting on the half-arch cantilever in which the particular section is located. They can be calculated by statics (see Fig. 10.5b) and are tabulated below:

Force and moment at section x, y

Due to.....	M_c	H_c	V_c	External loads
M	M_c	$H_c y$	$-V_c x$	M_0
N	0	$H_c \cos \theta$	$V_c \sin \theta$	N_0

The signs in this table are valid for both halves of the arch. In this connection it is important to note that for the right half arch, x and θ are negative, whereas y is positive by the sign convention given in Fig. 10.5a. It should be also be noted that external gravity loads result in negative M_0 (see Fig. 10.5b). N_0 at any section is the component, in the direction of the tangent to the arch axis, of all the external loads located between that section and the crown; gravity loads therefore result in positive N_0.

By substituting the values M and N from the table into Eqs. (10.13a) and (10.14a) and integrating for the left half from 0 to $+s/2$, and for the right half from 0 to $-s/2$, one obtains separately the crown displacements u and v for either half. If these are substituted in Eqs. (10.12), they result in three simultaneous equations for H_c, M_c, and V_c. If any particular action, for example that of H_c, results in equal displacements for both halves, these can obviously be omitted from Eqs. (10.12), since equal terms on both sides of an equation cancel. For example, H_c results in equal vertical displacements v_l and v_r, as is seen directly from Fig. 10.5c (the corresponding behavior of the right half is obtained by mirroring the entire figure about the vertical axis through the crown). Hence, the contribution of H_c to the vertical displacements can be omitted in the second of Eqs. (10.12). On the other hand, the horizontal crown displacements caused by H_c are equal in magnitude but opposite in sign, since the crown section of the left cantilever is displaced to the left (see Fig. 10.5b), that of the right cantilever to the right. Hence, these displacements must be entered in the first of Eqs. (10.12).

Table 10.1 gives the crown displacements caused by the various actions, separately for both halves. They must be substituted in Eqs. (10.12) to determine the three redundants. In this table the notation "cancels" designates

Table 10.1 Crown displacement of half arch of Fig. 10.5

All integrals $\int_0^{+s/2}$ for left half; $\int_0^{-s/2}$ for right half; multiply by $1/E$.

Caused by bending

	$+M_c$	$+H_c$	$+V_c$	$+M_0$
u	$\pm M_c \int \dfrac{y\,ds^*}{I}$	$\pm H_c \int \dfrac{y^2\,ds}{I}$	Cancels	$\pm \int \dfrac{M_0 y\,ds}{I}$
v	Cancels	Cancels	$\pm V_c \int \dfrac{x^2\,ds}{I}$	$\mp \int \dfrac{M_0 x\,ds}{I}$
ϕ	$\pm M_c \int \dfrac{ds}{I}$	$\pm H_c \int \dfrac{y\,ds^*}{I}$	Cancels	$\pm \int \dfrac{M_0\,ds}{I}$

Caused by arch shortening and temperature

	$+H_c$	$+V_c$	$+N_0$	$+t°$	$+\Delta t°$
u	$\pm H_c \int \dfrac{\cos^2\theta\,ds}{A}$	Cancels	$\pm N_0 \int \dfrac{\cos\theta\,ds}{A}$	$\mp \omega t° E \int \cos\theta\,ds$	Cancels
v	Cancels	$\pm V_c \int \dfrac{\sin^2\theta\,ds}{A}$	Cancels	Cancels	$\mp \Delta t° \omega E \int \dfrac{x\,ds}{t}$
ϕ	0	0	0	0	$\pm \Delta t° \omega E \int \dfrac{ds}{t}$

* These expressions are zero when x and y are measured from the elastic center.

displacements equal in sign and magnitude for both halves, which cancel if substituted in Eqs. (10.12). For simplicity, all values in the table are multiplied by E.

In addition to the displacements caused by the crown forces and the external loads (which include the weight of the arch), those resulting from temperature changes and shrinkage must be considered, just as in the case of the two-hinged arch. In fact, these influences are of more consequence for the fixed than for the two-hinged arch because the latter, being free to rotate at the supports, is better able to accommodate itself to changes in shape without excessive stress.

A uniform temperature increase of amount $t°$ causes an elongation $\Delta ds = \omega t°\,ds$ which, if substituted in Eqs. (10.14) and integrated over the respective half arches, gives the corresponding crown displacements. Likewise, a temperature differential between top and bottom fiber (Fig. 10.4d) results in a rotation $d\phi$, given by Eq. (10.8). This must be substituted in Eqs. (10.13) and similarly integrated to result in the corresponding crown displacements. These values are shown in Table 10.1. The general information as to range of temperature and method of accounting for shrinkage given in the article on two-hinged arches is valid for the fixed arch.

If the displacements of the table are substituted in Eqs. (10.12), the second equation contains only V_c as an unknown, and hence can be solved for this quantity. The first and third equations, however, contain both H_c and M_c, resulting in two simultaneous equations which are awkward to solve and, when solved, lead to rather lengthy expressions for these two quantities. This complication can be overcome by a suitable transformation of coordinates.

Consider the first term of the first row and the second term of the third row of the displacement table; if $\int y \, ds/I$ could be made zero, the first of Eqs. (10.12) would contain only H_c as an unknown, and the third only M_c. Obviously, for the location of the coordinate axes shown in Fig. 10.5a, this integral is not zero, since y is positive throughout. It is possible, however, to shift the origin to a position which will make this integral vanish. To visualize the meaning of this shift, assume the arch axis to be loaded by fictitious loads equal to $1/I$ at every point, as shown in Fig. 10.6a. Then, since y_1 is the lever arm of any of these forces about the origin, $\int y_1 \, ds/I = 0$ shows that the sum of the moments of all these forces about the x axis is to be zero; i.e., the axis is to pass through the center of these forces or is to be coincident with their resultant. The coordinate origin determined in this manner is known as the *elastic center* of the arch. By the general formula for the location of the resultant of parallel forces, the distance y_e of the elastic center from the center of the crown is then found from

$$y_e = \frac{\int_0^{s/2} (y \, ds/I)}{\int_0^{s/2} (ds/I)} \tag{10.15}$$

By symmetry these integrals need be extended only over the half arch.

Let y_1 designate the ordinate of any point on the arch axis measured from this new origin, as shown in Fig. 10.6a. Now, assume that the elastic center is joined to the crown section by a rigid, fictitious lever, as shown separately for each half in Fig. 10.6b. Instead of introducing the three unknown quantities

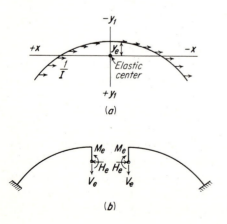

(a)

(b)   Fig. 10.6

M_c, H_c, and V_c at the crown, introduce three similar quantities M_e, H_e, and V_e at the ends of these fictitious levers, positive as shown. Then the previously tabulated expressions for u, v, and ϕ refer to the new origin (end of levers) if, in them, y is replaced by y_1, and M_c, H_c, and V_c by M_e, H_e, and V_e. Equations (10.12) are now valid for the ends of these levers. Since the levers are regarded as rigidly connected to their respective crown sections, and the two halves are fitted together to form the complete arch, the ends of the two levers cannot separate, no matter how the arch deforms. Thus the displacements u, v, and ϕ must be the same for the ends of both levers.

Since the origin is now chosen such that $\int y_1 \, ds / I = 0$, the first term of the first row and the second term of the third row in Table 10.1 vanish. (These terms are indicated by asterisks in the table.) Hence, if the remaining terms of the first, second, and third rows of the table are substituted, respectively, in the first, second, and third of Eqs. (10.12), each results in one equation with one unknown, H_e, V_e, and M_e in that order. Since in each individual expression the respective integral for the left half arch is added to that for the right half, integration can be extended directly from $-s/2$ to $+s/2$, rather than writing two integrals for each term. The three redundant quantities, referred to the elastic center, then become

$$H_e = \frac{-\displaystyle\int_{-s/2}^{s/2} \frac{M_0 y_1 \, ds}{I} - \int_{-s/2}^{s/2} \frac{N_0 \cos\theta \, ds}{A} + E t^\circ \omega l - E \, \Delta t^\circ \, \omega \int_{-s/2}^{s/2} \frac{y_1 \, ds}{t}}{\displaystyle\int_{-s/2}^{s/2} \frac{y_1^2 \, ds}{I} + \int_{-s/2}^{s/2} \frac{\cos^2\theta \, ds}{A}}$$

(10.16)

$$V_e = \frac{\displaystyle\int_{-s/2}^{s/2} (M_0 x \, ds / I) - \int_{-s/2}^{s/2} (N_0 \sin\theta \, ds / A)}{\displaystyle\int_{-s/2}^{s/2} (x^2 \, ds / I) + \int_{-s/2}^{s/2} (\sin^2\theta \, ds / A)}$$

(10.17)

$$M_e = \frac{-\displaystyle\int_{-s/2}^{s/2} (M_0 \, ds / I) - E \, \Delta t^\circ \, \omega \int_{-s/2}^{s/2} (ds / t)}{\displaystyle\int_{-s/2}^{s/2} (ds / I)}$$

(10.18)

where all integrals extend over the entire arch, as indicated.

Once these quantities are determined, it is more convenient to work with crown moments and forces (Fig. 10.5b) than with those referred to the elastic center (Fig. 10.6b). As is seen directly from this figure,

$$H_c = H_e \qquad V_c = V_e \qquad M_c = M_e - H_e y_e \tag{10.19}$$

Equations (10.16) to (10.18) account for all influences that act on symmetrical fixed arches. In any particular design it will often be found that certain of these influences are small compared with others and can be neglected. The analogous situation for the two-hinged arch was discussed in Art. 10.3. As in that case, the influence of rib shortening caused by axial forces N is small except for unusually flat and thick arches. Except for such cases, therefore, the second term in both numerator and denominator of Eqs. (10.16) and (10.17) can be omitted. The last term in the numerator of Eq. (10.16) will often be found to be small in comparison with other terms. Indeed, for an arch with constant cross section this term vanishes. It does so because the elastic center was chosen such that $\int y_1\, ds/I = 0$, which, for constant I, is the same as $\int y_1\, ds = 0$. This makes $\int y_1\, ds/t = 0$ if t is constant. It is seen, therefore, that the magnitude of that term in Eq. (10.16) depends essentially on the ratio of the depths at springing and crown and will usually be negligible. Finally, the term containing $\Delta t°$ in Eq. (10.18) must be included only in the rather rare case of a roof arch not insulated at the top.

Hence, the following simplified expressions are usually accurate enough for purposes of design.

$$H_e = \frac{-\int_{-s/2}^{s/2} (M_0 y_1\, ds/I) + Et°\omega l}{\int_{-s/2}^{s/2} (y_1{}^2\, ds/I)} \qquad (10.16a)$$

$$V_e = \frac{\int_{-s/2}^{s/2} (M_0 x\, ds/I)}{\int_{-s/2}^{s/2} (x^2\, ds/I)} \qquad (10.17a)$$

$$M_e = \frac{-\int_{-s/2}^{s/2} (M_0\, ds/I)}{\int_{-s/2}^{s/2} (ds/I)} \qquad (10.18a)$$

A word of warning should be included in connection with these simplified equations. Although under the specified conditions which were enumerated the omitted terms are usually negligibly small, there is no way of definitely ascertaining this fact except by computing them. It is therefore suggested, particularly for closely designed arches of significant spans, that the simplified equations be used for preliminary design but that in computing final stresses the influence of the omitted terms be at least estimated numerically and included in the final expressions if found significant.

10.5 EVALUATION OF INTEGRALS

The integrals which occur in the expressions for the redundants of both the two-hinged arch [Eqs. (10.11) and (10.11a)] and the fixed arch [Eqs. (10.16) to (10.18) and (10.16a) to (10.18a)] can be evaluated by direct integration only in a few simple cases. For this reason it is necessary, in general, to approximate the integrals with sums of a finite number of terms.

For this purpose the arch axis (not the span) is generally divided into a number of equal intervals Δs. A maximum of 10 intervals for the half arch is generally sufficient; 6 to 8 intervals may often be used for small, secondary structures. The pertinent quantities entering a particular integral, such as y_1, M_0, and I for $\int M_0 y_1 \, ds/I$, are then computed for the midpoint of each individual interval Δs. Then

$$\int \frac{M_0 y_1 \, ds}{I} \cong \sum \frac{M_{0n} y_{1n}}{I_n} \Delta s$$

where the subscript n refers to the values M_0, y_1, and I at the midpoint of the nth interval. The other integrals in Eqs. (10.11) and (10.16) to (10.18) are computed in a similar manner by summation. Attention must be paid to the signs of the various quantities. Thus, y_1 is negative if measured upward from the elastic center of the fixed arch (see Fig. 10.6a); x is negative if measured to the right of the crown of the fixed arch, but is positive throughout for the two-hinged arch, as is evident from Figs. 10.6a and 10.5a. The importance of this is evident, for instance, in connection with Eq. (10.4a). If an arch is symmetrically loaded, M_0 and N_0 are equal at corresponding points of the two half arches. On the other hand, for the left half arch, x and $\sin \theta$ are positive, whereas for the right half, they are negative. Each contribution to the sums of both numerator integrals made by an element on the left half is therefore canceled by the equal and opposite contribution made by the corresponding element on the right. Consequently, V_e, and thereby V_c, are zero for symmetrical loads, in analogy with a simple beam which has zero shear at midspan, if symmetrically loaded.

For fixed arches, the location of the elastic center [Eq. (10.15)] is likewise calculated by replacing the integrals with corresponding sums. The computation of these sums is best arranged in tabular form, as demonstrated in the design example of Art. 10.8.

Moments of inertia can be computed for the entire concrete section, including the tension portion, without regard to reinforcement.

10.6 PRELIMINARY ARCH DESIGN

The analysis discussed in the preceding articles can be carried out only if loads, as well as all dimensions of the arch, are definitely known. Before such an analysis can be made, it is therefore necessary to determine the shape and weight of the arch by a preliminary design. In general, only the span, the rise, and the loads superimposed on the arch are known. (In cases in which the designer is free to select the rise it is important to consider that the horizontal thrust decreases as the ratio of rise to span increases, with a consequent saving in cost of abutments.) It is then required to determine the shape of the arch axis and the dimensions of the cross sections at various points and, from these, to find the weight of the arch proper. Once these quantities are preliminarily determined, an exact stress analysis can be made. Should this analysis indicate the necessity for a considerable change in the assumed dimensions, a second, final analysis must be made, based on the corrected dimensions. Since such

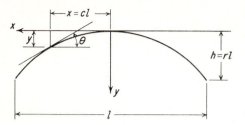

Fig. 10.7

computations are very time-consuming, it is desirable to determine the shape of the arch by preliminary design as accurately as possible, so that subsequent corrections are held to a minimum.

The axis of the arch is often made to coincide with the equilibrium or string polygon for dead load. If this is done, bending moments are caused by live load only. This polygon can be drawn only after the weight of the arch is determined. For roof arches, in which dead loads are mostly uniform along the axis, the parabola is a close approximation to the dead-load polygon, and, with the designations of Fig. 10.7, the equations of the axis and the slope are

$$y = 4rlc^2 \qquad\qquad\qquad\qquad\qquad\qquad\qquad\qquad\qquad\qquad (10.19a)$$

$$\tan \theta = 8rc \qquad\qquad\qquad\qquad\qquad\qquad\qquad\qquad\qquad\qquad (10.19b)$$

Next it is necessary to obtain an estimate of the maximum moments, thrusts, and shears for the purpose of determining the required cross section. To this end superimposed loads are determined, and a rough estimate is made of the weight of the arch proper. If temperature stresses need consideration, the limiting temperatures are also determined. With these data it is possible to determine M, R, and H from Tables 10.2 and 10.3.

For roof arches, the dead load is generally approximately uniformly distributed along the arch axis. The load per horizontal foot of span is therefore distributed as shown in Fig. 10.8. In order that Tables 10.2 and 10.3 may be used, this load can be divided into a uniform load g and a variable load of maximum intensity g' at the springing. The tables then allow one to compute separately the influences of each of these two loads, which must be added to result in the final M, R, and H. Live load must be so placed as to give maximum moments or forces; this will generally require partial loading of the span. The necessary formulas for partial loading are included in the tables; the load positions, determined by $\alpha = a/l$, are as shown in the sketches. Positions of loads and values of α which result with good accuracy in maximum moments at the

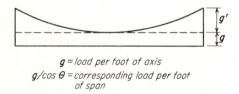

g = load per foot of axis
$g/\cos\theta$ = corresponding load per foot of span

Fig. 10.8

Table 10.2 Moments and reactions of two-hinged parabolic arches*

	Case I	Case II	Case III	Case IV	Case V	Case VI	Case VII	Case VIII — Uniform temperature change	Case IX — Temperature differential of $\Delta t°$
R	$\dfrac{gl}{2}$	$\dfrac{g'l}{6}$	pal	$\dfrac{pl}{2}(1-2\alpha)$	$\dfrac{pal}{2}(2-\alpha)$	$\dfrac{pa^2l}{2}$	$P(1-\alpha)$	None	None
R_r	$\dfrac{gl}{2}$	$\dfrac{g'l}{6}$	pal	$\dfrac{pl}{2}(1-2\alpha)$	$\dfrac{pa^2l}{2}$	$\dfrac{pal}{2}(2-\alpha)$	$P\alpha$	None	None
H	$\dfrac{gl^2}{8h}$	$\dfrac{g'l^2}{42h}$	$\dfrac{pl^2}{8h}\,\alpha^2(2\alpha^3 - 5\alpha^2 + 5)$	$\dfrac{-pl^2}{8h}(2\alpha^5 - 5\alpha^4 + 5\alpha^2 - 1)$	$\dfrac{pl^2}{16h}\,\alpha^2(2\alpha^3 - 5\alpha^2 + 5)$	$\dfrac{pl^2}{16h}\,\alpha^2(2\alpha^3 - 5\alpha^2 + 5)$	$\dfrac{5Pl}{8h}\,\alpha(\alpha^3 - 2\alpha^2 + 1)$	$\dfrac{15EI\cot^\circ}{8h^2}$	$\dfrac{5EI\omega\,\Delta t^\circ}{4t_c h}$
M_c	None	$-\dfrac{g'l^2}{338}$	$-\dfrac{pl^2}{8}\,\alpha^2(2\alpha^3 - 5\alpha^2 + 1)$	$\dfrac{pl^2}{8}\,\alpha^2(2\alpha^3 - 5\alpha + 1)$	....	....	$\dfrac{(1-\alpha)l}{8}\,P(5\alpha^3 - 5\alpha^2 - 5\alpha + 4)$	$-\dfrac{15EI\cot^\circ}{8h}$	$-\dfrac{5EI\omega\,\Delta t^\circ}{4t_c}$
$M_{l/4}$	None	$\dfrac{g'l^2}{234}$	....		$-\dfrac{pl^2}{64}(6\alpha^5 - 15\alpha^4 + 23\alpha^2 - 16\alpha + 2)$ †	$-\dfrac{pl^2}{64}\,\alpha^2(6\alpha^3 - 15\alpha^2 + 7)$ ‡	$\tfrac{1}{4}(R_m l - Hh)$ ¶ $\tfrac{1}{4}(Rl - Hh)$	$-\dfrac{45EI\cot^\circ}{32h}$	$-\dfrac{15EI\omega\,\Delta t^\circ}{16t_c}$

*According to Eqs. (10.19).

† $\dfrac{l}{4} \le \alpha l \le \dfrac{l}{2}$ ‡ $\dfrac{l}{2} \le \alpha l \le \dfrac{3l}{4}$ ¶ $\alpha l \le l/4$ $\alpha l \ge l/4$

Table 10.3 Moments and reactions of fixed parabolic arches*

	Case I	Case II	Case III	Case IV	Case V	Case VI	Case VII	Case VIII (Uniform temperature change)	Case IX (Temperature differential of $\Delta t°$)
R_l	$\dfrac{gl}{2}$	$\dfrac{g'l}{6}$	pal	$\dfrac{pl}{2}(1-2\alpha)$	$\dfrac{pl}{2}\alpha(\alpha^3 - 2\alpha^2 + 2)$	$\dfrac{pl}{2}(1+\alpha)(1-\alpha)^3$	$P(1-\alpha)^2(1+2\alpha)$	None	None
R_r	$\dfrac{gl}{2}$	$\dfrac{g'l}{6}$	pal	$\dfrac{pl}{2}(1-2\alpha)$	$\dfrac{pl}{2}\alpha^3(2-\alpha)$	$\dfrac{pl}{2}(1 - 2\alpha^3 + \alpha^4)$	$P\alpha^2(3-2\alpha)$	None	None
H	$\dfrac{gl^2}{8h}$	$\dfrac{g'l^2}{56h}$	$\dfrac{pl^2}{4h}\alpha^3(6\alpha^2 - 15\alpha + 10)$	$-\dfrac{pl^2}{8h}(12\alpha^5 - 30\alpha^4 + 20\alpha^3 - 1)$	$\dfrac{pl^2}{8h}\alpha^3(6\alpha^2 - 15\alpha + 10)$	$\dfrac{pl^2}{8h}(1-\alpha)^3(6\alpha^2 + 3\alpha + 1)$	$\dfrac{15}{4}\dfrac{Pl}{h}\alpha^2(1-\alpha)^2$	$\dfrac{45EI\omega t°}{4h^2}$	$-\dfrac{EI\omega \Delta t°}{t_c}$
M_{ls}	None	$-\dfrac{g'l^2}{210}$	$-\dfrac{pl^2}{2}\alpha^2(1-\alpha)^2(1-2\alpha)$	$\dfrac{pl^2}{2}\alpha^2(1-\alpha)^2(1-2\alpha)$	$-\dfrac{pl^2}{2}\alpha^2(1-\alpha)^3$	$\dfrac{pl^2}{2}\alpha^2(1-\alpha)^3$	$-\dfrac{Pl}{2}\alpha(1-\alpha)^2(2-5\alpha)$	$\dfrac{15EI\omega t°}{2h}$	$-\dfrac{EI\omega \Delta t°}{t_c}$
M_{rs}	None	$-\dfrac{g'l^2}{210}$	$-\dfrac{pl^2}{2}\alpha^2(1-\alpha)^2(1-2\alpha)$	$\dfrac{pl^2}{2}\alpha^2(1-\alpha)^2(1-2\alpha)$	$\dfrac{pl^2}{2}\alpha^3(1-\alpha)^2$	$-\dfrac{pl^2}{2}\alpha^3(1-\alpha)^2$	$\dfrac{Pl}{2}\alpha^2(1-\alpha)(3-5\alpha)$	$\dfrac{15EI\omega t°}{2h}$	$-\dfrac{EI\omega \Delta t°}{t_c}$
M_c	None	$-\dfrac{g'l^2}{560}$	$-\dfrac{pl^2}{4}\alpha^3(1-2\alpha)(2-\alpha)$	$\dfrac{pl^2}{4}\alpha^3(1-2\alpha)(2-\alpha)$	$-\dfrac{pl^2}{8}\alpha^3(1-2\alpha)(2-\alpha)$ †	$\dfrac{pl^2}{8}\alpha^3(1-2\alpha)(2-\alpha)$ ‡	$-\dfrac{Pl}{4}\alpha^2(5\alpha^2 - 10\alpha + 3)$	$\dfrac{15EI\omega t°}{4h}$	$-\dfrac{EI\omega \Delta t°}{t_c}$

* According to Eqs. (10.19).

† $0 \le \alpha l \le \dfrac{l}{2}$ ‡ $0 \le \alpha l \le \dfrac{l}{2}$

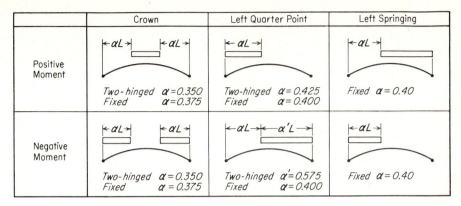

	Crown		Left Quarter Point	Left Springing
Positive Moment	Two-hinged $\alpha'=0.350$ Fixed $\quad\alpha'=0.375$		Two-hinged $\alpha=0.425$ Fixed $\quad\alpha=0.400$	Fixed $\alpha=0.40$
Negative Moment	Two-hinged $\alpha=0.350$ Fixed $\quad\alpha=0.375$		Two-hinged $\alpha'=0.575$ Fixed $\quad\alpha'=0.400$	Fixed $\alpha=0.40$

Fig. 10.9 Live-load positions for maximum moment.

crown, quarter point, or supports are given for both types of arches in Fig. 10.9.

To obtain simple expressions for Tables 10.2 and 10.3, it was assumed that

$$I = \frac{I_c}{\cos \theta}$$

Even considerable deviation from this relationship has relatively little influence and leads to satisfactory accuracy.

The length of the *half axis s* for parabolic arches is very nearly

$$s = \frac{l}{2} (1 + \tfrac{8}{3}r^2) \tag{10.20}$$

Once approximate values of R, H, and M are found for the critical sections from Table 10.2 or 10.3, the required cross-sectional dimensions are determined in the usual manner for bending plus axial thrust. Reinforcement is usually provided symmetrically near both faces. For reasons of appearance, the depth is often increased gradually from crown to support or held constant, but is hardly ever decreased toward the support.

With dimensions of the cross sections determined as required by strength and appearance, the weight of the arch is computed. If the new total dead load differs materially from that assumed in the beginning, the preliminary design procedure is repeated with adjusted loads, and dimensions are correspondingly corrected.

Since, at this stage, all loads are known, it is now possible to obtain the final location of the arch axis by passing a string polygon for dead load through the midpoints of the crown and springing sections. For this purpose, as well as for the subsequent exact analysis, the half axis is generally divided into six or ten equal sections (see Art. 10.5), and the dead load of each section applied at its center. A smooth curve which is tangent to this string polygon then represents the final location of the arch axis. Accurate values of M, N, and V are now determined by elastic analysis at the crown, quarter point, and springing and at such additional points as may be necessary. The difference between these accurate values and the approximate ones obtained from the preliminary design will require minor adjustments of cross-sectional dimensions, which can often be made merely by changing the reinforcement without alteration of overall size. Even if the thickness and width of the arch are changed in this stage of design, the elaborate elastic analysis need not be repeated for the adjusted sections unless these adjustments are quite sizable. This is due to the fact that the magnitudes of M, N, and V caused by external loads depend primarily on the ratios of the moments of inertia along the arch to that at the crown, rather than on their absolute values. Only the stresses due to rib shortening and temperature depend directly on the absolute size of the cross section and may have to be recomputed if the dimensions change significantly. The design of a simple roof arch in Art. 10.8 illustrates the above procedure.

10.7 HINGES IN REINFORCED-CONCRETE STRUCTURES

It is the function of hinges to transmit safely the thrust N and the shear force S and, at the same time, to offer little resistance to rotation. An ideal hinge should be entirely free to rotate, since only then will the bending moment at the hinge be zero, as assumed in the analysis of hinged structures. Such ideal hinges cannot be constructed, since friction or other inevitable restraining factors cannot be avoided. Practically, however, it is only necessary to ensure that the flexibility of the hinge be considerably greater than that of any other portion of the structure of which it is a part. This will result in very small moments in the hinge as compared with the moments in the structure proper, and therefore will lead only to negligible errors. The use of hinges is not restricted to arches. They are equally useful in rigid frames, in bridges as well as in buildings, particularly for long-span single-story frames. They reduce or, in the case of three-hinged spans, eliminate the stresses caused by rib shortening, by temperature, and by slight foundation movements. For these reasons a considerable reduction in the dimensions and weight of the structure and foundation can often be achieved by their use.

The main practical types of hinges for concrete structures are given in Fig. 10.10. The steel hinges shown there result in more perfect hinge action than the concrete hinges, but are considerably more expensive. Their use, today, is restricted to unusually heavy structures. Of the two types of concrete hinges, named after the French engineers who developed them, the Mesnager hinge ensures greater freedom of rotation and, except for very large thrusts, is the most satisfactory hinge presently in use.

To obtain maximum flexibility in a Mesnager hinge, it has been recommended that the hinge opening h should be at least be equal to the concrete

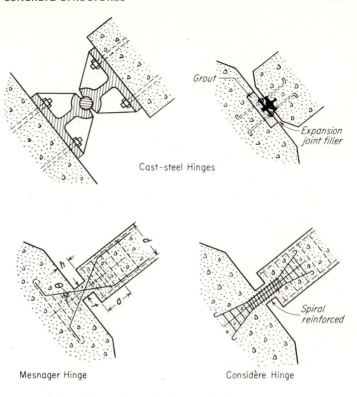

Fig. 10.10 Hinges for concrete structures.

thickness t. A value $h = 1.3t$ is recommended by some. However, hinges with with smaller openings have been used successfully. For the same purpose of maximum flexibility, the crossed steel bars should not be too thick and short. On the other hand, if their diameter were too small in relation to their length, they would be likely to buckle. For this reason the slenderness ratio L/r of the hinge bars should be between the limits of 20 and 40, L being the length of the inclined bar between concrete faces (that is, $h/\cos\theta$), and r the radius of gyration of the bar (that is, $D/4$ for round bars of diameter D). The length of embedment of the hinge bars on either side of the hinge should be at least equal to that prescribed for deformed bars in compression.

To find the required area of the hinge bars, it is assumed that they form a triangular truss which transmits the normal and shear forces N and S, as shown in Fig. 10.11. It is easily seen that the compression stress in bars 1 is equal to

$$f'_s = \frac{N}{A'_s \cos\theta} + \frac{S}{A'_s \sin\theta} \tag{10.21}$$

where A'_s is the total combined area of all hinge bars. The stress f'_s should not exceed ϕ times the yield stress f_y of the hinge bars when the full factored load

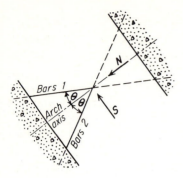

Fig. 10.11

is applied. It is recommended that a value ϕ not greater than 0.45 be used, for the following reason. Even though, by the above procedure, the bars are assumed to be subjected only to axial compression, like members of a truss, any rotation actually occurring at the hinge bends the bars and induces corresponding flexural stresses. Such rotations actually do take place during the lifetime of the structure and are caused primarily by changes in live load and in temperature. Rather than attempt to compute bending effects, which are complicated because of the interaction of the bars with the concrete, it is generally satisfactory to use a conservative capacity reduction factor in conjunction with assumed axial forces.

The inclined bars, transmitting their forces to the concrete by bond along the embedded length, exert a bursting force which must be resisted by lateral reinforcement (stirrups and ties), as shown in Fig. 10.10. Only the part of the lateral reinforcement within a distance $a = 8D$ from the face of the concrete is considered effective in resisting the bursting force. The stress in the lateral reinforcement can be computed from

$$f_s'' = \frac{(N/2)\tan\theta + Sa/jd}{0.005ab + A_s''} \tag{10.22}$$

where, in addition to previously defined quantities, A_s'' is the combined area of lateral stirrups or ties located within $a = 8D$ from the free face of the hinge, j can be taken as 0.9, d is shown in Fig. 10.10, and b is the horizontal width of the hinge, equal, in most cases, to the width of the arch rib. The stress f_s'' to be used in design may be taken as ϕf_y, where $\phi = 0.90$ for axial tension, as recommended by the ACI Code.

More detailed information on concrete hinges can be found in a paper by Ernst (Ref. 10.2). Most tests on steel bridges give adequate data on steel hinges of the type used for both steel and concrete arches and frames.

10.8 DESIGN OF TWO–HINGED, TIED ROOF ARCH

Arches of the type shown in Fig. 10.2 with 90-ft span are to be spaced at 20-ft intervals, to support a fireproof, insulated roof. The roof itself is to consist of precast slabs spanning between arches and weighing 40 psf. Four-ply tar-and-gravel roofing to cover the slabs weighs 5 psf. To account for the effects of wind and snow a live load of 35 psf is assumed.

The columns of Fig. 10.2 supporting the arch cannot develop sufficient rigidity to serve as fixed supports. A two-hinged arch is therefore indicated. In addition, if the horizontal thrust were transmitted to the supporting columns and frames, they would become excessively heavy and costly. For this reason a tied arch of the type shown in Fig. 10.12 will be used. A 15-ft rise has been selected, resulting in a rise-span ratio of $\frac{1}{8}$.

The sample design that follows is abbreviated for reasons of clarity and space. Computations of moments, etc., caused by small secondary loads, such as from the weight of the tie rod and the hangers, have been omitted, since they would complicate and lengthen the illustrative computations without adding new features. In actual design practice the influence of such loads should be included in the computations.

Preliminary design   The loads per linear foot of arch axis are
Dead load:

Concrete slabs:	40×20	=	800 lb per ft
Roofing:	5×20	=	100 lb per ft
Arch (assumed)		=	300 lb per ft
Total		=	1200 lb per ft

Live load:

Snow plus wind: $35 \times 20 =$ 700 lb per ft

Thus the factored dead load is $1.4 \times 1200 = 1680$ lb per ft, and the factored live load is $1.7 \times 700 = 1190$ lb per ft.

The load, uniform along the axis, results in distribution as shown in Fig. 10.8. With θ computed from Eq. (10.19b), $g' = 574$ lb per ft for dead plus live load, and $g' = 336$ lb per ft for dead load alone. From the information of Table 10.2, the following moments and forces are obtained for the trial design:

Maximum horizontal reaction (loading Fig. 10.13a):

$$H = \frac{gl^2}{8h} + \frac{g'l^2}{42h} = 201.0 \text{ kips}$$

Maximum negative crown moment (Fig. 10.13b; $\alpha = 0.35$ from Fig. 10.9):

$$M_c = -0.00296g'l^2 - 0.00725pl^2 = -78.0 \text{ ft-kips}$$

combined with

$$H = \frac{gl^2}{8h} + \frac{g'l^2}{42h} + 0.0685\frac{pl^2}{h} = 161.5 \text{ kips}$$

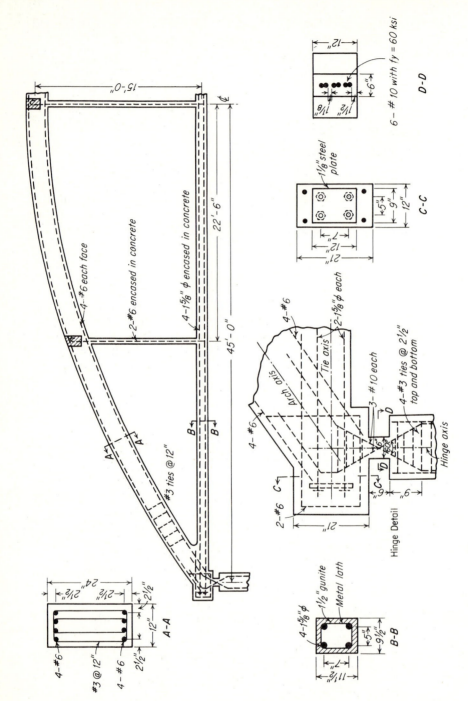

Fig. 10.12 Two-hinged tied roof arch.

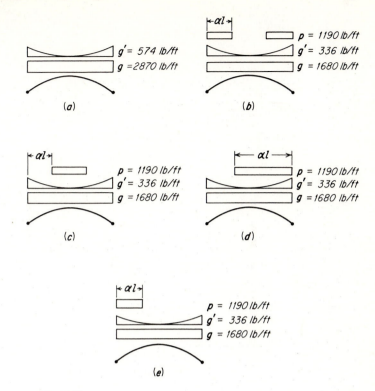

Fig. 10.13

Maximum positive crown moment (Fig. 10.13c; $\alpha = 0.35$):

$$M_c = -0.00296g'l^2 + 0.00725pl^2 = 61.8 \text{ ft-kips}$$

combined with

$$H = \frac{gl^2}{8h} + \frac{g'l^2}{42h} + 0.0564\frac{pl^2}{h} = 153.7 \text{ kips}$$

Maximum negative quarter-point moment (Fig. 10.13d; $\alpha = 0.575$):

$$M_{l/4} = 0.00428g'l^2 - 0.0164pl^2 = -146.3 \text{ ft-kips}$$

combined with

$$H = \frac{gl^2}{8h} + \frac{g'l^2}{42h} + 0.0760\frac{pl^2}{h} = 166.4 \text{ kips}$$

$$R_l = \frac{p\alpha^2 l}{2} + \frac{g'l}{6} + \frac{gl}{2} = 98.3 \text{ kips}$$

$V_{l/4} = $ vertical shear at quarter point

$\quad\quad = 98.3 - 22.5 \times 1.68 - 0.336 \times \frac{9.0}{8} = 56.7 \text{ kips}$

The inclination $\theta = 18°25'$ is determined from Eq. (10.19b). Then

$$
\begin{aligned}
N_{l/4} &= \text{axial thrust at quarter point} \\
&= V_{l/4} \sin \theta + H \cos \theta \\
&= 56.7 \times 0.316 + 166.4 \times 0.95 = 176 \text{ kips}
\end{aligned}
$$

Maximum positive moment at quarter point (Fig. 10.13e; $\alpha = 0.425$):

$$
M_{l/4} = 0.00428g'l^2 + 0.0164pl^2 = 169.7 \text{ ft-kips}
$$

combined with

$$
H = \frac{gl^2}{8h} + \frac{g'l^2}{42h} + 0.0435 \frac{pl^2}{h} = 145.5 \text{ kips}
$$

$$
R_l = pal\left(1 - \frac{\alpha}{2}\right) + \frac{gl}{2} + \frac{g'l}{6} = 116.4 \text{ kips}
$$

$$
V_{l/4} = 116.4 - 22.5 \times 2.87 - 0.336 \times \tfrac{9.0}{8} = 48.0 \text{ kips}
$$

$$
N_{l/4} = 48.0 \times 0.316 + 145.5 \times 0.95 = 153.3 \text{ kips}
$$

Maximum axial thrust at support (Fig. 10.13a):

$$
\begin{aligned}
H &= 201.0 \text{ kips (as previously computed)} \\
R_l &= 45 \times 2.87 + 15 \times 0.574 = 137.6 \text{ kips} \\
N_{l/4} &= R_1 \sin \theta + H \cos \theta \\
&= 137.6 \times 0.538 + 201 \times 0.834 = 241.8 \text{ kips}
\end{aligned}
$$

Maximum force in ties: $H = 201.0$ kips
Maximum crown moment: $M_c = -78.0$ ft-kips
Accompanying crown thrust: $N_c = 161.5$ kips
Maximum quarter-point moment: $M_{l/4} = 169.7$ ft-kips
Accompanying quarter-point thrust: $N_{l/4} = 153.3$ kips
Maximum thrust at support: $N_l = 241.8$ kips

In a structure of this type, in which the entire arch rib is located below an insulated roof, no temperature differential will arise in the rib. In addition, the arch and the tie are always at substantially the same temperature. Any temperature change in the enclosed space will therefore cause the arch and the tie to expand or contract by the same amount and, for this reason, will not cause any temperature stresses in the structure.

For simplicity of formwork and for good appearance, an arch of constant cross section will be used. It is consequently only necessary to design the most critical section, which is seen to be that at the quarter point. A concrete with $f'_c = 3$ ksi and a steel having $f_y = 40$ ksi will be used.

At the quarter point, the arch must provide axial strength $P_u = 153.3$ kips and moment resistance $M_u = 169.7$ ft-kips. The corresponding eccentricity of loading is $e = 169.7 \times 12/153.3 = 13.3$ in. A section 13×22 in. will be adopted for trial. In this case, $e/h = 13.3/22.0 = 0.605$. With

$$\gamma = \frac{17.0}{22.0} = 0.78$$

Graph 3 of the Appendix can be used without significant error.
For the given load and moment:

$$\frac{P_u}{\phi b h f_c'} = \frac{153.3}{0.70 \times 13 \times 22 \times 3} = 0.255$$

and

$$\frac{P_u}{\phi b h f_c'} \times \frac{e}{h} = 0.255 \times 0.605 = 0.154$$

For these values, Graph 3 indicates the required $\rho_t \mu$ to be 0.19. For the present
material strengths, $\mu = f_y/0.85f_c' = 40/0.85 \times 3 = 15.7$; hence

$$\rho_t = \frac{0.19}{15.7} = 0.0121$$

From this, $A_s = 0.0121 \times 13 \times 22 = 3.46$ in.² Eight No. 6 bars, four at each
face, provide 3.53 in.², and will be adopted tentatively. No change is necessary
in the estimated dead load of the arch.

The tie rod must resist $H_{max} = 201$ kips. To reduce the stretch of the tie,
a low design stress of $0.40 \times 60 = 24$ ksi will be used. Consequently, the
required tie area is $A_{tr} = 201/24 = 8.4$ in.² Since the final design is likely to
result in a smaller H_{max}, four $1\frac{5}{8}$-in. plain round bars with upset threaded ends
will be used, furnishing an area of 8.3 in.²

Final design   Since the preliminary design did not result in any change of the
weight of the arch rib, as originally assumed, the final design is carried out
for the same loads as before, that is, 1680 lb per ft dead load and 1190 lb per ft
live load.

For the purpose of drawing the dead-load string polygon and for subsequent computations, the length of the arch axis is divided into 20 equal segments Δs. From Eq. (10.20), with $l = 90$ ft and $r = \frac{1}{6}$,

$$\Delta s = l \frac{1 + \frac{8}{3}r^2}{20} = \frac{96.66}{20} = 4.83 \text{ ft}$$

The dead load per segment is $1.68 \times 4.83 = 8.1$ kips. To draw the string
polygon, it is simplest first to draw to a large scale the parabolic axis used
in the preliminary analysis, which is easily done by computing a sufficient
number of points by means of Eq. (10.19a). It is sufficient to draw half the
arch, as shown in Fig. 10.14. The half axis is next divided into 10 equal parts
of 4.83 ft each, and the weight of each such segment, 8.1 kips, is applied at its

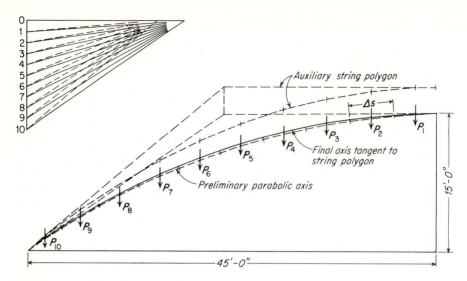

Fig. 10.14   Determination of final location of arch axis by means of dead-load string polygon.

midpoint. An accurate string polygon for these loads is then drawn through the midpoints of crown and support (not shown in Fig. 10.14 so as not to obscure the drawing). A curve tangent to this string polygon represents the final axis. Figure 10.14 shows that for this particular arch the final axis is very close to the original parabola. For subsequent computations the coordinates x and y of the 10 segments are accurately scaled off the drawing.

As is seen from Eq. (10.11a), the quantities $y\,\Delta s/I$ and $y^2\,\Delta s/I$ must be computed for the 10 segments of the half arch. For the tentative 13- $\times$ 22-in. section, $I = 11{,}600$ in.[4] The pertinent quantities are assembled in the first four columns of Table 10.4, which summarizes the numerical calculations.

For determining the dimensions of the arch, the tie rod, and the hinges, the forces and moments caused by the following loadings must be computed:

a. MAXIMUM VERTICAL REACTION R AND MAXIMUM FORCE IN TIE ROD H These are obtained for full live load, being placed over the entire span, so that the load per segment, dead plus live, is 13.9 kips. The corresponding reactions are $R_l = R_r = 139$ kips. The beam moments M_0 and the quantities $M_0 y\,\Delta s/I$ are computed at the midpoints of the 10 segments. They are entered in Table 10.4 in the columns headed "Case a." By virtue of symmetry, computations can be restricted to half the arch.

b. MAXIMUM POSITIVE QUARTER-POINT MOMENT   The preliminary design has shown that the positive quarter-point moment will govern the dimensions of the arch rib. Since a constant-cross-section arch was chosen, it is sufficient for our purposes to investigate only this one loading case. For a more complete investigation, the other loading cases of the preliminary design should also be computed. This is desirable even in the case of a constant-cross-section arch, in view of the approximate character of the preliminary design. It becomes mandatory if the final arch axis differs considerably from the original parabolic shape and, of course, if rib dimensions and reinforcement are varied along the span.

The exact location of the quarter point is at $x = 22.5$ ft. This point does not happen to coincide with any of the centers of the segments. Since moments and thrusts vary only very gradually in an arch subject only to distributed loads, it will be simpler and sufficiently accurate to compute the moment at that segment center located closest to the quarter point, i.e., at $x = 23.5$ ft. According to Fig. 10.9, live load should be placed from the left springing, up to $x = 0.425 \times 90 = 38.3$ ft. In the summation process it is quite inconvenient to place loads at locations other than over integer segments. For this reason the first nine segments from the left support (i.e., up to $x = 40.2$ ft) will be subjected to live load. The error so introduced is entirely negligible. Hence, these nine segments are loaded by 13.9 kips each, the remaining eleven segments by the dead load of 8.1 kips. The corresponding reactions are $R_l = 121.5$ kips and $R_r = 92.5$ kips.

In view of unsymmetrical loading, beam moments M_0 and quantities $M_0 y \, \Delta s/I$ must be computed for both halves of the arch. They are entered in Table 10.4 in the columns headed "Case b."

In Table 10.4, x and y are given in feet, $y \, \Delta s/I$ in $1/\text{in.}^2$, $y^2 \, \Delta s/I$ in $1/\text{in.}$, M_0 in ft-kips, and $M_0 y \, \Delta s/I$ in ft-kips per in.^{2} (It is advisable to select units so as not to obtain unwieldy figures, such as millions on the one hand, or small decimal fractions on the other.)

Table 10.4

| | | | | Case a | | Case b | | | |
| | | | | | | Left half | | Right half | |
x	y	$y \, \Delta s/I$	$y^2 \, \Delta s/I$	M_0	$M_0 y \, \Delta s/I$	M_0	$M_0 y \, \Delta s/I$	M_0	$M_0 y \, \Delta s/I$
1.9	1.30	0.078	1.2	264	21	231	18	176	14
6.0	3.90	0.235	10.9	777	182	672	158	523	123
10.2	6.20	0.373	27.6	1245	465	1066	398	834	311
14.5	8.30	0.498	49.6	1663	829	1410	702	1128	572
19.0	10.20	0.613	74.9	2038	1250	1706	1045	1399	858
23.5	11.80	0.708	100.2	2352	1662	1940	1372	1634	1160
28.2	13.10	0.786	123.2	2614	2050	2119	1670	1842	1450
32.9	14.10	0.847	143.2	2810	2380	2233	1890	2011	1705
37.7	14.70	0.882	155.5	2944	2590	2282	2010	2145	1890
42.5	14.95	0.897	160.8	3011	2710	2299	2060	2240	2010
Σ			847.1		14,139		11,323		10,093

From the general discussion of Eq. (10.11), it is evident that only the first term in the numerator and the first and last terms in the denominator of that equation need be considered for this particular roof arch. Hence, for determining H, the following equation is used:

$$H = \frac{\Sigma M_{0} y \,\Delta s / I}{\Sigma y^2 \,\Delta s / I + El/E_{tr} A_{tr}}$$

The denominator then becomes

$$2 \times 847.1 + (4 \times 10^6 \times 90 \times 12)/(30 \times 10^6 \times 8.3) = 1711$$

The maximum force in the tie rod (Case a) is

$$H_{max} = \frac{2 \times 14{,}139 \times 12}{1711} = 197 \text{ kips}$$

For the loading which results in maximum quarter-point moment (Case b),

$$H = \frac{(11{,}323 + 10{,}093)12}{1711} = 149 \text{ kips}$$

From Table 10.4 the simple beam moment M_0 at $x = 23.5$ ft is seen to be 1940 ft-kips. Hence the moment in the arch rib at that point is

$$M = M_0 - Hy = 1940 - 149 \times 11.80 = 177 \text{ ft-kips}$$

Since the point $x = 23.5$ ft is located at the center of the sixth arch segment, the vertical shear at that point is

$$V = 121.5 - 13.9 \times 5.5 = 45.0 \text{ kips}$$

At that point, from Eq. (10.19b), $\theta = 17°42'$. The axial thrust at that point then becomes

$$N = 45.0 \times 0.303 + 149 \times 0.95 = 156 \text{ kips}$$

It will be noticed that the values of H_{max}, M, and N are quite close to those obtained in the preliminary design. Such good coincidence obtains only when, as in this case, the final location of the axis is very close to that used in the preliminary design. Even then, significant differences between preliminary and final results are not infrequent; therefore the accurate analysis by summation should always be carried out.

For the revised forces, a section 12×24 in. will be tried. With $P_u = 156$ kips and $M_u = 177$ ft-kips, the equivalent eccentricity is

$$e = 177 \times 12/156 = 13.6 \text{ in.}$$

For the proposed section, $e/h = 13.6/24 = 0.567$, and with

$$\gamma = \frac{19.0}{24.0} = 0.79$$

Graph 3 of the Appendix will be used.
Then

$$\frac{P_u}{\phi bh f_c'} = \frac{156}{0.70 \times 12 \times 24 \times 3} = 0.258$$

and

$$\frac{P_u}{\phi bh f_c'} \times \frac{e}{h} = 0.258 \times 0.567 = 0.146$$

From the graph, $\rho_t \mu = 0.17$, and with $\mu = 15.7$ as before, the required steel ratio $\rho_t = 0.17/15.7 = 0.0108$. The corresponding steel area is

$$A_s = 0.0108 \times 12 \times 24 = 3.12 \text{ in.}^2$$

A total of eight No. 6 bars will be used, four at each face, providing an actual steel area of 3.53 in.^{2}

With four $1\frac{5}{8}$-in. round bars for the tie rod, as obtained in the preliminary design, the stress in the tie rods is $197/8.3 = 23.8$ ksi. The ends of the tie rods are upset and threaded. The long diameter of the nuts for bars of this size is $3\frac{1}{2}$ in. To accommodate the wrench for tightening the nuts, a 5-in.-minimum center-to-center spacing of tie bars is selected.

To transmit the pull of the tie to the concrete, an anchor plate $9 \times 12 \times 1\frac{1}{8}$ in. is provided. Actually, a considerable part of the pull will be transmitted to the concrete by bond over the embedded length of the tie bars; this will be discounted, however, and the anchor designed for the entire pull. The bearing surface is $9 \times 12 - 4 \times 2.1 = 99.6$ in.^{2}, and the bearing pressure immediately under the plate is $197,000/99.6 = 1980$ psi. According to the ACI Code, the allowable stress on the area of contact between the plate and the concrete is $2.0 \times 0.85 \phi f_c'$, where $\phi = 0.70$ for bearing. In the present case, the design stress is $2.0 \times 0.85 \times 0.70 \times 3000 = 3580$ psi, well above the actual value. Further, the Code requires that the stress on the total concrete area be not more than $0.85 \phi f_c'$, 1790 psi in the present case. The actual value is only $197,000/12 \times 21 = 782$ psi. The bending stresses in the anchor plate are investigated in the long direction for a strip 1 in. wide. The maximum negative moment of the portion projecting beyond the center line of the anchors is $1980 \times 2.5^2/2 = 6190$ in.-lb, while the maximum positive moment at the center line of the plate is $1980 \times 7^2/8 - 6190 = 5930$ in.-lb. With the section modulus equal to $1.125^2/6 = 0.210$ in.^{3}, the maximum bending stress is $6190/0.210 = 29,400$ psi. This is less than 85 per cent of the yield stress for the plate steel, and, since the calculations are based on factored loads, is completely satisfactory.

The hinges transmit only the vertical reaction of $V = 139$ kips, because the horizontal component is resisted by the tie. To reduce the number and size of the hinge bars so as to accommodate them in the limited width, a steel having $f_y = 60$ ksi will be specified. The design stress is taken as $0.45 \times 60 = 27$ ksi.

With an angle $\theta = 30°$ (Fig. 10.11), the required area of hinge bars is computed from Eq. (10.21); that is, $A_s' = 139/(27 \times 0.866) = 5.94$ in.[2] Six No. 9 bars are provided, furnishing an area of 6.00 in.[2] With a hinge opening of 6 in., as shown in Fig. 10.12, the free length of the hinge bars is $L = 6/0.866 = 6.93$ in., while the radius of gyration is $r = 1.128/4 = 0.281$. Hence, the ratio $L/r = 24.7$ is within the limits recommended in Art. 10.7.

The required lateral reinforcement is obtained from Eq. (10.22), where, in this case, $a = 8 \times 1.128 = 9.0$ in., $b = 12$ in., $N = 139$ kips, and $S = 0$. With a design stress $f_s'' = 0.90 \times 40 = 36$ ksi, the required area becomes

$$A_s'' = \frac{N}{2f_s''} \tan \theta - 0.005ab = 0.565 \text{ in.}^2$$

It is seen from Fig. 10.12 that within the length $a = 9$ in., four No. 3 closed ties are provided, which furnish an area of $8 \times 0.11 = 0.88$ in.[2]

The required development length for the 60-ksi No. 10 bars in compression is 25 in. beyond the faces of the concrete on either side of the hinge.

To avoid bending moments caused by eccentricities, the axes of the arch, the tie, and the hinge must be made to intersect at one common point, as shown in the hinge detail of Fig. 10.12.

10.9 SHELL ROOFS

Shell structures are by nature more complex than arches, and the mathematics of an exact analysis of a shell may become quite formidable. However, for specific shell shapes of great civil engineering utility, many of the unknowns to be found are negligibly small and may be neglected.

Shown in Fig. 10.15a is a shell of arbitrary shape. A typical element of the surface is reproduced to larger scale in Fig. 10.15b to d. The surface load w is shown, together with the system of internal forces which hold it in equilibrium. These forces are termed *stress resultants*, because they are equal to the integrals of normal or shearing unit stresses acting on the cut faces of the element. In general, the integration is carried out over the thickness, so that a particular stress resultant such as N_ϕ is expressed in units of force per unit of length along the line of the cut, e.g., pounds per inch.

Shown in Fig. 10.15b are the *membrane stress resultants*. These act in a plane tangent to the shell at a particular point. The normal membrane stress resultants N_ϕ and N_θ act perpendicular to the cut faces, while the membrane shears $N_{\phi\theta}$ and $N_{\theta\phi}$ act parallel to those faces. The *normal shears* Q_ϕ and Q_θ act perpendicular to the tangent plane, as shown in Fig. 10.15c. The *bending moments* M_ϕ and M_θ, as well as the *twisting moment* $M_{\phi\theta}$ and $M_{\theta\phi}$, are represented by vectors acting in the tangent plane, and are shown as double-headed vectors in Fig. 10.15b.

It is easily confirmed that there are 10 independent unknown stress resultants to be determined, which must be such as to equilibrate the applied

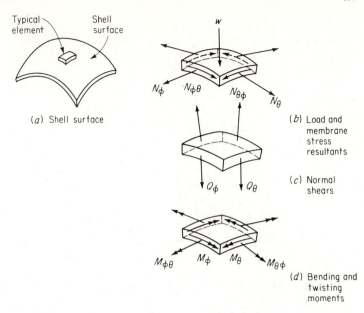

Typical
element

Shell
surface

(a) Shell surface

N_ϕ $N_{\phi\theta}$ $N_{\theta\phi}$ N_θ

w

(b) Load and
membrane
stress
resultants

Q_ϕ Q_θ

(c) Normal
shears

$M_{\phi\theta}$ M_ϕ M_θ $M_{\theta\phi}$

(d) Bending and
twisting
moments

Fig. 10.15   Shell stress resultants on typical element.

load w. Since there are but six independent equations of equilibrium available at any point (three of moment equilibrium and three of force equilibrium), it is clear that the shell problem is statically indeterminate and that, in general, one must consider conditions of compatibility as well as the force-displacement relations in order to obtain a solution (Refs. 10.3 to 10.5).

However, as previously noted, in specific instances many of these stress resultants may be negligibly small. In doubly curved shells such as spherical domes, for example, surface loads are held in equilibrium largely by membrane stresses alone throughout most of the shell.[1] For cylindrical shells the longitudinal bending moments and the twisting moments may usually be neglected. By assumptions such as these, justified in particular cases, shell analysis may be substantially simplified.

10.10 CYLINDRICAL SHELLS

One common form of shell surface is the cylindrical shell. This is a singly curved, developable surface, most commonly having a cross section which is a circular arc. The surface is bounded by two straight edges parallel to the axis of the cylinder and by two curved edges in planes perpendicular to the axis.

---

[1] The reason for this may be found in Castigliano's theorem of least work, which states that of all the possible values for the redundant forces of an indeterminate system, the true values are those which minimize the internal strain energy of the system.

Fig. 10.16 Cylindrical-shell roof.

Such shells are normally supported only by transverse diaphragms or ribs, the straight longitudinal edges being either free or built monolithically with edge beams. Figure 10.16 shows precast units of a cylindrical-shell roof in which no edge beams were employed. Figure 10.17 shows several of the characteristic types of cylindrical shells. Such shells may be used as single units (Fig. 10.17a) supported at the ends and free along the long edges or as multiple units (Fig. 10.17b) joined along the long edges. They may be continuous over one or more intermediate supports (Fig. 10.17c). If transverse spans are long, the use of a ribbed cylindrical shell (Fig. 10.17d) may be indicated. In the event that the longitudinal edges are continuously supported, the structural action is that of a barrel arch; such systems will not be considered here.

A typical circular cylindrical-shell surface is shown in Fig. 10.18a. A small element of that surface is reproduced to larger scale in Fig. 10.18b and c. While stress resultants corresponding to each of the ten shown in Fig. 10.15

may be present, it will be found for most cylindrical shells that the longitudinal bending moment M_x (M_θ in Fig. 10.15d) and the related normal shear Q_x, together with the twisting moments $M_{x\phi}$ and $M_{\phi x}$, are negligibly small. In such a case, the shell element is held in equilibrium mainly by the stress resultants shown in Fig. 10.18b and c, specifically the membrane thrusts N_x and N_ϕ, the membrane shears $N_{x\phi}$ and $N_{\phi x}$, the transverse normal shear Q_ϕ, and the transverse bending moment M_ϕ.

Note in Fig. 10.18b and c that, for clarity, the differential change in all stress resultants as the element width is traversed is not shown. For example,

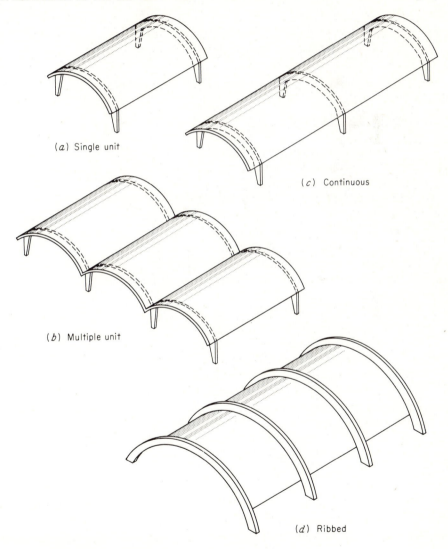

(a) Single unit

(c) Continuous

(b) Multiple unit

(d) Ribbed

Fig. 10.17   Cylindrical shells.

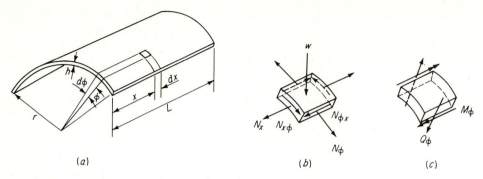

Fig. 10.18   Circular cylindrical-shell stress resultants. (*a*) Shell surface; (*b*) load and membrane stress resultants; (*c*) normal shears and bending moments.

the membrane shear force on the far side of the element corresponding to $N_{x\phi}$ is

$$N_{x\phi} + \frac{\partial N_{x\phi}}{\partial x}\,dx$$

In summing forces, when considering the equilibrium of the element, it is sometimes true that only the differential of such a force is considered, since the force itself may cancel.

By summing forces in the radial direction, it becomes apparent that the radial component of the load w must be resisted by the radial component of the transverse thrust N_ϕ plus the difference in the normal shears Q_ϕ acting on the faces of the element. The tangential component of the load must be resisted by the difference between the membrane shears $N_{x\phi}$ acting on the two opposite faces of the element, plus the difference in transverse thrust N_ϕ acting on opposite faces.

For cylindrical shells with a relatively high ratio L/r (see Fig. 10.18*a*), the important stresses can be estimated very closely by means of Lundgren's beam theory (Refs. 10.6 to 10.8), by which the shell is treated as a beam of curved cross section spanning between end supports. In Fig. 10.19*a*, the shell surface shown is considered to span as a beam in the direction L. The neutral axis of the curved cross section shown in Fig. 10.19*b* is found, and the moment of inertia of the section determined as usual. If $\overline{w}$ is the total load acting on the shell per unit of length in the direction of the span, the longitudinal bending moment $\overline{M}$ and shear force $\overline{V}$ can be found by ordinary means for simple or continuous beams. The longitudinal flexural stress is then

$$f_x = \frac{N_x}{h} = \frac{\overline{M}y}{I} \tag{10.23}$$

and has the distribution shown in Fig. 10.19*d*. If the shear stress is desired on a plane such as *a-a* (Fig. 10.19*b*), this may be found from the usual equations of mechanics. For example, if the area of the cross section outside of the plane considered is A, and the moment arm from its center of gravity to the neutral axis is y_a, as shown in Fig. 10.19b, then the static moment of that area is

$$m = Ay_a$$

and the shear stress associated with longitudinal bending is

$$v = \frac{N_{x\phi}}{h} = \frac{N_{\phi x}}{h} = \frac{\bar{V}m}{Ih} \qquad (10.24)$$

The shear stress distribution is as shown in Fig. 10.19e.

According to Lundgren's theory, the remaining three stress resultants are found from the equilibrium of a transverse slice of the shell as shown in Fig. 10.19f. The transverse moment M_ϕ must equilibrate the moment of the applied load w and that of the net tangential shear $\partial N_{x\phi}/\partial x$. Transverse thrust N_ϕ and normal shear Q_ϕ can likewise be found from equilibrium requirements

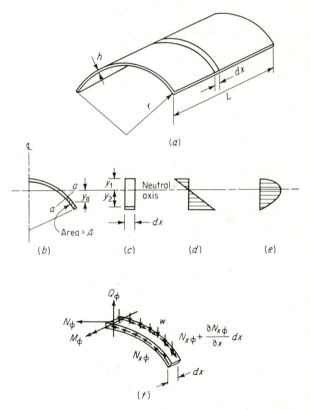

Fig. 10.19 Beam analysis of cylindrical shell. (a) Shell surface; (b) cross section; (c) elemental length; (d) $f_x = \bar{M}y/I$; (e) $v = \bar{V}m/Ih$; (f) forces for transverse analysis.

of the slice. Cutting planes at a number of locations along the arc may be chosen to obtain the variation of these three stress resultants with ϕ.

In practical cases it is useful to divide the arc of the cross section into finite segments as shown in Fig. 10.20. From Eq. (10.24)

$$N_{x\phi} = \frac{\bar{V}m}{I} \tag{10.24a}$$

The net shear force acting on the arch segment of length dx is

$$\frac{\partial N_{x\phi}}{\partial x} = \frac{\partial \bar{V}}{\partial x}\frac{m}{I} = \frac{\bar{w}m}{I}$$

Accordingly, the shear force acting on segment i is

$$s_i = \frac{\bar{w}m_i \Delta s}{I} \tag{10.25}$$

where m_i is calculated at the centroid of the segment. The segmental load w_i is easily obtained, after which M_ϕ, N_ϕ, and Q_ϕ are calculated directly by basic equilibrium relations for the slice.

The extension of the beam method of analysis to multiple shells, for which zero horizontal and rotational displacements are imposed at the lower edge, requires an analysis by indeterminate theory, but involves no real difficulty (Ref. 10.5).

Lundgren's beam method is based on the assumption that all points on a transverse cross section deflect the same amount vertically but do not deflect horizontally; i.e., the cross section of the shell retains its shape as load is applied. For relatively short cylindrical shells (low ratio L/r) this assumption is not acceptable. In such cases, the free edge tends to deflect downward and inward with respect to the crown. This relative movement tends to increase the longitudinal tensile strains at the free edge above the value predicted by beam theory; consequently, the tension along the free edge must be increased. But, for horizontal equilibrium, the compressive stresses above the neutral axis must be increased as well, although the internal moment of these forces must

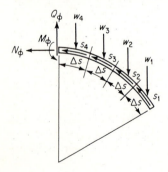

Fig. 10.20   Shell cross section.

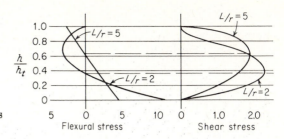

Fig. 10.21 Flexural and shear stress variations.

be the same as before. This indicates that the lever arm between the two must decrease, i.e., the center of compression must move down, as must the neutral axis.

Figure 10.21 shows the distribution of flexural and shear stress for single shells having L/r ratios of 5 and 2. While a nearly linear distribution of flexural stress is obtained for $L/r = 5$, it is clear that, for a short shell with $L/r = 2$, beam theory leads to stress predictions that are seriously in error. Comparative studies lead to the conclusion that beam theory should not be applied to single shells with L/r less than about 5, nor to interior units of multiple cylindrical shell with L/r less than about 2.

Shells which do not fall within these bounds may be analyzed by direct application of the theory of elasticity (Refs. 10.3 to 10.5). The solution is analogous to that of a conventional indeterminate frame by superposition, in which the structure is reduced to a determinate frame by the removal of redundant restraints, which are then reintroduced in turn and adjusted in value in order to satisfy the boundary conditions of the original frame. In the case of the shell, it is first assumed that the surface loads are transmitted to the supports by direct stress only. This is the membrane solution, equivalent to assuming the structure to be statically determinate. This assumption generally leads to displacements and reactions along the long edges which are inconsistent with the actual boundary conditions. To correct this situation, line loads must be applied along the long edges; the stresses resulting from these line loads are then superimposed on those obtained from the membrane solution to obtain the final stresses in the shell.

The mathematical labor associated with this procedure is greatly reduced by use of data and coefficients available in Manual of Engineering Practice No. 31—Design of Cylindrical Concrete Shell Roofs, published by the American Society of Civil Engineers in 1952 (Ref. 10.9). This publication includes a set of tables and charts which permit a solution for internal stresses for shells of common proportions and includes a valuable presentation of design theory and a number of detailed examples. In addition to this publication, the reader will find design data of interest in "Coefficients for Design of Cylindrical Concrete Shell Roofs," published by the Portland Cement Association (Ref. 10.10), which extends the Manual 31 tables, thus avoiding interpolation.

With regard to reinforcement of the shell, while the analysis is carried out assuming a homogeneous, elastic, isotropic material, the concrete has tensile strength so low that it is disregarded, and steel is placed to carry all the tension.

Transverse reinforcement must be provided to resist the bending moment M_ϕ. It may consist of relatively small bars, curved in the field to the proper shape, or may be welded wire mesh in the case of small shells. Transverse shear Q_ϕ and transverse thrust N_ϕ usually have low values and do not influence the design.

Longitudinal tensile stresses N_x require substantial reinforcement in the long direction of the shell. These bars may be curved to follow the lines of principal stress, although this leads to difficulty in the field, particularly since large bars would have to be bent. It is usually more convenient to use straight bars, parallel to the long edges. The steel area must be sufficient to provide for the total longitudinal tension force obtained from analysis. It is placed at the mid-thickness of the shell, as close to the edges as possible. The method of calculation is the same as for folded plates, treated in detail in Art. 10.11. If the main steel is concentrated in this way, at least a minimum amount of reinforcement, equal to 0.0035 times the concrete section, should be placed elsewhere in the tension zone to control cracking.

Additional bars are often required near the ends of the span to accommodate diagonal tensile stresses associated with $N_{x\phi}$. These are placed near the mid-thickness, generally at 45° with the generators of the surface. At least shrinkage and temperature steel should be provided elsewhere in the shell, the amount equal to 0.0018 times the gross concrete cross section.

The thickness of cylindrical shells is commonly dictated by requirements of transverse moment resistance, or by the practical need to provide sufficient room to include the necessary reinforcement with adequate concrete cover. Particular attention should be paid to the region in which the thin shell surface joins the more massive end diaphragms or supporting frames, where differential expansion and contraction may require additional steel as well as local thickening of the shell.

10.11 FOLDED PLATES

In an attempt to simplify formwork and yet retain many of the advantages characteristic of cylindrical shells, the folded-plate structure has evolved. These surfaces have a deep corrugated form similar to that of cylindrical shells, except that plane units are used, intersecting in "fold lines" parallel to the span direction. Folded-plate structures have been built, using very thin slabs, spanning clear distances greater than 200 ft. They are not ideal shells, because flexural stresses may have considerable influence on their dimensions and design, but they offer the sizable advantage of simplified formwork as compared with curved shell surfaces.

Folded-plate construction appears to have originated in Germany in the 1920's. Following the construction of a number of bunker-type structures

Fig. 10.22 Folded-plate roof.

using the folded-plate principle, a design theory was developed and published in Europe. The first technical publication on folded plates in the United States was a paper by Winter and Pei (Ref. 10.11) published in 1948. The past twenty years have seen numerous applications of the principles set forth. Figure 10.22 illustrates a folded-plate roof of 125-ft span which, in addition to carrying ordinary roof loads, carries the second floor as well from a system of cable hangers. The ground floor is thus kept free of columns.

The action of a folded-plate structure in carrying its loads is illustrated by Fig. 10.23. A surface load applied to the folded-plate structure shown in Fig. 10.23a would tend to deflect the individual plates, as indicated by the dotted lines for plate 1, if each plate were free to act independently, the plates spanning as one-way slabs between end diaphragms. The upper edges of plates 1 and 2, for example, would deflect in the directions CC' and CC'', respectively, as in Fig. 10.23b. The plates cannot act independently, however. The final location of the deflected edge C must lie at some singular position C_1. This necessitates that plates 1 and 2 undergo deflections Δ_1 and Δ_2, respectively, in their own planes. While the plates have negligible torsional and flexural stiffness

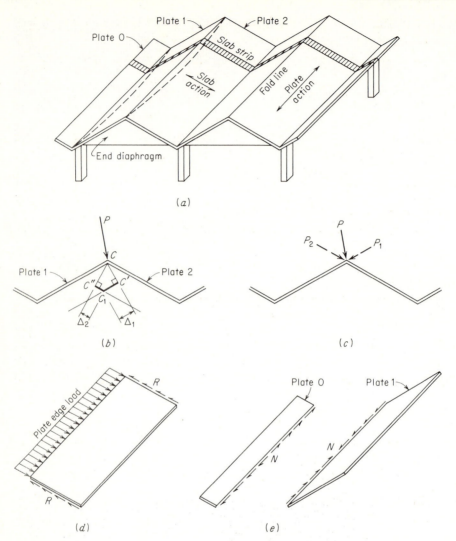

Fig. 10.23

in resisting the deflection components perpendicular to their own planes, they are extremely stiff in resisting in-plane deflections.[1] This resistance to bending about each plate's strong axis results in effective support for the slab along any fold line. A fold line cannot deflect without causing in-plane deflection of one or both of the adjacent plates, a type of deformation which the plates are able

---

[1] Note that the assumptions made here for folded-plate behavior are identical with those made for cylindrical shells, specifically that the stress resultants of consequence are N_x, $N_{x\phi}$, $N_{\phi x}$, Q_ϕ, M_ϕ, and N_ϕ, and that M_x, Q_x, $M_{x\phi}$, and $M_{\phi x}$ can be neglected.

to resist effectively. Consequently, the plates, rather than spanning in the long direction of the structure, act as slabs supported at the fold lines as well as the end diaphragms. For most folded plates, the ratio of plate length to width is sufficiently large, in fact, that it is satisfactory to consider all the surface load as carried by the plate in the direction perpendicular to the fold lines. This is termed *slab action*. A typical transverse slab strip is shown in Fig. 10.23*a*.

The reactions from such slab strips are applied as line loads to the fold lines. Because of the negligible torsional and out-of-plane flexural resistance of the plates, the only direction in which each plate can apply a reactive force to resist this line load is parallel to its own surface. The resultant line load P (see Fig. 10.23*c*) therefore resolves into components parallel to the two adjacent plates. The plates in turn carry this edge loading longitudinally between end diaphragms by plate action, as shown in Fig. 10.23*d*. Their behavior may be likened to that of deep, inclined-web girders, laterally braced by adjacent plates. Longitudinal membrane stresses N_x can be found as for any beam. These are termed the *free-edge* stresses.

If conditions are such that the free-edge stresses are the same on both sides of all fold lines, they are identical with the final stresses. This would be the case, for example, at the edges of an interior unit of a roof consisting of many identical units. If, on the other hand, the initial plate analysis indicates a stress difference on either side of a fold line, an incompatibility is indicated which cannot actually exist, because the strains on either side of a given fold line must be equal. This indicates the presence of longitudinal shears acting along the joint as shown in Fig. 10.23*e*.

For plate 0, the longitudinal stresses at the top and bottom edges, f_{t0} and f_{b0}, resulting from the application of an edge shear N acting as shown, are

$$(a) \quad f_{t0} = -\frac{2N}{A_0}$$

$$(b) \quad f_{b0} = +\frac{4N}{A_0}$$

with $+$ and $-$ indicating tension and compression, respectively, and A_0 indicating the cross-sectional area of the plate. Similarly, the same edge shear N applied to the lower edge of plate 1 in the opposite direction produces the following stresses in plate 1:

$$(c) \quad f_{b1} = -\frac{4N}{A_1}$$

$$(d) \quad f_{t1} = +\frac{2N}{A_1}$$

The edge shear force N must be of such magnitude as to eliminate completely any difference in longitudinal free-edge stresses at the fold line. The total required stress change is therefore equal to

$$f_{b0} - f_{b1} = \frac{4N}{A_0} + \frac{4N}{A_1}$$

The proportion of this change which occurs in plate 0 is

$$(e) \quad \frac{+4N/A_0}{4N/A_0 + 4N/A_1} = \frac{A_1}{A_0 + A_1} = K_0$$

while that in plate 1 is

$$(f) \quad \frac{4N/A_1}{4N/A_0 + 4N/A_1} = \frac{A_0}{A_0 + A_1} = K_1$$

The factors K_1 and K_2 are known as the *stress-distribution factors* and are applied to any difference in free-edge stresses to determine the stress change due to edge shear. Similar factors are found for all other joints.

Additionally, it is evident from Eqs. (a) to (d) that an edge shear causing a given stress change at the near edge of a plate will cause a stress change of one-half that amount, of opposite sign, at the far edge. Thus a *stress carry-over factor* of $-\frac{1}{2}$ is obtained. Stress-distribution factors and stress carry-over factors are applied to the free-edge stresses, in an iterative procedure exactly analogous to the familiar Cross method of moment distribution, to obtain N_x stresses corrected for the presence of edge shears.

Consider, for example, the folded-plate roof shown in Fig. 10.24a. A sawtooth cross section is shown having a modular width of 10 ft with a short cantilever plate at the outside edges. It is to carry a service live load of 40 psf of horizontal projection and a total calculated dead load of 55 psf of roof surface, on a simple span of 55 ft. Material strengths specified are $f_y = 50$ ksi and $f'_c = 4$ ksi.

Applying the usual load factors, the design live and dead loads are, respectively, $1.7 \times 40 = 68$ psf and $1.4 \times 55 = 77$ psf. The resolution of resultant forces for the analysis of slab and plate action is illustrated in Fig 10.24b and c. For plate BC, the total resultant applied load of 788 lb is resolved into components 682 lb normal to the surface (load intensity normal to the surface of $682/5.77 = 118$ plf) and 394 lb parallel to the surface. A similar calculation is made for each plate. A transverse strip of the roof is then analyzed as a continuous slab, assuming unyielding supports at each fold line, each slab span being loaded with the normal component of surface load to obtain transverse moment M_ϕ and transverse shear Q_ϕ. From the analysis, joint loads such as that shown for joint C in Fig. 10.24c are obtained. The end reaction from plate BC is 135 lb, acting normal to that plate, while that from plate CD is 284 lb. The resultant reaction of 365 lb is then resolved into components 410 and 319 lb parallel to plates BC and CD, respectively.

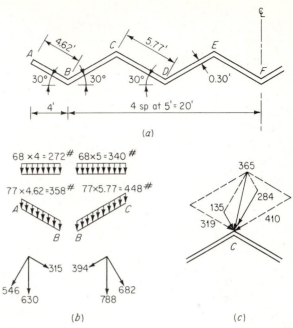

Fig. 10.24 Folded-plate example.

Loads at each edge of each plate are added algebraically to the parallel component of the surface load (394 lb for plate BC) to obtain the following line loads for each plate, acting as shown in Fig. 10.23d:

Plate AB	1260 plf
Plate BC	1580 plf
Plate CD	1350 plf
Plate DE	1650 plf
Plate EF	1650 plf

The section modulus for the exterior plate about its own neutral axis is $S_{ab} = bh^2/6 = 3.62 \times 55.5^2/6 = 1860$ in.³, while that for plate BC and all other interior plates is $S_{bc} = 3.62 \times 69.2^2/6 = 2880$ in.³ Following calculation of plate moments from the plate loads, these section moduli are used to obtain the free-edge stresses of Table 10.5. Stress-distribution factors are found from the cross-sectional areas of each plate, and the free-edge stresses are adjusted with successive balance and carry-over steps, as shown in Table 5, to obtain the final stresses $f_x = N_x/h$ given in the table.

In many cases the above analysis is a sufficient basis for design. In other cases, because of the geometry of the structure or the nature of the loads, it is necessary to correct for the relative displacement of the fold lines. This relative displacement amounts to support settlement for the slab strips. Although it will usually have little effect on the N_x stresses, there may be a significant change in slab moments M_ϕ and slab shears Q_ϕ.

Table 10.5   Balancing of free-edge stresses

Fold line	A	B		C		D		E		F
Stress-distribution factor	0	0.55	0.45	0.50	0.50	0.50	0.50	0.50	0.50	0.50
Free-edge stress	−3070	+3070	+2510	−2510	−2130	+2130	+2610	−2610	−2610	+2610
Balance	0	−310	+250	+190	−190	+240	−240	0	0	0
Carryover	+155	0	−95	−125	−120	+95	0	+120	0	0
	0	−52	+43	+2	+3	−47	+48	−60	+60	0
	+26	0	−1	−21	+23	+1	+30	−24	0	−30
	0	0	0	+22	−22	+15	−16	+12	−12	0
	0	0	−11	0	−7	+11	−6	+8	0	+6
	0	−6	+5	−3	+4	−8	+9	−4	+4	0
Final stress	−2889	+2702	+2701	−2445	−2445	+2437	+2437	−2558	−2558	+2586

There are several methods for determining the effect of joint translation. The method which is simplest conceptually is an iterative procedure in which the fold-line deflections for the initial plate stresses, calculated as described above, are determined. Slab moments and shears and joint loads resulting from such deflections are found, and the plate stresses resulting from these joint loads are calculated. The associated plate deflections are determined, and the cycle is repeated until sufficient accuracy is obtained (Ref. 10.2). While quite straightforward, the method has the disadvantage that it may converge very slowly, or actually diverge under certain circumstances.

Alternatively, a method analogous to the sidesway analysis of a multistory building is used. Each plate is given an arbitrary rotation in turn, and slab reactions, plate loads, plate stresses, and deflections are found for that rotation. A geometrical relation is established between plate deflection and plate rotation which permits the writing of a set of simultaneous equations. From these equations, constants are found which are applied to each of the arbitrary rotation solutions. Final moments and stresses are then found by superposition (Ref. 10.12). Other methods of solution are available, many particularly suited for use with digital computers. Additional information of interest in this respect will be found in Refs. 10.11 to 10.14.

Once the internal stress resultants for a folded plate are known, the design of the roof follows general procedures already established.

The slabs are designed for transverse bending moment M_ϕ just as for any other continuous one-way slab. Commonly, the transverse bending requirements control the thickness of folded plates. Bent bars are generally used, as shown in Fig. 10.25, for transverse reinforcement, although relatively small structures may utilize welded wire mesh to reduce placement costs. Particular attention must be given to proper detailing of the steel. At the valley fold line

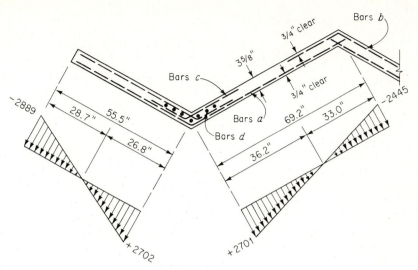

Fig. 10.25   Plate stresses and bar details.

between adjacent plates, it is generally necessary to bend up the positive reinforcement (bars a) so that it is continuous with the positive bars of the adjacent plate, because each plate is suspended, in effect, from the neighboring plate at its lowest point. Negative bars at the ridge fold line (bars b) may be continuous between adjacent plates, as shown. At the valley fold line, negative bars (bars c) should be carried to the *far* face of the adjacent slab before bending, to avoid spalling of the concrete at the top surface of the slab, at the fold line, when negative moment acts.

Large-diameter bars are usually required to provide for the longitudinal tensile stress N_x. They may be straight, or may be bent up at intervals along the plate span to help provide for diagonal tensile stresses. The longitudinal reinforcement is calculated to provide for the total tension in each plate resulting from the analysis (which has assumed the structure to be homogeneous and elastic). It is placed at the mid-thickness of each plate, as close to the tensile edge as possible. The steel centroid must be no farther from the tensile edge than the centroid of the triangular concrete tension zone. Elsewhere in the tension zone, a minimum steel area equal to 0.0035 times the concrete section should be provided, to control cracking.

Taking the stresses given in Table 10.5 for the preceding example for illustrative purposes,* the total longitudinal tension in plate BC is $3.62 \times 36.2 \times \frac{2701}{2} = 177{,}000$ lb. This requires a tensile steel area of $177{,}000/0.90 \times 50{,}000 = 3.94$ in.² Similar calculations for the adjacent plate AB indicate a tensile requirement there of 2.92 in.² The total bar area at fold

---

* These would be modified in a complete analysis due to the relative displacement of the fold lines.

line B must therefore be equal to $3.94 + 2.92 = 6.86$ in.^{2} Seven No. 9 bars will be used, providing an area of 7.00 in.^{2} They will be placed as shown (bars d) in Fig. 10.25. In order to provide room to lap-splice the main steel in its own plane (usually necessary because bar lengths greater than 60 ft are not commonly available) the bars will be spaced 4 in. on centers. Elsewhere in the tension zone, a minimum steel area of $0.0035 \times 3.62 \times 12 = 0.15$ in.^{2} per ft will be provided. Since the tension in the concrete is generally larger than $2\sqrt{f_c'}$, a maximum spacing of 3 times the slab thickness is required by Code. No. 3 bars at 9 in. spacing will be placed just above the transverse slab bars.

Diagonal tensile stresses associated with $N_{x\phi}$ may require placement of additional diagonal steel near the ends of the plate span. These bars may be designed using conventional equations. However, in calculating the nominal average shear stress, account should be taken of the influence of longitudinal fold-line shear forces (Fig. 10.23e) which may add to the plate shear caused by normal edge loads (Fig. 10.23d). Everywhere in the shell, at least shrinkage and temperature steel should be included, equal in amount to 0.0018 times the concrete cross-sectional area.

For other than very large folded plates, the thickness of the slabs may be governed by requirements of steel placement and concrete cover. In the preceding example, using $\frac{3}{4}$ in. cover top and bottom (implying that a waterproof membrane will protect the slab surface from the weather), and assuming that No. 3 bars will be used for transverse slab steel, the minimum possible thickness of the slab is

$$\tfrac{3}{4} + \tfrac{3}{8} + 1\tfrac{1}{8} + \tfrac{3}{8} + \tfrac{3}{4} = 3\tfrac{3}{8} \text{ in.}$$

only slightly less than the 3.62 in. thickness required by transverse bending.

10.12 HYBERBOLIC PARABOLOIDS

The possibilities of doubly curved shell roofs have been increasingly exploited in recent years. With both main curvatures in the same direction, say concave downward, such surfaces are termed *synclastic;* with the main curvatures in opposite directions, as in a saddle surface, they are termed *anticlastic.* The hyperbolic-paraboloid surface is an anticlastic form that has achieved great popularity owing to its economical use of material, relative simplicity of structural action, relatively easy forming, and interesting architectural effect.

With reference to Fig. 10.26, the surface is defined by two intersecting systems of straight-line generators, a and b. The lines AB and GH are both parallel to the plane YOZ, but they are not parallel to each other. The generators a intersect both AB and GH and are at the same time parallel to the plane XOZ. Similarly, the second set of generators b intersect HB and GA and are parallel to the plane YOZ. The intersection of the two planes XOZ and YOZ is usually, but not necessarily, a right angle.

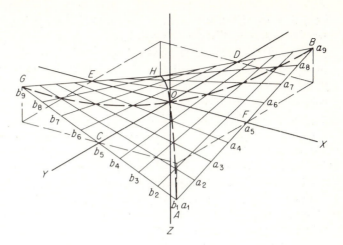

Fig. 10.26

In describing the surface, it is convenient to take as coordinate axes the two lines OF and OC, passing through the crown of the surface, and the intersection line OZ. The line OZ is always perpendicular to the plane XOY. The equation of the surface is

$$z = kxy \qquad\qquad\qquad\qquad\qquad\qquad (10.26)$$

where k is a constant representing the slope of the surface. It is evident that plane sections parallel to the two bisecting planes of the dihedral angle XOY are parabolic; these are termed the *principal parabolas*. They are respectively curved upward (GOB) and downward (AOH); the surface is anticlastic. Plane sections parallel to the plane XOY are hyperbolic; hence the name *hyperbolic paraboloid*.

The geometry of the surface described is most convenient in the building of forms. It is evident that the surface can be formed using straight joists, say 2 by 6's on edge, placed along either system of generators, over which can be laid plank or plywood sheathing thin enough to conform to the warped surface. Many different roof forms can be created either by use of the entire surface or by combining parts of it in various ways. For example, the surface of Fig. 10.26 can be used in its entirety, with a buttress support at A and H; Fig. 10.28 illustrates such an application. Alternatively, segments of the surface can be combined in a number of arrangements, a few of which are shown in Fig. 10.27. In most cases, stiffening ribs are required along the lines of intersection of the segments and along the free edges of the shell to accommodate the substantial thrusts developed.

In analyzing the hyperbolic paraboloid, it is convenient to work with X' and Y' axes, each rotated 45° in a horizontal plane from the original X or

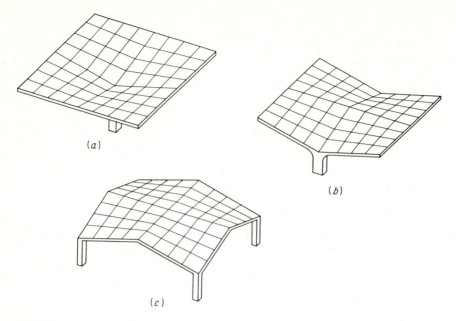

(a)

(b)

(c)

Fig. 10.27

Y axis; thus the axes are in the planes of the principal parabolas. The equation of the warped surface then becomes

$$z = 0.5k(x'^2 - y'^2) \tag{10.27}$$

The means by which a hyperbolic paraboloid carries its load may be apparent from its geometry. With two mutually perpendicular sets of parabolic strips, all of exactly the same curvature, it is reasonable to suppose that a load applied to a unit of surface area will be divided equally, half carried by each of the intersecting parabolic arch strips. The concave-downward strip carries its share by compression, while the concave-upward strip acts in tension. Since the parabolic axes correspond with the equilibrium polygons for a load distributed uniformly, no bending moments or transverse shears are developed; i.e., the direction of thrust must at any point be tangent to the axis of the parabolic strip. Figure 10.29 shows a parabolic compression strip, with a surface load $w/2$ producing a thrust at point (z,y'), with a horizontal component H and a vertical component $wL/4$. With x' a constant value, the equation of the curve is obtained from Eq. (10.27) and is

(a) $z = k_1 - 0.5ky'^2$

where k_1 is a new constant, simply defining the elevation of the vertex. Consider for convenience the parabola which passes through the origin (Fig.

Fig. 10.28 Hyperbolic paraboloid shell, St. Edmund's Episcopal Church, Elm Grove, Wisc.

10.26); k_1 is zero, and

(b) $z = -0.5ky'^2$

The slope of the parabola at any point is obtained by differentiation:

(c) $\dfrac{dz}{dy'} = -1.0ky'$

At $y' = L/2$, the slope is $-kL/2$. Since the slope of the resultant thrust must coincide with that of the parabola,

$$\frac{wL/4}{H} = \frac{kL}{2}$$

and

$$H = \frac{w}{2k} \qquad\qquad (10.28)$$

Equation (10.28) is an expression for the horizontal component of tensile or compressive thrust produced in the shell at any point due to a uniform load.

The above analysis is predicated upon the presence of a suitable restraining force, caused by continuity of each elemental unit of surface with those

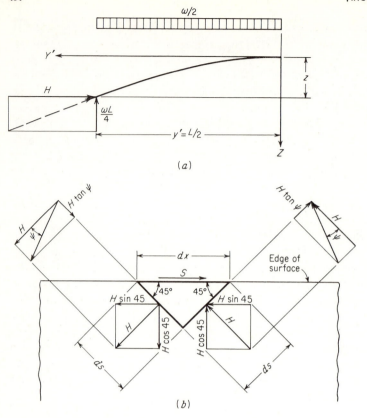

Fig. 10.29

adjacent. At the edges of the shell, a special situation obtains. The equilibrium of the small triangular element taken at the edge of the surface (Fig. 10.29b) demonstrates the need for edge members to take the thrust that is developed along the free edges. Consider first the horizontal component of thrust, and break that into components perpendicular and parallel to the edge; it is seen that the vectors normal to the edge are equal and act in opposite directions; hence they cancel, and no net force results normal to the free edge. The tangential components are additive, however. If S is the intensity of shear force per unit length along the edge,

$$2H \, ds \sin 45 = S \, dx$$

$$S = 2H \sin 45 \frac{ds}{dx} = 2H \sin^2 45$$

But $H = w/2k$, and $\sin^2 45 = 0.50$, so that

$$S = \frac{w}{2k} \tag{10.29}$$

In considering the effects of the vertical components of thrust, one must take separately the case of an element along a horizontal edge OF or OC (Fig. 10.26) and an element along a sloping edge AF or AC. Along the horizontal edges, since the slope of the axis of one parabola is equal to that of the intersecting parabola but of opposite sign, the vertical components cancel one another (see Fig. 10.29b). Consequently, the edge member must resist only a force parallel to its axis. Along the sloping edges, the vertical components do not cancel but rather add, and it is easily shown that the net vertical force is such that, when combined with the horizontal shear, it will produce a force acting along the sloping edge; again an edge member is required to resist axial load only.

The membrane analysis described provides the basis for the principal shell and edge-member reinforcement. In large hyperbolic paraboloids, careful attention must be given to the question of secondary stresses. These are due to several causes. There are relatively flat regions in most hyperbolic-paraboloid shells which give rise to significant local bending stresses. In addition, the membrane analysis results in an indicated strain incompatibility adjacent to edge members, which may be in compression or tension, as the case may be, while the adjacent monolithic shell surface is loaded primarily in shear. The result of this is the development of flexural and shearing stresses in the shell adjacent to the boundary members. Also, concentrated-load effects require special consideration, as well as the effects of temperature changes and support movements. The reader will find further information on hyperbolic-paraboloid shells in Refs. 10.5 and 10.15 to 10.17.

10.13 SPHERICAL DOMES

Another form of three-dimensional curved roof which has found frequent application is the spherical dome, generated by rotating an arc of a circle about a vertical axis through the center of the circle. The circular arc is termed a *meridian* (see Fig. 10.30a), and its plane a *meridian plane*. The movement of a point on the circular arc, as it rotates about the axis of the dome, describes a *parallel*, or a *circle of latitude*. Although simpler to describe geometrically than the hyperbolic-paraboloid surface, the dome is considerably more complicated to build, because in the case of the dome there are no straight generators. The forming joists as well as the sheathing must be curved to fit the surface, which adds considerably to the difficulty of construction and cost of such structures.

In spite of this, the appearance and economy of material associated with spherical domes has led to their choice in preference to other solutions in many cases. Spherical domes need not be completely closed, as shown in Fig. 10.30a, but may include "skylight openings" at the top (Fig. 10.30b) or arched openings along the lower edge (Fig. 10.30c). A spherical dome which has attracted considerable attention is the Kresge Auditorium at M.I.T., shown in Fig. 10.31, which is supported at only three points at ground level.

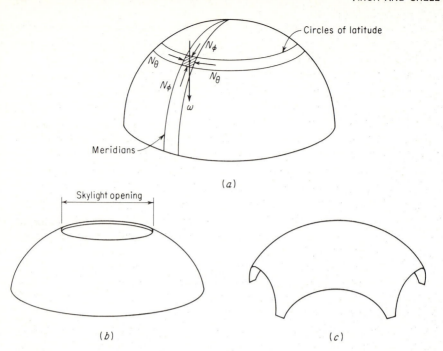

(a)

(b)   (c)

Fig. 10.30

Fig. 10.31   Spherical shell, Kresge Auditorium, at M.I.T.

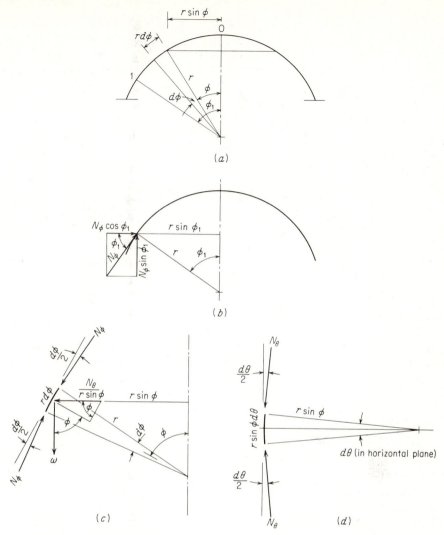

Fig. 10.32

As is the case for hyperbolic paraboloids, domes carry their load almost entirely by membrane stress tangent to the surface, over the greater part of the shell. As a consequence, the shell can be made very thin, even for long spans. Flexural stresses become significant only after axial strains of consider-able magnitude occur. In Fig. 10.32a, the surface of an annular ring formed by rotating $r\,d\phi$ about the axis is

(a) $r\,d\phi\,2\pi r \sin\phi$

The total surface of the shell above the level of point 1 is found by integrating this expression between the limits zero and ϕ to obtain

(b) $A = 2\pi r^2(1 - \cos \phi_1)$

If the load w_d per unit of surface area is constant (e.g., dead load), the total load above level 1 is then

(c) $W_d = 2\pi r^2 w_d(1 - \cos \phi_1)$

If the load w_s per unit of horizontal projected area is constant (e.g., snow load), the total load above the level 1 is

(d) $W_s = \pi r^2 w_s \sin^2 \phi_1$

With the total load on the portion of the shell above any specified circle of latitude known, the meridional thrust can be calculated from a consideration of the vertical equilibrium of that portion of the shell. In Fig. 10.32b, let W be the total load of the shell above the level concerned, and let N_ϕ be the meridional thrust per unit of length. Then

(e) $W = -2\pi r \sin \phi_1 N_\phi \sin \phi_1$

and

$$N = -\frac{W}{2\pi r \sin^2 \phi_1} \tag{10.30}$$

It is noted that the meridional thrust is always compressive. If the dome is hemispherical, the thrust reaches a maximum value of $W/2\pi r$ at $\phi_1 = 90°$.

The circumferential stress is obtained from the equilibrium of the small element shown shaded in Figs. 10.30a and 10.32c and d. Defining the component of the load in the direction of the shell normal,

$$w_z = w_d \cos \phi + w_s \cos^2 \phi$$

then the circumferential stress is

$$N_\theta = -N_\phi - w_z r \tag{10.31}$$

relating N_θ and N_ϕ. If one considers the particular case of a dome for which the load is uniformly distributed over the surface, one will find that the meridional thrust, always compressive, varies from a minimum of $-wr/2$ at the crown to $-wr$ if ϕ becomes as large as 90°. The circumferential stress varies from $-wr/2$ (compression) at the crown to wr (tension) at $\phi = 90°$. The transition point, at which circumferential stress is zero, occurs at $\phi = 51°50'$.

If the shell is terminated so that ϕ_1 is smaller than 90°, the meridional thrust will have an outward component which must be resisted by a ring girder subjected to an outward radial force of $N_\phi \cos \phi_1$ per unit length. The resulting ring tension is found from the hoop-tension expression and is

$$P_T = \frac{W \cos \phi_1}{2\pi \sin \phi_1} \tag{10.32}$$

Note that the ring tensile stress bears no direct relation to the circumferential stress present in the shell. In fact, as has been noted, for a uniformly loaded shell with ϕ_1 less than 51°50′, the shell circumferential stresses are compressive, although the stress in the ring girder must be tensile. Hence, one has an incompatibility of strains near the base ring, which results in local bending and transverse shearing stresses in the shell.

It should be kept in mind that the above pertains to an analysis of the spherical shell for membrane stress conditions. While the results will be close to the actual stresses over the greater part of uniformly loaded shells, particular attention must be directed to stresses resulting from concentrated loads, unbalanced loading, temperature and shrinkage effects, and discontinuity along the edges. For additional information with regard to these and related matters, the reader should consult Refs. 10.5 and 10.18.

10.14 CONCLUSION

Singly and doubly curved structures are becoming increasingly common as engineers become more familiar with the principles of their design and as builders gain experience with their construction. Great spans have been achieved and interesting architectural effects have been produced. An outstanding example is the ribbed shell designed by Nervi, shown in Fig. 10.33. Shells of different forms have been found suitable for a great variety of pur-

Fig. 10.33 Pallazzetto della Sport, Rome, designed by Pier Luigi Nervi.

poses, including the construction of convention halls, auditoriums, sports arenas, and column-free commercial buildings and stores. Shells are used with increasing frequency for the roofs of industrial buildings; all the forms discussed (but notably cylindrical shells and folded plates) have been used for such purposes.

Genuine saving of material can be accomplished by intelligent matching of form with loads and support conditions. It is expected (particularly as precasting techniques develop more fully) that cost economy may soon more closely follow economy of material.

The practicing engineer should not be alarmed at the mathematical complications of analysis, which in most cases are not as great as supposed. New techniques are available for the calculation of stresses, and the use of digital computers greatly reduces the amount of tedious computation. Also available to the structural engineer in general practice are the services of consultants who have made a special study of shell structures and, in collaboration with the principal engineer, can produce designs of great originality that can be built at a reasonable price.

REFERENCES

10.1. S. Timoshenko: "Strength of Materials," D. Van Nostrand Company, Inc., Princeton, N.J., 1956.

10.2. G. C. Ernst: Design of Hinges and Articulations in Reinforced Concrete Structures, *Trans. ASCE*, vol. 106, p. 862, 1941.

10.3. S. P. Timoshenko and S. Woinowsky-Krieger: "Theory of Plates and Shells," 2d ed., McGraw-Hill Book Company, New York, 1959.

10.4. W. Flügge: "Stresses in Shells," Springer-Verlag, Berlin, 1962.

10.5. D. P. Billington: "Thin Shell Concrete Structures," McGraw-Hill Book Company, New York, 1965.

10.6. H. Lundgren: "Cylindrical Shells," vol. 1, Danish Technical Press, Copenhagen, 1951.

10.7. J. Chinn: Cylindrical Shell Analysis Simplified by Beam Method, *J. ACI*, May, 1959.

10.8. M. S. Ketchum, A. L. Parme, H. W. Connor, A. Siev, A. Tedesko, and A. Zweig: Discussion of Ref. 10.7, *J. ACI*, December, 1959.

10.9. Design of Cylindrical Concrete Shell Roofs, American Society of Civil Engineers Manual of Engineering Practice 31, 1952.

10.10. "Coefficients for Design of Cylindrical Concrete Shell Roofs," Portland Cement Association, Chicago, 1959.

10.11. G. Winter and M. Pei: Hipped Plate Construction. *J. ACI*, January, 1947.

10.12. H. Simpson: Design of Folded Plate Roofs, *J. Struct. Div. ASCE*, January, 1958.

10.13. D. Yitzhaki: "Prismatic and Cylindrical Shell Roofs," Haifa Science Publishers, Haifa, Israel, 1958.

10.14. K. S. Lo and A. C. Scordelis: Finite Segment Analysis of Folded Plates, *J. Struct. Div. ASCE*, May, 1969.

10.15. Elementary Analysis of Hyperbolic Paraboloid Shells, *Portland Cement Assoc. Bull.* ST85, 1960.

10.16. A. L. Parme: Shells of Double Curvature, *Proc. ASCE*, vol. 123, p. 989, 1958.

10.17. F. Candella: General Formulas for Membrane Stresses in Hyperbolic Paraboloidical Shells, *J. ACI*, October, 1960.

10.18. Design of Circular Domes, *Portland Cement Assoc. Bull.* ST55 (undated publication).

11
Prestressed Concrete

11.1 INTRODUCTION

Modern structural engineering tends to progress toward more economic structures through gradually improved methods of design and the use of higher-strength materials. This results in a reduction of cross-sectional dimensions and consequent weight savings. Such developments are particularly important in the field of reinforced concrete, where the dead load represents a substantial part of the total design load. Also, in multistory buildings, any saving in depth of members, multiplied by the number of stories, can represent a substantial saving in total height, load on foundations, length of heating and electrical ducts, plumbing risers, and wall and partition surfaces.

Significant savings can be achieved by the use of high-strength concrete and steel in conjunction with present-day design methods, which permit an accurate appraisal of member strength. However, as was discussed in Art. 2.9, there are limitations to this development, due mainly to the interrelated problems of cracking and deflection at service loads. The efficient use of high-strength steel is limited by the fact that the amount of cracking (width and number of cracks) is proportional to the strain, and therefore the stress, in the steel. Although a moderate amount of cracking is normally not objectionable in structural concrete, excessive cracking is undesirable in that it exposes the reinforcement to corrosion, it may be visually offensive, and it may trigger a premature failure by diagonal tension. The use of high-strength materials is further limited by deflection considerations, particularly when refined analysis is used. The slender members which result may permit deflections which are functionally or visually unacceptable. This is further aggravated by cracking, which reduces the flexural stiffness of members.

These undesirable characteristics of ordinary reinforced concrete have been largely overcome by the development of prestressed concrete. A prestressed-concrete member may be defined as one in which there have been introduced internal stresses of such magnitude and distribution that the stresses resulting from the given external loading are counteracted to a desired degree. Concrete is basically a compressive material, with its strength in tension a low and unreliable value. Prestressing applies a precompression to the member which reduces or eliminates undesirable tensile stresses that would otherwise be present. Cracking under service loads can be minimized or even avoided entirely. Deflections may be limited to an acceptable value; in fact, members may be designed to have zero deflection under the combined effects of service load and prestress force.

The first suggestions for prestressing seem to have been made between 1886 and 1908 by the Americans P. H. Jackson and G. R. Steiner, the Austrian J. Mandl, and the German J. Koenen. The use of high-strength steel was first suggested by the Austrian F. von Emperger in 1923, while at about the same time, the American R. H. Dill proposed "full prestressing" to eliminate cracks completely. These proposals remained mainly on paper; the practical development of prestressed-reinforced-concrete structures is chiefly due to E. Freyssinet and Y. Guyon (France), E. Hoyer (Germany), and G. Magnel (Belgium). Circular prestressing of cylindrical tanks and pipes, originated by W. H. Hewitt in 1923, was the first important application of the principles of prestressing in the United States. Important contributions have been made by T. Y. Lin in the design of many types of prestressed-concrete structures in the United States since 1950.

It is interesting to contrast the development of prestressed concrete in Europe, where a high ratio of material to labor costs prevails, with that in the United States, where the ratio is reversed. In Europe, many sophisticated designs have been executed which minimize the material used, but require a great amount of highly qualified labor to build and to prestress. Long-span bridges, two-dimensional floor systems, shells, and even space trusses have resulted. In the United States, development has been in the direction of relatively small precast building components, well suited to mass production. A few types of standardized precast members have accounted for the bulk of prestressed-concrete construction in the United States over the past twenty years. This difference is less apparent now than previously. European industry now produces standardized precast building components, utility poles, railroad ties, and other items using techniques perfected in the United States, while engineers here have made use of European experience in the design of special structures of major span.

11.2 DESIGN BASIS

It has already been pointed out, in Art. 3.1, that the design of concrete structures may be based either on providing sufficient strength, which would be utilized fully only if the expected loads were increased by an overload factor, or on keeping material stresses within permissible limits when actual service loads act. In the case of ordinary reinforced-concrete members, strength design is recommended and widely used. Members are proportioned on the basis of strength requirements, and then checked for satisfactory service-load behavior, notably with respect to deflection and cracking. Proportions are modified if necessary.

In the case of prestressed-concrete members, present practice is to proportion members so that stresses in the concrete and steel at actual service loads are within permissible limits. These limits are a fractional part of the actual capacities of the materials. There is some logic to this approach, since the main purpose of prestressing is to improve the performance of members at service loads. Furthermore, service-load requirements often control the amount of prestress force used. Design based on service loads may usually be carried out assuming elastic behavior of both the concrete and the steel, since stresses are relatively low in each and cracking is not extensive.

Regardless of the starting point chosen for the design, a structural member must be satisfactory at all stages of its loading history. Accordingly, prestressed members designed on the basis of permissible stresses must also be checked to ensure that sufficient strength is provided should overloads occur.

11.3 SOURCES OF PRESTRESS FORCE

Prestress force can be applied to a concrete beam in many ways. Perhaps the most obvious method of precompressing is to use jacks reacting against abutments, as shown in Fig. 11.1a. Such a scheme has been employed for large projects. Many variations are possible, including replacing the jacks with compression struts after the desired stress in the concrete is obtained or the use of inexpensive jacks which remain in place in the structure, in some cases with a cement grout used as the hydraulic fluid. The principal difficulty associated with such a system is that even a slight movement of the abutments will drastically reduce the prestress force.

In most cases the same result is more conveniently obtained by tying the jack bases together with wires or cables, as shown in Fig. 11.1b. These wires or cables may be external, located on each side of the beam; more usually they are passed through a hollow conduit embedded in the concrete beam. Usually, one end of the prestressing tendon is anchored, and all the force is

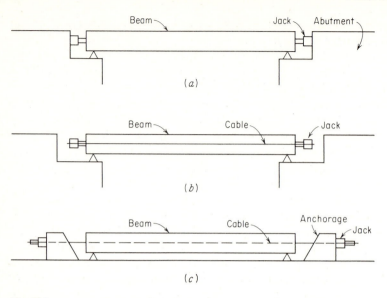

Beam   Jack Abutment

(a)

Beam   Cable   Jack

(b)

Beam   Cable   Anchorage   Jack

(c)

Fig. 11.1

applied at the other end. After attainment of the desired amount of prestress force, the wires are wedged against the concrete, and the jacking equipment is removed for reuse. Note that, in this type of prestressing, the entire system is self-contained and is independent of relative displacement of the supports.

Another method of prestressing which is widely used is illustrated by Fig. 11.1c. The prestressing strands are tensioned between massive abutments in a casting yard prior to placing the concrete in the beam forms. The beam is poured around the tensioned strands, and after the concrete has attained sufficient strength, the jacking pressure is released. This transfers the prestressing force to the concrete by bond and friction along the strands, chiefly at the outer ends.

Other means for introducing the desired prestressing force have been attempted on an experimental basis. Thermal prestressing can be achieved by preheating the steel by electrical or other means. Anchored against the ends of the concrete beam while in the extended state, the steel cools and tends to contract. The prestress force is developed through the restrained contraction. The use of expanding cement in concrete members has been tried with varying success. The volumetric expansion, restrained by steel strands or by fixed abutments, produces the prestress force.

Most of the patented systems for applying prestress in current use are variations of those shown in Fig. 11.1b and c. Such systems can generally be classified as *pretensioning* or *posttensioning* systems. In the case of pretensioning, the wires are stressed before the concrete is placed, as in Fig. 11.1c. This system is well suited for mass production, since casting beds can be made several hundred feet long, the entire length cast at once, and individual beams cut to the desired length from the long casting.

In posttensioned construction, shown in Fig. 11.1b, the wires are tensioned after the concrete is placed and has acquired its strength. Usually, a hollow conduit or sleeve is provided in the beam, through which the wires are passed. In some cases, hollow box-section beams are used, as are external posttensioning wires. The jacking force is usually applied against the ends of the hardened concrete, eliminating the need for massive abutments.

A large number of particular systems, steel elements, jacks, and anchorage fittings have been developed in this country and abroad, many of which differ from each other only in minor details (Refs. 11.1 to 11.3). As far as the designer of prestressed-concrete structures is concerned, it is unnecessary, and perhaps even undesirable, to specify in detail the technique that is to be followed and the equipment to be used. It is frequently best to specify only the magnitude and line of action of the prestress force. The contractor is then free, in bidding the work, to receive quotations from several different prestressing subcontractors, with resultant cost savings. It is evident, however, that the designer must have some knowledge of the details of the various systems contemplated for use, so that in selecting cross-sectional dimensions, any one of several systems can be accommodated.

As noted in Art. 2.10, the magnitude of prestress force in concrete members is not constant, but diminishes with time at a gradually decreasing rate. The reduction of the initial prestress force P_i to the final effective value P_e is due to a number of causes. The most significant are elastic shortening of the concrete, concrete creep under sustained load, concrete shrinkage, relaxation of the stress in the steel, frictional loss between the tendons and the concrete during the stressing operation, and loss due to slip of the steel strands as stress is transferred from the prestressing jacks to the anchorages at the ends of the beam. The approximate magnitudes of the various losses and means for predicting stress loss will be discussed in detail in Art. 11.6.

11.4 PRESTRESSING STEELS

Early attempts at prestressing concrete were unsuccessful because steel of ordinary structural strength was used. The low prestress obtainable in such rods was quickly lost due to shrinkage and creep in the concrete.

Such changes in length of concrete have much less effect on prestress force if that force is obtained using highly stressed steel wires or cables. In Fig. 11.2a, a concrete member of length L is prestressed using steel bars of ordinary strength stressed to 20,000 psi. With $E_s = 29 \times 10^6$ psi, the unit strain ϵ_s required to produce the desired stress in the steel of 20,000 psi is

$$\epsilon_s = \frac{\Delta L}{L} = \frac{f_s}{E_s} = \frac{20,000}{29 \times 10^6} = 6.9 \times 10^{-4}$$

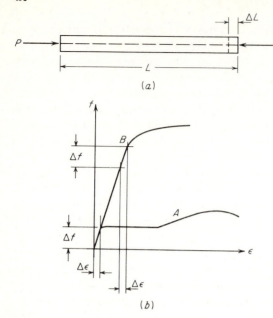

Fig. 11.2

But the long-term strain in the concrete due to shrinkage and creep alone may be of the order of 8.0×10^{-4}, which would be sufficient to completely relieve the steel of all stress.

Alternatively, suppose that the beam is prestressed using high-tensile wire stressed to 150,000 psi. The elastic modulus of steel does not vary greatly, and the same value of 29×10^6 psi will be assumed. Then in this case the unit strain required to produce the desired stress in the steel is

$$\epsilon_s = \frac{150,000}{29 \times 10^6} = 51.7 \times 10^{-4}$$

If shrinkage and creep strain are the same as before, the net strain in the steel after these losses is

$$\epsilon_s' = (51.7 - 8.0)10^{-4} = 43.7 \times 10^{-4}$$

and the corresponding stress after losses is

$$f_s = \epsilon_s' E_s = 43.7 \times 10^{-4} \times 29 \times 10^6 = 127,000 \text{ psi}$$

This represents a stress loss of about 15 per cent, compared with 100 per cent loss in the beam using ordinary steel. It is apparent that the amount of stress lost because of shrinkage and creep is independent of the original stress in the steel. Therefore, the higher the original stress, the lower will be the percentage loss. This is illustrated graphically by the stress-strain curves of Fig. 11.2b. Curve A is representative of ordinary reinforcing rods, with a yield stress between 40,000 and 60,000 psi, while curve B represents high-tensile wire, with an ultimate stress of about 250,000 psi. The stress change Δf resulting from a certain change in strain $\Delta \epsilon$ is seen to have much less effect when high steel-stress levels are attained. Prestressing of concrete is therefore practical only when steels of very high strength are used.

Prestressing steel is most commonly used in the form of individual wires, stranded cable made up of seven wires, and alloy-steel bars. The physical properties of these have been discussed in Art. 1.12, and typical stress-strain curves appear in Fig. 1.14.

The tensile stress permitted by the ACI Code in prestressing wires, strands, or bars is dependent upon the stage of loading. When the jacking force is first applied, a stress of $0.80f_{pu}$ is allowed, where f_{pu} is the ultimate strength of the steel. In pretensioning tendons, immediately after transfer of prestress force to the concrete, and in posttensioning tendons, immediately after anchoring, the permissible stress is $0.70f_{pu}$. The justification for a higher allowable stress during the stretching operation is that the steel stress is known quite precisely at this stage. Hydraulic jacking pressure and total steel strain are quantities that are easily measured. In addition, if an accidentally deficient tendon should break, it could easily be replaced at small cost; in effect, the tensioning operation is a performance test of the material. The lower value of $0.70f_{pu}$ applies after elastic shortening of the concrete and anchorage slip have taken place, when service loads may be applied. The steel stress is further reduced during the life of the member due to shrinkage and creep in the concrete, and relaxation in the steel.

The strength and other characteristics of prestressing wire, strands, and bars vary somewhat among manufacturers, as do methods of grouping tendons and anchoring them. Typical information is given for illustration in Table 11 of the Appendix and in Refs. 11.1 and 11.2.

11.5 CONCRETE FOR PRESTRESSED CONSTRUCTION

Ordinarily, concrete of substantially higher compressive strength is used for prestressed structures than for those constructed of ordinary reinforced concrete. Most prestressed construction in the United States at the present time is designed for a compressive strength between 4000 and 6000 psi. There are several reasons for this:

1. High-strength concrete normally has a higher modulus of elasticity (see Fig. 1.4). This means a reduction in initial elastic strain under application of prestress force and a reduction in creep strain, which is approximately proportional to elastic strain. This results in a reduction in loss of prestress.
2. In posttensioned construction, high bearing stresses result at the ends of beams where prestressing force is transferred from the tendons to anchor-

age fittings, which bear directly against the concrete. This problem can be met by increasing the size of the anchorage fitting or by increasing the bearing capacity of the concrete by increasing its compressive strength. The latter is usually more economical.

3. In pretensioned construction, where transfer by bond is customary, the use of high-strength concrete will permit the development of higher bond stresses.
4. A substantial part of the prestressed construction in the United States is precast, with the concrete mixed, placed, and cured under carefully controlled conditions which facilitate obtaining higher strengths.

The strain characteristics of concrete under short-time and sustained loads assume an even greater importance in prestressed structures than in reinforced-concrete structures, because of the influence of strain on loss of prestress force. Strains due to stress, together with volume changes due to shrinkage and temperature changes, may have considerable influence on prestressed structures. In this connection, it is suggested that the reader review Arts. 1.7 to 1.10, which discuss in some detail the compressive and tensile strengths of concrete under short-time and sustained loads and the changes in concrete volume that occur owing to shrinkage and temperature change.

As is the case for prestressing steels, the allowable stresses in the concrete, according to the ACI Code, are dependent upon the stage of loading. These stresses are given in Table 11.1. Here f_{ci}' is the compressive strength of the concrete at the time of initial prestress, and f_c' the specified compressive strength of the concrete.

Table 11.1 Permissible stresses in concrete in prestressed flexural members

---

1. Flexural stresses immediately after transfer, before losses, shall not exceed the following:
   a. Compression... $0.60f_{ci}'$
   b. Tension in members without bonded auxiliary reinforcement (unprestressed or prestressed) in the tension zone.................................... $3\sqrt{f_{ci}'}$

   Where the calculated tension stress exceeds this value, reinforcement shall be provided to resist the total tension force in the concrete computed on the assumption of an uncracked section.
2. Stresses at service loads, after allowance for all prestress losses, shall not exceed the following:
   a. Compression... $0.45f_c'$
   b. Tension in precompressed tensile zone................................ $6\sqrt{f_c'}$
   c. Tension in precompressed tensile zone in members where computations based on the transformed cracked section and on bilinear moment-deflection relationships show that immediate and long-term deflections comply with restrictions stated elsewhere in the Code.............................. $12\sqrt{f_c'}$
3. The permissible stresses may be exceeded when it is shown experimentally or analytically that performance will not be impaired.

11.6 LOSS OF PRESTRESS

The prestressing force is not constant, but decreases with time, due to a variety of causes. Most losses take place rapidly at first, then at a gradually decreasing rate, approaching a limit value asymptotically. In practice, only initial and final prestress values need be considered. Initial prestress P_i is the force existing immediately after transfer of load from the jacks to the concrete, and is lower than the jacking force because of elastic shortening of the concrete, frictional losses, and slip of the tendon at the anchorages. Final prestress P_e is obtained after all losses have occurred, including the time-dependent losses due to shrinkage and creep of the concrete and relaxation of the steel. Although a member must be satisfactory at all stages in its life history, it is usually sufficient to investigate prestressed beams under initial prestress plus self-weight, final prestress plus service load, and ultimate load. The most important causes of prestress loss will be discussed briefly.

a. Elastic shortening of the concrete In pretensioned members, as the tendon force is transferred from the fixed abutments to the concrete beam, elastic instantaneous compressive strain will take place in the concrete, tending to reduce the stress in the bonded prestressing steel. The steel stress loss is

$$\Delta f_{s,\text{elastic}} = E_s \frac{f_c}{E_c} = n f_c \tag{11.1}$$

where f_c is the concrete stress, at the level of the steel centroid, immediately after prestress is applied. If the strands are placed with significantly different effective depths, the stress loss in each should be calculated separately.

In posttensioned members, if all the strands are tensioned at one time, there will be no loss due to elastic shortening, because this shortening will occur as the jacking force is applied and before the prestressing force is measured. On the other hand, if various strands are tensioned sequentially, the stress loss in each strand will vary, being a maximum in the first strand tensioned and zero in the last strand. In most cases, it is sufficiently accurate to calculate the loss in the first strand and to apply one-half that value to all strands.

b. Creep of concrete Shortening of concrete under sustained load has been discussed in Art. 1.7. It may be expressed in terms of either the creep coefficient C_c or the unit creep strain δ_t. Creep shortening may be several times the initial elastic shortening, and it is evident that it will result in loss of prestress force. For pretensioned members, the stress loss in the steel due to concrete creep may be found most conveniently by applying the creep coefficient to the loss due to elastic shortening. For posttensioned members in which there is normally no loss due to elastic shortening, the stress loss due to creep must be calculated separately, based on the concrete stress f_c after the tendons are anchored. In this case

$$\Delta f_{s,\text{creep}} = C_c n f_c = E_s f_c \delta_t \tag{11.2}$$

Values of C_c and δ_t for different concrete strengths, for average conditions of humidity, are given in Table 1.1.

c. Shrinkage of concrete It is apparent that a decrease in the length of a member due to shrinkage of the concrete will be just as detrimental as length changes due to stress, creep, or other causes. As discussed in Art. 1.10, the shrinkage strain ϵ_{sh} may vary between about 0.0002 and 0.0007. A typical value of 0.0003 may be used in lieu of specific data. The steel stress loss resulting from shrinkage is

$$\Delta f_{s,\text{shrink}} = \epsilon_{sh} E_s \tag{11.3}$$

d. Relaxation of steel The phenomenon of relaxation, similar to creep, was discussed in Art. 1.12. Loss in stress due to relaxation will vary depending upon the stress in the steel, and may be estimated using Eq. (1.3). Since a sustained prestress of about 65 per cent of the ultimate strength of the steel is common, it is apparent that some allowance must be made for relaxation.

It is interesting to observe that the largest part of the relaxation loss occurs shortly after the steel is stretched. For stresses of $0.80f_s'$ and higher, even a very short period of loading will produce substantial relaxation, and this in turn will reduce the relaxation that will occur later at a lower stress level. The relaxation rate can thus be artificially accelerated by temporary over-tensioning. This technique has sometimes been employed to reduce losses.

e. Frictional losses Losses due to friction, as the tendon is stressed in post-tensioned members, are usually separated for convenience into two parts: curvature friction and wobble friction. The first is due to intentional bends in the tendon profile as specified, and the second to the unintentional variation of the tendon from its intended profile. It is apparent that even a "straight" tendon duct will have some unintentional misalignment, so that wobble friction must always be considered in posttensioned work. Usually, curvature friction must be considered as well. The force at the jacking end of the tendon, P_0, required to produce the force P_x at any point x along the tendon, can be found from the expression

$$P_0 = P_x \epsilon^{KL + \mu\alpha} \tag{11.4a}$$

where ϵ = base of natural logarithms
$\quad\quad L$ = tendon length from jacking end to point x
$\quad\quad \alpha$ = angular change of tendon from jacking end to point x, radians
$\quad\quad K$ = wobble friction coefficient, lb per lb per ft
$\quad\quad \mu$ = curvature friction coefficient

Table 11.2   Friction coefficients for posttensioned tendons

	Wobble coefficient K	Curvature coefficient μ
Grouted tendons in metal sheathing:		
Wire tendons	0.0010–0.0015	0.15–0.25
High-strength bars	0.0001–0.0006	0.08–0.30
7-wire strand	0.0005–0.0020	0.15–0.25
Mastic-coated unbonded tendons:		
Wire tendons	0.0010–0.0020	0.05–0.15
7-wire strand	0.0010–0.0020	0.05–0.15
Pregreased unbonded tendons		
Wire tendons	0.0003–0.0020	0.05–0.15
7-wire strand	0.0003–0.0020	0.05–0.15

There has been much research on frictional losses in prestressed construction, particularly with regard to the values of K and μ. These vary appreciably, depending on construction methods and materials used. The values in Table 11.2 may be used as a guide.

If one accepts the approximation that the normal pressure on the duct causing the frictional force results from the undiminished initial tension all the way around the curve, the following simplified expression for loss in tension is obtained:

$$P_0 = P_x(1 + KL + \mu\alpha) \tag{11.4b}$$

where α is the angle between the tangents at the ends. The ACI Code permits the use of the simplified form if the value of $(KL + \mu\alpha)$ is not greater than 0.3.

The loss of prestress for the entire tendon length can be computed by segments, with each segment assumed to consist of either a circular arc or a length of tangent.

f. Slip at the anchorages   As the load is transferred to the anchorage device in posttensioned construction, a slight inward movement of the tendon will occur as the wedges "seat" themselves and as the anchorage itself deforms under stress. The amount of movement will vary greatly, depending on the type of anchorage and on construction techniques. The amount of movement due to "seating" and stress deformation associated with any particular type of anchorage is best established by test. Once this amount ΔL is determined, the stress loss is easily calculated using the equation

$$\Delta f_{s,\text{slip}} = \frac{\Delta L}{L} E_s \tag{11.5}$$

It is significant to note that the amount of slip is nearly independent of the cable length. For this reason, the stress loss will be large for short tendons and

Table 11.3 Summary of losses in prestress, per cent of initial prestress

Losses	Pretensioned beams		Posttensioned beams	
Before transfer	Shrinkage	3%		
At transfer	Elastic strain in concrete	3%	Elastic strain in concrete (only if multiple strands)	1%
			Anchor slip	2%
			Bending of beam (usually negligible)	
			Friction	2%
After transfer	Shrinkage	4%	Shrinkage	4%
	Concrete creep	7%	Concrete creep	4%
	Steel relaxation	3%	Steel relaxation	3%
Total		20%		16%

relatively small for long tendons. The practical consequence of this is that it is most difficult to posttension short tendons with any degree of accuracy.

g. Summary of losses   As indicated in the foregoing discussion, the magnitude of losses in prestressed construction will vary widely, depending on conditions. Table 11.3 presents a summary of losses for pretensioned and posttensioned beams, and gives an order of magnitude of the various amounts. Individual losses may vary substantially from the values given; however, for ordinary cases in which detailed analysis is not justified, the figures may provide a basis for design. Often, in practice, a lump sum loss of 35,000 psi for pretensioned members and 25,000 psi for posttensioned work is used to estimate time-dependent losses.

11.7 ELASTIC FLEXURAL ANALYSIS

A convenient starting point in the study of prestressed members is to think of the prestressing forces as a system of external forces acting on the concrete member, which must be in equilibrium under the action of those forces. Figure 11.3a shows a simple-span prestressed beam with curved tendons, typical of many posttensioned installations. The portion of the beam to the left of a vertical cutting plane x-x is taken as a free body, with forces acting as shown in Fig. 11.3b. The force P at the left end is exerted on the concrete through the tendon anchorage, while the force P at the cutting plane x-x results from combined shear and normal force acting on the concrete surface at that location. The direction of P is tangent to the curve of the tendon at each location. Note the presence of the force N, acting on the concrete from the tendon, due to tendon curvature. This force will be distributed in some manner along the length

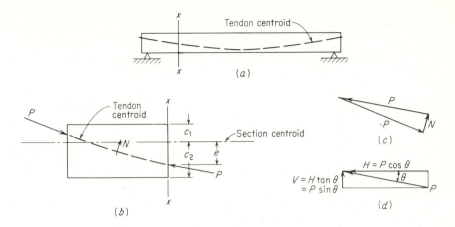

Fig. 11.3

of the tendon, the exact distribution depending upon the tendon profile. Its resultant, and the direction in which the resultant acts, can be found from the force diagram of Fig. 11.3c.

It is convenient in working with the prestressing force P to divide it into its components in the horizontal and vertical directions. The horizontal component (Fig. 11.3d) is $H = P \cos \theta$, and the vertical component is $V = H \tan \theta = P \sin \theta$, where θ is the angle of inclination of the tendon centroid at the particular section. Since the slope angle is normally quite small, the cosine of θ is very close to unity, and it is sufficient for most calculations to take $H = P$.

In developing elastic equations for flexural stress, the effects of prestress force, dead-load moment, and live-load moment are calculated separately, and the separate stresses are superimposed. When the initial prestress force P_i is applied with an eccentricity e below the neutral axis of the cross section with area A and top and bottom fiber distances c_1 and c_2, respectively, it causes the compressive stress $-P_i/A$ and the bending stresses $+P_i e c_1/I$ and $-P_i e c_2/I$ in the top and bottom fibers, respectively (compressive stresses are designated as negative, tensile stresses as positive), as shown in Fig. 11.4a. Then, at the top fiber, the stress is

$$f_{1pi} = -\frac{P_i}{A} + \frac{P_i e c_1}{I} = -\frac{P_i}{A}\left(1 - \frac{ec_1}{r^2}\right) \tag{11.6a}$$

and at the bottom fiber

$$f_{2pi} = -\frac{P_i}{A} - \frac{P_i e c_2}{I} = -\frac{P_i}{A}\left(1 + \frac{ec_2}{r^2}\right) \tag{11.6b}$$

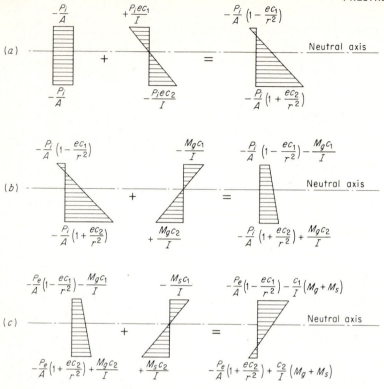

Fig. 11.4 Stress distribution in beams. (a) Effect of prestress; (b) effect of prestress and girder dead load; (c) effect of prestress, girder dead load, and service load.

where r is the radius of gyration of the concrete section. Normally, as the eccentric prestress force is applied, the beam deflects upward. The beam self-weight w_g then causes additional moment M_g to act, and the net top and bottom fiber stresses become

$$f_{1gi} = f_{1pi} - \frac{M_g c_1}{I} \tag{11.7a}$$

$$f_{2gi} = f_{2pi} + \frac{M_g c_2}{I} \tag{11.7b}$$

as shown in Fig. 11.4b. At this stage time-dependent losses due to shrinkage, creep, and relaxation commence, and the prestress force gradually reduces from P_i to P_e. It is usually acceptable to assume that all such losses occur prior to the application of service loads, since the concrete stresses at service loads will be critical after losses, not before. Accordingly, the stresses in the top and bottom fiber, with P_e and beam load acting, become

$$f_{1ge} = -\frac{P_e}{A}\left(1 - \frac{ec_1}{r^2}\right) - \frac{M_g c_1}{I} \tag{11.8a}$$

$$f_{2ge} = -\frac{P_e}{A}\left(1 + \frac{ec_2}{r^2}\right) + \frac{M_g c_2}{I} \tag{11.8b}$$

When full service loads (dead load in addition to self-weight of the beam, plus service live load) are applied, the stresses are

$$f_{1s} = f_{1ge} - \frac{M_s c_1}{I} \tag{11.9a}$$

$$f_{2s} = f_{2ge} + \frac{M_s c_2}{I} \tag{11.9b}$$

as shown in Fig. 11.4c.

It is necessary, in reviewing the adequacy of a beam (or in designing a beam on the basis of permissible stresses), that the stresses in the extreme fibers remain within specified limits under any combination of loadings that can occur. Normally, the stresses at the section of maximum moment, in a properly designed beam, will range between the distributions shown in Fig. 11.5 as the beam passes from the unloaded stage (P_i plus self-weight) to the loaded stage (P_e plus full service loads). In the figure, f_{ci} and f_{ti} are the permissible compression and tension stresses, respectively, in the concrete immediately after transfer, and f_{cs} and f_{ts} are the permissible compression and tension stresses at service loads (see Table 11.1).

In calculating the section properties A, I, etc., to be used in the above equations, it is relevant that, in posttensioned construction, the tendons are usually grouted in the conduits after tensioning. Before grouting, stresses should be based on the net section with holes deducted. After grouting, the transformed section should be used, with holes considered filled with concrete, and with the steel replaced with an equivalent area of concrete. However, it is satisfactory, unless the holes are quite large, to compute section properties on the basis of the gross concrete section. Similarly, while in pretensioned

Fig. 11.5   Stress limits.   (a) Unloaded   (b) Loaded

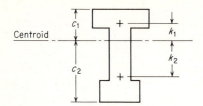

Fig. 11.6 Location of kern points.

beams the properties of the transformed section should be used, it makes little difference if calculations are based on properties of the gross concrete section.[1]

It is useful to establish the location of the upper and lower *kern points* of a cross section. These are defined as the limiting points inside which the prestress-force resultant may be applied without causing tension anywhere in the cross section. Their locations are obtained by writing the expression for the tensile fiber stress due to application of an eccentric prestress force acting alone and setting this expression equal to zero to solve for the required eccentricity. In Fig. 11.6, to locate the upper kern-point distance k_1 from the neutral axis, let the prestress-force resultant P act at that point. Then the bottom fiber stress is

$$f_2 = -\frac{P}{A}\left(1 - \frac{ec_2}{r^2}\right) = 0$$

Set

$$1 - \frac{ec_2}{r^2} = 0$$

and solve for

$$e = k_1 = \frac{r^2}{c_2} \qquad\qquad\qquad\qquad (11.10a)$$

Similarly, the lower kern-point distance k_2 is

$$k_2 = \frac{r^2}{c_1} \qquad\qquad\qquad\qquad (11.10b)$$

The region between these two limiting points is known as the *kern*, or in some cases the *core*, of the section.

11.8 SERVICE-LOAD DESIGN FOR FLEXURE

As in reinforced concrete, problems in prestressed concrete can be separated generally as review problems and design problems. For the former, with the

---

[1] The ACI Code states: "In calculations of section properties prior to bonding of tendons, areas of the open ducts shall be deducted. The transformed area of bonded tendons and reinforcing steel may be included in pretensioned members and in posttensioned members after grouting."

applied loads, the concrete cross section, and the amount and point of application of the prestress force resultant known, Eqs. (11.6) to (11.9) permit the direct calculation of the resulting stresses. However, if the dimensions of a section and the amount and location of the prestress force are to be found, given the loads and limiting stresses, the problem is more complicated because of the several interrelated variables.

Some designers prefer to assume a section, calculate the required prestress force and eccentricity for what will probably be the controlling load stage, and then check the stresses at all stages, using the preceding equations, revising the section if necessary. If the section is to be chosen from a limited number of standardized shapes, this procedure is probably the best. For the more general case, in which the designer is free to choose concrete dimensions as well as prestress force and eccentricity to meet particular job requirements, the procedure described herein is preferred (Refs. 11.3 and 11.4).

a. Beams with variable eccentricity   Concrete stress limits for the initial stage (P_i plus self-weight of beam) and service-load stage are normally established by specification (see Table 11.1, for example).

Let f_{ci} = permissible concrete compressive stress at initial stage

   f_{ti} = permissible concrete tensile stress at initial stage

   f_{cs} = permissible concrete compressive stress at service-load stage

   f_{ts} = permissible concrete tensile stress at service-load stage

The concrete dimensions, prestress force, and eccentricity must be chosen so that these limits on stress are not violated.

Figure 11.7 shows concrete stress distributions, at the maximum-moment section, that obtain at several stages of loading. A prestress force P_i, applied

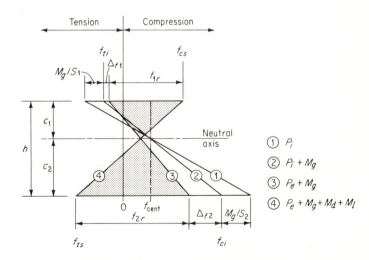

Fig. 11.7   Concrete stress distributions at various loading stages.

with eccentricity e below the neutral axis, and acting alone, would produce distribution 1. However, since the beam deflects upward as P_i is introduced, the self-weight of the beam normally causes bending moment M_g to act at once. The associated stress changes at the top and bottom of the beam are, respectively, M_g/S_1 and M_g/S_2, where S_1 and S_2 are the section moduli with respect to the top and bottom fibers. When these are superimposed on the effect of the eccentric prestress force, distribution 2 is obtained. At this stage of loading, the tension at the top surface is limited by f_{ti}, and the compression at the bottom surface by f_{ci}, as shown.

For reasons given earlier, it may be assumed that all time-dependent losses occur prior to the application of full service load. As the prestress force diminishes from P_i to P_e, the tension at the top surface is reduced by the amount Δf_1 and the compression at the bottom is reduced by Δf_2. When P_e acts with M_g, stress distribution 3 is obtained. Finally, as the additional dead load and the live load are applied, with associated moments M_d and M_l, the stress changes f_{1r} at the top fiber and f_{2r} at the bottom fiber are caused. Concrete stress distribution 4 results. At this stage, the tension at the bottom is limited by f_{ts}, and the compression at the top by f_{cs}, as shown. Thus, the requirements for section moduli for the beam are

$$(a) \quad S_1 \geq \frac{M_d + M_l}{f_{1r}}$$

$$(b) \quad S_2 \geq \frac{M_d + M_l}{f_{2r}}$$

It is necessary, as is clear from Fig. 11.7, to evaluate Δf_1 and Δf_2 in order to find the section moduli.[1] Defining the *effectiveness ratio* R as

$$(c) \quad R = \frac{P_e}{P_i}$$

the loss in prestress force is $(1 - R)P_i$. Then the change in top and bottom fiber stresses as losses occur is $(1 - R)$ times the corresponding stresses due to initial prestress acting alone (distribution 1 of Fig. 11.7); hence

$$(d) \quad \Delta f_1 = (1 - R)\left(f_{ti} + \frac{M_g}{S_1}\right)$$

$$(e) \quad \Delta f_2 = (1 - R)\left(-f_{ci} + \frac{M_g}{S_2}\right)$$

representing a reduction in tension at the top of the member and a reduction in compression at the bottom. The stress range available at the top surface, as

[1] In writing statements of inequality, it is convenient to designate tensile stress as larger than zero and compressive stress smaller than zero. Thus $+450 > -1350$, $-600 > -1140$, etc. Stress changes, for example, M_g/S_1 and Δf_1, are dealt with as absolute quantities.

moment $M_d + M_l$ is applied, is

$$f_{1r} = f_{ti} - \Delta f_1 - f_{cs}$$

$$(f) \qquad = f_{ti} - (1 - R)\left(f_{ti} + \frac{M_g}{S_1}\right) - f_{cs}$$

$$= Rf_{ti} - \frac{(1 - R)M_g}{S_1} - f_{cs}$$

and that the bottom surface is

$$f_{2r} = f_{ts} - f_{ci} - \Delta f_2$$

$$(g) \qquad = f_{ts} - f_{ci} - (1 - R)\left(-f_{ci} + \frac{M_g}{S_2}\right)$$

$$= f_{ts} - Rf_{ci} - \frac{(1 - R)M_g}{S_2}$$

Thus, the minimum values of S_1 and S_2 are, respectively,

$$(h) \quad S_1 \geq \frac{M_d + M_l}{Rf_{ti} - (1 - R)M_g/S_1 - f_{cs}}$$

$$(i) \quad S_2 \geq \frac{M_d + M_l}{f_{ts} - Rf_{ci} - (1 - R)M_g/S_2}$$

which, after simple algebraic transformation, become

$$S_1 \geq \frac{(1 - R)M_g + M_d + M_l}{Rf_{ti} - f_{cs}} \tag{11.11a}$$

$$S_2 \geq \frac{(1 - R)M_g + M_d + M_l}{f_{ts} - Rf_{ci}} \tag{11.11b}$$

The cross section must be chosen to satisfy the two requirements of Eq. (11.11). Since $I = S_1c_1 = S_2c_2$, the neutral axis of the section must be such that

$$(j) \qquad \frac{c_1}{c_2} = \frac{S_2}{S_1}$$

or in terms of the total section depth $h = c_1 + c_2$,

$$(k) \qquad \frac{c_1}{h} = \frac{S_2}{S_1 + S_2}$$

Referring again to Fig. 11.7, the concrete centroidal stress under initial conditions is

$$f_{cent} = f_{ti} - \frac{c_1}{h}(f_{ti} - f_{ci}) \tag{11.12}$$

The required initial prestress force is found by multiplying this stress by the cross-sectional area of the concrete, A.

$$P_i = A \times |f_{\text{cent}}| \tag{11.13}$$

That prestress force, applied with eccentricity e, must produce a bending moment $P_i e$ such that, at the top fiber,

$$\frac{P_i e}{S_1} = (f_{ti} - f_{\text{cent}}) + \frac{M_g}{S_1}$$

From this, the required eccentricity e_m, at the maximum-moment section, is

$$e = e_m = (f_{ti} - f_{\text{cent}}) \frac{S_1}{P_i} + \frac{M_g}{P_i} \tag{11.14a}$$

The same result will be obtained from consideration of requirements at the bottom fiber, leading to the alternative expression

$$e = e_m = (f_{\text{cent}} - f_{ci}) \frac{S_2}{P_i} + \frac{M_g}{P_i} \tag{11.14b}$$

Equations (11.11) to (11.14) permit the selection of the optimum concrete section, prestress force, and eccentricity to satisfy exactly the specified limit stresses.

b. Beams with constant eccentricity   The preceding development was based on conditions at the maximum-moment section and on the assumption that the effect of the beam dead load is immediately superimposed on the effect of prestressing. Elsewhere in the member, the moment M_g is less, and at the supports of simple spans becomes zero. Consequently, if both P_i and e were constant everywhere in the span, the stress limits f_{ti} and f_{ci} would be exceeded wherever M_g was less than its maximum value. In many designs, e is reduced near the supports, partially for this reason (see Art. 11.11). If this is inconvenient (e.g., for pretensioned mass-produced members), the maximum e which can be used is governed by conditions at the supports, where no beam-load moment exists. In this case, the extremities of stress distribution 1 of Fig. 11.7 must fall at the stress limits f_{ti} and f_{ci}, and several of the preceding equations must be modified. The extreme fiber stress changes as losses occur are

$$(l) \quad \Delta f_1 = (1 - R)(f_{ti})$$
$$(m) \quad \Delta f_2 = (1 - R)(-f_{ci})$$

Accordingly, the stress ranges f_{1r} and f_{2r} become

$$(n) \quad f_{1r} = R f_{ti} - f_{cs}$$
$$(o) \quad f_{2r} = f_{ts} - R f_{ci}$$

The section must be sufficient to provide for stress changes due to beam-load moment M_g, as well as M_d and M_l. The required section moduli are

$$S_1 \geq \frac{M_g + M_d + M_l}{Rf_{ti} - f_{cs}} \tag{11.15a}$$

$$S_2 \geq \frac{M_g + M_d + M_l}{f_{ts} - Rf_{ci}} \tag{11.15b}$$

The centroidal concrete stress f_{cent} and the initial prestress force P_i may be found from Eqs. (11.12) and (11.13) as before. However, the bending moment $P_i e$, due to initial prestress, must now be such that at the top fiber

$$\frac{P_i e}{S_1} = (f_{ti} - f_{cent})$$

giving the required (constant) eccentricity

$$e = (f_{ti} - f_{cent}) \frac{S_1}{P_i} \tag{11.16a}$$

The same result will be obtained from requirements at the bottom fiber, in terms of f_{ci}:

$$e = (f_{cent} - f_{ci}) \frac{S_2}{P_i} \tag{11.16b}$$

There is a significant difference between the type of beam considered first (variable eccentricity) and the present case (constant eccentricity). Previously, the section modulus was governed mainly by the superimposed dead- and live-load moments. Almost all the self-weight of the beam was carried "free" (i.e., without increasing the section modulus or prestress force) simply by increasing the eccentricity along the span by the amount M_g/P_i. In the present case, the constant eccentricity is controlled by conditions at the support, where no beam-load moment acts, and this saving is impossible. In spite of this, many beams are made with constant eccentricity because of the practical advantages during production.

c. Symmetrical cross sections Often, for practical reasons, a symmetrical cross section is chosen, even though the limit stresses indicate otherwise. In such cases, $S_1 = S_2 = S$, and the stress change at the top and bottom fiber will be identical as the transverse load is applied. As a result, only three of the four limit stresses will be achieved, in general. The fourth must be only within the stated limit. In such circumstances, there is no unique combination of P_i and e that will serve, but rather a range of values. The best choice will normally be that which minimizes the prestress force required.

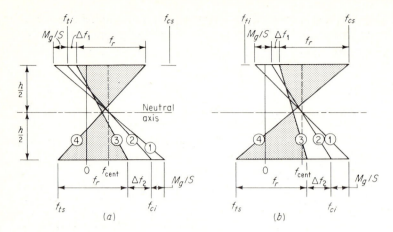

Fig. 11.8 Symmetrical section, bottom-fiber stress controlling.

If S_2, found from Eqs. (11.11), is larger than S_1 (i.e., if the bottom fiber stress controls), and a symmetrical section is chosen, then $S = S_2$. The full permissible stress range is not utilized at the top of the member, as is clear from Fig. 11.8. Two possible solutions represented by (a) and (b) both satisfy stress limits, as do all solutions between these extreme cases. However, the first requires the least prestress force (indicated by the least concrete centroidal stress) and would normally be preferred. The prestress force and eccentricity may be found from Eqs. (11.12) to (11.14) as usual.

If the top fiber stress controls (i.e., if S_1 is greater than S_2), then once again an infinite number of solutions is possible. The two extreme cases are represented by Fig. 11.9a and b. In this case, the second solution would normally be preferred, because of the lower prestress force. Equation (11.12)

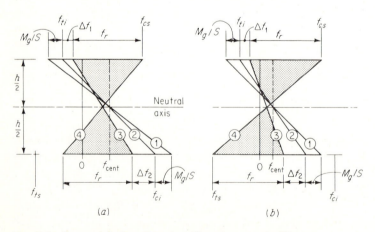

Fig. 11.9 Symmetrical section, top-fiber stress controlling.

cannot be applied presently, because the stress limit f_{ci} is not achieved exactly. In this case, the concrete centroidal stress is conveniently found from the service-load stress distribution 4:

$$f_{cent} = \frac{1}{R}\left[f_{ts} - \frac{c_2}{h}\,(f_{ts} - f_{cs})\right]$$ (11.17)

The initial prestress force may be found from Eq. (11.13) as usual, and the required eccentricity from Eq. (11.14a). Equation (11.14b) is not valid because the stress limit f_{ci} is never realized.

d. Example: design of a beam with variable eccentricity   A simply supported prestressed beam is to be designed to carry a superimposed dead load of 500 plf and a service live load of 1000 plf, in addition to its own weight on a 40-ft span. Normal-weight concrete having a specified strength f_c' of 6000 psi is to be used. At the time of tensioning, the initial compressive strength f_{ci}' will be approximately 4200 psi. The prestress force is to be applied using alloy-steel bars having strength f_s' of 145 ksi, the eccentricity of which may be varied along the span as desired. Permissible stresses recommended by ACI are to be adopted. It is estimated that losses will result in an effectiveness ratio R of 0.85.

According to the recommendations quoted in Table 11.1, the permissible concrete stresses are

Compression at transfer: $f_{ci} = -0.60 \times 4200 = -2520$ psi
Tension at transfer: $f_{ti} = +3\sqrt{4200} = +195$ psi
Compression at service load: $f_{cs} = -0.45 \times 6000 = -2700$ psi
Tension at service load: $f_{ts} = +6\sqrt{6000} = +465$ psi

Since the design is to be based on stress limits at service loads, moments must be found corresponding to the calculated dead load and the specified service live load, without increase by load factors. Estimating the beam weight to be 350 plf,

$M_g = \frac{1}{8} \times 0.350 \times 1600 = 70$ ft-kips

The combined superimposed dead and live loads produce the moment

$M_d + M_l = \frac{1}{8} \times 1.500 \times 1600 = 300$ ft-kips

The minimum required section moduli may be found from Eq. (11.11):

$$S_1 \geq \frac{(0.15 \times 70 + 300) \times 12,000}{0.85 \times 195 + 2700} = 1300 \text{ in.}^3$$

$$S_2 \geq \frac{(0.15 \times 70 + 300) \times 12,000}{465 + 0.85 \times 2520} = 1440 \text{ in.}^3$$

These values are sufficiently alike so that a symmetrical cross section will be chosen; for simplicity in constructing forms, a rectangular member is used. The required section modulus is

$$S = \frac{bh^2}{6} = 1440 \text{ in.}^3$$

A member having $b = 12$ in. and $h = 26.8$ in. is just adequate. (In practice, the dimensions would be rounded, probably to the next larger inch; however, for the sake of clarity in this example and for comparison with later results, theoretical dimensions will be retained.) For this member, the cross-sectional area $A = 321$ in.^{2} and the weight is 335 plf, sufficiently close to the assumed value so that no revision is necessary. The concrete centroidal stress just after transfer is found from Eq. (11.12):

$$f_{cent} = 195 - \tfrac{1}{2}(195 + 2520) = -1162 \text{ psi}$$

and the corresponding initial prestress force, from Eq. (11.13), is

$$P_i = 321 \times 1.162 = 373 \text{ kips}$$

Equation (11.14) gives the required steel eccentricity at midspan:

$$e_m = (195 + 1162) \times \frac{1440}{373,000} + \frac{70 \times 12,000}{373,000} = 7.50 \text{ in.}$$

For alloy-steel bars with $f_s' = 145$ ksi, the permitted stress immediately after transfer is $0.70 \times 145 = 102$ ksi. Accordingly, the steel area required is

$$A_s = \tfrac{373}{102} = 3.65 \text{ in.}^2$$

Six $\tfrac{7}{8}$-in.-diameter bars provide 3.61 in.^{2}, sufficiently close for design purposes. They will be placed in two layers, as shown in Fig. 11.10. The steel centroid will be curved upward, following the ordinates of a parabola, reducing the eccentricity to zero over the supports.

As a check on the work, stresses are next calculated for the load stages of interest, using Eqs. (11.6) to (11.9). Results appear in Fig. 11.10b, with the notation of Fig. 11.7. The slight differences between specified and calculated stresses are due to rounding errors, and are of no practical consequence. Note that the compressive stress at full service load is well below the allowable value of 2700 psi; this results

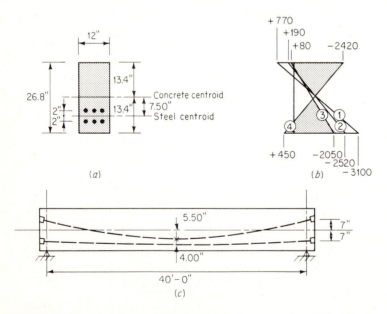

(a)   (b)

(c)

Fig. 11.10   Beam with curved tendons. (a) Beam cross section; (b) concrete stresses; (c) beam profile.

from the choice of a symmetrical cross section with minimum prestress force. It can be concluded that the design is satisfactory with respect to all specified stress limits.

e. Example: design of a beam with constant eccentricity   The beam of the preceding example is to be redesigned using straight tendons in order to simplify construction. All other specifications remain unchanged.

In this case, anticipating the need for a somewhat larger member, the beam weight is estimated to be 400 plf. Accordingly,

$$M_g = \tfrac{1}{8} \times 0.400 \times 1600 = 80 \text{ ft-kips}$$

The moment due to superimposed dead and live load is 300 ft-kips as before. With constant eccentricity imposed, the required section moduli must be found using Eq. (11.15).

$$S_1 \geq \frac{(80 + 300) \times 12,000}{0.85 \times 195 + 2700} = 1590 \text{ in.}^3$$

$$S_2 \geq \frac{(80 + 300) \times 12,000}{465 + 0.85 \times 2520} = 1750 \text{ in.}^3$$

Again a rectangular section is chosen. In the present case,

$$S = \frac{bh^2}{6} = 1750$$

which requires $b = 13$ in. and $h = 28.4$ in. The cross-sectional area of the member is 369 in.², and the weight is 385 plf, very close to the assumed value. The centroidal stress of -1162 psi is found from Eq. (11.12) as before, without change. However, the larger concrete section requires a greater prestress force to produce that stress. In this case,

$$P_i = 369 \times 1.162 = 429 \text{ kips}$$

The constant eccentricity can be found from Eq. (11.16):

$$e = (195 + 1162) \times \frac{1750}{429,000} = 5.53 \text{ in.}$$

and the required steel area is

$$A_s = \tfrac{429}{102} = 4.20 \text{ in.}^2$$

Seven $\tfrac{7}{8}$-in.-diameter bars will provide exactly this amount. They will be placed in the pattern shown in Fig. 11.11a. A spacing between bars of 5 in. will be maintained to provide room at the ends of the member for the bar anchorages and the prestress jacks.

Once again, stresses at the several load stages are checked using Eqs. (11.6) to (11.9), with the results shown in Fig. 11.11b. It is seen that all limit stresses are closely met, except for the top fiber compression under full service load. This is below the allowable value, as before, because of the use of a symmetrical section with minimum prestress.

In a comparison of the present results with those of the preceding example, the advantage of curved tendons becomes obvious. The additional eccentricity permitted through the use of curved tendons results in a saving of 15 per cent in both

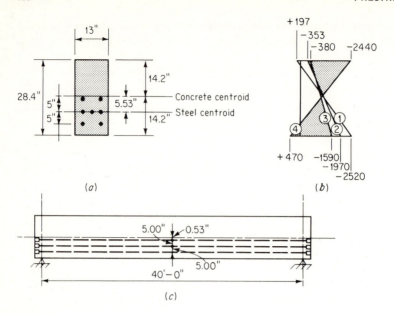

Fig. 11.11 Beam with straight tendons. (*a*) Beam cross section; (*b*) concrete stresses; (*c*) beam profile.

concrete and steel. For longer spans, with a higher ratio of beam load to superimposed load, the saving would be considerably larger. For major spans, curved tendons are almost always used.

11.9 PARTIAL PRESTRESSING VS. FULL PRESTRESSING

Early in the development of prestressed concrete, the goal of prestressing was the complete elimination of concrete tensile stress at service load. The limit stress f_{ts} was taken as zero. Certain difficulties result from this restrictive definition of prestressing. The highly prestressed beams which are produced in some cases exhibit undesirable deformational characteristics. In particular, they tend to camber upward due to the eccentric prestress force, a displacement which is only partially counteracted by the usual gravity loads producing downward deflection. This tendency is aggravated by creep in the concrete, which magnifies the upward displacement due to the sustained prestress force, but influences to a much lesser degree the downward deflection due to live load, which may be only intermittently applied. In addition, should highly prestressed beams be overloaded and fail, they do so in a less ductile manner than members with a lesser degree of prestress. Furthermore, they may lack the desirable capacity to absorb the energy of impact loading (Ref. 11.3).

In recent years, a broader view of prestressing has prevailed. It is argued that, since concrete tension and cracking is permitted in ordinary reinforced-concrete construction without detrimental effect, it may also be allowed in prestressed concrete. By this means, the required prestress force may be reduced, and undesirable deflections controlled more easily. *Partially pre-stressed beams*, in which some concrete tension is permitted under service loads, are permitted by many specifications. Partial prestressing may be obtained by reducing the stress in the tendons or by leaving a part of the total tensile reinforcement unstressed.

Sufficient tensile steel area must be provided (although perhaps at less than its full allowable stress) to produce the necessary flexural strength (see Art. 11.12). Analysis, confirmed by tests, indicates that the flexural strength of a member is not much affected by reduction in the amount of prestress force, provided that the total area of tensile reinforcement is not reduced.

This is illustrated by the curves of Fig. 11.12. An ordinary reinforced-concrete beam, or a beam with unstressed tendons, will display the behavior indicated by curve *a*. As the load is applied, the beam stiffness is at first governed by the uncracked, transformed concrete section. When the bottom fiber stress exceeds the modulus of rupture, the slope of the load-deflection curve is reduced. Additional nonlinearity occurs near the failure load due to further cracking, destruction of bond, and nonlinear material characteristics. For partially prestressed beams, curves *b*, cracking is delayed, but may occur prior to attainment of full service load. After cracking, the slope of the curve is about the same as for the beam with unstressed tendons. For the fully pre-stressed member exhibiting curve *c*, the bottom-fiber stress is zero at full service load, and cracking does not occur until additional stress equal to the modulus of rupture is caused by further loading. Finally, the overprestressed member illustrated by curve *d* fails at or before the cracking load, as the highly strained steel ruptures. Brittle behavior such as this is highly undesirable. Note that in all cases the failure load is about the same, regardless of the amount of prestress initially introduced. These curves illustrate clearly that the main purposes served by prestressing are (1) the control of cracking and (2) the limitation of deflections (Ref. 11.3).

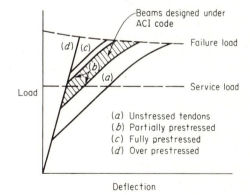

Fig. 11.12   Load-deflection curves for beams having varying amounts of prestress with steel area constant. (*Adapted from Ref. 11.3.*)

In many cases in design, it is useful to distinguish between normal service live load and temporary abnormal loading. The service loads specified by building codes are infrequently attained during the life of the typical member, and a *characteristic load*, often substantially less, is the normal condition. The Joint European Committee on Concrete (Ref. 11.5) establishes three classes of prestressed beams:

Class 1: *Fully prestressed*, in which no tensile stress is allowed in the concrete at service load.

Class 2: *Partially prestressed*, in which occasional temporary cracking is permitted under infrequent high loads.

Class 3: *Partially prestressed*, in which there may be permanent cracks provided that their width is suitably limited.

The Draft British Standard Code (Ref. 11.6) includes similar categories.

The 1971 ACI Code permits concrete tension of $6\sqrt{f'_c}$ at full service load, slightly less than the usual modulus of rupture, as shown in Table 11.1. Provided that explicit calculation of deflections, due to both immediate and sustained loads, shows that deflections are within stated limits, the nominal tensile stress may be increased to $12\sqrt{f'_c}$. Both cases amount to partial prestressing, by the definition given earlier. In the first case, no cracking should result, since the limit of tension is slightly below the usual modulus of rupture. However, nominal tension at the higher stress limit would certainly be accompanied by some cracking. Stresses based on the assumption of an uncracked section are, to that extent, fictitious, and should be regarded only as a guide to the actual conditions in the beam. If the higher stress limit is used, the required deflection calculation should be based on a bilinear load-deflection curve (or the equivalent), as shown in Fig. 11.12, with slopes related to the flexural stiffness of the uncracked and cracked transformed sections.

11.10 SHAPE SELECTION

The examples and problems up to this point have treated only beams of rectangular cross section. These are important, not only in demonstrating the basic principles of prestressing, but also because rectangular sections are often used in practice for slabs, small beams, and designs in which only a few beams of given dimensions are required. Forming costs are least for rectangular beams. However, for improved material economy, it is desirable to use more sophisticated sections, such as I, T, or box shapes. While in ordinary reinforced-concrete design, diagonal tensile stresses restrict the reduction of web thickness, the introduction of prestress force permits much more slender sections to be used.

Several common shapes are shown in Fig. 11.13. Some of these are standardized and mass-produced, using reusable steel forms. Others are indi-

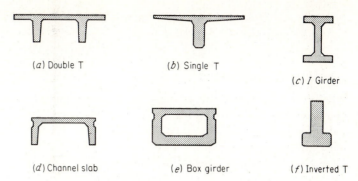

(a) Double T   (b) Single T

(c) I Girder

(d) Channel slab   (e) Box girder   (f) Inverted T

Fig. 11.13   Typical beam cross sections.

vidually proportioned for large and important works. The double T (Fig. 11.13*a*) is probably the most widely used cross section in U.S. prestressed construction. A flat surface is provided, usually 4 ft in width. Slab thickness and web depth vary, depending upon requirements. Spans to 60 ft are not unusual. The single T (Fig. 11.13*b*) is more appropriate for longer spans, to 120 ft, and heavier loads. The I section (Fig. 11.13*c*) is widely used for bridge spans and roof girders up to about 120 ft, while the channel slab (Fig. 11.13*d*) is suitable for floors in the intermediate-span range. The box girder (Fig. 11.13*e*) is advantageous for bridges of intermediate and major span. The inverted-T section (Fig. 11.13*f*) provides a bearing ledge to carry the ends of precast deck members spanning in the perpendicular direction.

As indicated, the cross section may be symmetrical or unsymmetrical. An unsymmetrical section is a good choice (1) if the available stress ranges f_{1r} and f_{2r} at the top and bottom surface are not the same; (2) if the beam must provide a flat, useful surface as well as offering load-carrying capacity; (3) if the beam is to become a part of composite construction, with cast-in-place slab acting together with precast web; or (4) if the beam must provide supporting surfaces as in Fig. 11.13*f*. In addition, T sections provide increased flexural strength, since the internal arm of the resisting couple at maximum design load is greater than for rectangular sections.

Generally speaking, I, T, and box sections with relatively thin webs and flanges are more efficient than members with thicker parts (see Ref. 11.7 for quantitative discussion of section efficiency). However, several factors limit the gain in efficiency which may be obtained in this way. These include the instability of very thin overhanging compression parts, the vulnerability of thin parts to breakage in handling (in the case of precast construction), and the practical difficulty of depositing concrete in very thin elements. The designer must also recognize the need for providing adequate spacing and concrete protection for tendons and anchorages, the importance of construction depth limitations, and the need for lateral stability if the beam is not braced by other members against buckling.

11.11 TENDON PROFILES

The preceding discussion has dealt mainly with critical sections (usually mid-span sections) in establishing steel centroid location. Most present prestressed construction uses members that are essentially prismatic and employs a nearly constant prestressing force from one end of the beam to the other, although curved or draped tendons are often used. In determining the best location for the centroid of such steel through the length of the beam, it is necessary to ensure that the limiting stress states, such as are shown in Fig. 11.5, are nowhere violated under any anticipated loading. For a member which is properly designed at the critical sections, it is sufficient elsewhere to ensure that the tendon eccentricity is such that the allowable tensile stress in the concrete is not exceeded.

   With no transverse loading applied to a beam, the center of compression at any cross section will coincide with the center of tension, i.e., the centroid of the prestressing steel. When the beam load is initially applied, the center of compression will move upward an amount M_g/P_i. Similarly, when the total service load is applied, the center of compression will be $(M_g + M_s)/P_e$ above its original location.

   The upper limit position of the compressive resultant at service load, y_1 above the neutral axis, is such that it produces a tensile stress f_{ts} at the bottom fiber of the section. With the compressive resultant at that limiting point, the bottom-fiber stress is

$$-\frac{P_e}{A} + \frac{P_e e c_2}{I} = f_{ts}$$

from which

$$e = y_1 = \frac{I}{P_e c_2}\left(\frac{P_e}{A} + f_{ts}\right) = \frac{A r^2}{P_e c_2}\left(\frac{P_e}{A} + f_{ts}\right)$$

$$y_1 = \frac{r^2}{c_2}\left(1 + \frac{A f_{ts}}{P_e}\right)$$

(11.18)

Similarly, the lower limit position of the compressive resultant, y_2 below the neutral axis, is such that, when only beam load acts with P_i, they produce the allowable tensile stress f_{ti} at the top fiber, and its location relative to the neutral axis is

$$y_2 = \frac{r^2}{c_1}\left(1 + \frac{A f_{ti}}{P_i}\right)$$

(11.19)

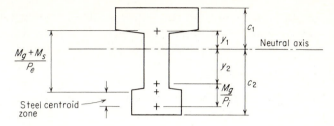

Fig. 11.14

With reference to Fig. 11.14, the location of the steel centroid must therefore be such that it is no less than $(M_g + M_s)/P_e$ below the level defined by y_1, and no more than M_g/P_i below the level defined by y_2. In this way the possible range of steel-centroid locations is established. The procedure should be repeated at sufficient intervals along the member to establish cable-centroid envelope curves. Note that, ideally, the upper and lower locations of the steel centroid, found in this way, will coincide. Only for this condition will the full efficiency of the cross section be obtained; i.e., the full allowable stress range will be utilized for both the upper and lower fibers of the beam. Usually, this optimum condition is obtained at only one cross section.

A saving in design time is possible if it is noted that the ordinates of the cable envelope curves, measured downward from the limit lines defined by y_1 and y_2, are directly proportional to the ordinates of the respective moment diagrams. For a prismatic beam, for example, the dead load is uniform, and the lower envelope curve will vary parabolically from one end of the span to the other. If the live load is uniformly distributed, the upper envelope curve will likewise be a parabola, passing through established points at midspan and at the two supports. Limiting profiles can be drawn as shown in Fig. 11.15. Figure 11.15a illustrates, by the shaded area, the possible range of tendon

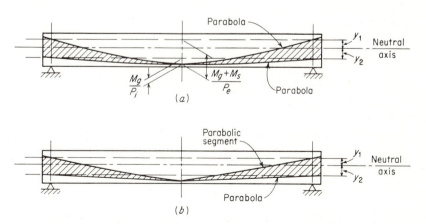

Fig. 11.15 Tendon profile zones.

profiles for a uniformly loaded prismatic beam such that the midspan section is fully utilized, with upper and lower tendon envelopes coinciding at midspan. Any tendon profile falling completely within the shaded zone will be satisfactory from the point of view of flexural stress. Figure 11.15b shows the tendon zone for a prismatic beam with a single concentrated live load at midspan. The lower tendon envelope is a continuous parabola as before, while the upper curve consists of two parabolic segments with ordinates from the upper limit line proportional to those of the combined dead- and live-load moment diagrams.

From a practical point of view, it is not always wise to adopt any arbitrary shape. In pretensioned construction, cable depressors and cable templates are used to obtain a linear segmental profile. In general, it is desirable to minimize the number of changes in slope. In posttensioned construction, curved-tendon alignment can easily be obtained, but in general, points of reversal of curvature are to be avoided, since these greatly increase losses in tension due to friction between the tendon and the duct during the prestressing operation. (If such points of contraflexure cannot be avoided, there is some advantage in tensioning from both ends of the tendons, rather than from one end only.) The best tendon profile is that which follows the natural curve taken by the cable when it is suspended between the two end anchorages and allowed to sag until it passes through a single specified low point. To avoid displacing the conduit during concreting, it must be wired to the vertical web reinforcement. In many cases, in both pretensioned and posttensioned members, it is advantageous to provide zero eccentricity at simple supports, thus matching prestress moment with that due to transverse loads.

11.12 FLEXURAL STRENGTH

In an ordinary reinforced-concrete beam, the stress in the tensile steel and the compressive force in the concrete increase in proportion to the applied moment up to and somewhat beyond service load, the distance between the two internal force resultants remaining essentially constant. In contrast to this behavior, in a prestressed beam, increased moment is resisted by a proportionate increase in the distance between the compressive and tensile resultant forces, the compressive resultant moving upward as the load is increased. The magnitude of the internal forces remains nearly constant up to and usually somewhat beyond service loads.

This situation is changed drastically upon tensile flexural cracking of the prestressed beam. When the beam cracks, there is a sudden increase in the stress in the steel as the tension which was formerly carried by the concrete is transferred to it. After cracking, the prestressed beam behaves essentially as an ordinary reinforced-concrete beam. The compressive resultant cannot continue to move upward indefinitely, and increasing moment must be accompanied by a nearly proportionate increase in steel stress and compressive force. The strength of a prestressed beam can therefore be predicted by the methods developed for ordinary reinforced-concrete beams, with minor modifications to account for the differing nature of the stress-strain relationship of prestressing steel as compared with ordinary reinforcing steel.

In Art. 2.10 it was shown that underreinforced prestressed beams will fall when the steel stress reached its ultimate tensile value f_{pu}, at which

$$M'_u = A_{ps}f_{pu}\left(d - \frac{A_{ps}f_{pu}}{1.7f'_c b}\right) = A_{ps}f_{pu}d\left(1 - 0.59\rho_p \frac{f_{pu}}{f'_c}\right) \qquad (2.81)$$

Whether the beam is underreinforced or overreinforced is determined by calculating the concrete strain in the extreme fiber at the instant of steel rupture. If the concrete strain ϵ_c is less than the crushing strain (usually taken equal to 0.003) when the steel has reached its breaking strain ϵ'_s, then the beam is underreinforced. The concrete strain can be found from the expression

$$\epsilon_c = (\epsilon'_s - \epsilon_{se}) \frac{a/\beta_1}{d - a/\beta_1} \qquad (2.82)$$

where

$$a = \frac{A_{ps}f_{pu}}{0.85f'_c b} \qquad (2.80)$$

If ϵ_c, found in this way, exceeds the crushing strain of the concrete, the beam is said to be overreinforced, in which case the steel stress at failure of the beam is some value $f_{ps} < f_{pu}$. In this case the strength of the beam can be found by successive approximations as described in detail in Art. 2.10.

The above discussion and that of Art. 2.10 pertain to prestressed beams in which the steel is bonded to the concrete continuously throughout its length. This will normally be the case in pretensioned construction and will be the case in posttensioned construction if the tendons are grouted after tensioning, as is usually the case. If a posttensioned beam is not grouted, it is difficult to determine the exact steel stress at failure load. The stress will be uniformly distributed along the entire length of the tendons. Characteristically, there will be a small number of rather wide cracks. These have the effect of reducing the failure load as compared with that of an otherwise identical bonded beam.

According to procedures contained in the ACI Code Commentary, if the ratio $\rho_p f_{ps}/f'_c$ is not more than 0.30, the strength of prestressed beams can be found from the expression

$$M_u = \phi A_{ps}f_{ps}\left(d - \frac{a}{2}\right) \qquad (11.20)$$

where

$$a = \frac{A_{ps}f_{ps}}{0.85f'_c b}$$

This may be stated in the alternative form

$$M_u = \phi A_{ps} f_{ps} d \left(1 - \frac{0.59\rho_p f_{ps}}{f'_c}\right)$$ (11.20a)

Equation (11.20a) is seen to be equivalent to Eq. (2.81), with the substitution of f_{ps} for f_{pu} and the inclusion of the understrength factor ϕ (see Art. 3.4). Where a specific stress-strain curve for the prestressing steel used is not available, use of the following values of f_{ps} is permissible.

For bonded members: $$f_{ps} = f_{pu} \left(1 - \frac{0.5\rho_p f_{pu}}{f'_c}\right)$$ (11.21a)

For unbonded members: $$f_{ps} = f_{se} + 10,000 + \frac{f'_c}{100\rho_p}$$ (11.21b)

but not to exceed the yield strength f_{py} and also not to exceed $f_{se} + 60,000$, where f_{se} is the effective steel prestress after losses (not to be less than $0.5f_{pu}$).

If $\rho_p f_{ps}/f'_c$ exceeds 0.30, the ultimate moment of the member should not be taken as greater than

$$M_u = \phi(0.25f'_c bd^2)$$ (11.22)

For flanged members such as I and T beams in which the neutral axis falls outside the flange,[1] the method of determining flexural strength is exactly analogous to that used for ordinary reinforced-concrete I and T beams. The total steel area is divided into two parts for computational purposes. The first part, at a stress f_{ps}, balances the compression in the overhanging portion of the flange, and the remainder, at stress f_{ps}, balances the compression in the web. In applying the criterion $\rho_p f_{ps}/f'_c$ to flanged members, ρ_p is taken as the steel ratio of only that portion of the total tension-steel area which is required to develop the compression in the web, expressed in terms of the web area. If $\rho_p f_{ps}/f'_c$ does not exceed 0.30, then

$$M_u = \phi \left[A_{pw} f_{ps} d \left(1 - \frac{0.59 A_{pw} f_{ps}}{b_w d f'_c}\right) + 0.85 f'_c (b - b_w) h_f \left(d - \frac{h_f}{2}\right) \right]$$ (11.23)

where

$$A_{pf} = 0.85 \frac{f'_c}{f_{ps}} (b - b_w) h_f$$

and

$$A_{pw} = A_{ps} - A_{pf}$$

If $\rho_p f_{ps}/f'_c$ exceeds 0.30, then

$$M_u = \phi \left[0.25 f'_c b_w d^2 + 0.85 f'_c (b - b_w) h_f \left(d - \frac{h_f}{2}\right) \right]$$ (11.24)

[1] This is usually the case where the flange thickness is less than $1.4 d\rho_p f_{ps}/f'_c$.

If, after a prestressed beam is designed by elastic methods at service loads, inadequate strength exists to resist factored loads, it is possible to include nonprestressed reinforcement in combination with prestressed steel to improve the factor of safety. Such nonprestressed steel may be considered to contribute to the tension force in a member at design moment an amount equal to its area times its yield strength.

It is required by Code that the total amount of prestressed and unprestressed reinforcement be adequate to develop a design load in flexure at least equal to 1.2 times the cracking load, calculated on the basis of a modulus of rupture of $7.5 \sqrt{f_c'}$.

To control cracking in members with unbonded tendons, some bonded reinforcement should be added, uniformly distributed over the tension zone near the extreme tensile fiber. According to the Code, the minimum amount of such bonded reinforcement is the larger of

$$A_s = \frac{N_c}{0.5 f_y} \qquad \text{or} \qquad A_s = 0.004A$$

where N_c is the tensile force in the concrete at load $D + 1.2L$, A is the area of that part of the cross section between the flexural tension face and the center of gravity of the gross section, and f_y is not to exceed 60 ksi.

Example The posttensioned, bonded, prestressed beam shown in cross section in Fig. 11.16 is stressed using wires of strength $f_{pu} = 275,000$ psi. If the concrete strength is $f_c' = 5000$ psi, what is the design strength of the member?

With $f_{pu} = 275,000$ psi, the stress in the steel at design load is, approximately,

$$f_{ps} = f_{pu} \left(1 - \frac{0.5 \rho_p f_{pu}}{f_c'} \right)$$

The tensile-steel ratio is

$$\rho_p = \frac{A_{ps}}{bd} = \frac{1.75}{18 \times 24.5} = 0.00397$$

Then

$$f_{ps} = 275(1 - 0.5 \times 0.00397 \times \tfrac{275}{5}) = 245 \text{ ksi}$$

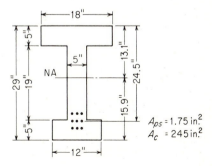

Fig. 11.16

It is next necessary to determine the location of the neutral axis. Since

$$1.4\rho_p \frac{f_{ps}}{f_c'} = 1.4 \times 24.5 \times 0.00397 \times \frac{245}{5} = 6.68 \text{ in.} > 5.0 \text{ in.}$$

it is concluded that the neutral axis is below the underside of the flange, and equations for flanged members will be used. Then

$$A_{pf} = \frac{0.85 f_c'(b - b_w)h_f}{f_{ps}} = 0.85 \times \frac{5}{245} \times 13 \times 5 = 1.13 \text{ in.}^2$$

$$A_{pw} = A_{ps} - A_{pf} = 1.75 - 1.13 = 0.62 \text{ in.}^2$$

The ratio

$$\rho_p \frac{f_{ps}}{f_c'} = \frac{0.62}{5 \times 24.5} \times \frac{245}{5} = 0.248$$

is less than 0.30, and so the design strength of the member is found from Eq. (11.23):

$$M_u = 0.90 \left[0.62 \times 245 \times 24.5 \left(1 - \frac{0.59 \times 0.62 \times 245}{5 \times 24.5 \times 5} \right) \right.$$
$$\left. + 0.85 \times 5 \times 13 \times 5 \times 22 \right]$$

$$= 8340 \text{ in.-kips}$$

As already noted, the satisfaction of stress limits at service loads does not automatically ensure that strength will be adequate at factored loads. Rectangular sections usually have adequate strength, but this is not so for I sections. This suggests the advantage of an alternative approach to design in which the beam is proportioned to have the required strength at factored load. The required concrete section and steel area having been found in this way, the appropriate amount of prestress force is introduced to obtain the desired control of cracking and deflection at service load (Refs. 11.8 and 11.9). The main difficulty with this approach is that, at the outset of the design, neither the steel stress at failure nor the moment arm of the resisting couple is known. However, Abeles has pointed out (Ref. 11.8) that, for I sections, the internal lever arm at failure is always nearly equal to the distance between the steel centroid and the centroid of the compressive flange, and that the steel stress can be approximated as $0.90f_{pu}$ without serious error, except when the steel percentage is unusually large.

11.13 SHEAR AND DIAGONAL TENSION

In prestressed-concrete beams at service loads, there are two factors which greatly reduce the intensity of diagonal tensile stress, as compared with that which would exist if no prestress force were present. The first of these results from the combination of longitudinal compressive stress and shearing stress.

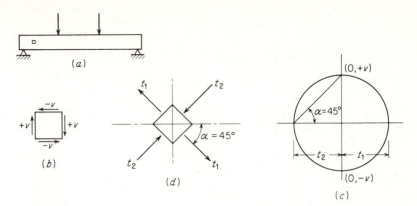

Fig. 11.17

An ordinary tensile-reinforced-concrete beam under load is shown in Fig. 11.17a. The stresses acting on a small element of the beam taken near the neutral axis and near the support are shown in (b). It is found by means of Mohr's circle of stress (c) that the principal stresses act at 45° to the axis of the beam (d) and are numerically equal to the shear-stress intensity; thus

(a) $t_1 = t_2 = v$

Now suppose that the same beam, with the same loads, is subjected to a precompression stress in the amount c, as shown in Fig. 11.18a and b. From Mohr's circle (c) the principal tensile stress is

(b) $t_1 = -\dfrac{c}{2} + \sqrt{v^2 + \left(\dfrac{c}{2}\right)^2}$

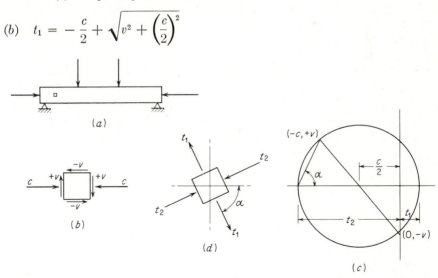

Fig. 11.18

and the direction of the principal tension with respect to the beam axis is

(c) $\tan 2\alpha = \dfrac{2v}{c}$

as shown in (d).

Comparison of Eq. (a) with (b) and Fig. 11.17c with 11.18c shows that, with the same shear-stress intensity, the principal tension in the prestressed beam is much reduced.

The second factor working to reduce the intensity of the diagonal tension at service loads results from the slope of the tendons. Normally, this slope is such as to produce a shear due to the prestress force which is opposite in direction to the load-imposed shear. The magnitude of this "countershear" is $V_p = P_e \sin \theta$, where θ is the slope of the tendon at the section considered (see Fig. 11.3).

It is important to note, however, that in spite of these characteristics of prestressed beams at service loads, an investigation of diagonal tensile stresses at service loads does not ensure an adequate factor of safety against failure. In Fig. 11.18c, it is evident that a relatively small decrease in compressive stress and increase in shear stress, which may occur when the beam is overloaded, will produce a disproportionate, large increase in the resulting principal tension. In addition to this effect, if the countershear of inclined tendons is used to reduce design shear, its contribution does not increase directly with load, but much more slowly (see Art. 11.12). Consequently, a small increase in total shear may produce a large increase in the net shear for which the beam must be designed. For these two reasons, it is usual to base design for diagonal tension in prestressed beams on conditions at factored load rather than at service load. The study of principal stresses in the uncracked prestressed beam is significant only in predicting the load at which the first diagonal crack forms.

At loads near failure, a prestressed beam usually is extensively cracked and behaves much like an ordinary reinforced-concrete beam. Accordingly, many of the procedures and equations developed in Art. 3.5 for the design of web reinforcement for nonprestressed beams may be applied to prestressed beams also. The nominal shear stress at factored loads is

$$v_u = \frac{V_u}{\phi b_w d} \qquad\qquad\qquad (3.23)$$

For prestressed members, where the position of the tendon centroid often varies along the length, d need not be taken less than $0.80h$, where h is the total depth of the section. The required spacing of vertical-web reinforcement[1]

[1] Because of the characteristically flatter slope of diagonal cracking in prestressed beams compared with reinforced-concrete beams, inclined web reinforcement is not permitted.

can be found as before:

$$s = \frac{A_v f_y}{(v_u - v_c)b_w} \tag{3.26a}$$

where v_c is the shear stress carried by the concrete at factored loads.

According to the ACI Code, wherever v_u is greater than one-half of v_c, a minimum amount of web steel must be provided, equal in area to

$$A_v = \frac{A_{ps} f_{pu}}{80 \; f_y} \frac{s}{d} \sqrt{\frac{d}{b_w}} \tag{11.25}$$

where A_{ps} is the area of prestressed reinforcement, and f_{pu} and f_y are, respectively, the ultimate strength of the prestressing steel and the yield strength of the web reinforcement. A maximum spacing of $\frac{3}{4}h$, not to exceed 24 in., is specified. If the quantity $(v_u - v_c)$ exceeds $4\sqrt{f_c'}$, this is reduced by half. Under no condition is the excess shear $(v_u - v_c)$ permitted to exceed $8\sqrt{f_c'}$. The Code specifies further that sections located at a distance less than $h/2$ from a support face may be designed for the shear computed at $h/2$.

The shear stress v_c resisted by the concrete after cracking has occurred is taken equal to the shear stress which caused the first diagonal crack. Two types of diagonal cracks have been observed in tests of prestressed-concrete beams:

1. *Flexure-shear cracks*, occurring at nominal shear stress v_{ci}, start as nearly vertical flexural cracks at the tension face of the beam, then spread diagonally upward (under the influence of diagonal tension), toward the compression face. These are common in beams with a low value of prestress force.
2. *Web-shear cracks*, occurring at nominal shear stress v_{cw}, start in the web due to high diagonal tension, then spread diagonally both upward and downward. These are often found in beams with thin webs and in beams with high prestress force.

On the basis of extensive tests, it was established that the shear causing flexure-shear cracking can be found from the following expression:

$$(a) \quad v_{ci} = 0.6\sqrt{f_c'} + \frac{V_{cr,d+l}}{b_w d}$$

where $V_{cr,d+l}$ is the shear force, due to total load, at which the flexural crack forms at the section considered, and $0.6\sqrt{f_c'}$ represents an additional shear required to transform the flexural crack into an inclined crack. While dead loads are generally uniformly distributed, the live loads may have any distribution. Consequently, it is convenient to separate the total shear into V_d caused by the beam load and superimposed dead load (without load factor), and V_{cr}, the additional shear force, due to live load only, corresponding to flexural cracking. Thus,

$$(b) \quad v_{ci} = 0.6\ \sqrt{f_c'} + \frac{V_d + V_{cr}}{b_w d}$$

The shear V_{cr} may then be found conveniently from

$$(c) \quad V_{cr} = \frac{V_l}{M_l}\ M_{cr}$$

where V_l/M_l, the ratio of live-load shear to moment at the section, remains constant as the load is increased to cracking load, and

$$M_{cr} = \frac{I}{c_2}\ (6\ \sqrt{f_c'} + f_{2pe} - f_{2d}) \tag{11.26}$$

In the last equation, c_2 is the distance from the concrete centroid to the tension face, f_{2pe} is the compressive stress at the tension face resulting from effective prestress force acting alone, and f_{2d} is the bottom-fiber stress due to beam load and superimposed dead load.[1] The first term in the parentheses is a conservative estimate of the modulus of rupture. The stress due to dead load is subtracted here because dead load is considered separately in Eq. (b). Thus, Eq. (b) becomes

$$v_{ci} = 0.6\ \sqrt{f_c'} + \frac{V_d + V_l M_{cr}/M_l}{b_w d} \tag{11.27}$$

Tests indicate that v_{ci} need not be taken less than $1.7\ \sqrt{f_c'}$. If the live load is uniformly distributed as well as the dead load, Eq. (11.27) can be replaced by

$$v_{ci} = 0.6\ \sqrt{f_c'} + \frac{V_u}{M_u}\ \frac{M_{cr}}{b_w d} \tag{11.27a}$$

where the cracking moment is calculated from

$$M_{cr} = \frac{I}{c_2}\ (6\ \sqrt{f_c'} + f_{2pe}) \tag{11.26a}$$

The shear causing web-shear cracking can be found from an exact principal-stress calculation, in which the principal tensile stress is set equal to the direct tensile capacity of the concrete (conservatively taken as $4\ \sqrt{f_c'}$ according to the ACI Code). Alternatively, the Code permits the use of the approximate expression

$$v_{cw} = 3.5\ \sqrt{f_c'} + 0.3 f_{pc} + \frac{V_p}{b_w d} \tag{11.28}$$

[1] All stresses are used with absolute value here, consistent with ACI practice.

in which f_{pc} is the compressive stress in the concrete, after losses, at the centroid of the concrete section (or at the junction of the web and the flange when the centroid lies in the flange), and V_p is the vertical component of the effective prestress force.

After calculating v_{ci} and v_{cw}, then v_c, the shear resistance provided by the concrete, is taken equal to the smaller of the two values.

Calculating M_{cr}, v_{ci}, and v_{cw} for a prestressed beam is a tedious matter, because many of the parameters vary along the member axis. In practical cases, the required quantities may be found at discrete intervals along the span, such as at $l/2$, $l/3$, $l/6$, and at $h/2$ from the support face, and stirrups spaced accordingly.

To shorten the calculation required, the ACI Code includes, as a conservative alternative to the above procedure, an equation for finding the concrete shear resistance v_c directly.

$$v_c = 0.6\sqrt{f_c'} + 700\frac{V_u d}{M_u} \tag{11.29}$$

in which M_u is the bending moment occurring simultaneously with shear force V_u, but $V_u d/M_u$ is not to be taken greater than 1.0. When this equation is used, v_c need not be taken less than $2\sqrt{f_c'}$ and must not be taken greater than $5\sqrt{f_c'}$. Note that, for uniformly loaded, simply supported members,

$$\frac{V_u}{M_u} = \frac{l-2x}{x(l-x)} \tag{11.30}$$

where x is the distance from the support center line along a beam of span l. While Eq. (11.29) is temptingly easy to use, and may be adequate for uniformly loaded members of minor importance, its use is apt to result in highly uneconomical designs for I beams of medium and long span and for composite construction (Ref. 11.10).

Example The unsymmetrical I beam shown in Fig. 11.16 carries an effective prestress force of 288 kips and supports a superimposed dead load of 345 plf and service live load of 900 plf, in addition to its own weight of 255 plf, on a 50-ft simple span. At the maximum-moment section, the effective depth to the main steel is 24.5 in. (eccentricity 11.4 in.). The wires are deflected upward 15 ft from the support, and eccentricity is reduced linearly to zero at the support. If concrete having strength $f_c' = 5000$ psi and stirrups with $f_y = 40,000$ psi are used, and if the prestressed wires have strength $f_{pu} = 275$ ksi, what is the required stirrup spacing at a point 10 ft from the support?

For a cross section of the given dimensions, it is easily confirmed that $I = 24,200$ in.^{4} and $r^2 = I/A = 99$ in.^{2} At a distance 10 ft from the support center line, the tendon eccentricity is

$$e = 11.4 \times \tfrac{10}{15} = 7.6 \text{ in.}$$

corresponding to an effective depth d from the compression face of 20.7 in. According to the Code, the larger value of $d = 0.80 \times 29 = 23.2$ in. will be used. Calculation of v_{ci} is based on Eqs. (11.26) and (11.27). The bottom-fiber stress due to effective prestress acting alone is

$$f_{2pe} = -\frac{P_e}{A}\left(1 + \frac{ec_2}{r^2}\right) = -\frac{288,000}{245}\left(1 + \frac{7.6 \times 15.9}{99}\right) = -2600 \text{ psi}$$

The moment and shear at the section due to combined beam load and superimposed dead load are, respectively,

$$M_{d,10} = \frac{w_d x}{2}(l - x) = 0.600 \times 5 \times 40 = 120 \text{ ft-kips}$$

$$V_{d,10} = w_d\left(\frac{l}{2} - x\right) = 0.600 \times 15 = 9 \text{ kips}$$

and the bottom-fiber stress due to these loads is

$$f_{2d} = \frac{120 \times 12,000 \times 15.9}{24,200} = 950 \text{ psi}$$

Then, from Eq. (11.26),

$$M_{cr} = \frac{24,200(425 + 2600 - 950)}{15.9 \times 12} = 264,000 \text{ ft-lb}$$

The ratio of live-load shear to moment at the section is

$$\frac{V_l}{M_l} = \frac{l - 2x}{x(l - x)} = \frac{30}{400} = 0.075 \text{ ft}^{-1}$$

Equation (11.27) is then used to determine the shear at which flexure-shear cracks may be expected to form:

$$v_{ci} = 43 + \frac{9000 + 0.075 \times 264,000}{5 \times 23.2} = 291 \text{ psi}^1$$

The lower limit of $1.7\sqrt{5000} = 120$ psi does not control.

Calculation of v_{cw} is based on Eq. (11.28). The slope θ of the tendons at the section under consideration is such that $\sin\theta \cong \tan\theta = 11.4/(15 \times 12) = 0.063$. Consequently, the vertical component of the effective prestress force is $V_p = 0.063 \times 288 = 18.1$ kips. The concrete compressive stress at the section centroid is

$$f_{pc} = 288,000/245 = 1170 \text{ psi}$$

Equation (11.28) may now be used to find the shear at which web-shear cracks should occur.

$$v_{cw} = 248 + 0.3 \times 1170 + \frac{18,100}{5 \times 23.2} = 756 \text{ psi}$$

Thus, in the present case,

$$v_c = v_{ci} = 291 \text{ psi}$$

[1] The same result may be obtained for the present case by use of Eqs. (11.26a) and (11.27a).

$$M_{cr} = \frac{24,200}{15.9 \times 12}(425 + 2600) = 383,000 \text{ in.-lb}$$

$$v_{ci} = 43 + \frac{0.075 \times 383,000}{5 \times 23.2} = 291 \text{ psi}$$

At the section considered, the total shear at factored loads is

$$V_u = 1.4 \times 0.600 \times 15 + 1.7 \times 0.900 \times 15 = 35.6 \text{ kips}$$

and from Eq. (3.23),

$$v_u = \frac{35,600}{0.85 \times 5 \times 23.2} = 362 \text{ psi}$$

Using No. 3 U stirrups, for which $A_v = 2 \times 0.11 = 0.22$ in.^{2}, the required spacing is found from Eq. (3.26a) to be

$$s = \frac{0.22 \times 40,000}{(362 - 291) \times 5} = 25 \text{ in.}$$

Equation (11.25) is then applied to establish a maximum-spacing criterion:

$$0.22 = \frac{1.75}{80} \times \frac{275}{40} \frac{s}{23.2} \sqrt{\frac{23.2}{5}} = 0.014s$$

$$s = 15.7 \text{ in.}$$

The other criteria for maximum spacing, $\frac{3}{4} \times 29 = 22$ in., and 24 in., do not control here. Open U stirrups similar to those shown in Fig. 3.8c will be used, at a spacing of 15 in.

For comparison, the concrete shear will be calculated on the basis of Eq. (11.29). The ratio V_u/M_u is 0.075, and

$$v_c = 43 + 700 \times 0.075 \times 23.2/12 = 144 \text{ psi}$$

The lower and upper limits are $2\sqrt{5000} = 142$ psi and $5\sqrt{5000} = 353$ psi, and do not control here. Thus, on the basis of v_c obtained from Eq. (11.29), the required spacing of No. 3 U stirrups is

$$s = \frac{0.22 \times 40,000}{(362 - 144) \times 5} = 8 \text{ in.}$$

For the present case, an I-section beam of intermediate span, more than 3 times the web steel is required at the location investigated, if the alternative expression giving v_c directly is used.

11.14 BOND

There are two separate sources of bond stress in prestressed concrete beams: (1) flexural bond, which, in pretensioned construction, exists between the tendons and the concrete, and in grouted posttensioned construction between the tendons and the grout and between conduit (if any) and concrete; and (2) prestress transfer bond, generally applicable to pretensioned members only.

Flexural bond   The magnitude of bond stress due to flexure will depend significantly on whether or not the beam has cracked. If the beam is uncracked (often the case at service loads), flexural-bond stress is low. When the beam cracks, a sudden increase in bond stress results. The beam behaves very much as an ordinary reinforced-concrete beam, and flexural-bond stress can be calculated with the equations of Chap. 3. While bond stresses after cracking are an order of magnitude higher than those before cracking, the beam will not fail due to these stresses if adequate development length is provided for the steel past the point of maximum stress. According to the ACI Code, pretensioning strand must be bonded beyond the critical section for a development length, in inches, not less than

$$l = (f_{ps} - \tfrac{2}{3}f_{se})d_b \qquad\qquad\qquad (11.31)$$

where d_b is the nominal strand diameter in inches, and the stresses are expressed in kips per square inch. Investigation may be restricted to those cross sections nearest each end of the member that are required to develop their full strength under the specified design load. Equation (11.31) for minimum bonded length is an empirical one, based on tests of stranded tendons. It does not apply to plain wires; the length for smooth wire could be considerably greater due to the absence of mechanical interlock.

Prestress transfer bond   In the end regions of pretensioned beams (and in some cases for posttensioned beams), the prestress force from the tendons is transferred to the concrete by bond alone. Thus there is a length of transfer at each end of the tendons that serves the purpose of anchorage. The stress condition in the end zones of such beams is exactly the reverse of that in ordinary reinforced construction. In ordinary reinforced construction, the concrete is poured around an unstressed bar. As the bar is brought into tension, its diameter decreases at the stressed end, owing to Poisson's effect, creating tension normal to the surface of contact between the rod and the concrete. In prestressed beams, on the other hand, the wire is stretched before the concrete is poured around it. As the exposed wire end is cut loose from the anchorage, the stress at the outer end becomes zero. As a consequence, the wire expands as it attempts to regain its original diameter, causing compression normal to the surface of contact with the concrete. The stress in the wires is zero at the free end and increases progressively with distance from that end, approaching its full value at some finite distance, termed the *transfer length*.

   The development length given by Eq. (11.31) is also a measure of transfer length. Slightly rusted tendons have an appreciably shorter transfer length than clean ones. Gentle release of the strands or wires during transfer of prestress to the concrete results in shorter transfer length than does abruptly cutting the strands.

   In shear design, if the critical section $h/2$ from the support face is closer to the end of the member than the development length, the reduced prestress force should be used in calculating v_{cw}. For practical purposes, the prestress force may be assumed to vary linearly from zero at the end of the tendon to its full value at a distance from the end equal to the length given by Eq. (11.31).

11.15 END-ZONE STRESSES

In all the foregoing analysis, consideration has been given to the resultant stress distribution in the concrete without regard for the fact that the prestress force is actually applied as a concentrated load, or as a group of concentrated loads, on the end of the member. It has been shown mathematically and experimentally that the effect of such concentration of the load becomes insignificant at a distance from the end of the member approximately equal to the member depth. In pretensioned construction, in which the prestress is normally transferred to the concrete by bond over a length which may about equal the member depth, no problem arises, because of the gradual stress input. However, in posttensioned beams, in which, in many cases, the prestress force is transferred through end anchorages and is applied in full value at the extreme end of the beam, the resulting stress concentrations require special consideration.

Two types of stress may be troublesome: (1) direct bearing compression under the anchorages and (2) transverse tension leading to longitudinal splitting of the member at the level of the steel centroid. It is common in posttensioned I beams to use a solid end block of rectangular cross section as shown in Fig. 11.19. This may extend from the end of the beam a distance equal to about $\frac{3}{4}$ to 1 times the beam depth, and reduces local stresses near the anchorages. According to the Code, the end zone must be designed to develop the full tensile strength f_{pu} of the prestressing tendons.

The maximum permitted concrete bearing stress f_{cp} may be found from

$$f_{cp} = 0.6 f'_{ci} \sqrt[3]{\frac{A_2}{A_1}} \tag{11.32}$$

but not to exceed f'_{ci}. In this equation, f'_{ci} is the concrete compressive strength at the time of tensioning, A_1 is the bearing area of the anchor plate, and A_2 is the maximum possible area of the portion of the anchorage surface that is geometrically similar to, and concentric with, the area of the anchor plate.

Determination of transverse end-block tension is a complex problem. Theoretical treatments are highly complicated, and to obtain workable methods one must introduce simplifications and assumptions which leave the solution at least open to question. Fortunately, one can afford to be quite conservative in reinforcing the end zone against local stress concentration. This will not significantly affect the overall economy of the construction, because even a

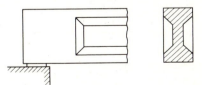

Fig. 11.19   End block for posttensioned beam.

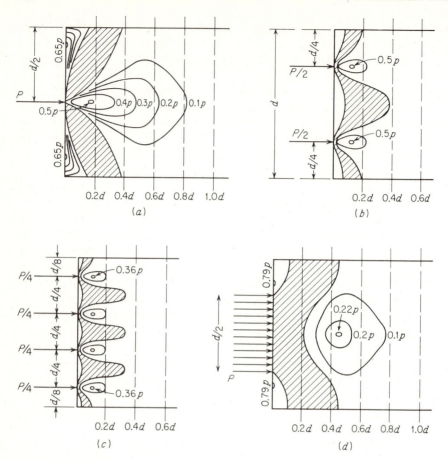

Fig. 11.20 Isobar charts. (*After Ref. 11.7.*)

conservative analysis will usually require only a small amount of reinforcement to be placed in a very limited region.

Practically, the problem is to estimate the magnitude and location of the local tensile stresses, which are greatest in the direction at right angles to the prestressing force, and which may be sufficient to cause cracking or splitting of the end of the member. Guyon has published an important mathematical study of end-zone stresses based on elastic analysis, and has presented a set of diagrams for the transverse tension resulting from concentrated longitudinal compressive forces (Ref. 11.7). Examples of such charts, in the form of *isobars*, i.e., curves of equal transverse stress, are shown in Fig. 11.20. These diagrams express the transverse unit tensile stress in terms of the average longitudinal compressive stress $p = P/A$. It is evident from these diagrams that there are two separate zones of transverse tension resulting in the end block. The first, defined as the *bursting* zone, lies deep within the block, along the line of action of the load; the second, the *spalling* zone, is at the corners of the end block. These tensile zones are separated from each other by a continuous band, shown crosshatched, in which the transverse stress is compressive.

It will be found that in many cases the transverse tensile stress is higher than the direct tensile strength of the concrete, and some form of tensile reinforcement is indicated. Usually, this is in the form of ordinary bars, bent in the form of closed stirrups or hoops, used at a design stress of $0.90f_y$. The reinforcement to accommodate spalling tension should be located as close to the end face of the beam as is practical. Guyon observes that, for most practical tendon groupings, the total transverse spalling tension will not exceed $0.03P$, where P is the total longitudinal load, and he recommends that steel be placed in the spalling zone to accommodate that amount of tension. In the bursting zone the amount of steel required to carry tension will vary considerably, depending on the pattern of loading. This total tension can be evaluated from diagrams such as Fig. 11.20. Reference 11.7 includes such diagrams covering many specific loading situations. Although these charts resulted from elastic analysis, they are commonly used as the basis for strength design. This is justifiable, since in tension the concrete is nearly elastic up to the cracking load, and since elastic behavior is obtained in compression, as well, because transverse concrete compressive stress is normally quite low.

Example   Figure 11.21a and b shows the end zone of a rectangular posttensioned beam 12 in. wide and 15 in. deep, for which end reinforcement is to be designed. Assuming the tendon area of 1.31 in.^{2} to be stressed to f_{pu} of 275 ksi, the end zone must be designed for a total force of $1.31 \times 275 = 360$ kips. Anchorage is such that this may be considered uniformly distributed over the center half of the depth. The direct tensile strength of the concrete is 300 psi.

The average compressive stress in the concrete is $p = \frac{360}{180} = 2$ ksi. From Fig. 11.20d, the transverse tension along the middepth of the beam is obtained, and plotted against distance from the end in Fig. 11.21c. The tensile strength of 300 psi is exceeded in the region from $0.32d$ to $0.64d$, a horizontal distance of 4.8 in. While the tensile stress varies from 300 to 440 psi in that length, use of an average value of 400 psi over the entire distance will not seriously affect the total tension, which is

$0.400 \times 4.8 \times 12 = 23.0$ kips

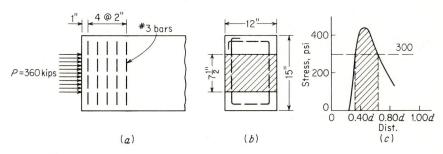

(a)   (b)   (c)

Fig. 11.21

No. 3 stirrups having the form shown in Fig. 11.21b and an area of $0.11 \times 2 = 0.22$ in.^{2} will be used at a design stress $\phi f_y = 0.90 \times 40 = 36$ ksi. The number required to resist the bursting tension is

$$N = 23.0/(36 \times 0.22) = 2.9$$

Three stirrups will be used for this purpose. The spalling tension can be obtained from Fig. 11.20d also, but it is sufficient to estimate the total spalling tension as

$$0.03P = 0.03 \times 360 = 10.8 \text{ kips}$$

The number of No. 3 stirrups required to resist this force is

$$N = 10.8/(36 \times 0.22) = 1.4$$

Two additional stirrups will be specified. The total of five No. 3 stirrups meets requirements of spalling and bursting tension and is arranged as shown in Fig. 11.21a.

11.16 DEFLECTION

Deflection of the slender, relatively flexible beams that are made possible by prestressing must be predicted with care. Many members, satisfactory in all other respects, have proved to be unserviceable because of excessive deformation. In some cases, the absolute amount of deflection is excessive. Often, it is the differential deformation between adjacent members (e.g., precast roof-deck units) that has caused problems. More often than not, any difficulties which occur are associated with upward deflection due to the sustained prestress load. Such difficulties are easily avoided by proper consideration in design.

When the prestress force is first applied, a beam will normally camber upward, due to the combined effect of initial eccentric prestress force and its own weight. With the passage of time, concrete shrinkage and creep will cause a gradual reduction of prestress force. In spite of this, the upward deflection will increase, due to the differential creep, affecting the highly stressed bottom fibers more than the top. With the application of superimposed dead and live loads, this upward deflection will be partially or completely overcome, and zero or downward deflection obtained. Clearly, in computing deformation, careful attention must be given to the duration of loading. In many practical cases, the most significant deflection is that occurring under dead load plus the sustained portion of the live load (characteristic load).

The calculation of deflection of a prestressed member may be based on the moment diagrams associated with the various loads acting upon it. For statically determinate beams, the ordinates of the moment diagram resulting from eccentric prestress force are directly proportional to the eccentricity of the steel centroid with respect to the concrete centroid. Deflections due to this and other moments may be found using the ordinary methods of moment area or conjugate beam (Refs. 7.1 and 7.2). Creep may be allowed for with sufficient accuracy by applying the creep coefficients of Table 1.1 to initial elastic deformations.

Since members are mostly or completely uncracked at characteristic load levels of interest, calculations may usually be based on properties of the uncracked concrete section. If the bottom-fiber tension exceeds the modulus of rupture at the load at which deflection is to be found, calculations should be based on a bilinear load-deformation curve or the equivalent, with slopes related to the moment of inertia of the transformed uncracked and cracked sections (Ref. 11.12).

Example   The 40-ft simply supported T beam shown in Fig. 11.22 is prestressed with a force of 314 kips, using a parabolic tendon which has eccentricity 3 in. above the concrete centroid at the supports and 7.9 in. below the centroid at midspan. After time-dependent losses have occurred, this prestress is reduced to 267 kips. In addition to its own weight of 330 plf, the girder must carry a short-term superimposed live load of 900 plf. Estimate the deflection at all critical stages of loading. The creep coefficient $C_c = 2.0$, $E_c = 4 \times 10^6$ psi, and modulus of rupture = 580 psi.

It is easily confirmed that the stress in the bottom fiber when the beam carries the maximum load to be considered is 80 psi compression. All deflection calculations may therefore be based on the moment of inertia of the gross concrete section, $I = 15,800$ in.[4] It is convenient to calculate the deflection due to prestress and that due to girder load separately, superimposing the results later. For the eccentricities shown by the tendon profile of Fig. 11.22b, the application of $P_i = 314$ kips causes the moments shown in Fig. 11.22c (moments are stated in in.-lb units). Applying the second moment-area theorem by taking moments of the M/EI diagram between midspan and the support, about the support, produces the desired vertical displacement between those two points as follows:

$$\Delta_{pi} = \frac{-(3.42 \times 10^6 \times 240 \times \frac{2}{3} \times 240 \times \frac{5}{8}) + (0.942 \times 10^6 \times 240 \times 120)}{4 \times 10^6 \times 15,800}$$

$$= -0.87 \text{ in.}$$

the minus sign indicating upward deflection due to initial prestress alone. The downward deflection due to the self-weight of the girder is calculated by the well-known equation

$$\Delta_g = \frac{5wl^4}{384EI} = \frac{5 \times 330 \times 40^4 \times 12^4}{384 \times 12 \times 4 \times 10^6 \times 15,800} = +0.30 \text{ in.}$$

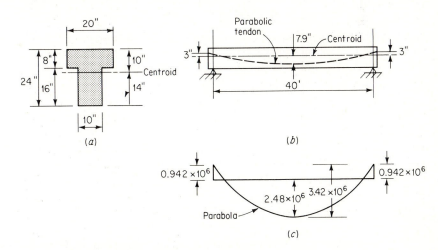

Fig. 11.22   T beam for deflection calculation. (a) Cross section; (b) tendon profile; (c) moment from prestress force.

Superimposing these two results, the net upward deflection when initial prestress and girder load act together is

$$\Delta_{pi} + \Delta_g = -0.87 + 0.30 = -0.57 \text{ in.}$$

Shrinkage and creep of the concrete cause gradual reduction of prestress force from $P_i = 314$ kips to $P_e = 267$ kips, and reduce the bending moment due to prestress proportionately. However, concrete creep acts to increase both the upward-deflection component due to prestress force and the downward-deflection component due to girder load. Consequently, the net deflection after these changes take place is upward and equal to

$$\Delta_{pe} + \Delta_g = (-0.87 \times \tfrac{267}{314} + 0.30)(1 + 2.0) = -1.33 \text{ in.}$$

where the term in the second parentheses is $(1 + C_c)$, the creep coefficient being equal to 2.0.[1] In spite of loss of prestress, the upward deflection is considerably larger than before. Finally, as the 900-plf short-term superimposed load is applied, the net deflection is

$$\Delta_{pe} + \Delta_g + \Delta_s = -1.33 + 0.30 \times 900/330 = -0.51 \text{ in.}$$

A net deflection upward of about 1/1000 of the span is thus obtained when the member carries its superimposed load.

REFERENCES

11.1. T. Y. Lin: "Prestressed Concrete Structures," 2d ed., John Wiley & Sons, Inc., New York, 1963.

11.2. N. Khachaturian and G. Gurfinkel: "Prestressed Concrete," McGraw-Hill Book Company, New York, 1969.

11.3. P. W. Abeles: "An Introduction to Prestressed Concrete," vols. 1 and 2, Frederick Ungar Publishing Co., New York, 1969.

11.4. A. H. Nilson: Flexural Design Equations for Prestressed Concrete Members, *J. Prestressed Concrete Inst.*, vol. 14, no. 1, pp. 62–71, February, 1969.

11.5. CEB-FIP Joint Committee: "International Recommendations for the Design and Construction of Prestressed Concrete Structures," Cement and Concrete Association, London, June, 1970.

11.6. British Standards Institution: Draft British Standard Code of Practice for the Structural Use of Concrete, London, 1969.

11.7. Y. Guyon: "Prestressed Concrete," vols. 1 and 2, John Wiley & Sons, Inc., New York, 1960.

11.8. P. W. Abeles: Design of Partially-prestressed Concrete Beams, *J. ACI*, vol. 64, no. 10, pp. 669–677, October, 1967.

11.9. A. H. Nilson: Discussion of Ref. 11.8, *J. ACI*, vol. 65, no. 4, pp. 345–347, April, 1968.

11.10. J. G. MacGregor and J. M. Hanson: Proposed Changes in Shear Provisions for Reinforced and Prestressed Concrete Beams, *J. ACI*, vol. 66, no. 4, pp. 276–288, April, 1969.

11.11. Deflection of Prestressed Concrete Members, Report by ACI Committee 435, *J. ACI*, vol. 60, no. 12, pp. 1697–1728, December, 1963.

[1] This calculation is not strictly correct, because the creep takes place as the prestress force gradually reduces from P_i to P_e, while it is assumed here that creep results from a constant force P_e. The difference is not important in most cases. More rigorous methods are given in Ref. 11.11.

12
Reinforced-concrete Bridges

12.1 TYPES OF BRIDGES

Reinforced concrete is particularly adaptable for use in highway bridges be-cause of its durability, rigidity, and economy, as well as the comparative ease with which a pleasing architectural appearance can be secured. For very short spans, from about 10 to 25 ft, one-way-slab bridges (Fig. 12.1) are economical. For somewhat longer spans, concrete girder spans (Fig. 12.2) may be used.

Probably most highway spans of medium length, from 40 to 90 ft, presently use composite steel-concrete construction (Fig. 12.3) or composite

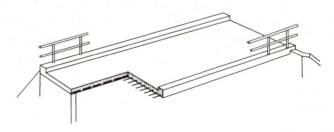

Fig. 12.1 Slab bridge.

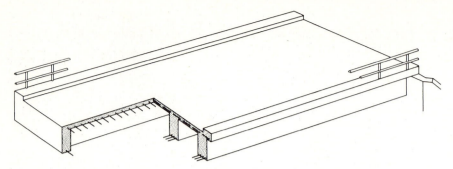

Fig. 12.2   Deck-girder bridge.

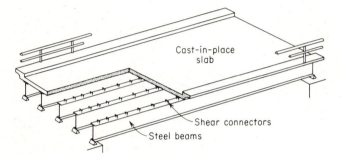

Fig. 12.3   Composite steel-concrete bridge.

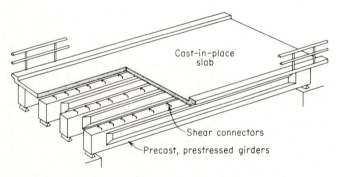

Fig. 12.4   Composite prestressed-concrete bridge.

prestressed-concrete construction (Fig. 12.4). In composite construction with structural steel, the concrete deck is made to act integrally with supporting steel stringers by means of devices called *shear connectors*, welded to the top flange of the steel section and embedded in the slab. Although such a bridge is not strictly a reinforced-concrete structure, the design of this type of bridge will be discussed in some detail in this section because of its widespread use. Prestressed-concrete bridges frequently make use of composite section characteristics also. Commonly, the girders are precast and placed into final location

Fig. 12.5 Prestressed box girders of 80-ft span for Bay Area Rapid Transit System, San Francisco.

by crane, eliminating the necessity for obstructing traffic by falsework. The deck slab is then cast in place, bonded, and tied to the precast sections by steel dowels.

Hollow box girders, usually prestressed, are often used for intermediate and long-span concrete bridges. Spans up to about 80 ft are precast in one piece and lifted into position, as shown in Fig. 12.5. Longer spans of similar cross section, carrying two lanes of highway traffic as well as shoulders and walkways, have either been cast in place or precast in short segments, which are posttensioned after positioning.

Bridge spans as long as 320 ft (Fig. 12.6) have been attained using prestressed girders. Other possibilities for long-span concrete bridges are the various forms of arches, including the barrel arch (Fig. 12.7) and the three-hinged arch (Fig. 12.8).

This chapter will be devoted to an introductory discussion of the design and detailing of highway bridges. It is the present practice in the United States to proportion concrete highway bridges by elastic stress methods at service loads, rather than by strength methods at factored loads. Undoubtedly, this results from the view that, where repetitive loads act, primary attention should be given to actual stresses obtained at normal load conditions.

Fig. 12.6 Oneida Lake Bridge, New York; NYS Dept. of Pub. Works; 320-ft span, pre-stressed-concrete girders.

Fig. 12.7 Fox River Valley Bridge, Illinois; Vogt, Ivers, and Assoc., Engineers; five 178-ft barrel-arch spans.

12.2 LIVE LOADS

Truck loadings The live loads specified by the American Association of State Highway Officials (Ref. 12.2)—whose specifications control the design of most American highway bridges—consist of standard, idealized trucks or of live loads which are equivalent to a series of trucks. Two systems of loadings are provided: the H loadings, as shown in Fig. 12.9, and the HS loadings, as shown in Fig. 12.10. The H loadings represent a two-axle truck; the HS loadings represent a two-axle tractor plus a single-axle semitrailer.

Fig. 12.8 Salgintobel Bridge, Switzerland, by Robert Maillart: three-hinged arch span of 295 ft.

As shown in the figures, the highway loadings are divided into several classes. The number of the loading indicates the gross weight in tons of the truck or tractor. The gross weight is divided between the front and rear axles, as shown in Figs. 12.9 and 12.10.

Lane loadings, which are to be used when they produce greater stress than the corresponding truck loadings, are shown in Fig. 12.11. In general, the lane loadings, which are designed to approximate the effect of a series of trucks, govern the design of the longer-span bridges.

Other roadway loadings The possibility that the bridge may be required to carry electric railways, railroad freight cars, military vehicles, or other extraordinary vehicles should be considered in design. The class of such traffic shall determine the loads to be used.

Sidewalk loadings Sidewalk floors, stringers, and their immediate supports are usually designed for a live load of at least 85 psf of sidewalk area.

Selection of loadings The AASHO specifications state that for truck highways or for other highways which carry heavy truck traffic the minimum live load shall be the HS15 loading. It is common practice at present to design bridge

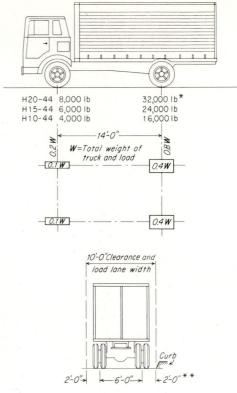

Fig. 12-9 Standard H truck loadings.

STANDARD H TRUCKS

* In the design of timber floors for H20 loading, one axle load of 24,000 lb or two axle loads of 16,000 lb each, spaced 4 ft apart, whichever produces the greater stress, may be used instead of the 32,000-lb axle shown.

** For slab design the center line of the wheel shall be assumed to be 1 ft from the face of the curb.

spans on major highways for the HS20 loading, using the lighter loadings only for structures on secondary roads, on which the occurrence of heavy traffic will be negligible.

12.3 APPLICATION OF LOADINGS

Some of the more important rules for applying the selected AASHO loading follow.

1. The lane loading of standard trucks shall be assumed to occupy a width of 10 ft. They shall be assumed to occupy any position within their design traffic lane which will produce the maximum stress. The width of the design traffic lanes shall be determined from the formula

$$W = \frac{W_c}{N} \tag{12.1}$$

where W_c = roadway width between curbs, exclusive of median strip

N = number of design traffic lanes as shown in the following table

W = width of design traffic lane

W_c, ft	N
20–30	2
Over 30–42	3
Over 42–54	4
Over 54–66	5
Over 66–78	6
Over 78–90	7
Over 90–102	8
Over 102–114	9
Over 114–126	10

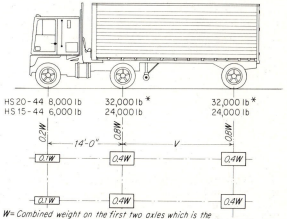

W= Combined weight on the first two axles which is the
same as for the corresponding H truck
V= Variable spacing–14 feet to 30 feet inclusive.
Spacing to be used is that which produces
maximum stresses.

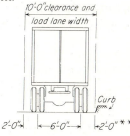

STANDARD H-S TRUCKS

Fig. 12-10 Standard HS truck loading.

* In the design of timber floors for HS20 loading, one axle load of 24,000 lb or two axle loads of 16,000 lb each, spaced 4 ft apart, whichever produces the greater stress, may be used instead of the 32,000-lb axle shown.

** For slab design the center line of the wheel shall be assumed to be 1 ft from the face of the curb.

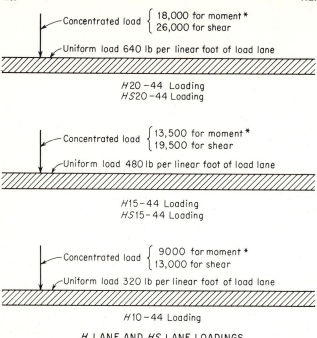

H LANE AND HS LANE LOADINGS

Fig. 12.11 Equivalent lane loadings.

* For continuous spans, another concentrated load of equal weight shall be placed in one other span in the series in such a position as to produce maximum negative moment.

2. Each 10-ft lane loading or single standard truck shall be considered as a unit, and fractional load-lane widths or fractional trucks shall not be used.
3. The number and position of the lane loading or truck loading shall be as specified above. The type of loading to be used—lane loading or truck loading—shall be such as to produce maximum stress subject to the reduction specified below.

Reduction in load intensity Where maximum stresses are produced in any member by loading any number of traffic lanes simultaneously, the following percentages of the resultant live-load stresses shall be used in view of improbable coincident maximum loading:

Number of lanes	Per cent
1 or 2	100
3	90
4 or more	75

12.4 IMPACT

Live-load stresses due to truck loading (or equivalent lane loading) are increased to allow for vibration and the sudden application of the load. The increase is computed by the formula

$$I = \frac{50}{l + 125} \tag{12.2}$$

where I = impact fraction of live-load stress
 l = loaded length, ft
The maximum impact allowance to be used shall be 30 per cent.

12.5 DISTRIBUTION OF LOADS

When a concentrated load is placed on a reinforced-concrete slab, the load is distributed over an area larger than the actual contact area. For example, if a concentrated load with a bearing area of 1 ft² were placed on the slab of a through-girder bridge, it is reasonable to assume that, on account of the stiffness of the slab, the strips of slab at right angles to the girder and adjacent to the 1-ft strip in direct contact with the load would assist in carrying the load. Similarly, with the beams of a T-beam bridge spaced fairly closely, a concentrated load placed directly over one of the beams would not be carried entirely by that beam, for the concrete slab would be sufficiently rigid to transfer part of the load to adjacent beams.

No distribution is assumed in the direction of the span of the member. The effect of any such distribution would be comparatively small.

The following recommendations for the distribution of loads are taken from the specifications of the American Association of State Highway Officials.

12.6 DISTRIBUTION OF WHEEL LOADS ON CONCRETE SLABS

The pertinent rules for the distribution of wheel loads on concrete slabs and some additional design requirements are as follows:

Span lengths For simple spans, the span length shall be the distance center to center of supports but shall not exceed clear span plus thickness of slab.

The following effective span lengths shall be used in calculating the distribution of loads and bending moments for slabs continuous over more than two supports:

Slabs monolithic with beam (without haunches): S = clear span
Slabs supported on steel stringers: S = distance between edges of flanges plus one-half the stringer flange width
Slabs supported on timber stringers: S = clear span plus one-half the thickness of the stringer

Edge distance of wheel load In designing slabs, the center line of the wheel load shall be assumed to be 1 ft from the face of the curb.

Bending moment Bending moment per foot width of slab shall be calculated according to the methods given under Cases 1 and 2 below. In both cases,

S = effective span length as defined under the above heading, ft
E = width of slab over which wheel load is distributed, ft
P = load on one rear wheel of truck
P_{15} = 12,000 lb for H15 loading
P_{20} = 16,000 lb for H20 loading

Case 1: main reinforcement perpendicular to traffic (spans 2 to 24 ft inclusive)
The live-load moment for simple spans shall be determined by the following formulas (impact not included):

HS20 loading:

$$\frac{S+2}{32} P_{20} = \text{moment in ft-lb per foot width of slab} \tag{12.3a}$$

HS15 loading:

$$\frac{S+2}{32} P_{15} = \text{moment in ft-lb per foot width of slab} \tag{12.3b}$$

In slabs continuous over three or more supports, a continuity factor of 0.8 shall be applied to the above formulas for both positive and negative moments.

Case 2: main reinforcement parallel to traffic Distribution of wheel loads $E = 4 + 0.06S$; maximum 7.0 ft. Lane loads are distributed over a width of $2E$. Longitudinally reinforced slabs shall be designed for the appropriate HS loading.

For simple spans, the maximum live-load moment per foot width of slab, without impact, is closely approximated by the following formulas:

HS20 loading:

Spans up to and including 50 ft: L.L.M. = $900S$ ft-lb   (12.4a)
Spans 50 to 100 ft: L.L.M. = $1000(1.30S - 20.0)$ ft-lb   (12.4b)

HS15 loading:
Use three-fourths of the values obtained from the formulas for HS20 loading.

Moments in continuous spans shall be determined by suitable analysis using the truck or appropriate lane loading.

Edge beams (longitudinal) Edge beams shall be provided for all slabs having main reinforcement parallel to traffic. The beam may consist of a slab section additionally reinforced, a beam integral with and deeper than the slab, or an integral reinforced section of slab and curb. It shall be designed to resist a live-load moment of $0.10PS$, where P is the wheel load in pounds (P_{15} or P_{20}) and S is the span length in feet. The moment as stated is for a freely supported span. It may be reduced 20 per cent for continuous spans unless a greater reduction results from a more exact analysis.

Distribution reinforcement   Reinforcement shall be placed in the bottoms of all slabs transverse to the main steel reinforcement to provide for the lateral distribution of the concentrated live loads, except that reinforcement will not be required for culverts or bridge slabs when the depth of the fill over the slab exceeds 2 ft. The amount shall be the percentage of the main-reinforcement steel required for positive moment as given by the following formulas:

For main reinforcement parallel to traffic:

$$\text{Percentage} = \frac{100}{\sqrt{S}} \qquad \text{maximum} = 50 \text{ per cent} \qquad (12.5a)$$

For main reinforcement perpendicular to traffic:

$$\text{Percentage} = \frac{220}{\sqrt{S}} \qquad \text{maximum} = 67 \text{ per cent} \qquad (12.5b)$$

where S is the effective span length in feet.

Shear and bond stress in slabs   Slabs designed for bending moment in accordance with the foregoing shall be considered satisfactory in bond and shear.

12.7 DISTRIBUTION OF WHEEL LOADS TO STRINGERS, LONGITUDINAL BEAMS, AND FLOOR BEAMS

The following rules govern the distribution of wheel loads affecting longitudinal and transverse concrete beams supporting concrete slabs. The AASHO specification contains additional recommendations for steel and timber flooring which are not reproduced here.

Position of loads for shear   In calculating end shears and end reactions in transverse floor beams and longitudinal beams and stringers, no lateral or longitudinal distribution of the wheel shall be assumed for the wheel or axle load adjacent to the end at which the stress is being determined. For loads in other positions on the span, the distribution for shear shall be determined by the method prescribed for moment.

Bending moment in stringers and longitudinal beams   In calculating bending moments in longitudinal beams or stringers, no longitudinal distribution of the wheel loads shall be assumed. The lateral distribution shall be determined as follows:

1. *Interior stringers* supporting concrete floors shall be designed for loads determined in accordance with the following table, in which S is the average spacing of stringers in feet:

Floor system	One traffic lane, fraction of a wheel load to each stringer	Two or more traffic lanes, fraction of a wheel load to each stringer
Concrete slab on steel I-beam stringers and prestressed-concrete girders	$S/7.0$ ($S_{max} = 10$ ft)*	$S/5.5$ ($S_{max} = 14$ ft)*
Concrete slab on concrete stringers	$S/6.0$ ($S_{max} = 6$ ft)*	$S/5.0$ ($S_{max} = 10$ ft)*
Concrete box girder	$S/8.0$ ($S_{max} = 12$ ft)*	$S/7.0$ ($S_{max} = 16$ ft)*

* If S exceeds the value in parentheses, the load on each stringer shall be the reaction of the wheel loads, under the assumption that the flooring between the stringers acts as a simple beam.

2. The live load supported by *outside stringers* shall be the reaction of the truck wheels, under the assumption that the flooring acts as a simple beam between stringers.
3. The *combined load capacity* of all the beams in a span shall not be less than the total live and dead load in the panel.

Bending moment in floor beams (transverse)   In calculating bending moments in floor beams, no transverse distribution of the wheel loads shall be assumed. If longitudinal stringers are omitted and the concrete floor is supported directly on floor beams, the fraction of wheel load allotted to each floor beam is $S/6$, where S equals the spacing of beams in feet. If S exceeds 6 ft, the load on the beam shall be the reaction of the wheel loads, under the assumption that the flooring between beams acts as a simple beam.

12.8 ABUTMENTS

Bridge abutments serve to transmit the load from the superstructure to the foundation and act as retaining walls to hold back the earth fill behind them.

Bridge seats   The bridge seats of abutments which support the fixed ends of concrete highway bridges are quite frequently built as horizontal surfaces without parapets or backwalls, as shown in Fig. 12.12. The slab or the beams of the deck rest directly on the bridge seat; in deck-girder bridges, transverse

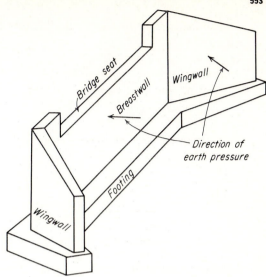

Fig. 12.12 Details of abutment without backwalls.

diaphragm walls are constructed between the beams to hold back the earth above the bridge seat. These diaphragms are thin walls, 6 or 8 in. in thickness, with a nominal amount of reinforcement.

Figure 12.13 shows a modified form of the construction described above in which a very low backwall is used, primarily for the purpose of counteracting the tendency of the abutment to move inward under the deck. The dowels b and c shown in the abutment and in the deck construction serve to tie the deck, the abutment, and the approach slab together.

Another type of bridge seat is shown in Fig. 12.14. This type is particularly adapted to the fixed ends of deck-girder bridges. The bridge seat is constructed with notches into which the bridge beams are built. Figure 12.14

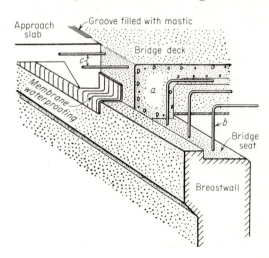

Fig. 12.13 Details of bridge seat with low backwall.

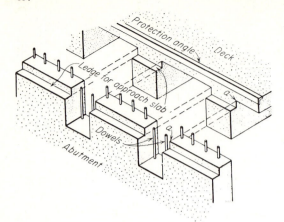

Fig. 12.14 Details of bridge seat with pockets for beams.

is a diagrammatic sketch and does not show the final relative positions of the deck and abutment; the two parts of the figure actually fit together, as shown by the dotted lines. Ledges are provided in the abutment and at the top of the deck beams to support the approach slab, as shown in the figure. Dowels are placed in the abutment in such a manner as to project up into the beams and deck slab.

A third type of bridge-seat construction is shown in Fig. 12.15. This is the usual type with backwall, used at the expansion end of the bridge. Two arrangements of joints are shown in the figure. The backwall is extended upward to the roadway surface in (*a*); it stops at the bottom of the roadway slabs in (*b*). Type *b* is usually preferred, in that only one joint is required in the roadway surface, as compared with two joints in type *a*.

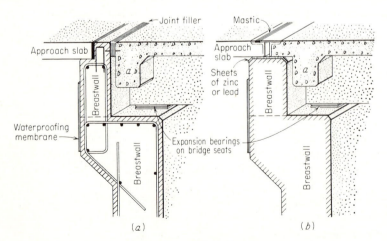

Fig. 12.15 Details of bridge seats with backwalls.

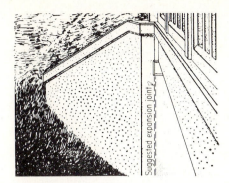

Fig. 12.16   Vertical expansion joint in breast wall.

Breast walls   The two most common types of breast walls are the cantilever retaining wall of reinforced concrete and the type that acts as a vertical slab supported horizontally by the deck and by the foundation. Buttressed breast walls are rarely used in short-span concrete bridges. Gravity walls of plain concrete are sometimes used, especially where the height of the abutment is not great.

The cantilever breast wall is generally used at the expansion end of the bridge. The type that is supported horizontally by the deck and the footing can be used only at the fixed end; in such cases it is always advisable to have at least a low parapet or backwall, as shown in Fig. 12.13, in order to obtain the direct horizontal resistance which is assumed to be furnished by the deck.

Wing walls   Gravity or cantilever-type wing walls may be used. It is advisable to construct vertical expansion joints at the junction of the wing walls and breast wall to avoid unsightly cracks which are apt to form along these planes. Sometimes, instead of an expansion joint, a vertical groove is constructed at the joint; any crack that may develop will then be inconspicuous. The expansion joint or the groove should preferably be placed at the end of the bridge seat, as shown in Fig. 12.16.

12.9 MISCELLANEOUS CONSTRUCTION DETAILS

Diaphragms   In deck-girder bridges, thin transverse walls called *diaphragms* are constructed between the beams at the ends of the bridge, as shown at (*a*) in Figs. 12.13 and 12.15. These diaphragms serve two purposes: first, they furnish lateral support for the beams, and second, when no parapet or backwall is built above the bridge seat, they prevent the earth fill from spilling out on the bridge seat. At the fixed ends, the diaphragms may rest directly on the bridge seat, as shown in Fig. 12.13; at the expansion ends, the diaphragms should be kept clear of the bridge seat, as shown in Fig. 12.15, so as not to develop friction which would interfere with the freedom of longitudinal movement. The diaphragms are usually from 6 to 8 in. thick, with a nominal amount of reinforcement.

Intermediate diaphragms should be constructed between the beams of deck-girder bridges with spans greater than 40 ft; the intermediate diaphragms are placed at the center or at the third points of the span.

Fixed bearings   Fixed bearings can safely be used at both ends of spans up to 40 ft. Two such bearings are illustrated in Figs. 12.13 and 12.14. The dowels which serve to tie together the deck construction and the abutment are not designed to resist any appreciable amount of bending moment, and hence no joint rigidity can be assumed in the design of the deck or abutment.

Expansion bearings   For spans greater than 40 ft, an expansion bearing should be provided at one end of the bridge. A suitable form of expansion bearing for spans up to 50 ft is shown in Fig. 12.15. A pair of steel plates is placed under each girder; the lower plate is anchored to the concrete in the abutment, and the upper plate is fastened to the underside of the girder. Frequently, plain or laminated elastomeric bearings (natural rubber or neoprene) are placed between the plates to facilitate movement.

For the shorter spans and light loadings, frequently the only expansion device is a layer of tar paper inserted between the bridge seat and the deck. Obviously, this is not desirable where provision for expansion is of any importance, because of the questionable efficiency of the construction.

In spans greater than 50 ft, the deflection of the deck may rotate the ends of the girders so much that the bearing pressure may become concentrated on a narrow strip along the front edge of the bridge seat and may cause local damage if flat bearing plates are used. To counteract this tendency, one plane and one curved plate may be used, as shown in Fig. 12.17. Alternatively, a hinge or pin arrangement or elastomeric pads may be provided.

Deck joints   The number of expansion joints in a bridge deck should be kept as low as possible. In single-span bridges of the type under discussion, one such joint usually suffices; that joint is at one end of the bridge. A suitable detail for the joint is shown in Fig. 12.15b; in this detail an angle is anchored to the end of the deck slab, and a similar angle is anchored to the end of the approach slab. A tread plate is fastened, by means of rivets or tap screws,

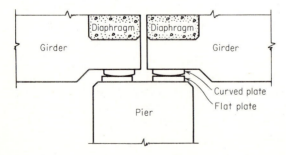

Fig. 12.17   Expansion bearings.

to the top of the angle at the end of the deck, and this plate bears on the angle at the end of the approach slab. The tread plate does not cover the latter angle completely, sufficient provision for expansion of the deck slab being made. The groove between the end of the tread plate and the edge of the approach slab is usually filled with mastic, as shown in Fig. 12.15b, in order to make the joint watertight. The mastic is soft enough to permit movement of the tread plate, but firm enough to resist displacement.

Drainage   Surface water on bridge decks should be disposed of as quickly and as directly as possible. This is accomplished by crowning the roadway about $\frac{1}{8}$ in. per ft and pitching the gutters to drain into inlets. In single-span bridges built on a grade, no special provision for longitudinal drainage is ordinarily necessary. The water is carried by the transverse crown to the curbs or gutters and then to the low end of the bridge, where it can easily be deflected away from the roadway.

  If the bridge is horizontal longitudinally, a longitudinal camber can be built into the bridge deck by elevating the middle portion of the span when the forms are erected. This camber will serve to carry the water in the gutters to either end of the span. A camber of about $\frac{1}{10}$ in. per ft is satisfactory for concrete decks.

DESIGN OF A SLAB BRIDGE

12.10   DATA AND SPECIFICATIONS

A slab bridge similar to that shown in Fig. 12.1 is to be designed according to the following data:

Clear span	15 ft
Clear width	26 ft
Live loading	HS20
Wearing surface	30 psf
Concrete strength f_c'	3000 psi
Grade 40 reinforcement	

By AASHO specifications, an allowable concrete stress of

$$f_c = 0.40 f_c' = 1200 \text{ psi}$$

and an allowable steel stress of 20,000 psi will be used.

  The effective span of the slab is assumed as 15 ft + 1 ft = 16 ft, and a total thickness of 12 in. is selected for trial. The total dead load per square foot is then 180 psf, and the dead-load moment is

$$\tfrac{1}{8} \times 180 \times 16^2 = 5760 \text{ ft-lb}$$

The load on each rear wheel is 16,000 lb.

$$E = 4 + 0.06S = 4 + 0.06 \times 16 = 4.96 \text{ ft}$$

The load on a unit width of slab is therefore

$$16,000/4.96 = 3230 \text{ lb}$$

and the live-load moment is

$$\frac{PS}{4} = \frac{3230 \times 16}{4} = 12,900 \text{ ft-lb}[1]$$

The impact coefficient is

$$I = \frac{50}{l + 125} = \frac{50}{16 + 125} = 0.355$$

which is greater than the specified maximum of 0.300. The latter will therefore be used, giving an impact moment of

$$0.300 \times 12,900 = 3870 \text{ ft-lb}$$

The total moment due to dead, live, and impact effects is 22,530 ft-lb.

Consistent with the AASHO specification, the elastic design equations of Art. 3.14 will be used. For the stresses specified, Eq. (3.42) gives

$$k = \frac{n}{n + r} = \frac{10}{10 + 16.7} = 0.375$$

and from Eq. (2.19), $j = 0.875$. The minimum required depth may then be found from Eq. (2.17).

$$d = \sqrt{\frac{2M}{f_c kjb}} = \sqrt{\frac{2 \times 22,530 \times 12}{1200 \times 0.375 \times 0.866 \times 12}} = 10.7 \text{ in.}$$

With $t = 12$ in., less 1 in. protective covering under the main reinforcement, and less one-half of an assumed bar diameter of 1 in., an effective depth of 10.5 in. is obtained, sufficiently close to the calculated requirement. A 12-in. slab will be used.

The required main reinforcement, from Eq. (2.15), is

$$A_s = \frac{M}{f_s jd} = \frac{22,530 \times 12}{20,000 \times 0.875 \times 10.5} = 1.47 \text{ in.}^2 \text{ per ft}$$

which is furnished by No. 8 bars 6 in. on centers ($A_s = 1.57$ in.² per ft). Fifty per cent of these bars will be stopped 15 diam beyond the point at which they are no longer required for moment, or 8 in. short of the face of supports.

[1] Note that this calculation of live-load moments results in a somewhat lower value than the approximate $900S = 14,400$ ft-lb.

The remainder will be carried a distance $l = A_s f_s / 2\Sigma_0 u = 9$ in. into the support, in accordance with the AASHO specification. As specified in Art. 12.6, the amount of transverse reinforcement required for proper distribution of concentrated loads (also to provide for shrinkage) is

$$\frac{A_s}{\sqrt{S}} = \frac{1.47}{\sqrt{16}} = 0.37 \text{ in.}^2$$

No. 5 bars 10 in. on centers are placed directly on top of the longitudinal reinforcement.

To facilitate screeding off the slab, the required curbs will not be made monolithic. Required edge beams will consist of an edge slab section additionally reinforced. The dead load carried by the edge beam is

$$\frac{22 \times 24}{144} 150 = 550 \text{ plf}$$

and the dead-load moment is

$$\tfrac{1}{8} \times 550 \times 16^2 = 17,600 \text{ ft-lb}$$

The specified live-load moment is

$$0.10 P_{20} S = 0.10 \times 16,000 \times 16 = 25,600 \text{ ft-lb}$$

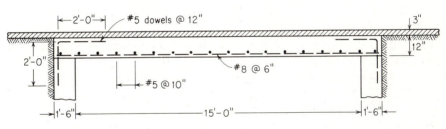

(a) Longitudinal section

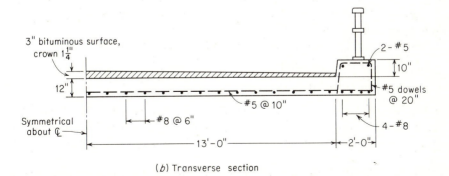

(b) Transverse section

Fig. 12.18 Details of slab bridge.

giving a total moment of 43,200 ft-lb. The resisting moment of the given section is

$$M = \tfrac{1}{2}f_ckjbd^2 = 600 \times 0.375 \times 0.875 \times 24 \times 10.5^2 \times \tfrac{1}{12} = 43,400 \text{ ft-lb}$$

which is just adequate. The steel required is

$$A_s = \frac{43,200 \times 12}{20,000 \times 0.875 \times 10.50} = 2.82 \text{ in.}^2$$

which will be provided by four No. 8 bars.

The crown of the roadway is obtained by varying the thickness of the bituminous wearing surface. No expansion joints are necessary for a span of this length, and a fixed bearing is constructed at each end by placing vertical dowels in the abutment so that they will project up into the concrete deck slab. These are extended horizontally 24 in. to control the cracking that may occur near the edge of the support. The bridge seat is to be coated with a bituminous mixture before the slab is poured. Details are shown in Fig. 12.18.

DESIGN OF A T-BEAM OR DECK-GIRDER BRIDGE

12.11 DATA AND SPECIFICATIONS

A T-beam or deck-girder bridge similar to that shown in Fig. 12.2 is to be designed for the following conditions:

Clear span	48 ft
Clear width	29 ft
Live loading	HS20
Concrete strength f_c	3000 psi
Grade 40 reinforcement	

The design is to meet the AASHO specifications. The bridge will consist of six parallel girders supporting a floor slab, as shown in Fig. 12.19.

12.12 SLAB DESIGN

Since the slab will be poured monolithically with the girders and will be fully continuous, the span will be taken as the clear distance between girders. Under the assumption that the girders will be 14 in. in width, the clear span will be 4 ft 4 in. For a total thickness of slab of 6 in. (including a $\tfrac{3}{4}$-in. wearing surface), with a 15-psf allowance for possible future protective covering, the total dead load is 90 psf. A coefficient of $\tfrac{1}{10}$ for positive and negative dead-load moments will be assumed in the absence of definite specification values. The dead-load positive and negative moments are

$$\tfrac{1}{10} \times 90 \times 4.33^2 = 169 \text{ ft-lb}$$

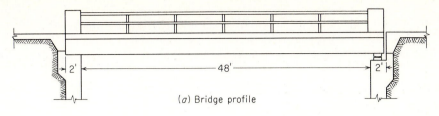

(*a*) Bridge profile

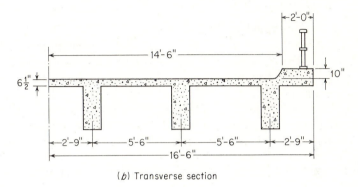

(*b*) Transverse section

Fig. 12.19 Deck-girder bridge.

For live-load moment computations, Case 1 of Art. 12.6 applies:

$$0.80 \frac{S+2}{32} P_{20} = \frac{0.80 \times 6.33 \times 16,000}{32} = 2530 \text{ ft-lb}$$

Since the loaded length is small, the impact coefficient is 0.30, and the impact moment is 760 ft-lb. The total positive and negative moments are 3459 ft-lb. Then

$$d = \sqrt{\frac{2 \times 3459 \times 12}{1200 \times 0.375 \times 0.866 \times 12}} = 4.19 \text{ in.}$$

With the addition of 1 in. of protective concrete below the center of an assumed $\frac{3}{4}$-in. bar, and with the $\frac{3}{4}$-in. wearing surface (not considered structurally effective), the total thickness is 6.32 in. A $6\frac{1}{2}$-in. slab will be used, with an effective depth of 4.37 in. The dead-load estimate need not be revised.

$$A_s = \frac{3459 \times 12}{20,000 \times 0.875 \times 4.37} = 0.54 \text{ in.}^2$$

which is furnished by No. 6 bars 10 in. on centers. To avoid excessive bending of bars, and since short spans are involved, straight bars will be used in the slab, with No. 6 bars at 10 in. top and bottom. This will provide a surplus of steel in some areas, but the cost of the additional steel will probably be offset by the savings in fabrication and handling of the bent bars which would otherwise be required. Temperature and distribution stresses in the perpendicular direction are provided for by placing five No. 5 bars directly on top of the main slab steel in each slab panel. Complete details of the slab are shown in Fig. 12.24.

12.13 INTERIOR GIRDERS

The interior girders are T beams with flange width equal to the center-to-center distance of girders. The required stem dimensions are governed by either the maximum moment or maximum shear. The bridge seats will be assumed 2 ft in width, and the effective span length center to center of bearings taken as 50 ft.

Dead-load moments   The weight of the slab per foot of beam is

$$96 \times 5.5 = 528 \text{ plf}$$

The stem section below the slab is assumed as 14×30 in., which adds weight of 437 plf, making the total 965 plf. The dead-load moment at the center of the span is

$$M_d = \tfrac{1}{8} \times 965 \times 50^2 = 302{,}000 \text{ ft-lb}$$

To determine the points at which some of the horizontal steel may be bent up, it is necessary to compute the moment at some sections between the point of maximum moment and the support. At 10 ft from the support, the dead-load moment is 192,800 ft-lb; at 20 ft, 289,000 ft-lb.

Live-load moments   The absolute maximum live-load moment will occur with an HS20 truck on the bridge in the position[1] shown in Fig. 12.20. With the distribution of loads as specified in Art. 12.7, each interior girder must support $5.50/5.0 = 1.10$ wheel loads per wheel. Therefore the load from

[1] For a discussion of the position of moving loads for absolute maximum moment, see any elementary text in structural theory.

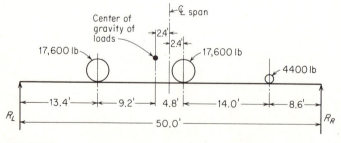

Fig. 12.20

the rear wheel is $1.1 \times 16{,}000 = 17{,}600$ lb, and that from the front wheel is $1.1 \times 4000 = 4{,}400$ lb.

$$R_L = [17{,}600(36.6 + 22.6) + 4400 \times 8.6]\tfrac{1}{50} = 21{,}600 \text{ lb}$$
$$M_{\max} = 21{,}600 \times 27.4 - 17{,}600 \times 14 = 346{,}000 \text{ ft-lb}$$

Ten feet from the left support, the maximum live-load moment occurs with the rear trailer wheels at that point and the front wheels 38 ft from the left support. With this position of the loads,

$$R_L = [17{,}600(40 + 26) + 4400 \times 12]\tfrac{1}{50} = 24{,}400 \text{ lb}$$
$$M_{10} = 24{,}400 \times 10 = 244{,}000 \text{ ft-lb}$$

Similarly,

$$M_{20} = 330{,}000 \text{ ft-lb}$$

Impact moments In computing impact moments for a girder such as this, it is usual to consider the whole span as the loaded length. The impact coefficient is therefore $50/(50 + 125) = 0.285$, and the impact moments are as follows:

$$M_{\max} = 98{,}700 \text{ ft-lb}$$
$$M_{20} = 94{,}100 \text{ ft-lb}$$
$$M_{10} = 69{,}600 \text{ ft-lb}$$

Maximum total moments The sum of the maximum dead load, live load, and impact moments is 746,700 ft-lb. The total moments at the 20- and 10-ft points are 713,100 and 506,400 ft-lb, respectively.

Dead-load shears The maximum dead-load shear at the end of the beam is $965 \times 25 = 24{,}100$ lb. Ten feet from the support, the shear is 14,450 lb; 20 ft from the support, it is 4800 lb.

Live-load shears The absolute maximum shear occurs with the truck on the span in the position shown in Fig. 12.21. The maximum live-load shear is 30,600 lb. The maximum shears at the 10-, 20-, and 25-ft points are 24,300, 16,400, and 12,700 lb, respectively.

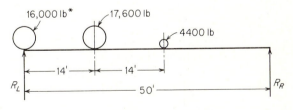

Fig. 12.21

*See Art 12.7 for prescribed distribution of loads for shear computations.

Impact shears   For loaded lengths of 50, 40, 30, and 25 ft the impact shears are as follows:

End shear	8700 lb
10-ft section	6900 lb
20-ft section	4700 lb
Midspan section	3600 lb

Total shears   The total shears are as follows:

End shear	63,400 lb
10-ft section	45,650 lb
20-ft section	25,900 lb
Midspan section	16,300 lb

Determination of cross section and steel area   According to the AASHO specification, the nominal shear stress at service load is to be calculated from the equation $v = V/b'jd$, where b' = web width. In beams with web reinforcement, the nominal shear stress v may not exceed $0.075f'_c$. In sizing the web area for the present design, a maximum shear of $0.06f'_c = 180$ psi will be used. In this case,

$$b'd = \frac{V}{vj} = \frac{63,400}{180 \times 0.875} = 403 \text{ in.}^2$$

If b' is taken as 14 in., the required d is 28.8 in. If three rows of No. 11 bars are assumed, with 2 in. clear between rows and $2\frac{1}{2}$ in. clear below the bottom row to allow for stirrups and concrete protection, a total depth of 35.42 in. is obtained. A 36-in. total depth will be used, resulting in an effective depth of 29.4 in. The depth of the stem below the slab is then $36 - 6\frac{1}{2} = 29\frac{1}{2}$ in. This is sufficiently close to the assumed value so that the dead-load moments need not be revised. The required tensile-steel area is then found:

$$A_s = \frac{M}{f_s(d - t/2)} = \frac{746,700 \times 12}{20,000(29.4 - 5.75/2)} = 16.90 \text{ in.}^2$$

which will be furnished by eleven No. 11 bars.

A check of the maximum stress in the concrete, using Eq. (3.52b), shows a maximum compression of 1190 psi, practically equal to the allowable stress of 1200 psi.

The lower four tensile bars, somewhat more than one-third the area, will be extended into the support. The upper two layers will be cut off 15 diam or one-twentieth of the span beyond the point at which they are no longer required, in accordance with the specification.[1] The critical extension in this

---

[1] Note that the stringent ACI requirements for extra stirrup reinforcement where flexural reinforcement is terminated in a tension zone (see Art. 3.8), surprisingly, have no counterpart in AASHO specifications.

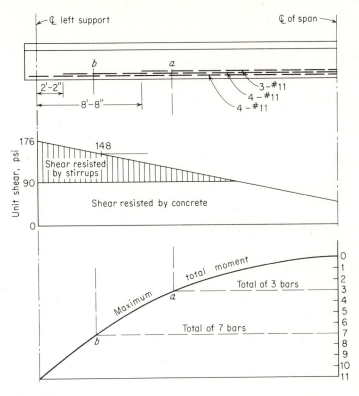

Fig. 12.22

case is one-twentieth of the span, or 30 in. Cutoff points are found graphically as in Fig. 12.22.

In beams designed under the AASHO specification, the nominal bond stress u is to be checked at critical locations by the expression $u = V/\Sigma_0 jd$, where Σ_0 is the sum of bar perimeters at the section. Checking the bond at the center line of supports, where the shear is greatest and the total bar perimeter is least, four No. 11 bars provide a perimeter of 17.7 in., and

$$u = \frac{V}{\Sigma_0 jd} = \frac{63,400}{17.7 \times 0.875 \times 29.4} = 139 \text{ psi}$$

which is well below the value of $4.8 \sqrt{f_c'}/D = 4.8 \sqrt{3000}/1.410 = 187$ psi permitted by specification.

Web reinforcement   The portion of the beam through which web reinforcement is required is determined by computing the unit shears at various points on the beam. This is shown graphically in Fig. 12.22. The concrete resists a unit shear of $0.03f_c' = 90$ psi, and the remainder is taken up by stirrups. The required spacing is to be calculated from the expression

$$s = \frac{A_v f_v jd}{V - V_c} = \frac{A_v f_v}{(v - v_c)b'}$$

No. 5 U stirrups will be used. In the region through which web reinforcement is required, the specification calls for a maximum spacing of $d/2 = 15$ in. The excess shear corresponding to that spacing is

$$v - v_c = \frac{A_v f_v}{sb'} = \frac{0.61 \times 20,000}{15 \times 14} = 58 \text{ psi}$$

and the total shear stress is 148 psi. This is attained 5 ft 3 in. from the support. The maximum spacing in the region in which stirrups are not required to carry shear is $3d/4 = 22\frac{1}{2}$ in. This will apply at a distance of 16 ft 3 in. from the support. At the support, the required spacing is

$$s = \frac{A_v f_v}{(v - v_c)b'} = \frac{0.61 \times 20,000}{86 \times 14} = 10.1 \text{ in.}$$

The first stirrup will be placed $s/2 = 5$ in. from the support center line; this will be followed by six stirrups at 10 in., nine at 15 in., and five at 20 in., filling out the distance to the center of the span.

12.14 EXTERIOR GIRDERS

The exterior girders are identical in cross section with the interior girders, because the raised curb section will be poured separately and cannot be counted on to participate in carrying loads.

Moments   In addition to the 965-plf dead load obtained for the interior girders, the safety curb adds 250 plf, producing a total dead load of 1215 plf and a maximum dead-load moment at the center of the span of 380,000 ft-lb. A portion of each wheel load which rests on the exterior slab panel is supported by the exterior girder. That portion is obtained by placing the wheels as close to the curb as the clearance diagram will permit and treating the exterior slab panel as a simple beam. The position is shown in Fig. 12.23, and the proportion

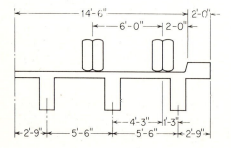

Fig. 12.23

of the load is $4.25/5.50 = 0.773$. The longitudinal position of the load which will produce the absolute maximum bending moment is the same as for the interior girders. The absolute maximum live-load moment can therefore be obtained by direct proportion:

$$M_{max} = (0.773/1.10)(346,000) = 243,000 \text{ ft-lb}$$

The impact moment is $0.285 \times 243,000 = 69,300$ ft-lb, and the total maximum moment is 692,300 ft-lb.

Shears   The maximum dead-load shear is

$$V_D = 1215 \times 25 = 30,400 \text{ lb}$$

The maximum live-load shear is proportional to the maximum live-load shear in the interior girders:

$$V_L = (0.773/1.10)(30,600) = 21,500 \text{ lb}$$

The impact shear at the support is 6100 lb. The total shear at the support is then 58,000 lb. Shears at other points in the span are found similarly.

Determination of cross section and reinforcement   Once the shears and moments due to dead and live loads and impact are obtained, the design of the exterior girders would follow along the lines of that for the interior girders. The specification stipulates that in no case shall an exterior stringer have less carrying capacity than an interior stringer. This controls in the present design, and the exterior girders will duplicate the interior girders.

12.15   MISCELLANEOUS DETAILS

Diaphragms   A transverse diaphragm (see Figs. 12.13 and 12.15) will be built between the girders at either end of the bridge. The chief function of these diaphragms is to furnish lateral support to the girders; with some abutment details, the diaphragms also serve to prevent the backfill from spilling out onto the bridge seats. A similar diaphragm will be built between the girders at midspan. Such intermediate diaphragms are required for all spans in excess of 40 ft and serve to ensure that all girders act together in resisting the loads.

Fixed bearing   A fixed bearing similar to that in Fig. 12.13 is provided at one end. Vertical dowels are placed in the breast wall of the abutment and are bent so as to project into the longitudinal beams or deck slab. Horizontal dowels are embedded in the deck and approach slabs. The diaphragm rests directly on the bridge seat.

Expansion bearing   An expansion bearing is provided at one end of the span. Its details are similar to Fig. 12.15*b*. The bottom of the diaphragm is flush with the bottoms of the beams and is not in contact with the bridge seat.

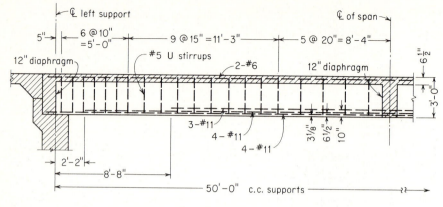

(*a*) Details of interior girder

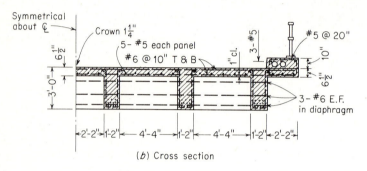

(*b*) Cross section

Fig. 12.24   Details of T-beam bridge.

Waterproofing and drainage   All joints will be filled with a mastic compound to prevent water from seeping through the joints. Removal of surface water will be accomplished by crowning the roadway $1\frac{1}{4}$ in. and pitching the gutters toward the ends by providing a camber of $2\frac{1}{2}$ in. at the center. Besides facilitating drainage, this camber serves to prevent the appearance of sag that would be evident if the girders were at the same level throughout the span.

Full details are shown in Fig. 12.24.

COMPOSITE-BRIDGE CONSTRUCTION

12.16   GENERAL DESCRIPTION

In addition to girder bridges constructed entirely of reinforced concrete, frequently bridges are constructed of longitudinal steel girders with reinforced-concrete decks. If the deck merely rests on the girders, it does not represent an integral part of the transverse cross section. That is, since no provision is made to transfer longitudinal shears from the girder to the deck, the latter does not

Fig. 12.25   Continuous bridge of composite construction.

assist in carrying longitudinal bending moments. In contrast, steel-concrete composite construction (Fig. 12.3), as used in many present-day bridges, consists essentially of three elements: (1) longitudinal steel girders, (2) a reinforced-concrete deck, and (3) devices called *shear connectors* which are welded to the top flanges of the beams and which protrude into the slab to tie slab and beam together and to force them to act as a unit in resisting girder moments and shears (Ref. 12.3). Figure 12.25 shows a typical example of a completed bridge of this type. Simple spans up to about 125 ft and continuous spans much longer have been built in this manner.[1]

The principal advantage of this type of construction is that it utilizes the concrete deck, which in normal construction does not contribute to longitudinal strength, to assist in resisting longitudinal bending. The resulting composite section is usually shallower and lighter—but nonetheless stiffer—than a comparable section of ordinary concrete-slab–steel-stringer construction in which there is no provision for composite action.

The dimensions of the concrete slab in composite construction are usually determined by the transverse flexural requirements. Principal reinforcement is in the direction transverse to the axis of the bridge; in the longitudinal direction, only distribution reinforcement is used. The slab design is carried out in the same manner as for noncomposite construction.

The steel beams, spanning longitudinally between abutments or piers, may be rolled sections, rolled sections with cover plates, or built-up plate girders.

[1] Composite construction is used increasingly for buildings as well as for bridges. See Ref. 12.4.

Unsymmetrical steel sections, such as a rolled beam with a cover plate on the lower flange only, can be used to advantage. For longer spans, variable-depth sections are often used. The design may be continuous over several supports.

12.17 DESIGN OF COMPOSITE SECTIONS

The steel stringers alone must support their own dead weight, as well as the dead weight of the formwork and fresh concrete of the deck, unless intermediate shoring is used. The usual principles of steel design apply to the design of the stringers to resist noncomposite dead loads.

A portion of the dead load and all the live load are resisted by composite action. The resisting moment of the composite section is usually computed from the properties of the transformed section obtained by substituting an equivalent amount of steel for the concrete slab.

In transforming the concrete slab into equivalent steel, the depth of the transformed area is held constant, but the width is reduced to $1/n$ times its actual value, where $n = E_s/E_c$. The value of n to be used with concrete of given compressive strength may be taken from the following tabulation:

f_c'	n
2000 to 2400 psi	15
2500 to 2900 psi	12
3000 to 3900 psi	10
4000 to 4900 psi	8
5000 or more psi	6

If loads are long-term (such as the composite portion of the dead load), the effects of creep and plastic flow are approximated by reducing E_c to one-third its normal value. Thus, when stresses due to long-term loads are dealt with, the substituted steel area is only one-third what it would otherwise have been.

AASHO recommends that the effective width of the compression-concrete flange be taken not larger than (1) one-fourth the span length of the beam, (2) the distance center to center of beams, or (3) 12 times the least thickness of the slab. For beams having a flange on one side only, the effective flange width shall not exceed one-twelfth the span length of the beam, nor 6 times the thickness of the slab, nor one-half the distance center to center of the next beam.

In composite design, the section must be selected largely by trial. Although some tables are available giving the section properties for certain combinations of slab and steel beam (Ref. 12.3), and although methods have been developed to reduce the labor of finding an appropriate section (Ref.

12.5), it is not practical to tabulate all desirable combinations of slab width and thickness, steel-beam section, and cover-plate size. Direct design of composite sections is therefore not possible, and the design process is actually a review of an assumed section to determine if the stresses in the concrete and steel are within the allowable range. Stresses due to each type of loading are computed using the appropriate section properties, and these stresses are added algebraically to obtain final values.

12.18 SHEAR CONNECTORS

Shear connectors are essential to the development of composite action. They must transfer the horizontal shear, with extremely small deformation, so that the structure behaves as a unit. They must also be capable of resisting any tendency for the slab to separate vertically from the steel flange due to buckling or other causes.

The two types of shear connector most widely used in U.S. practice are shown in Fig. 12.26. Short pieces of steel channel may be used, as in Fig. 12.26a, set transverse to the beam axis and welded to the beam along the heel and toe of the channel. According to specification, welds must be at least $\frac{3}{16}$-in. fillets. Stud shear connectors (Fig. 12.26b) are commonly available in diameters of $\frac{5}{8}$, $\frac{3}{4}$, and $\frac{7}{8}$ in. They are granular-flux-filled, and are resistance-welded to the steel beam with special equipment. According to AASHO specification, the clear depth of concrete cover over the tops of all shear connectors must be not less than 2 in. The connectors must penetrate at least 2 in. above the bottom of the slab, and the clear distance between the edge of the girder flange and the edge of the connector must be not less than 1 in.

Connectors of both types are most economically installed in the fabricating shop. However, the advantages of this procedure may be offset by the possibility of damage during shipment and the difficulty in working along the top of the beams during erection.

According to AASHO specification, shear connectors must be designed for fatigue (Ref. 12.6), and then the number of connectors obtained checked to ensure that sufficient strength has been provided to develop the capacity of the member.

a. Design for fatigue The horizontal shear per unit length along the interface between concrete slab and steel beam may be found by the usual equation of

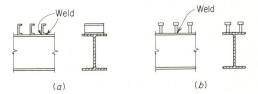

Fig. 12.26 Typical shear connectors.   (a)   (b)

structural mechanics:

$$(a) \quad S = \frac{VQ}{I}$$

where S = horizontal shear, lb/in.

V = external shear force at section, lb

Q = static moment of transformed compression-concrete area about neutral axis of entire section, in.[3]

I = moment of inertia of entire transformed section, in.[4]

In designing for fatigue, it is the variation of horizontal shear, as live load and impact load are applied, that is significant. Consequently, Eq. (a) is restated in terms of the change in horizontal shear that occurs as live and impact loads are applied:

$$S_r = \frac{V_r Q}{I} \tag{12.6}$$

where S_r = range of horizontal shear, lb/in.

V_r = range of external shear force at section, due to live and impact loads, lb

and other terms are as already defined. The range of shear may be taken as the difference in the maximum and minimum shear envelopes, excluding dead loads.

The allowable range of horizontal shear, Z_r, per individual connector, has been established by extensive fatigue testing (Ref. 12.6). The following values are accepted by AASHO:

For channels:

$$Z_r = Bw \tag{12.7}$$

where Z_r = allowable range of horizontal shear per channel, lb

w = length of channel shear connector measured in a transverse direction on flange of girder, in.

B = 4000 for 100,000 cycles; 3000 for 500,000 cycles; 2400 for 2,000,000 cycles

For welded studs:

$$Z_r = \alpha d^2 \tag{12.8}$$

where Z_r = allowable range of horizontal shear per stud, lb

d = diameter of stud, in. (height H must not be less than $4d$)

α = 13,000 for 100,000 cycles; 10,600 for 500,000 cycles; 7850 for 2,000,000 cycles

The allowable range of horizontal shear of all the connectors at one transverse girder section, ΣZ_r, divided by the pitch p, or spacing of connector groups along the beam, must at least equal the range of horizontal shear, S_r. Thus the maximum pitch is

$$p = \frac{\Sigma Z_r}{S_r} \tag{12.9}$$

According to specification, the pitch is not to exceed 24 in.

b. Design for strength According to specification, the number of connectors provided on the basis of fatigue must be checked to ensure that sufficient connectors have been provided to develop the strength of the composite girders. The number N of shear connectors needed between the point of maximum positive moment and the end supports is equal to

$$N = \frac{P}{\phi S_u} \tag{12.10}$$

where S_u = ultimate strength of one shear connector, lb
ϕ = capacity reduction factor, equal to 0.85
P = force to be developed, lb
The force P is equal to the smaller of the following:

$$P_1 = A_s f_y \tag{12.11a}$$

where A_s = total area of steel section, including cover plates, in.²
f_y = yield strength of steel beam, psi
or

$$P_2 = 0.85 f_c' bc \tag{12.11b}$$

where b = effective width of slab, in.
c = slab thickness, in.
The ultimate strength S_u of the shear connectors has been established by testing, and the following expressions are recommended:

For channels:

$$S_u = 550 \left(h + \frac{t}{2} \right) w \sqrt{f_c'} \tag{12.12}$$

where h = average thickness of channel flange, in.
t = thickness of channel web, in.
w = length of channel measured in a transverse direction on flange of girder, in.

For welded studs:

$$S_u = 930 d^2 \sqrt{f_c'} \tag{12.13}$$

where d is the diameter of the stud, in inches (height H must not be less than $4d$).

DESIGN OF A COMPOSITE BRIDGE

12.19 DATA AND SPECIFICATIONS

A composite bridge similar to that shown in Fig. 12.3 is to be designed for the following conditions:

Span center to center of supports	70 ft
Clear width	28 ft
Live loading	HS20
Wearing surface	30 psf
Concrete strength f_c'	3000 psi
A36 steel beam	
Grade 40 reinforcement	

Welded-stud shear connectors are specified, and a single cover plate will be used on the bottom flange and discontinued where no longer required. The bridge will consist of five parallel girders, as shown in Fig. 12.27. The design is to be based on 2×10^6 cycles of loading.

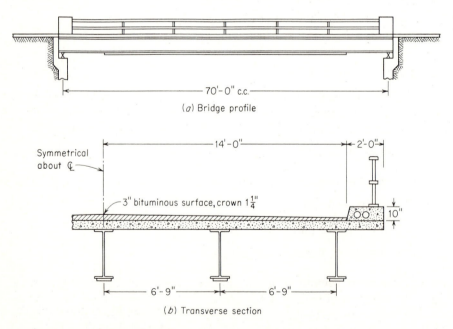

(a) Bridge profile

(b) Transverse section

Fig. 12.27   Composite girder bridge.

12.20 SLAB DESIGN

The design of the slab in the transverse direction follows the procedure presented in Art. 12.12. For the given loads and spans, a slab of $6\frac{1}{2}$-in. total thickness is obtained, reinforced with No. 6 bars 8 in. on centers, top and bottom, in the direction transverse to the supporting beams. In the direction parallel to the girders, six No. 6 bars are equally spaced in each slab panel. Details of the slab are shown in Fig. 12.29.

12.21 INTERIOR GIRDERS

The effective width of flange, in accordance with AASHO, is taken equal to the beam spacing of 81 in.

Dead-load moment on steel beams The noncomposite portion of the dead load includes the weight of the concrete slab, the estimated girder and cover-plate weight, and the estimated weight of transverse diaphragms. The total weight is estimated at 800 plf, and the noncomposite dead-load moment is

$$M_{DS} = \tfrac{1}{8} \times 800 \times 70^2 = 490,000 \text{ ft-lb}$$

Dead-load moment on composite beam The composite dead load includes the weight of bridge rails, safety curbs, and bituminous pavement. These sum up to 300 plf, and the composite dead-load moment is

$$M_{DC} = \tfrac{1}{8} \times 300 \times 70^2 = 184,000 \text{ ft-lb}$$

Live-load moment The live-load distribution factor in this case is

$$6.75/5.5 = 1.23 \text{ wheel loads per wheel}$$

The absolute maximum live-load moment will occur when an HS20 truck is located with one of its inner 16,000-lb wheels 2.35 ft from the span center line, and it will occur under that wheel. The live-load moment at that point is

$$M_{\text{max}} = 494,000 \times 1.23 = 608,000 \text{ ft-lb}$$

Impact moment The impact coefficient in this case is

$$50/(70 + 125) = 0.256$$

and the maximum impact moment is

$$M_{\text{max}} = 608,000 \times 0.256 = 156,000 \text{ ft-lb}$$

Selection of the cross section The selection of the cover-plated cross section is necessarily a matter of trial. A section made up of a 36-in.-wide flange 150-lb beam plus a single 10- $\times$ $1\frac{1}{4}$-in. cover plate on the bottom flange, as shown in Fig. 12.28, will be tried. Section properties are calculated in Table 12.1.

Table 12.1 Properties of section with cover plate

	I_o	A	y_b NA to bottom	Ay_b	d CG to NA	Ad^2
Steel only:						
W36 × 150	9012	44.16	19.17	848	4.07	730
Pl 10 × 1¼		12.50	0.63	8	14.47	2630
	9012	56.66		856		3360
						9012
						$I = 12{,}372$

$$\bar{y} = \frac{856}{56.66} = 15.10 \qquad S_b = \frac{12{,}372}{15.10} = 822$$

$$S_{ts} = \frac{12{,}372}{21.99} = 564$$

	I_o	A	y_b	Ay_b	d	Ad^2
Composite with $n = 30$:						
Steel	12,372	56.66	15.10	856	6.00	2040
Slab	62	17.50	40.34	706	19.24	6480
	12,434	74.16		1562		8,520
						12,434
						$I = 20{,}954$

$$\bar{y} = \frac{1562}{74.16} = 21.10 \qquad S_b = \frac{20{,}954}{21.10} = 990$$

$$S_{ts} = \frac{20{,}954}{15.99} = 1310 \qquad S_{tc} = \frac{20{,}954}{22.49} = 929$$

	I_o	A	y_b	Ay_b	d	Ad^2
Composite with $n = 10$:						
Steel	12,372	56.66	15.10	856	12.20	8450
Slab	186	52.50	40.34	2120	13.04	8940
	12,558	109.16		2976		17,390
						12,558
						$I = 29{,}948$

$$\bar{y} = \frac{2976}{109.16} = 27.30 \qquad S_b = \frac{29{,}948}{27.30} = 1095$$

$$S_{ts} = \frac{29{,}948}{9.79} = 3060 \qquad S_{tc} = \frac{29{,}948}{16.29} = 1830$$

The stresses due to various loads are shown in Table 12.2. It is seen that the bottom fiber is stressed to about 90 per cent of its allowable stress of 20,000 psi, while the top fiber is understressed by about 25 per cent. Since the lightest 36-in. section has been selected, the only alternative would be to use a 33-in. rolled section, which would require a heavier cover plate and which would save little if any steel. Consequently, the 36-in.-wide-flange 150-lb beam with

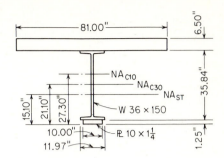

Fig. 12.28 Girder cross section.

cover plate $10 \times 1\frac{1}{4}$ in. will be used. The concrete fiber stress is well within the allowable limit.

The cover plate on the lower flange will be discontinued where it is no longer needed to resist stress. Section properties for the beam without cover plate are computed as shown in Table 12.3. Once the section moduli of the beam without cover plate have been found, the point at which the cover plate can be discontinued without exceeding allowable stresses in the remaining section is found by trial to be 12 ft 9 in. from the support center line. In accordance with specifications, the plate will be continued past that point $1\frac{1}{2}$ times the cover-plate width, or 15 in. in the present case. It will be welded across its end.

The shear connectors are designed for the stress change resulting from application of live and impact load (the minimum shear when dead loads are excluded being zero for the simple span). At the left support, maximum live-load shear is obtained with one 16,000-lb wheel at the support and 16,000- and 4000-lb wheels at 14 and 28 ft from the support, respectively. After applying the live-load distribution factor of 1.23 and the impact coefficient of 0.256, maximum shear $V_{l+i} = 48,200$ lb at this location. Equation (12.6) will be used to calculate the variation of horizontal shear there. Section properties applicable at this location are those of the section without cover plate, with slab transformed on the basis of $n = 10$.

$$S_r = \frac{48,200 \times 52.50 \times 9.69}{19,938} = 1230 \text{ lb per in.}$$

For the present design, $\frac{3}{4}$-in. studs are selected, and tentatively, three studs

Table 12.2 Total stresses

Load	Moment	f_{sb}	f_{st}	$3nf_c$ or nf_c	f_c
M_{DS}	490,000	7170	10,400		
M_{DC}	184,000	2230	1,690	2380	80
M_{LL+I}	764,000	8370	2,990	5000	500
Total		17,770	15,080		580

Table 12.3   Properties of section without cover plate

	I_o	A	y_b NA to bottom	Ay_b	d CG to NA	Ad^2
Steel only			$S_b = 503$		$S_t = 503$	
Composite with $n = 30$:						
Steel	9012	44.16	17.92	791	5.98	1580
Slab	62	17.50	39.09	685	15.19	4040
	9074	61.66		1476		5620
						9074
						$I = 14,694$

$$\bar{y} = \frac{1476}{61.66} = 23.90 \qquad S_b = \frac{14,694}{23.90} = 616$$

$$S_{ts} = \frac{14,694}{11.94} = 1230 \qquad S_{tc} = \frac{14,694}{18.44} = 798$$

	I_o	A	y_b NA to bottom	Ay_b	d CG to NA	Ad^2
Composite with $n = 10$:						
Steel	9012	44.16	17.92	791	11.48	5820
Slab	186	52.50	30.09	2050	9.69	4920
	9198	96.66		2841		10,740
						9,198
						$I = 19,938$

$$\bar{y} = \frac{2841}{96.66} = 29.40 \qquad S_b = \frac{19,938}{29.40} = 678$$

$$S_{ts} = \frac{19,938}{6.44} = 3090 \qquad S_{tc} = \frac{19,938}{12.94} = 1540$$

per transverse row are used, with studs 4 in. high. The allowable shear range for the group is found using Eq. (12.8), with $\alpha = 7850$:

$$\Sigma Z_r = 3 \times 7850 \times 0.75^2 = 13,260 \text{ lb}$$

From Eq. (12.9), the required pitch at this location is

$$p = 13,260/1230 = 10.8 \text{ in.}$$

Similar calculations at 7, 14, 21, 28, and 35 ft, using appropriate section properties at each location, indicate required spacings at those points of 12.2, 15.7, 18.5, 22.4, and 28.5 in., respectively. The maximum spacing permitted by AASHO is 24 in. The stud spacing pattern selected, starting at the left support, is

$$
\begin{aligned}
13 \text{ spaces at } 10 \text{ in.} &= 10 \text{ ft } 10 \text{ in.} \\
12 \text{ spaces at } 15 \text{ in.} &= 15 \text{ ft } \ 0 \text{ in.} \\
5 \text{ spaces at } 22 \text{ in.} &= \ 9 \text{ ft } \ 2 \text{ in.}
\end{aligned}
$$

A strength check must now be made to ensure that a sufficient number of connectors has been provided between the maximum-moment section and the support to develop the horizontal stress resultant for the composite beam. From Eqs. (12.11a) and (12.11b),

$$P_1 = 56.66 \times 36,000 = 2,040,000 \text{ lb}$$
$$P_2 = 0.85 \times 3000 \times 81 \times 6.50 = 1,340,000 \text{ lb}$$

P_2 controls here, being the smaller of the two values. The ultimate strength of one stud connector, from Eq. (12.13), is

$$S_u = 930 \times 0.75^2 \sqrt{3000} = 28,700 \text{ lb}$$

and accordingly, from Eq. (12.10), the required number of connectors to develop the force in the concrete slab is

$$N = \frac{1,340,000}{0.85 \times 28,700} = 55$$

A total of 31 transverse groups of three studs each has been selected as a result of the preceding design, based on fatigue. The resulting total of 93 studs is more than sufficient for strength.

The spacing of the fillet welds joining the cover plate to the lower flange can likewise be increased as the shear-transfer requirement becomes less. The design method follows that for spacing of the welded studs.

12.22 EXTERIOR GIRDERS

The design of the two exterior girders does not differ materially from that for the interior girders, and details will not be presented here. Dead and live loads will differ somewhat; transverse positioning of live load for maximum effect will follow the procedures outlined for the design of exterior girders for the deck-girder bridge (Art. 12.14).

12.23 MISCELLANEOUS DETAILS

Diaphragms Transverse diaphragms will be provided at the center line of supports and at the third points of the span. These will consist of C18 $\times$ 42.7 members, field-bolted to vertical ribs welded to the girder webs, as shown in Fig. 12.29. In addition to providing safety against overturning of the girders such as might be caused by lateral forces on the roadway, these transverse beams provide some distribution of vertical loads between girders, helping to ensure that the bridge acts as an integral unit.

Bearings A hinged bearing will be provided at one end of each girder, and a rocker bearing at the other, to provide for expansion and contraction of the

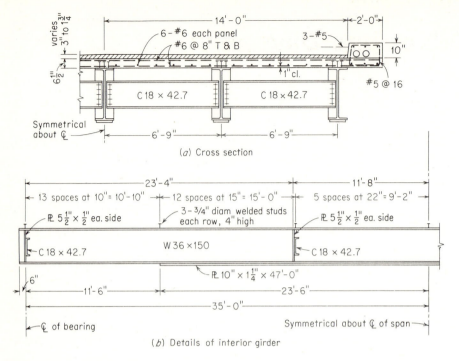

Fig. 12.29 Details of composite bridge.

70-ft span. A joint detail similar to that shown in Fig. 12.15b will be used between the bridge decks and the approach slab.

Drainage The bituminous surface will be crowned to provide lateral drainage, with a depth of 3 in. at the bridge center line and a depth of $1\frac{3}{4}$ in. at the curbs. A longitudinal camber of 3 in. will ensure that no water stands on the bridge.

Full details of the composite bridge are shown in Fig. 12.29.

REFERENCES

12.1. Standard Specifications for Highway Bridges, 10th ed., American Association of State Highway Officials, Washington, D.C., 1969.
12.2. Interim Specifications, American Association of State Highway Officials, Committee on Bridges and Structures, Washington, D.C., 1970.
12.3. I. M. Viest, R. S. Fountain, and R. C. Singleton: "Composite Construction in Steel and Concrete," McGraw-Hill Book Company, New York, 1958.
12.4. ACI-ASCE Committee 333: Tentative Recommendations for Design of Composite Beams and Girders for Buildings, *J. ACI*, December, 1960, p. 623.
12.5. R. S. Fountain and I. M. Viest: Selection of the Cross Section for a Composite T Beam, *J. Struct. Div. ASCE*, pp. 1313-1 to 1313-29, July, 1957.
12.6. R. G. Slutter and J. W. Fisher: Fatigue Strength of Shear Connectors, Highway Res. Board, *Highway Res. Record 147*, Washington, D.C., 1966.

Appendix

Tables and Diagrams

Table 1   Designations, areas, perimeters, and weights of standard bars

Bar No.*	Diameter, in.	Cross-sectional area, in.^{2}	Perimeter, in.	Unit weight per foot, lb
2	$\frac{1}{4} = 0.250$	0.05	0.79	0.167
3	$\frac{3}{8} = 0.375$	0.11	1.18	0.376
4	$\frac{1}{2} = 0.500$	0.20	1.57	0.668
5	$\frac{5}{8} = 0.625$	0.31	1.96	1.043
6	$\frac{3}{4} = 0.750$	0.44	2.36	1.502
7	$\frac{7}{8} = 0.875$	0.60	2.75	2.044
8	$1 = 1.000$	0.79	2.14	2.670
9	$1\frac{1}{8} = 1.128†$	1.00	3.54	3.400
10	$1\frac{1}{4} = 1.270†$	1.27	3.99	4.303
11	$1\frac{3}{8} = 1.410†$	1.56	4.43	5.313
14	$1\frac{3}{4} = 1.693†$	2.25	5.32	7.650
18	$2\frac{1}{4} = 2.257†$	4.00	7.09	13.600

* Based on the number of eighths of an inch included in the nominal diameter of the bars. The nominal diameter of a deformed bar is equivalent to the diameter of a plain bar having the same weight per foot as the deformed bar. Bar No. 2 is available in plain rounds only. All others are available in deformed rounds.

† Approximate to the nearest $\frac{1}{8}$ in.

Table 2 Areas of groups of standard bars, in square inches

Bar No.	\(Number of bars\)												
	2	3	4	5	6	7	8	9	10	11	12	13	14
4	0.39	0.58	0.78	0.98	1.18	1.37	1.57	1.77	1.96	2.16	2.36	2.55	2.75
5	0.61	0.91	1.23	1.53	1.84	2.15	2.45	2.76	3.07	3.37	3.68	3.99	4.30
6	0.88	1.32	1.77	2.21	2.65	3.09	3.53	3.98	4.42	4.86	5.30	5.74	6.19
7	1.20	1.80	2.41	3.01	3.61	4.21	4.81	5.41	6.01	6.61	7.22	7.82	8.42
8	1.57	2.35	3.14	3.93	4.71	5.50	6.28	7.07	7.85	8.64	9.43	10.21	11.00
9	2.00	3.00	4.00	5.00	6.00	7.00	8.00	9.00	10.00	11.00	12.00	13.00	14.00
10	2.53	3.79	5.06	6.33	7.59	8.86	10.12	11.39	12.66	13.92	15.19	16.45	17.72
11	3.12	4.68	6.25	7.81	9.37	10.94	12.50	14.06	15.62	17.19	18.75	20.31	21.87
14	4.50	6.75	9.00	11.25	13.50	15.75	18.00	20.25	22.50	24.75	27.00	29.25	31.50
18	8.00	12.00	16.00	20.00	24.00	28.00	32.00	36.00	40.00	44.00	48.00	52.00	56.00

Table 3 Perimeters of groups of standard bars, in inches

Bar No.	\(Number of bars\)												
	2	3	4	5	6	7	8	9	10	11	12	13	14
4	3.1	4.7	6.2	7.8	9.4	11.0	12.6	14.1	15.7	17.3	18.8	20.4	22.0
5	3.9	5.9	7.8	9.8	11.8	13.7	15.7	17.7	19.5	21.6	23.6	25.5	27.5
6	4.7	7.1	9.4	11.8	14.1	16.5	18.8	21.2	23.6	25.9	28.3	30.6	33.0
7	5.5	8.2	11.0	12.7	16.5	19.2	22.0	24.7	27.5	30.2	33.0	35.7	38.5
8	6.3	9.4	12.6	15.7	18.9	22.0	25.1	28.3	31.4	34.6	37.7	40.9	44.0
9	7.1	10.6	14.2	17.7	21.3	24.8	28.4	31.9	35.4	39.0	42.5	46.0	49.6
10	8.0	12.0	16.0	20.0	23.9	27.9	31.9	35.9	39.9	43.9	47.9	51.9	55.9
11	8.9	13.3	17.7	22.2	26.6	31.0	35.4	39.9	44.3	48.7	53.2	57.6	62.0
14	10.6	16.0	21.3	26.6	31.9	37.2	42.6	47.9	53.2	58.5	63.8	69.2	74.5
18	14.2	21.3	28.4	35.5	42.5	49.6	56.7	63.8	70.9	78.0	85.1	93.2	100.3

Table 4 Areas of bars in slabs, in square inches per foot

Spacing, in.	\(Bar No.\)								
	3	4	5	6	7	8	9	10	11
3	0.44	0.78	1.23	1.77	2.40	3.14	4.00	5.06	6.25
3½	0.38	0.67	1.05	1.51	2.06	2.69	3.43	4.34	5.36
4	0.33	0.59	0.92	1.32	1.80	2.36	3.00	3.80	4.68
4½	0.29	0.52	0.82	1.18	1.60	2.09	2.67	3.37	4.17
5	0.26	0.47	0.74	1.06	1.44	1.88	2.40	3.04	3.75
5½	0.24	0.43	0.67	0.96	1.31	1.71	2.18	2.76	3.41
6	0.22	0.39	0.61	0.88	1.20	1.57	2.00	2.53	3.12
6½	0.20	0.36	0.57	0.82	1.11	1.45	1.85	2.34	2.89
7	0.19	0.34	0.53	0.76	1.03	1.35	1.71	2.17	2.68
7½	0.18	0.31	0.49	0.71	0.96	1.26	1.60	2.02	2.50
8	0.17	0.29	0.46	0.66	0.90	1.18	1.50	1.89	2.34
9	0.15	0.26	0.41	0.59	0.80	1.05	1.33	1.69	2.08
10	0.13	0.24	0.37	0.53	0.72	0.94	1.20	1.52	1.87
12	0.11	0.20	0.31	0.44	0.60	0.78	1.00	1.27	1.56

Table 5 Design of rectangular beams and slabs

$k = n/(n + r); j = 1 - k/3; \rho_e = n/2r(n + r); K = \frac{1}{2}f_c kj$ or $\rho f_s j$

n and f_c'	f_s	f_c	k	j	ρ_e	K
	18,000	1000	0.357	0.881	0.0099	157
		1125	0.385	0.872	0.0120	189
10	20,000	1000	0.333	0.889	0.0083	148
(2500)		1125	0.360	0.880	0.0101	178
	24,000	1000	0.294	0.902	0.0061	133
		1125	0.319	0.894	0.0075	160
	18,000	1200	0.375	0.875	0.0125	197
		1350	0.403	0.866	0.0151	235
9	20,000	1200	0.351	0.883	0.0105	186
(3000)		1350	0.377	0.874	0.0128	223
	24,000	1200	0.310	0.897	0.0078	167
		1350	0.336	0.888	0.0095	201
	18,000	1600	0.416	0.861	0.0185	286
		1800	0.444	0.852	0.0222	341
8	20,000	1600	0.390	0.870	0.0156	272
(4000)		1800	0.419	0.860	0.0188	324
	24,000	1600	0.345	0.884	0.0116	246
		1800	0.375	0.875	0.0141	295
	18,000	2000	0.438	0.854	0.0243	374
		2250	0.467	0.844	0.0292	443
7	20,000	2000	0.412	0.863	0.0206	355
(5000)		2250	0.441	0.853	0.0248	423
	24,000	2000	0.368	0.877	0.0154	323
		2250	0.396	0.868	0.0186	387

Table 6   Review of rectangular beams and slabs

$k = \sqrt{2\rho n + (\rho n)^2} - \rho n; \ j = 1 - \frac{1}{3}k$

	n = 7		n = 8		n = 9		n = 10	
ρ	k	j	k	j	k	j	k	j
0.0010	0.112	0.963	0.119	0.960	0.125	0.958	0.132	0.956
0.0020	0.154	0.949	0.164	0.945	0.173	0.942	0.180	0.940
0.0030	0.185	0.938	0.196	0.935	0.207	0.931	0.217	0.928
0.0040	0.210	0.930	0.223	0.926	0.235	0.922	0.246	0.918
0.0050	0.232	0.923	0.246	0.918	0.258	0.914	0.270	0.910
0.0054	0.240	0.920	0.254	0.915	0.267	0.911	0.279	0.907
0.0058	0.247	0.918	0.262	0.913	0.275	0.908	0.287	0.904
0.0062	0.254	0.915	0.269	0.910	0.283	0.906	0.296	0.901
0.0066	0.261	0.913	0.276	0.908	0.290	0.903	0.303	0.899
0.0070	0.268	0.911	0.283	0.906	0.298	0.901	0.311	0.896
0.0072	0.271	0.910	0.287	0.904	0.301	0.900	0.314	0.895
0.0074	0.274	0.909	0.290	0.903	0.304	0.899	0.318	0.894
0.0076	0.277	0.908	0.293	0.902	0.308	0.897	0.321	0.893
0.0078	0.280	0.907	0.296	0.901	0.311	0.896	0.325	0.892
0.0080	0.283	0.906	0.299	0.900	0.314	0.895	0.328	0.891
0.0082	0.286	0.905	0.303	0.899	0.317	0.894	0.331	0.890
0.0084	0.289	0.904	0.306	0.898	0.321	0.893	0.334	0.889
0.0086	0.292	0.903	0.308	0.897	0.324	0.892	0.338	0.887
0.0088	0.295	0.902	0.311	0.896	0.327	0.891	0.341	0.886
0.0090	0.298	0.901	0.314	0.895	0.330	0.890	0.344	0.885
0.0092	0.300	0.900	0.317	0.894	0.332	0.889	0.347	0.884
0.0094	0.303	0.899	0.320	0.893	0.335	0.888	0.350	0.883
0.0096	0.306	0.898	0.323	0.892	0.338	0.887	0.353	0.882
0.0098	0.308	0.897	0.325	0.892	0.341	0.886	0.355	0.882
0.0100	0.311	0.896	0.328	0.891	0.344	0.885	0.358	0.881
0.0104	0.316	0.895	0.333	0.889	0.349	0.884	0.364	0.879
0.0108	0.321	0.893	0.338	0.887	0.354	0.882	0.369	0.877
0.0112	0.325	0.892	0.343	0.886	0.359	0.880	0.374	0.875
0.0116	0.330	0.890	0.348	0.884	0.364	0.879	0.379	0.874
0.0120	0.334	0.889	0.353	0.882	0.369	0.877	0.384	0.872
0.0124	0.339	0.887	0.357	0.881	0.374	0.875	0.389	0.870
0.0128	0.343	0.886	0.362	0.879	0.378	0.874	0.394	0.867
0.0132	0.347	0.884	0.366	0.878	0.383	0.872	0.398	0.867
0.0136	0.351	0.883	0.370	0.877	0.387	0.871	0.403	0.866
0.0140	0.355	0.882	0.374	0.875	0.392	0.869	0.407	0.864
0.0144	0.359	0.880	0.378	0.874	0.396	0.868	0.412	0.863
0.0148	0.363	0.879	0.382	0.873	0.400	0.867	0.416	0.861
0.0152	0.367	0.878	0.386	0.871	0.404	0.865	0.420	0.860
0.0156	0.371	0.876	0.390	0.870	0.408	0.864	0.424	0.859
0.0160	0.374	0.875	0.394	0.869	0.412	0.863	0.428	0.857
0.0170	0.383	0.872	0.403	0.867	0.421	0.860	0.437	0.854
0.0180	0.392	0.869	0.412	0.863	0.430	0.857	0.446	0.851
0.0190	0.400	0.867	0.420	0.860	0.438	0.854	0.455	0.848
0.0200	0.407	0.864	0.428	0.857	0.446	0.851	0.463	0.846

Table 7a   Resisting moment for slab sections 12 in. wide, in foot-kips
$f_y = 40$ ksi

f_c'	ρ						*Effective depth d, in.*							
		3.0	3.5	4.0	4.5	5.0	5.5	6.0	6.5	7.0	8.0	9.0	10.0	12.0
	.002	0.6	0.9	1.1	1.4	1.8	2.1	2.6	3.0	3.5	4.5	5.7	7.1	10.2
	.003	0.9	1.3	1.7	2.1	2.6	3.2	3.8	4.5	5.2	6.7	8.5	10.5	15.2
	.004	1.3	1.7	2.2	2.8	3.5	4.2	5.0	5.9	6.8	8.9	11.3	13.9	20.1
3000	.005	1.6	2.1	2.8	3.5	4.3	5.2	6.2	7.3	8.5	11.1	14.0	17.3	24.9
	.006	1.9	2.5	3.3	4.2	5.1	6.2	7.4	8.7	10.1	13.2	16.7	20.6	29.6
	.007	2.1	2.9	3.8	4.8	6.0	7.2	8.6	10.1	11.7	15.2	19.3	23.8	34.3
	.008	2.4	3.3	4.3	5.5	6.7	8.2	9.7	11.4	13.2	17.3	21.9	27.0	38.9
	.009	2.7	3.7	4.8	6.1	7.5	9.1	10.8	12.7	14.8	19.3	24.4	30.1	43.4
	.010	3.0	4.1	5.3	6.7	8.3	10.0	11.9	14.0	16.3	21.2	26.9	33.2	47.8
	.011	3.3	4.4	5.8	7.3	9.0	10.9	13.0	15.3	17.7	23.2	29.3	36.2	52.1
	.002	0.6	0.9	1.1	1.4	1.8	2.2	2.6	3.0	3.5	4.6	5.8	7.1	10.2
	.003	1.0	1.3	1.7	2.1	2.7	3.2	3.8	4.5	5.2	6.8	8.6	10.6	15.3
	.004	1.3	1.7	2.2	2.8	3.5	4.3	5.1	5.9	6.9	9.0	11.4	14.1	20.2
	.005	1.6	2.1	2.8	3.5	4.4	5.3	6.3	7.4	8.6	11.2	14.1	17.5	25.2
	.006	1.9	2.6	3.3	4.2	5.2	6.3	7.5	8.8	10.2	13.3	16.9	20.8	30.0
	.007	2.2	3.0	3.9	4.9	6.0	7.3	8.7	10.2	11.8	15.5	19.6	24.2	34.8
4000	.008	2.5	3.4	4.4	5.6	6.9	8.3	9.9	11.6	13.4	17.6	22.2	27.4	39.5
	.009	2.8	3.8	4.9	6.2	7.7	9.3	11.0	13.0	15.0	19.6	24.9	30.7	44.2
	.010	3.0	4.1	5.4	6.9	8.5	10.2	12.2	14.3	16.6	21.7	27.4	33.9	48.8
	.011	3.3	4.5	5.9	7.5	9.3	11.2	13.3	15.6	18.1	23.7	30.0	37.0	53.3
	.012	3.6	4.9	6.4	8.1	10.0	12.1	14.5	17.0	19.7	25.7	32.5	40.1	57.8
	.013	3.9	5.3	6.9	8.8	10.8	13.1	15.6	18.3	21.2	27.7	35.0	43.2	62.2
	.014	4.2	5.7	7.4	9.4	11.6	14.0	16.6	19.5	22.7	29.6	37.5	46.2	66.6
	.015	4.4	6.0	7.9	10.0	12.3	14.9	17.7	20.8	24.1	31.5	39.9	49.2	70.9
	.002	0.6	0.9	1.1	1.4	1.8	2.2	2.6	3.0	3.5	4.6	5.8	7.1	10.3
	.003	1.0	1.3	1.7	2.2	2.7	3.2	3.8	4.5	5.2	6.8	8.6	10.6	15.3
	.004	1.3	1.7	2.3	2.9	3.5	4.3	5.1	6.0	6.9	9.0	11.4	14.1	20.3
	.005	1.6	2.2	2.8	3.6	4.4	5.3	6.3	7.4	8.6	11.2	14.2	17.6	25.3
	.006	1.9	2.6	3.4	4.3	5.2	6.3	7.6	8.9	10.3	13.4	17.0	21.0	30.2
	.007	2.2	3.0	3.9	4.9	6.1	7.4	8.8	10.3	11.9	15.6	19.7	24.4	35.1
	.008	2.5	3.4	4.4	5.6	6.9	8.4	10.0	11.7	13.6	17.7	22.4	27.7	39.9
5000	.009	2.8	3.8	5.0	6.3	7.8	9.4	11.2	13.1	15.2	19.9	25.1	31.0	44.7
	.010	3.1	4.2	5.5	6.9	8.6	10.4	12.3	14.5	16.8	22.0	27.8	34.3	49.4
	.011	3.4	4.6	6.0	7.6	9.4	11.4	13.5	15.9	18.4	24.0	30.4	37.5	54.1
	.012	3.7	5.0	6.5	8.3	10.2	12.3	14.7	17.2	20.0	26.1	33.0	40.8	58.7
	.013	4.0	5.4	7.0	8.9	11.0	13.3	15.8	18.6	21.5	28.1	35.6	43.9	63.3
	.014	4.2	5.8	7.5	9.5	11.8	14.2	16.9	19.9	23.1	30.1	38.1	47.1	67.8
	.015	4.5	6.1	8.0	10.2	12.5	15.2	18.1	21.2	24.6	32.1	40.6	50.2	72.3
	.016	4.8	6.5	8.5	10.8	13.3	16.1	19.2	22.5	26.1	34.1	43.1	53.3	76.7
	.017	5.1	6.9	9.0	11.4	14.1	17.0	20.3	23.8	27.6	36.0	45.6	56.3	81.1

Table 7b   Resisting moment for slab sections 12 in. wide, in foot-kips

$f_y = 60$ ksi

| f'_c | ρ | \multicolumn{13}{c}{Effective depth d, in.} |
		3.0	3.5	4.0	4.5	5.0	5.5	6.0	6.5	7.0	8.0	9.0	10.0	12.0
3000	.002	0.9	1.3	1.7	2.1	2.6	3.2	3.8	4.5	5.2	6.7	8.5	10.5	15.2
	.003	1.4	1.9	2.5	3.2	3.9	4.7	5.6	6.6	7.7	10.0	12.7	15.6	22.5
	.004	1.9	2.5	3.3	4.2	5.1	6.2	7.4	8.7	10.1	13.2	16.7	20.6	29.6
	.005	2.3	3.1	4.1	5.1	6.4	7.7	9.1	10.7	12.4	16.3	20.6	25.4	36.6
	.006	2.7	3.7	4.8	6.1	7.5	9.1	10.8	12.7	14.8	19.3	24.4	30.1	43.4
	.007	3.1	4.2	5.5	7.0	8.7	10.5	12.5	14.7	17.0	22.2	28.1	34.7	49.9
	.008	3.5	4.8	6.3	7.9	9.8	11.8	14.1	16.5	19.2	25.0	31.7	39.1	56.3
	.009	3.9	5.3	7.0	8.8	10.9	13.1	15.6	18.4	21.3	27.8	35.2	43.4	62.6
	.010	4.3	5.8	7.6	9.6	11.9	14.4	17.1	20.1	23.3	30.5	38.6	47.6	68.6
	.011	4.7	6.3	8.3	10.5	12.9	15.6	18.6	21.8	25.3	33.1	41.9	51.7	74.4
4000	.002	1.0	1.3	1.7	2.1	2.7	3.2	3.8	4.5	5.2	6.8	8.6	10.6	15.3
	.003	1.4	1.9	2.5	3.2	3.9	4.8	5.7	6.7	7.7	10.1	12.8	15.8	22.7
	.004	1.9	2.6	3.3	4.2	5.2	6.3	7.5	8.8	10.2	13.3	16.9	20.8	30.0
	.005	2.3	3.2	4.1	5.2	6.5	7.8	9.3	10.9	12.6	16.5	20.9	25.8	37.2
	.006	2.8	3.8	4.9	6.2	7.7	9.3	11.0	13.0	15.0	19.6	24.9	30.7	44.2
	.007	3.2	4.3	5.7	7.2	8.9	10.7	12.8	15.0	17.4	22.7	28.7	35.5	51.1
	.008	3.6	4.9	6.4	8.1	10.0	12.1	14.5	17.0	19.7	25.7	32.5	40.1	57.8
	.009	4.0	5.5	7.2	9.1	11.2	13.5	16.1	18.9	21.9	28.6	36.2	44.7	64.4
	.010	4.4	6.0	7.9	10.0	12.3	14.9	17.7	20.8	24.1	31.5	39.9	49.2	70.9
	.011	4.8	6.6	8.6	10.9	13.4	16.2	19.3	22.7	26.3	34.3	43.4	53.6	77.2
	.012	5.2	7.1	9.3	11.7	14.5	17.5	20.9	24.5	28.4	37.1	46.9	57.9	83.4
	.013	5.6	7.6	9.9	12.6	15.5	18.8	22.4	26.2	30.4	39.8	50.3	62.1	89.5
	.014	6.0	8.1	10.6	13.4	16.6	20.0	23.8	28.0	32.5	42.4	53.6	66.2	95.4
	.015	6.3	8.6	11.2	14.2	17.6	21.2	25.3	29.7	34.4	45.0	56.9	70.2	101.2
5000	.002	1.0	1.3	1.7	2.2	2.7	3.2	3.8	4.5	5.2	6.8	8.6	10.6	15.3
	.003	1.4	1.9	2.5	3.2	4.0	4.8	5.7	6.7	7.8	10.1	12.8	15.9	22.8
	.004	1.9	2.6	3.4	4.3	5.2	6.3	7.6	8.9	10.3	13.4	17.0	21.0	30.2
	.005	2.3	3.2	4.2	5.3	6.5	7.9	9.4	11.0	12.8	16.7	21.1	26.0	37.5
	.006	2.8	3.8	5.0	6.3	7.8	9.4	11.2	13.1	15.2	19.9	25.1	31.0	44.7
	.007	3.2	4.4	5.7	7.3	9.0	10.9	12.9	15.2	17.6	23.0	29.1	35.9	51.7
	.008	3.7	5.0	6.5	8.3	10.2	12.3	14.7	17.2	20.0	26.1	33.0	40.8	58.7
	.009	4.1	5.6	7.3	9.2	11.4	13.8	16.4	19.2	22.3	29.1	36.9	45.5	65.5
	.010	4.5	6.1	8.0	10.2	12.5	15.2	18.1	21.2	24.6	32.1	40.6	50.2	72.3
	.011	4.9	6.7	8.8	11.1	13.7	16.6	19.7	23.1	26.8	35.1	44.4	54.8	78.9
	.012	5.3	7.3	9.5	12.0	14.8	17.9	21.3	25.1	29.1	37.9	48.0	59.3	85.4
	.013	5.7	7.8	10.2	12.9	15.9	19.3	22.9	26.9	31.2	40.8	51.6	63.7	91.8
	.014	6.1	8.3	10.9	13.8	17.0	20.6	24.5	28.8	33.4	43.6	55.2	68.1	98.1
	.015	6.5	8.9	11.6	14.7	18.1	21.9	26.1	30.6	35.5	46.3	58.6	72.4	104.3
	.016	6.9	9.4	12.3	15.5	19.2	23.2	27.6	32.4	37.5	49.0	62.1	76.6	110.3
	.017	7.3	9.9	12.9	16.4	20.2	24.4	29.1	34.1	39.6	51.7	65.4	80.8	116.3

Table 8 Development length, in inches

For No. 11 bars or smaller: $l_d = 0.04A_b f_y/\sqrt{f_c'}$ but $\geq 0.0004 d_b f_y$

For No. 14 bars: $l_d = 0.085 f_y/\sqrt{f_c'}$

For No. 18 bars: $l_d = 0.11 f_y/\sqrt{f_c'}$

(Minimum length 12 in. in all cases.)

Bar No.	f_y	f_c'							
		3000		4000		5000		6000	
		Basic l_d	Top bars $1.4 l_d$	Basic l_d	Top bars $1.4 l_d$	Basic l_d	Top bars $1.4 l_d$	Basic l_d	Top bars $1.4 l_d$
2	40	12	12	12	12	12	12	12	12
	50	12	12	12	12	12	12	12	12
	60	12	12	12	12	12	12	12	12
3	40	12	12	12	12	12	12	12	12
	50	12	12	12	12	12	12	12	12
	60	12	13	12	13	12	13	12	13
4	40	12	12	12	12	12	12	12	12
	50	12	14	12	14	12	14	12	14
	60	12	17	12	17	12	17	12	17
5	40	12	14	12	14	12	14	12	14
	50	13	18	13	18	13	18	13	18
	60	15	21	15	21	15	21	15	21
6	40	13	18	12	17	12	17	12	17
	50	16	22	15	21	15	21	15	21
	60	19	27	18	25	18	25	18	25
7	40	18	25	15	21	14	20	14	20
	50	22	31	19	27	18	25	18	25
	60	26	37	23	32	21	29	21	29
8	40	23	32	20	28	18	25	16	23
	50	29	40	25	35	22	31	20	29
	60	35	48	30	42	27	38	24	34
9	40	29	41	25	35	23	32	21	29
	50	37	51	32	44	28	40	26	36
	60	44	61	38	53	34	48	31	43
10	40	37	52	32	45	29	40	26	37
	50	46	65	40	56	36	50	33	46
	60	56	78	48	67	43	60	39	55
11	40	46	64	39	55	35	49	32	45
	50	57	80	49	69	44	62	40	56
	60	68	96	59	83	53	74	48	68
14	40	62	87	54	75	48	67	44	61
	50	78	109	67	94	60	84	55	77
	60	93	130	81	113	72	101	66	92
18	40	80	113	70	97	62	87	57	80
	50	100	141	87	122	78	109	71	99
	60	121	169	104	146	93	131	85	119

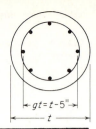

$gt = t - 5''$

Table 9 Weights, areas, moments of inertia of circular columns and moments of inertia of column verticals arranged in a circle 5 in. less than the diameter of column

Diameter of column h, in.	Weight per foot, lb	Area, in.²	I, in.⁴	A_s where $\rho_g = 0.01$*	I_s, in.⁴†
12	118	113	1,018	1.13	6.92
13	138	133	1,402	1.33	10.64
14	160	154	1,886	1.54	15.59
15	184	177	2,485	1.77	22.13
16	210	201	3,217	2.01	30.40
17	237	227	4,100	2.27	40.86
18	265	255	5,153	2.55	53.87
19	295	284	6,397	2.84	69.58
20	327	314	7,854	3.14	88.31
21	361	346	9,547	3.46	110.7
22	396	380	11,500	3.80	137.2
23	433	416	13,740	4.16	168.4
24	471	452	16,290	4.52	203.9
25	511	491	19,170	4.91	245.5
26	553	531	22,430	5.31	292.7
27	597	573	26,090	5.73	346.7
28	642	616	30,170	6.16	407.3
29	688	661	34,720	6.61	475.9
30	736	707	39,760	7.07	552.3

* For other values of ρ_g, multiply the value in the table by $100\rho_g$.

† The bars are assumed transformed into a thin-walled cylinder having the same sectional area as the bars. Then $I_s = A_s(\gamma t)^2/8$.

Table 10 Size and pitch of spirals, ACI Code

Diameter of column, in.	Out to out of spiral, in.	f'_c			
		2500	3000	4000	5000
$f_y = 40,000$:					
14, 15	11, 12	$\frac{3}{8}$-2	$\frac{3}{8}$-1$\frac{3}{4}$	$\frac{1}{2}$-2$\frac{1}{2}$	$\frac{1}{2}$-1$\frac{3}{4}$
16	13	$\frac{3}{8}$-2	$\frac{3}{8}$-1$\frac{3}{4}$	$\frac{1}{2}$-2$\frac{1}{2}$	$\frac{1}{2}$-2
17-19	14-16	$\frac{3}{8}$-2$\frac{1}{4}$	$\frac{3}{8}$-1$\frac{3}{4}$	$\frac{1}{2}$-2$\frac{1}{2}$	$\frac{1}{2}$-2
20-23	17-20	$\frac{3}{8}$-2$\frac{1}{4}$	$\frac{3}{8}$-1$\frac{3}{4}$	$\frac{1}{2}$-2$\frac{1}{2}$	$\frac{1}{2}$-2
24-30	21-27	$\frac{3}{8}$-2$\frac{1}{4}$	$\frac{3}{8}$-2	$\frac{1}{2}$-2$\frac{1}{2}$	$\frac{1}{2}$-2
$f_y = 60,000$:					
14, 15	11, 12	$\frac{1}{4}$-1$\frac{3}{4}$	$\frac{3}{8}$-2$\frac{3}{4}$	$\frac{3}{8}$-2	$\frac{1}{2}$-2$\frac{3}{4}$
16-23	13-20	$\frac{1}{4}$-1$\frac{3}{4}$	$\frac{3}{8}$-2$\frac{3}{4}$	$\frac{3}{8}$-2	$\frac{1}{2}$-3
24-29	21-26	$\frac{1}{4}$-1$\frac{3}{4}$	$\frac{3}{8}$-3	$\frac{3}{8}$-2$\frac{1}{4}$	$\frac{1}{2}$-3
30	27	$\frac{1}{4}$-1$\frac{3}{4}$	$\frac{3}{8}$-3	$\frac{3}{8}$-2$\frac{1}{4}$	$\frac{1}{2}$-3$\frac{1}{4}$

Table 11   Areas, strengths, and initial tensioning load for prestressing steels

Type	Minimum tensile strength f_{pu}, ksi	Nominal diameter, in.	Nominal area, in.^{2}	Ultimate strength $A_{ps}f_{pu}$, kips	Initial tension $0.70A_{ps}f_{pu}$, kips
Grade 250	250	0.250	0.0356	8.9	6.2
seven-wire		0.313	0.0578	14.5	10.2
strand		0.375	0.0799	20.0	14.0
		0.437	0.1089	27.2	19.0
		0.500	0.1438	36.0	25.2
Grade 270	270	0.250	0.0356	9.6	6.7
seven-wire		0.313	0.0578	15.6	10.9
strand		0.375	0.0799	21.6	15.1
		0.437	0.1089	29.4	20.6
		0.500	0.1438	38.8	27.2
Stress-relieved	250	0.192	0.0290	7.25	5.08
solid wire	235	0.276	0.0598	14.05	9.84
Alloy-steel	145	0.750	0.442	64	45
bars (regular)		0.875	0.601	87	61
		1.000	0.785	114	80
		1.125	0.994	144	101
		1.250	1.227	178	125
		1.375	1.485	215	151
Alloy-steel	160	0.750	0.442	71	50
bars (special)		0.875	0.601	96	67
		1.000	0.785	126	88
		1.125	0.994	159	111
		1.250	1.227	196	137
		1.375	1.485	238	167

Table 12a   Coefficients for slabs with variable moment of inertia*

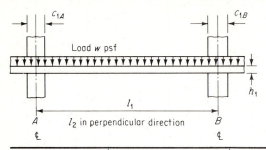

Column dimension		Uniform load F.E.M. = coef. $(wl_2l_1{}^2)$		Stiffness factor†		Carry-over factor	
$\dfrac{c_{1A}}{l_1}$	$\dfrac{c_{1B}}{l_1}$	M_{AB}	M_{BA}	k_{AB}	k_{BA}	COF_{AB}	COF_{BA}
0.00	0.00	0.083	0.083	4.00	4.00	0.500	0.500
	0.05	0.083	0.084	4.01	4.04	0.504	0.500
	0.10	0.082	0.086	4.03	4.15	0.513	0.499
	0.15	0.081	0.089	4.07	4.32	0.528	0.498
	0.20	0.079	0.093	4.12	4.56	0.548	0.495
	0.25	0.077	0.097	4.18	4.88	0.573	0.491
0.05	0.05	0.084	0.084	4.05	4.05	0.503	0.503
	0.10	0.083	0.086	4.07	4.15	0.513	0.503
	0.15	0.081	0.089	4.11	4.33	0.528	0.501
	0.20	0.080	0.092	4.16	4.58	0.548	0.499
	0.25	0.078	0.096	4.22	4.89	0.573	0.494
0.10	0.10	0.085	0.085	4.18	4.18	0.513	0.513
	0.15	0.083	0.088	4.22	4.36	0.528	0.511
	0.20	0.082	0.091	4.27	4.61	0.548	0.508
	0.25	0.080	0.095	4.34	4.93	0.573	0.504
0.15	0.15	0.086	0.086	4.40	4.40	0.526	0.526
	0.20	0.084	0.090	4.46	4.65	0.546	0.523
	0.25	0.083	0.094	4.53	4.98	0.571	0.519
0.20	0.20	0.088	0.088	4.72	4.72	0.543	0.543
	0.25	0.086	0.092	4.79	5.05	0.568	0.539
0.25	0.25	0.090	0.090	5.14	5.14	0.563	0.563

* Applicable when $c_1/l_1 = c_2/l_2$. For other relationships between these ratios, the constants will be slightly in error.

† Stiffness is $K_{AB} = k_{AB}E(l_2h_1{}^3/12l_1)$ and $K_{BA} = k_{BA}E(l_2h_1{}^3/12l_1)$.

Table 12b   Coefficients for slabs with variable moment of inertia*

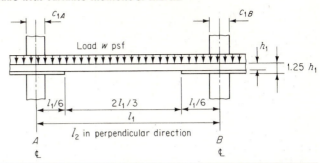

Column dimension		Uniform load F.E.M. = coef. $(wl_2l_1^2)$		Stiffness factor†		Carry-over factor	
$\dfrac{c_{1A}}{l_1}$	$\dfrac{c_{1B}}{l_1}$	M_{AB}	M_{BA}	k_{AB}	k_{BA}	COF_{AB}	COF_{BA}
0.00	0.00	0.088	0.088	4.78	4.78	0.541	0.541
	0.05	0.087	0.089	4.80	4.82	0.545	0.541
	0.10	0.087	0.090	4.83	4.94	0.553	0.541
	0.15	0.085	0.093	4.87	5.12	0.567	0.540
	0.20	0.084	0.096	4.93	5.36	0.585	0.537
	0.25	0.082	0.100	5.00	5.68	0.606	0.534
0.05	0.05	0.088	0.088	4.84	4.84	0.545	0.545
	0.10	0.087	0.090	4.87	4.95	0.553	0.544
	0.15	0.085	0.093	4.91	5.13	0.567	0.543
	0.20	0.084	0.096	4.97	5.38	0.584	0.541
	0.25	0.082	0.100	5.05	5.70	0.606	0.537
0.10	0.10	0.089	0.089	4.98	4.98	0.553	0.553
	0.15	0.088	0.092	5.03	5.16	0.566	0.551
	0.20	0.086	0.094	5.09	5.42	0.584	0.549
	0.25	0.084	0.099	5.17	5.74	0.606	0.546
0.15	0.15	0.090	0.090	5.22	5.22	0.565	0.565
	0.20	0.089	0.094	5.28	5.47	0.583	0.563
	0.25	0.087	0.097	5.37	5.80	0.604	0.559
0.20	0.20	0.092	0.092	5.55	5.55	0.580	0.580
	0.25	0.090	0.096	5.64	5.88	0.602	0.577
0.25	0.25	0.094	0.094	5.98	5.98	0.598	0.598

* Applicable when $c_1/l_1 = c_2/l_2$. For other relationships between these ratios, the constants will be slightly in error.

† Stiffness is $K_{AB} = k_{AB}E(l_2h_1^3/12l_1)$ and $K_{BA} = k_{BA}E(l_2h_1^3/12l_1)$.

Table 12c   Coefficients for columns with variable moment of inertia

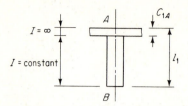

Slab depth c_{1A}/l_1	Uniform load $F.E.M. = coef. (wl_2l_1{}^2)$		Stiffness factors		Carry-over factors	
	M_{AB}	M_{BA}	k_{AB}	k_{BA}	COF_{AB}	COF_{BA}
0.00	0.083	0.083	4.00	4.00	0.500	0.500
0.05	0.100	0.075	4.91	4.21	0.496	0.579
0.10	0.118	0.068	6.09	4.44	0.486	0.667
0.15	0.135	0.060	7.64	4.71	0.471	0.765
0.20	0.153	0.053	9.69	5.00	0.452	0.875
0.25	0.172	0.047	12.44	5.33	0.429	1.000

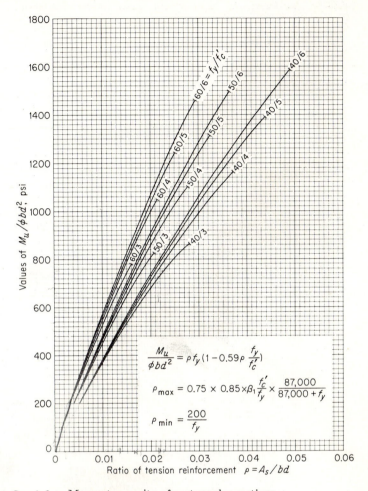

Graph 1a Moment capacity of rectangular sections.

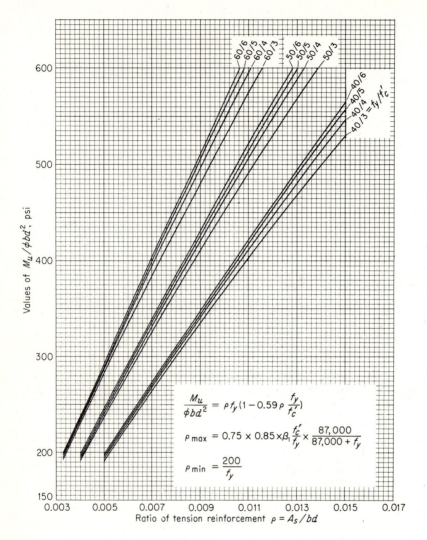

Graph 1b   Moment capacity of rectangular sections.

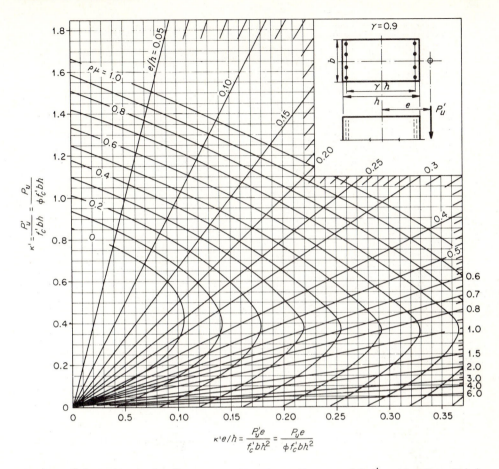

Graph 2   Column interaction diagram—rectangular section: $\gamma = 0.9$, $f'_c = 4$ ksi, $f_y = 60$ ksi. (*Adapted from "Ultimate Strength Design of Reinforced Concrete Columns," Interim Report of ACI Comm. 340, by Noel J. Everard and Edward Cohen, ACI Publ. SP-7, 1964.*)

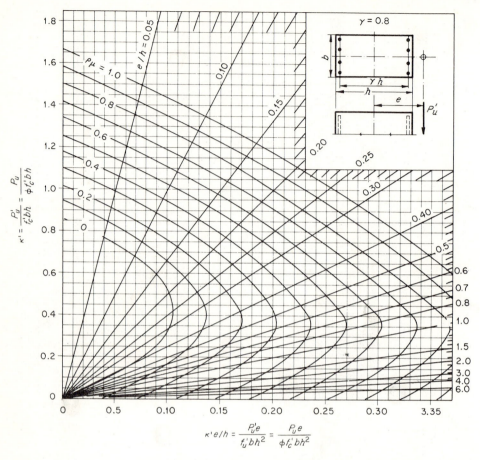

Graph 3   Column interaction diagram—rectangular section: $\gamma = 0.8, f_c' = 4$ ksi, $f_y = 60$ ksi. (*Adapted from same source as Graph 2.*)

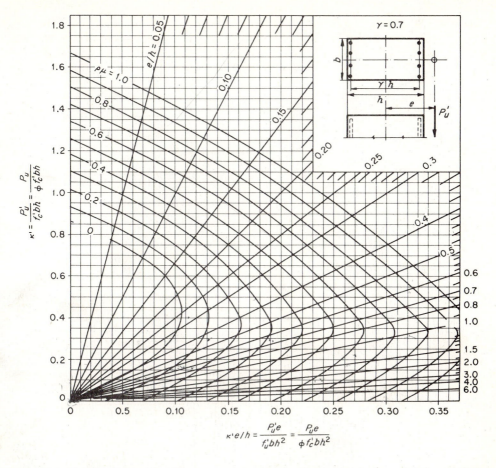

Graph 4 Column interaction diagram—rectangular section: $\gamma = 0.7, f'_c = 4\,\text{ksi}, f_y = 60\,\text{ksi}$. (*Adapted from same source as Graph 2.*)

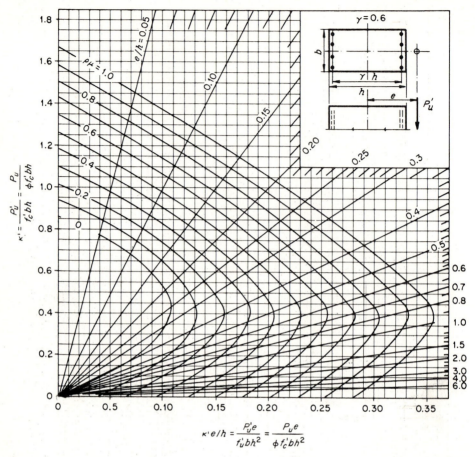

Graph 5   Column interaction diagram—rectangular section: $\gamma = 0.6$, $f'_c = 4$ ksi, $f_y = 60$ ksi. (*Adapted from same source as Graph 2.*)

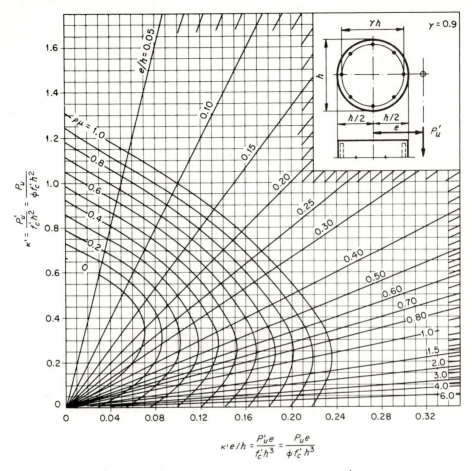

$$\kappa' e/h = \frac{P'_u e}{f'_c h^3} = \frac{P_u e}{\phi f'_c h^3}$$

Graph 6 Column interaction diagram—circular section: $\gamma = 0.9$, $f'_c = 4$ ksi, $f_y = 60$ ksi. (*Adapted from same source as Graph 2.*)

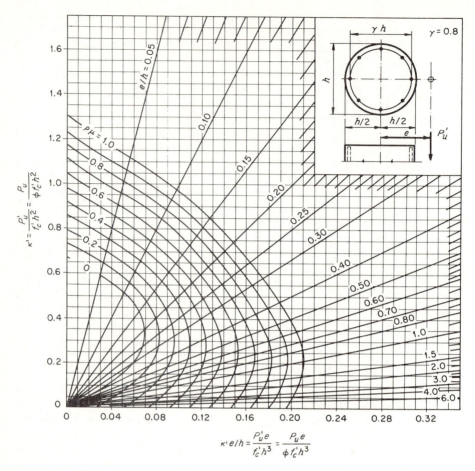

Graph 7 Column interaction diagram—circular section: $\gamma = 0.8$, $f'_c = 4$ ksi, $f_y = 60$ ksi. (*Adapted from same source as Graph 2.*)

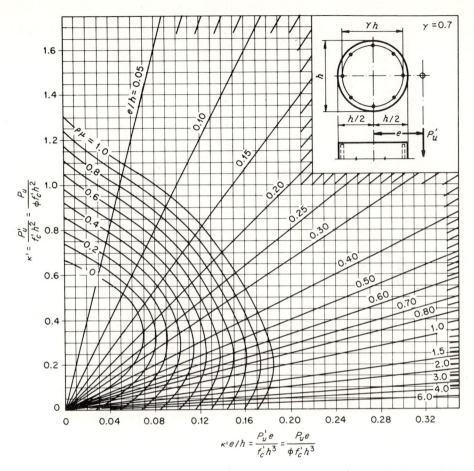

Graph 8 Column interaction diagram—circular section: $\gamma = 0.7$, $f_c' = 4$ ksi, $f_y = 60$ ksi. (*Adapted from same source as Graph 2.*)

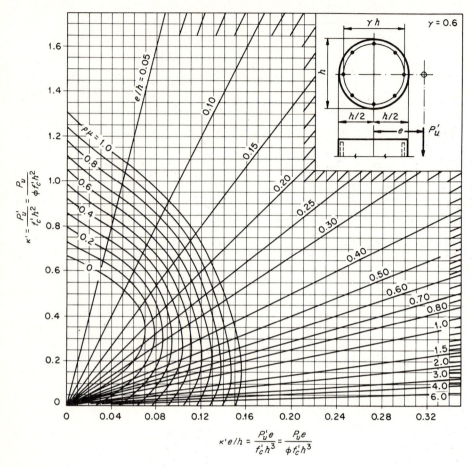

$$\kappa'e/h = \frac{P'_u e}{f'_c h^3} = \frac{P_u e}{\phi f'_c h^3}$$

Graph 9 Column interaction diagram—circular section: $\gamma = 0.6$, $f'_c = 4$ ksi, $f_y = 60$ ksi. (*Adapted from same source as Graph 2.*)

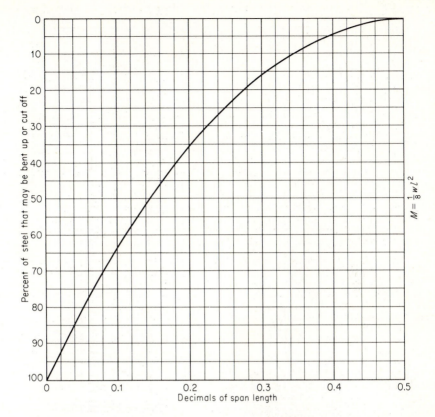

Graph 10 Locations of points at which bars may be bent up for simply supported beams uniformly loaded.

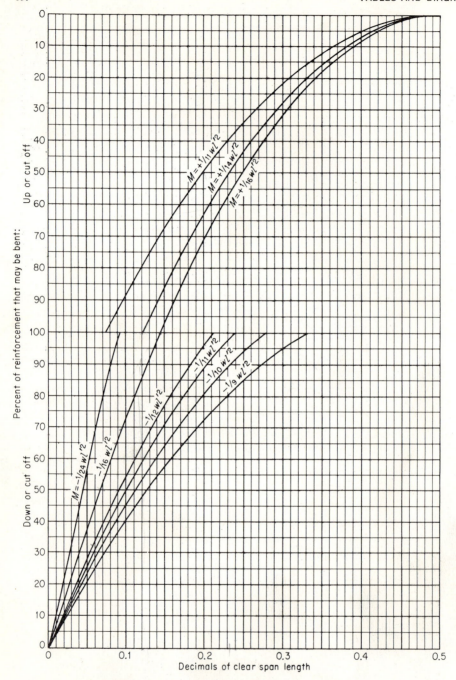

Graph 11 Approximate locations of points at which bars may be bent up or down or cut off for continuous beams uniformly loaded and built integrally with their supports according to the coefficients of the ACI Code.

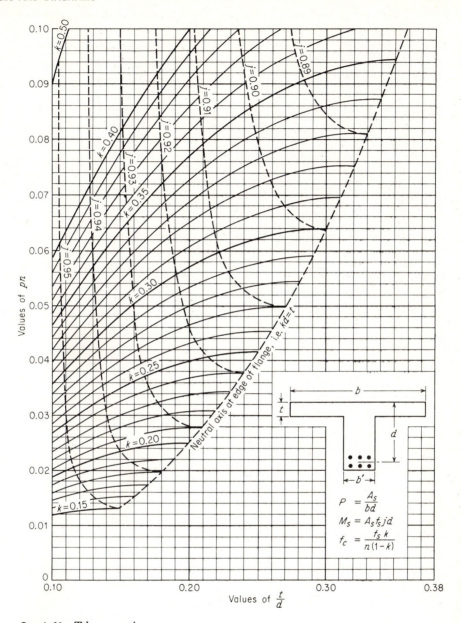

Graph 12 T-beam review.

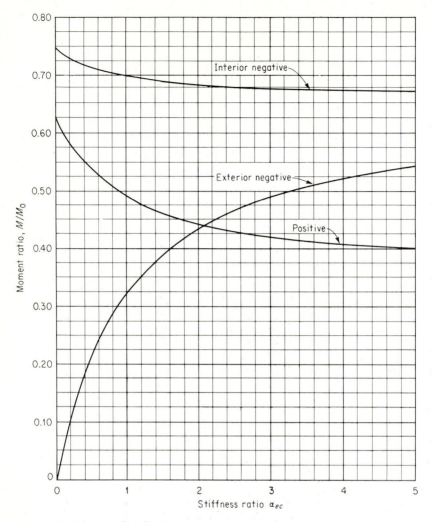

Graph 13   Moment-distribution ratios for two-way slabs.

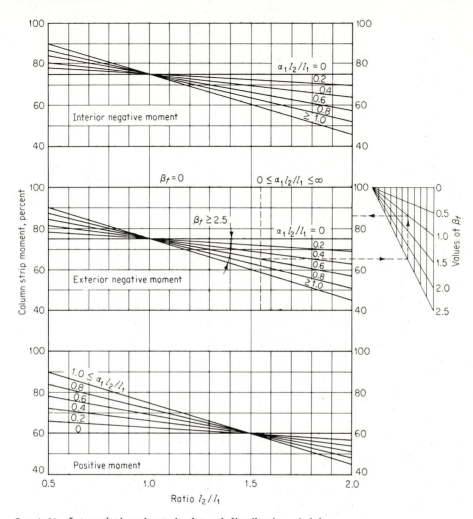

Graph 14 Interpolation charts for lateral distribution of slab moments.

Index